# Palgrave Studies in the History of Economic Thought

Palgrave Studies in the History of Economic Thought publishes contributions by leading scholars, illuminating key events, theories and individuals that have had a lasting impact on the development of modern-day economics. The topics covered include the development of economies, institutions and theories.

More information about this series at
http://www.palgrave.com/gp/series/14585

Ajit Sinha
Editor

# A Reflection on Sraffa's Revolution in Economic Theory

palgrave
macmillan

*Editor*
Ajit Sinha
Thapar School of Liberal Arts and Sciences
Patiala, India

ISSN 2662-6578 ISSN 2662-6586 (electronic)
Palgrave Studies in the History of Economic Thought
ISBN 978-3-030-47205-4 ISBN 978-3-030-47206-1 (eBook)
https://doi.org/10.1007/978-3-030-47206-1



Cover illustration: Evgenii Emelianov / Alamy Stock Photo

This Palgrave Macmillan imprint is published by the registered company Springer Nature Switzerland AG.
The registered company address is: Gewerbestrasse 11, 6330 Cham, Switzerland

# Praise for *A Reflection on Sraffa's Revolution in Economic Theory*

"Piero Sraffa's elusive classic *Production of Commodities by Means of Commodities* was published sixty years ago with the aim of reconstructing the foundations of Economics on solid theoretical grounds purging the Classical and Neoclassical Economics of logical and philosophical weaknesses. It set off a debate which has still not been settled. In the meantime, the global economy has been through the crises of stagflation, globalisation, the Great Recession and lately Covid-19. Nostrums and solutions have come—Keynesianism, Monetarism, New Classical Economics—and gone. The Soviet system of economic planning collapsed and has been forgotten.

Practical remedies to real life problems having failed, it may be the time for going back to fundamentals. Here is a splendidly rich collection of contributions, with criticisms and replies that Ajit Sinha has brought together which will merit serious study by senior economists as much as the next generation of young students. Here the Sraffa system is analysed in its full complexity and dissected by scholars who were there when Sraffa published the book as well as subsequent three generations of students. Here are Geoffrey Harcourt, Bertram Schefold, Roberto Scazzieri, Christina Marcuzzo, Anna Carabelli, Annalisa Rosselli, John King, Guglielmo Chiodi, names which have been known to us as experts in theory, methodology and philosophy of economics.

Sraffa was not just an economic theorist. The book casts light on his influence on Wittgenstein (Sinha), his discussions with Keynes (Carabelli) and his friendship with Gramsci (Naldi) and Kalecki (King). The availability of the Sraffa archives has enriched our understanding of his thinking and the debates he carried on with himself and his correspondents. This book captures the spirit of these archives which previous work on Sraffa has missed out on. The debate here is creative as there is also a contribution relating Sraffa's theory to the problem of non-renewable natural capital (Verger).

Ajit Sinha has prepared a feast for students and scholars of economics, philosophy and politics to enjoy."

—Lord Meghnad Desai, *Professor Emeritus, London School of Economics*

# Preface

Sraffa's contribution to economics is mainly thought of in terms of his specific contributions to particular theoretical issues in Economics—his early contributions (1925, 26)[1] were acknowledged as a devastating critique of Marshall's theoretical framework of partial equilibrium in a perfectly competitive market under the condition of ceteris paribus; the latter paper also proved to be seminal in the development of the theories of 'imperfectly competitive' markets; he is also noted for his polemic with Hayek on Hayek's *Production and Prices* (1931)[2] in which he developed an ingenious idea of 'a commodity's own rate of interest' (Sraffa 1932),[3] which Keynes utilised in his *General Theory* (1936)[4] and is also used in the inter-temporal general equilibrium analysis; and most importantly, he is credited to have presented a definitive critique of the orthodox marginal productivity theory of distribution of income in his magnum opus, *Production of Commodities by Means of Commodities* (1960).[5] Along with

[1] 'Sulle relazioni tra costo e quantita prodotta. *Annali di Economia*, Vol. 2: pp. 277–328, 1925 and 'The laws of returns under competitive conditions', *The Economic Journal*, vol. 36: pp. 535–50, 1926.

[2] *Prices and Production*, London: George Routledge & Sons ltd., 1931.

[3] 'Dr. Hayek on Money and Capital', *Economic Journal*, vol. 42 (165), pp. 42–53.

[4] *The General Theory of Employment, Interest and Money*, London: A Harvest/ HBJ Books. In a footnote to Chapter 17 of the *General Theory*, Keynes wrote, 'This relationship was first pointed out by Mr. Sraffa, *Economic Journal*, March 1932, p.50'. Sraffa, however, found Keynes's use of this concept contradictory and inconsistent as is evident from the marginal comments in his personal copy of the *General Theory*.

[5] *Production of Commodities by Means of Commodities: Prelude to a Critique of Economic Theory*, Cambridge: Cambridge University Press.

these important specific contributions to economic theory, he is also credited to have given a new interpretation of 'Classical Economics' by providing a highly original interpretation of Ricardo's 'labour theory of value' in his universally acknowledged magisterial editing of *Works and Correspondence of David Ricardo*[6] and, in some sense, connecting it to the theory developed in his *Production of Commodities by Means of Commodities.*

Even though the list is highly impressive and would be good enough to qualify Sraffa to be inducted in the 'Economics Hall of Fame', this sort of listing of specific contributions does not do justice to the overall nature and significance of his contribution to Economics. My own study of Sraffa's archives convinced me of this conclusion, which I tried to communicate to the larger intellectual audience in my book, *A Revolution in Economic Theory: The Economics of Piero Sraffa* (Sinha 2016).[7] It seems to me that Sraffa wanted to cut the foundation of Economics from its old moorings either in an essentialist problematic in which the theory searches for the ultimate cause of a phenomenon or the mechanical causation in which the theory searches for an equilibrium of forces as the solution of the problem. Sraffa seems to want to bring economic theory to the modern era of the twentieth century quantum mechanics, which eschews determinacy and classical causation in favour of indeterminacy and structural relations. Moreover, he wants to build economics as a science based on observable data and therefore does not want to allow human psychology any role in economic theory as human psychology is unobservable. This is a manifesto for a revolution in economic theory.

While Sraffa's archive at the Wren Library, Trinity College, University of Cambridge, was officially opened up to public in the early 1990s, it was not available in print or electronic format and the use of its contents was kept on a tight leash for more than a decade by the literary executor of the time. For many years there were only a handful of Sraffa-scholars who had an opportunity to study it partially or completely and were allowed to quote from it. Lately, however, the situation has eased and increasingly more and more scholars interested in Sraffa's work are coming up with their interesting findings and alternative interpretations of the methodological and philosophical underpinnings of Sraffa's economics. In a way,

[6] 'Introduction', in *Works and Correspondence of David Ricardo*, vol. I, Cambridge: Cambridge University Press, 1951.

[7] *A Revolution in Economic Theory: The Economics of Piero Sraffa*, Cham: Palgrave Macmillan.

Sraffa's economics is currently going through a churning. In this milieu, Rachel Sangster of Palgrave Macmillan approached me in early 2019 to consider editing a volume that would reflect on Sraffa's economics in a fresh way by taking account of the insights from his archives—the present volume is the outcome of that—it, indeed, stands as an example of the current renaissance of Sraffa's economics that is well underway.

I thank all the contributors for accepting my request to contribute to this volume in a short period of time. All papers, except for one, have comments but not all have replies to the comments as those comments were largely explanatory and complimentary and therefore authors did not see any reason to 'reply'. Naldi's paper on a particular aspect of the relationship between Sraffa and Gramsci throws a rare light on a facet of Sraffa's personal life and personality—the nature of the paper did not require any comment.

Patiala, India — Ajit Sinha

# Contents

# Notes on Contributors

**Carlo Benetti** is Emeritus Professor of Economics at the University of Paris Nanterre. His research interests are money and prices; reproduction, disequilibrium, and crises; Marxian and Classical theories.

**Anna Carabelli** is a full professor of Political Economy, Università degli Studi del Piemonte Orientale, Italy. She graduated from Bocconi University in Milan and took her PhD at the University of Cambridge. Her research interests are in the field of methodology, epistemology, history of economic thought, philosophy of measurement with particular reference to Keynes, Hayek and Sraffa. Other research interests are in the field of happiness and material conditions, and of international monetary relations (the morality of debt, international reserves, 'policy space', Bretton Woods 2 and Washington consensus).

**Jean Cartelier** is Emeritus Professor at the University of Paris-Nanterre. He has recently published L'intrus et l'absent, Essai sur le place du travail et du salariat dans la théorie économique (Presses Universitaires de Paris-Ouest 2016) and Money, Markets, and Capital, The Case for a Monetary Analysis (London, Routledge 2018).

**Guglielmo Chiodi** is a former Professor of Economics at the University of Rome "La Sapienza", Italy. He did his post-graduate studies at the University of Cambridge (GB). He has been Visiting Scholar at the University of Lund, Sweden, and Visiting Professor at the European University Institute in Fiesole, Italy. He is the author of several articles in journals and books on monetary theory, the theory of value and distribution, with special regard to Classical, Sraffian and Marxian economic theory and, more recently, he has been concerned with the relationship between ethics and economics.

**John B. Davis** is Professor Emeritus of Economics, Marquette University, USA, and Professor Emeritus of Economics, University of Amsterdam, Netherlands. His research is on the history and methodology of economics. He is a former editor of the Journal of Economic Methodology and the Review of Social Economy.

**Ghislain Deleplace** is Emeritus Professor at the University Paris 8. He has published Private Money and Public Currencies (with M.T. Boyer-Xambeu and L. Gillard; Sharpe, 1994), Money in Motion. The Post Keynesian and Circulation Approaches (co-edited with E.J. Nell; Macmillan, 1996), Ricardo on Money: A Reappraisal (Routledge, 2017), Histoire de la pensée économique (Dunod, 3rd edition 2018).

**Geoffrey Colin Harcourt** is Emeritus Reader in the History of Economic Theory, Cambridge, 1998; Emeritus Fellow, Jesus College, Cambridge, 1998; Professor Emeritus, Adelaide, 1988; and currently Honorary Professor, School of Economics, UNSW Sydney, 2010–. He has published 33 books and over 400 articles, chapters in books, and reviews. His research interests include post-Keynesian theory, applications and policy, intellectual biography and history of economic theory.

**John King** is Emeritus Professor at La Trobe University and Honorary Professor at Federation University Australia. His principal research interests are in the history of heterodox economic thought, especially Marxian political economy and Post-Keynesian economics. Recent publications include The Distribution of Wealth (2016; with Michael Schneider and Mike Pottenger), A History of American Economic Thought (2018; with Samuel Barbour and James Cicarelli) and The Alternative Austrian Economics, dealing with socialist economic thought in Austria between 1900 and the present day (2019).

**Maria Cristina Marcuzzo** is Professor of Economics at the University of Rome, "La Sapienza", Italy and Fellow of the Italian Academy of Lincei. Former President of the European Society for the History of Economics and the Italian Society for the History of Political Economy, she has been Visiting Professor in several universities in Europe, Japan, Mexico and United States. She has worked on classical monetary theory, the Cambridge

School of Economics, Keynesian economics, and more recently, Keynes's investments in financial markets. She has published about 100 articles in journals and books, plus authoring or editing 20 volumes.

**Nuno Martins** has a PhD in Economics by the University of Cambridge, and a licenciatura in Economics by the Universida de Católica Portuguesa (Porto), where he also obtained his habilitation in History of Economic Thought. He is currently full professor at the Universida de Católica Portuguesa where he teaches History of Economic Thought and Social Philosophy and Ethics. He also taught at the University of Cambridge and the University of the Azores. He is a member of the Cambridge Social Ontology Group (CSOG) interdisciplinary research network, and of the editorial board of various academic journals.

**Nerio Naldi** is Associate professor of Economics, Department of Statistics at the University of Rome, La Sapienza.

**Goddanti Omkarnath** is Professor, School of Economics, University of Hyderabad, India. His teaching and research interests in Classical Political Economy and Capital Theory have been prompted by an abiding interest in the complex structure of Indian economy for which neo-classical economics seemed to be singularly inappropriate. Trained in Jawaharlal Nehru University, New Delhi, he was earlier a faculty of Centre for Development Studies, Thiruvananthapuram (Kerala). He is the author, among others, of *Economics – A Primer for India* (Orient Blackswan, 2016).

**Carlo Panico** is Professor at the Universidad Nacional Autonoma de Mexico. He was Professor of Political Economy at the University Federico II of Naples (Italy). His main research interest regards the effects of monetary events on income distribution and growth. He has published articles and monographs on how money can be integrated in the foundations of the economic discipline, that is in the theory of value, distribution and production. He has also contributed to the history of economic thought and to the institutional organization of economic policy.

**Murray Patterson** holds the Chair in Ecological Economics at Massey University and was the Founding Director of the New Zealand Centre for

Ecological Economics from 2003 to 2011. He is widely published in a number of fields including ecological economics, energy analysis, theories of value, environmental valuation, environmental policy and policy modelling. Much of this research is of an applied and interdisciplinary nature, working at the interface of policy and sustainability concerns. He has over 30 years' professional experience as policy analyst, economist and planner in a number of government and university settings.

**Annalisa Rosselli** is Professor of History of Economic Thought at the University of Rome Tor Vergata, Italy and Fellow of the Italian Academy of Lincei. She graduated in mathematics at the University of Florence and did her post-graduate studies at the London School of Economics and at the University of California, Los Angeles. She was President of the Italian Association for the History of Political Economy from 2009 to 2012, of the European Society for the History of Economic Thought from 2012 to 2014 and of the Italian Economic Society from 2016 to 2019. She is author of several publications and reports on issues of gender equality. She has authored or co-edited six books and published many articles in journals and books on history of monetary theory, Classical Political Economy, Cambridge School, Keynesian policies in the 1950s.

**Roberto Scazzieri** is Professor of Economic Analysis, University of Bologna and Fellow, Lincei Academy, Rome. He is Senior Member, Gonville and Caius College, and Life Member, Clare Hall, Cambridge. He has been *Professor Attaché* at Interdisciplinary Research Centre 'Beniamino Segre', Lincei Academy, founding Scientific Director of the Institute of Advanced Study, University of Bologna and founding Managing Editor of the journal *Structural Change and Economic Dynamics*.

**Bertram Schefold** was born on 28.12.1943 in Basel (Switzerland), studied at Munich, Hamburg and Basel, 1962–1967; diploma in mathematics, physics and philosophy. Doctoral studies in Cambridge (GB) and Basel. Ph.D. in economics 1971. He was Research Associate at Harvard University and became Full Professor Faculty of Economics, Johann Wolfgang Goethe-University, Frankfurt, March 74. Emeritus, teaching full-time as Senior Professor since 2012. More than 40 books and 350 articles on economic theory, history of economic thought, energy policy and general economic policy.

**Ajit Sinha** is Professor of Economics at the Thapar School of Liberal Arts and Sciences. His main area of research has been the history of economic theory. His book *A Revolution in Economic Theory: The Economics of Piero Sraffa* opens new avenues for advancing classical theory and has been credited to have 'put paid' to the established interpretation of Sraffa. He has held visiting positions at Cambridge University, Collège de France, University of Trento, Delhi School of Economics and several other universities. The second edition with a long 'Afterword' of his first book, *Theories of Value from Adam Smith to Piero Sraffa*, and his latest book, *Essays on Theories of Value in the Classical Tradition*, were both published in 2018.

**Albert E. Steenge** is Emeritus Professor of Economics at the University of Groningen (Honorary) and at the University of Twente. He has published widely on topics in public economics, environmental economics and multi-sectoral and inter-industry economics. His current interests include the economics of unplanned events (disasters) and tourism. He has been Visiting Scholar and Professor at various universities and institutes, including the Institute of Economics of the Hungarian Academy of Sciences, the University of Cambridge (UK), MIT (Cambridge, Mass.), the University of Bologna and the University of Macerata. He is a Life Member of Clare Hall, Cambridge, UK.

**K. Vela Velupillai** is a retired economics Professor, now residing in Stockholm, Sweden. He was, during his working years, a Professor at NUI Galway, Ireland, University of Trento, Italy, Queen's University in Belfast and Fellow of Girton College, Cambridge. He was educated at the Universities of Kyoto, Lund and Cambridge.

**Roberto Veneziani** is Professor of Economics at the School of Economics and Finance, Queen Mary University of London. His research interests include topics of dynamic models of cyclical growth, egalitarian principles, distribution of resources between generations, sustainable development, and normative principles in economics. He has focused mainly on dynamic recursive optimisation models with heterogeneous agents. He is also interested in the history of economic thought and in political economy from a mathematical perspective. He has published widely in economics, philosophy and political science.

**Ragupathy Venkatachalam** is a lecturer in Economics at the Institute of Management Studies, Goldsmiths, University of London, UK. He is a member of the Algorithmic Social Sciences Research Unit (ASSRU) at the

University of Trento, Italy. His research covers economic dynamics, computable economics, history of economic thought and classical behavioural economics.

**Yoann Verger** is Professor agrégé in Social Sciences. He holds a Ph.D. in economics titled "Sraffa and ecological economics: links and possibilities" (2016).

**Naoki Yoshihara** is Professor of Economics at Department of Economics, University of Massachusetts Amherst. He won the First Encouragement Prize by the Japan Society for Political Economy (2010) and the Distinguished Achievement Award in Political Economy for the Twenty-first Century by the World Association for Political Economy (2011). His main research fields are welfare economics, analytical approach to classical and Marxian economics. He has published in variety of topics including distributive justice, cooperative games, implementation theory, axiomatic theory of Marxian exploitation, and unequal exchange in international economies.

**Stefano Zambelli** is Professor of Economics at the Department of Economics and Management, University of Trento, Italy. He is a founding member of the Algorithmic Social Sciences Research Unit (ASSRU) at the University of Trento. His research focuses on Algorithmic production theory, capital theory and aggregation, computable economics and endogenous business cycle theory.

# List of Figures

# List of Tables

CHAPTER 1

# Interpreting the Nature of Sraffa's Equations: A Critique of Garegnani's Interpretation

*Ajit Sinha*

## 1 Introduction

In this chapter I argue that Sraffa's (1960) idea of ***a system of basic goods*** makes a decisive break from all previous economic theories. Classical economics was rooted in the idea of vertically integrated industries as silos where production could be conceptually reduced to a direct labouring activity of Man against Nature. In this context, only 'final' goods exchange, and fluctuations in prices of any particular commodity due to changes in its market conditions of demand and supply do not affect the price of any other commodity. Within this framework, the classicists developed a long-term theory of 'equilibrium' or 'natural' prices that could be determined by the given conditions of production and distribution of income alone. However, this long-term result could only be ensured by introducing a mechanism of gravitation of 'disequilibrium' or 'market' prices towards equilibrium or 'natural' prices. This required hypotheses of rational human behaviour, rooted in what Adam Smith called 'self-love', and constant returns to scale or linear technique of production.

A. Sinha (✉)
Thapar School of Liberal Arts and Sciences, Patiala, India



A. Sinha (ed.), *A Reflection on Sraffa's Revolution in Economic Theory*, Palgrave Studies in the History of Economic Thought,
https://doi.org/10.1007/978-3-030-47206-1_1

The revolution of 1870s in economic theory, which replaced classical economics by putting human subjectivity such as utility and demand at the centre of economic theory, did not however do away with the idea of conceptually reducing production to a long chain of productive activity all the way back to the primordial state of direct relation between Man and Nature. During the winter of 1927–28, Sraffa set up a task for himself to rehabilitate the classical paradigm by removing human psychology or human subjectivity from economic analysis. In this context, he zeroed in on the explanation of the rate of profits of the moderns. Sraffa discovered that the attempts by Jevons, Menger, Böhm-Bawerk, Wicksell and others to find a measure of capital independently of prices in terms of time of 'waiting' or 'roundabout way of production' could not be sustained because if capital goods were produced as commodities and industries bought and sold these capital goods to use them as their inputs then no matter how far back we go in time there always will remain a commodity residue—one can never reach the primordial state. One consequence of this finding was that if wages become zero then the rate of profits of the system must reach a finite maximum and not infinity as would be the case if no commodity residue remained. Later in early 1940s, Sraffa developed a hypothesis that the maximum rate of profits of any system of basic goods, where basic goods are produced inputs that are used directly or indirectly in the production of all goods, is an objective property of the system of production and therefore it must be independent of prices. The search for a confirmation of this hypothesis led him to develop the Standard system and the proof of its uniqueness, and the Standard commodity as the unit of measure for wages and prices. One consequence of this development was to show that the classical condition of the uniform rate of profits associated with their 'natural prices' does not require the notion of equilibrium of quantities supplied with their effectual demands, and thus its associated assumptions regarding human behaviour and constant returns can also be discarded—a uniform rate of profits is a structural property of the system of basic goods, and the determination of prices is independent of human subjectivity or demand considerations. Another consequence of this analysis was that the distribution of income in terms of rates of profits and wages could be separated from price determination, which Sraffa maintained was the core of classical paradigm—profit is a non-price phenomenon (see Sinha 2016 for relevant references).

Unfortunately, the leading commentators of Sraffa's book such as Garegnani and his associates did not see that Sraffa's rehabilitation of classical economics was not just a restatement of it. They failed to notice that Sraffa had carefully removed the classical notions of 'equilibrium', 'effectual demand', mechanism of gravitation of 'market prices' to 'natural prices' and constant returns from his analysis. Garegnani maintained that Sraffa's equations of basic goods *assume* classical equilibrium—they simply represent the end or the rest state of classical gravitation of 'market prices' to 'natural prices'. Sraffa's main contribution to classical economics, according to Garegnani (1991), was to show that the 'natural prices' and the uniform rate of profits of the classical economics are determined *simultaneously* and not *sequentially* as the classicists had tried to do since they did not possess the knowledge of matrix algebra to determine the two simultaneously. In this chapter I show that Garegnani's position becomes incoherent once one removes the assumption of constant returns or linear technique from the classical gravitation mechanism and Sraffa's production equations. However, it is well known that Sraffa was quite explicit and emphatic that he makes no such assumption.

## 2 Adam Smith's Basic Propositions on 'Natural' and 'Market' Prices

Adam Smith (1981 [1776]) wanted to find a Standard of measure for the values of commodities so that the *real* wealth of a nation could be compared over periods of time independently of fluctuations in the prices of commodities. He thought that if the ultimate cause of value could be discovered then that could provide us with the Standard that will remain invariant in the face of apparent changes in prices. This led Adam Smith to think of Man's primordial state when he must have had to act *directly* against Nature to wrest his basic needs of survival from it. For Adam Smith this primordial act of Man against Nature is both an act of *production* as well as an act of *exchange*. Expenditure of labour in the process of production is also a sacrifice in terms of 'toil and trouble', which is a payment of price for the product appropriated from Nature. Thus all prices or all economic values must be measured by this 'originary' or the ultimate price, which measures the 'real' value of the commodity as opposed to the 'nominal' value measured by the money-commodity such as gold or silver:

> Labour was the first price, the original purchase-money that was paid for all things. It was not by gold or by silver, but by labour, that all the wealth of the world was originally purchased; and its value, to those who possess it and

> who want to exchange it for some new production, is precisely equal to the quantity of labour which it can enable them to purchase or command. (Smith 1981 [1776], p. 48; *WN*, I.v.2)

After having determined the Standard by which to measure the values of commodities in a commodity's ability to command certain length of time of the labourer or his 'sacrifice of labour', Adam Smith developed a theory of value in terms of accounting of the income generated in production by 'adding up' the 'ordinary' or 'average' wages, profits and rent, which he considered were known data at any given point of time:

> These ordinary or average rates may be called the natural rates of wages, profit and rent, at the time and place in which they commonly prevail. (ibid., p. 72, *WN*, I.vii.3)

On the question of how to take account of the raw materials and other means of production that get used up in the process of production, Adam Smith seems to evoke an image of vertically integrated industries where every industry that produces a commodity produces all its raw materials and means of production as a long chain of productive activity which culminates in some form of the originary act of production, and thus they all can be reduced to a long series of wages, profits and rent in the final analysis:

> In the price of corn, for example, one part pays the rent of the landlord, another pays the wages or maintenance of the labourers and labouring cattle employed in producing it, and the third pays the profit of the farmer. These three parts seem either immediately or ultimately to make up the whole price of corn. A fourth part, it may perhaps be thought, is necessary. In the price of corn, for example, one part pays the rent of the landlord, another pays the wages or for replacing the stock of the farmer, or for compensating the wear and tear of his labouring cattle, and other instruments of husbandry. But it must be considered that the price of any instrument of husbandry, such as labouring horse, is itself made up of the same three parts; the rent of the land upon which he is reared, the labour of tending and rearing him, and the profits of the farmer who advances such a rent of this land, and the wages of this labour. Though the price of the corn, therefore, may pay the price as well as the maintenance of the horse, the whole price still resolves itself either immediately or ultimately into the same three parts of rent, labour, and profit. (ibid., p. 68; *WN*, I.vi.11)

The reason why I think it invokes an image of a vertically integrated industry is because if all the commodity-producing industries were buying and selling their raw materials and means of production from each other, then the reduction of any production chain to the primordial state would require that there must exist an independent industry that uses no produced raw materials and means of production, and its product is of such a nature that it must be directly or indirectly used by all industries. This obviously is a very difficult assumption to maintain in a capitalist economy. I will come back to this presently.

In Book 1, Chapter vii of the *Wealth of Nations* Adam Smith introduces his notion of the 'centre of gravitation'. As I mentioned above, Smith claims that at any given point of time the 'average' or 'expected' rates of profits, wages and rent in an economy are known data. Given this, if the price of a commodity in the market turns out to be such that it generates wages, profits and rent such that their rates turn out to be the known 'average' or 'expected' rates then that price is defined as its 'natural price'. Then the question arises: under what circumstances a commodity's price would turn out to be such that it generates the 'expected' rates of wages, profits and rent? The answer to this question Smith gives is that in the market a schedule of prices and quantity demanded of the commodity exists—at higher prices the quantity demanded is low and at lower prices the quantity demanded is high. If this schedule is a continuous one then there must exist a quantity that is demanded at the given 'natural price'. If the quantity supplied of the commodity turns out to be exactly the same as the quantity demanded at the 'natural price' then the commodity's 'market price' will coincide with its 'natural price' and it will generate the 'expected' rates of returns. However, if the quantity supplied of the commodity turns out to be lower (higher) than the quantity demanded at the 'natural price' then the 'market price' of the commodity will turn out to be higher (lower) than its 'natural price'. In this scenario at least one income category must receive a rate of return higher (lower) than its 'natural' or 'expected' returns. This would generate a movement of resources into (out of) the industry thus changing its output over time bringing it in conformity with its quantity demanded at its 'natural price'. Hence the 'natural price' of a commodity works as the centre of gravitation for the 'market price'—at this price there is no incentive for resources to move in or out of the industry and thus the ratio of quantity demanded and quantity supplied (in this case equal to one) remains stable:

> If at any time it [the supply] exceeds the effectual demand, some of the component parts of its price must be paid below their natural rate. If it is rent, the interest of the landlords will immediately prompt them to withdraw a part of their land; and if it is wages or profit, the interest of the labourers in the one case, and of the employers in the other, will prompt them to withdraw a part of their labour or stock from this employment. The quantity brought to market will soon be no more than sufficient to supply the effectual demand. All the different parts of its price will rise to their natural rate, and the whole price to natural price. If, on the contrary, the quantity brought to the market should at any time fall short of the effectual demand, some of the component parts of its price must rise above its natural rate. … The natural price, therefore, is, as it were, the central price, to which the prices of all commodities are continually gravitating. (ibid., pp. 74–75, *WN*, vii.13–15)

There are two theoretical points to note in this context: (i) Adam Smith's industries are apparently vertically integrated and hence are independent of each other. Otherwise, adjustments of 'market prices' to the 'natural prices' of raw materials and means of production would affect the cost prices of all other commodities and hence in turn their 'natural prices'. Since Adam Smith does not consider this effect, it implies an implicit assumption of vertically integrated industries the cost conditions of which are immune to price adjustments of any other industry. And (ii) though the given rates of returns and the production technique are sufficient data or information to determine the 'natural price', the level of the demand schedule or the point of 'effectual demand' is not taken as a given datum—all we know is that a demand schedule exists and therefore a particular quantity demanded must correspond to the 'natural price' and that when the quantity supplied does not correspond to that 'effectual demand' quantity then a market mechanics comes into play that brings the quantity of supply to correspond with the quantity of effectual demand—this movement in quantity supplied happens over a period of time and several production cycles. Therefore, implicit in Adam Smith's discussion on the gravitation mechanism is the assumption of given technique with constant returns to scale, since it is assumed that the 'natural prices' do not change when industries adjust their supplies. This is a highly Newtonian world view where *independent* industries interact with each other due to the pull or push created by rational human agents' 'self-love' to do the best for themselves, which brings them together in an equilibrium or a state of 'repose'.

Though Ricardo ([1821] 1951) criticised Adam Smith's 'adding up' theory of value and replaced it with his proposition that the 'natural price' of a commodity is directly determined by the total direct and indirect labour input needed in the production of a commodity, he concurred with Adam Smith that, in the final analysis, all inputs in production can be reduced to labour and Nature alone. He also accepted Adam Smith's basic proposition regarding the adjustment of 'market prices' to 'natural prices'. Marx ([1867], [1894] 1977, 1991), though not very comfortable with the idea of reducing production process to a relation between Man and Nature in its final analysis, nevertheless, remained within the Ricardian tradition of maintaining labour as the 'ultimate cause' of value or 'prices of production' and also accepted Adam Smith's explanation of the adjustment of 'market prices' to 'natural prices'. It is now widely accepted that Adam Smith's mechanism of the gravitation of 'market prices' to 'natural prices' implicitly assumes fixed technique with constant returns to scale. This is not a position only held by economists of neoclassical orientation such as Marshall (1890 [1949]), Samuelson (1978) or Hollander (1992), but even the young Sraffa (1925) and the young Garegnani (1959) also interpreted it in the same vein:

> It can be said that all classical writers accept implicitly, as an obvious fact, that cost is independent of quantity, and they do not bother to discuss the contrary hypothesis. (Sraffa 1925, p. 325)

> We can here remember that Smith and Ricardo's theory of price is *founded on the assumption of constant returns to scale* for manufactures, while the position of the margin in agriculture is given since it is treated as broadly determined by the level of accumulation and population. (Garegnani 1959, p. 29, f.n. 2, emphasis added)

## 3 Why Was There a Change in Garegnani's Position on the Constant Returns Assumption?

A problem for Garegnani however arose when in 1960, Sraffa, in the Preface to his book, boldly proclaimed that the propositions of his book do not rest on an assumption of constant returns to scale or on an assumption of returns to scale of any kind as his system does not allow for changes in outputs. He actually made it quite clear that if the reader thinks in terms of equilibrium of demand and supply then he or she would suppose that his arguments are built on the tacit assumption of constant returns but that would be a mistake:

> Anyone accustomed to think in terms of the equilibrium of demand and supply may be inclined, on reading these pages, to suppose that the argument rests on the tacit assumption of constant returns in all industries. … In fact, however, no such assumption is made. … The investigation is concerned exclusively with such properties of an economic system as do not depend on changes in the scale of production or the proportions of 'factors'. (Sraffa 1960, p. v)

Just after this proclamation, Sraffa in the next paragraph goes on to state: 'This standpoint, which is that of the old classical economists from Adam Smith to Ricardo, has been submerged and forgotten since the advent of the "marginal" method' (ibid., p. v). Garegnani and almost all the Sraffians interpreted Sraffa's reference to 'this standpoint' as an endorsement of classical economics in almost its entirety including the mechanism of gravitation of 'market prices' to 'natural prices' as its central theoretical pillar. Instead of relating the reference of 'this standpoint' to the propositions of the book or what Sraffa characterised as 'the argument in these pages', the Sraffians related it directly as an allusion to an absence of constant returns to scale assumption in classical economics. From this perspective, it became a problem for Garegnani as to how to square Adam Smith's gravitation mechanism, which obviously rests on quantity adjustments and constant returns, with Sraffa's bold claim that his propositions do not assume any kind of returns to scale, let alone constant returns?

To tackle this problem, Garegnani dropped his original proposition that the classical economics of Adam Smith and Ricardo rests on the assumption of constant returns to scale in manufacturing and went on to claim that the 'core' of classical theory, just like Sraffa's in his interpretation, was to determine prices and the rate of profits from given outputs, technique of production and wages (see Garegnani 1984, 1987, 1990). So the first question that arises in this context is: what are those 'given outputs'? Are they empirical outputs? The answer Garegnani gives is that no, the given outputs are the outputs that the system needs to have to ensure equilibrium of the system or the 'natural prices'. In other words, these are the outputs that the system would arrive at after the supplies have completely adjusted to their effectual demands. These outputs in general will be different from empirical outputs at any point of time:

> … no economist had previously supposed the economy to ever actually *be* in equilibrium position, or more generally in a position of rest, except by fluke: *gravitation around such positions and not achievement of them* being what was always thought relevant for the positions of the economy in the focus of the analysis. (Garegnani 2012, 1429–30, original emphasis)

## 4 How Do We Know 'Given Outputs'?

Thus it becomes quite clear that to know the 'given outputs' of the theory one cannot rely on observation of the outputs produced; one needs to know the effectual demands by other means. This marks the first major difference between Garegnani's description of classical economics and Adam Smith's original formulation. As we have seen above, Adam Smith defines 'natural price' as '[w]hen the price of any commodity is neither more or less than what is sufficient to pay the rent of the land, the wages of the labour, and the profits of the stock employed in raising, preparing, and bringing it to market, according to their natural rates, the commodity is then sold for what may be called its natural price' (*WN*, I.vii.4, p. 72) and then goes on to add, '[t]he market price of every particular commodity is regulated by the proportion between the quantity which is actually brought to market, and the demand of those who are willing to pay the natural price of the commodity, or the whole value of the rent, labour and profit, which must be paid in order to bring it thither. Such people may be called the effectual demanders, and their demand the effectual demand' (*WN*, I.vii.8, p. 73). Here the 'natural price' is a price that can be calculated by given 'natural rates' of rent, wages and profits along with the technique that helps reduce raw materials and means of production costs to their 'natural' wages, profits and rent. In Adam Smith's context, the knowledge of effectual demand quantities is not required for the determination of the 'natural prices'. The constant returns to scale assumption along with the gravitation mechanism take care of the fact that the system would come to rest when supplies completely adjust to their effectual demands, no matter what those quantities of effectual demands happen to be. In Garegnani's description of the classical theory, one needs an extra information about the precise quantities of effectual demands as it does not assume constant returns to scale and therefore cannot allow any adjustments in 'given quantities'.

Before getting to the question of how does one know the precise quantities of the given effectual demands, let us inquire into what does 'given effectual demands' entail? Garegnani argues that in classical economics of Adam Smith and Ricardo, '[o]utputs are therefore naturally determined *separately* from prices, i.e. can be taken as given in determining the latter' (Garegnani 2007, p. 52, emphasis in original—all the page numbers are from Kurz 2013). This implies that effectual demands are ***independent*** of prices as they are not only determined separately from prices but this information is used as an input in determining the prices. In this case, the effectual demand can be drawn as a vertical straight line on the quantity axis in the price-quantity plane. Thus if the actual quantity supplied falls short of the effectual demand then the price must rise to infinity and if the actual supply turns out to be higher than the given effectual demand then price must fall to zero. Hence the classical gravitation mechanism becomes inoperative in this case. For classical gravitation mechanism to be operational, one needs to allow some relation between prices and quantity demanded—the role of price as a rationing devise is essential for the gravitation mechanism to work as Adam Smith explains, '[w]hen the quantity of any commodity which is brought to market falls short of the effectual demand, all those who are willing to pay the whole value of the rent, wages and profit, which must be paid in order to bring it thither, cannot be supplied with the quantity they want. Rather than want it altogether, some of them will be willing to give more. A competition will immediately begin among them, and the market price will rise more or less above the natural price, according as either the greatness or the deficiency, or the wealth and wanton luxury of the competitors, happen to animate more or less the eagerness of the competition' (*WN*, I.vii.9, pp. 73–74). Thus when quantity supplied falls short of the given effectual demand then the 'market price' would turn out to be finite but higher than whatever its 'natural price' happens to be. Let's call it point 'A' on the price quantity plane. Now, as we go on to regularly increase the quantity supplied on the horizontal axis, the 'market price' would regularly fall on the vertical axis and there will come a point where the 'market price' will meet the vertical-straight-line effectual demand when quantity supplied would become equal to the given effectual demand. Let's call this point 'B', which is, by definition the 'natural price' of the commodity. In this case, the 'natural price' gets determined by the curvature or the elasticity of the historical demand curve, which is drawn by simply plotting the 'market prices' by regularly increasing the quantity supplied till it meets the known effectual demand quantity. This, of course, is quite far from Adam Smith's description of the gravitation mechanism; nonetheless, Garegnani's above proposition logically leads to this conclusion.

However, after a few pages, Garegnani puts forward another proposition, which is that the given effectual demand is not a vertical straight line but just a point in the price-quantity space. In other words, the knowledge of effectual demand quantity comes with the knowledge of 'natural price' of the commodity as well: 'The second difference is that the natural price—corresponding to an equilibrium price in modern terms—far from being an unknown to be determined by those "supply and demand" as in neoclassical theory—is there a *given* for the very definition of the demand, the single quantity' (Garegnani 2007, p. 55, original emphasis). If, however, the knowledge of the given effectual demand comes with the knowledge of 'natural price' as well then what remains of the problem of *determining* the natural price and the rate of profits given 'outputs', 'technique' and 'wages'?

Now, let us look at Garegnani's reasoning behind his idea that the vector of 'effectual demand' quantities can be taken as known data. He argues that effectual demands are determined by mainly three factors that the classical economics takes as given. They are: '(1) the level of aggregate income and activity; (2) the technical conditions of production (governing, among other things, the outputs of means of production); (3) the distribution of the social product among the social classes (and therefore, in terms of the classical theories, the level of the independent distributive variable), since different classes generally spend their income on different commodities' (Garegnani 1990, p. 129). It is, however, clear that all the three items in Garegnani's list can be known only if we know the vector of outputs. Otherwise, what does it mean to know 'the level of aggregate income and activity'? Similarly, how can one know what would be the total demand for the commodities that function as inputs unless we know what has been the total utilisation of inputs in the system and what are the savings decisions of the capitalists (assuming workers don't save)? Moreover, the levels of aggregate incomes of either the workers or the capitalists cannot be known unless the vector of outputs is known. In other words, Garegnani's argument runs in a circle: the vector of outputs is supposed to be equal to the vector of effectual demands, which in turn is supposed to depend on the vector of outputs!

So why all this rigmarole? Because the idea of 'given outputs' obfuscates the basic question, which relates to the nature of the 'given technique' rather than 'given output'. Now, it is the data of *observed* inputs and outputs that we can definitely take as 'given'. The problem is that the observed output data at any time are supposed to not coincide with the quantities of 'effectual demands'. Classical economists argue that given the 'rational behaviour' of the agents, there is an inherent tendency in this system that actual outputs tend to adjust to effectual demands. Thus this adjustment process requires us to know how changes in outputs must relate to changes in inputs. The classical economists such as Adam Smith and Ricardo (1821 [1951]) assumed that the prices determined on the basis of either the 'adding up' theory or the labour theory of value do not change due to such supply adjustments. In other words, they assumed that the proportions between inputs and outputs remain constant when such supply adjustments take place. This implies an assumption of constant returns to scale. At one point Garegnani seems to reluctantly accept it: 'Smith and Ricardo could therefore leave physical returns to scale quite naturally aside when dealing with relative prices in a given position of the economy, *with the kind of comparatively small output changes generally involved in that specific analysis*' (Garegnani 2007, p. 53, emphasis added). But then again reverts back to an untenable position: 'the classical separate determination of prices and outputs does not exclude at all the dependence of normal prices on normal outputs because of variable physical returns: so much so that rigour asks us to speak of outputs being data when determining prices' (ibid., p. 54). Strangely, Garegnani never refers to inputs while talking about given outputs. But to write Sraffa's equations he needs precise input data to go along with his given 'effectual demands' or 'normal' output data. Where does he get that from, without the knowledge of returns to scale?

## 5 Why Is There No Room for the Classical Gravitation in Sraffa's System?

The revolution of 1870s that came to dominate economics had rejected the classical idea that labour is the ultimate cause of value. Instead, they argued that the ultimate cause of value is scarcity, which is fundamentally a subjective condition of the intensity of our desire for something in relation to its availability. If something is not freely available in the amount that will satiate us then we are willing to pay a price for it, which can be a sacrifice of our comfort or sacrifice of anything that we possess which has positive utility for us—there is nothing special about loss of comfort or leisure (i.e., labour) as a sacrifice for acquiring something of value. In this context, forgoing consumption is no different from forgoing leisure, and therefore, if forgoing leisure (i.e., labour) must receive a return for it (i.e., wages) then forgoing consumption, which is how savings and capital investment can be interpreted, must also receive a return as interest or profits. Now the question is, how do we measure the sacrifice of consumption? Let us suppose a farmer 'A' sacrifices one quintal of consumption of wheat just harvested and uses it as seed for production of wheat in the next harvest cycle and another farmer 'B' sacrifices one quintal of wheat to use as seed for production of wheat and then another harvest cycle to turn it into beer. Should the two farmers receive the same return on their equal sacrifice of consumption of one quintal of wheat? The answer is no; because farmer 'B' has sacrificed one quintal of wheat for two time periods whereas farmer 'A' has done it only for one. Therefore, farmer 'B' must receive higher return. What we have noticed here is that the notion of sacrifice of consumption has also a time dimension. This gave rise to the idea that capital could also be measured in the time dimension as 'time of waiting' by going back and back in the production cycle of any commodity till we hit upon the primordial state. This was the approach taken up by Jevons (1957 [1871]), Menger (2007 [1871]), Böhm-Bawerk (1959 [1884]) and Wicksell (1934) that had become highly influential in the profession as the alternative to the classical (and Marx's) explanation of profits in terms of some kind of deduction from what legitimately belonged to the workers. In the late 1920s Sraffa had set himself a task of demolishing the theories that rooted economic calculations or the cause of prices and profits in human subjectivity or human psychology.

Sraffa soon realised that in a commodity-producing society where means of production are produced by separate industries and bought and sold by each other in the manner as any final or consumption goods are then it is impossible to trace back production of any commodity to its primordial state—*production of commodities is always by means of commodities.* No matter how far back we go in the chain of production some *commodity residue* will always remain—the road to the primordial stage is theoretically blocked forever! Though it is true that by going back and back in the production chain one can always reduce the commodity residue to negligible proportion and thus ignore it in the calculation of a long chain of labouring activity, it so happens that at what stage the commodity residue becomes negligible depends upon the rate of wages—if wages are relatively high then the commodity residue will become negligible more quickly than when wages are comparatively low; and if wages are zero then the commodity residue will never become negligible. This reveals a fundamental mistake in understanding the relationship between wages and profits when we root our theory of production in the idea of primordial relation of Man to Nature—if we could reduce the production chain to the primordial stage then we could reduce all capital investment to a long series of only wage advances and in this case if wages go to zero then the rate of profits must become infinity; however, if there must remain a commodity residue, no matter how far back we go in the production chain, then when wages go to zero the rate of profits must reach a finite maximum. This theoretical insight had a momentous implication for Sraffa's theory—later Sraffa credited Marx for this insight:

> The notion of a Maximum rate of profits corresponding to a zero wage has been suggested by Marx, directly through an incidental allusion to the possibility of a fall in the rate of profits 'even if the workers could live on air'; but more generally owing to his emphatic rejection of the claim of Adam Smith and others after him that the price of every commodity 'either immediately or ultimately' resolves itself entirely (that is, to say, without leaving any commodity residue) into wage, profit and rent—a claim which necessarily presupposed the existence of 'ultimate' commodities produced by pure labour without means of production except land, and which therefore was incompatible with a fixed limit to the rise in the rate of profits. (Sraffa 1960, Appendix D, p. 94)

The fact that there always remains a commodity residue rules out the possibility of conceptualising industries as vertically integrated independent silos which produce all their means of production themselves and only exchange their final commodities or consumption goods. Once the idea of independence of industries is rejected, we realise that social production relies on a complex web of interconnected industries that produce at least one good that goes directly or indirectly into the production of all goods. An interconnected web of all such 'basic goods' form a social system of production, where removal of one such industry would amount to a complete cessation of the whole economy. It is in this context that we can understand why the industries get structurally constrained such that its productivity or the maximum rate of profits of the system becomes a physical property of the system of production.

Sraffa's theoretical story begins with a subsistence system, which is similar to Adam Smith's 'early and rude state of society' or Marx's 'simple commodity production'. The characteristic of this system is that in a self-replacing state it produces outputs exactly equal to what it uses as inputs—there is no net output production in the sense that all the income received by the labourers appear as necessary consumption similar to feed for the horses. So suppose such a system is given by:

$$
\begin{array}{c}
90\,\text{t.iron} + 120\,\text{t.coal} + 60\,\text{qr.wheat} \rightarrow 180\,\text{t.iron} \\
50\,\text{t.iron} + 125\,\text{t.coal} + 150\,\text{qr.wheat} \rightarrow 285\,\text{t.coal} \\
40\,\text{t.iron} + 40\,\text{t.coal} + 200\,\text{qr.wheat} \rightarrow 410\,\text{qr.wheat} \\
\hline
180\,\text{t.iron} + 285\,\text{t.coal} + 410\,\text{qr.wheat} \rightarrow 180\,\text{t.iron} + 285\,\text{t.coal} + 410\,\text{qr.wheat}
\end{array}
$$

This system must solve the problem of pricing, i.e., the problem of determining the exchange ratios at which the three commodities must exchange with each other so that the system can reproduce itself. In price terms this system can be represented by:

$$
\begin{array}{c}
90p_i + 120p_c + 60p_w = 180p_i \\
50p_i + 125p_c + 150p_w = 285p_c \\
40p_i + 40p_c + 200p_w = 410p_w \\
\hline
180p_i + 285p_c + 410p_w = 180p_i + 285p_c + 410p_w
\end{array} \qquad \text{(i)}
$$

The condition of 'subsistence' that the aggregate values of all inputs must be equal to the values of their outputs reduces this system of equations to only two independent equations and thus given any commodity as the measuring Standard, say $p_w = 1$, we can uniquely determine the values of $p_i$ and $p_c$. Thus the exchange ratios in this system are uniquely and completely determined by the objective input-output data—no extra information from outside is needed.

It should be noted that in this context there is no need to assume equilibrium of the industrial demands and supplies or a self-replacing state—as long as the equations remain intact the solution of prices would remain the same. As a matter of fact, Harrod (1961) in his review of Sraffa's book had suggested that Sraffa was implicitly assuming equilibrium of demand and supply in his price equations to which Sraffa (1962) responded that '(...[even when] the system ceased to be in self-replacing state) the exchange ratio would remain the same but the ratio between the excess productions of the two commodities would be changed, so that the two would not be equal' (p. 477; for a recent debate on this controversy, see Sinha 2016). This brings out the fundamental methodological difference between the classical gravitation idea and Sraffa's position. In the classical or Marx's simple commodity production system if demand and supplies are not equal then 'market prices' would diverge from the labour-value ratios which brings forth the movement of labour in search for higher returns to per unit of labour that in turn brings about supplies in conformity with demands and 'market prices' to 'natural prices' or labour-value ratios. In Sraffa's context, however, prices have no degree of freedom to be anywhere else than what the equations determine them to be irrespective of the system being in equilibrium or not—no dual concept of 'market prices' and 'natural prices' could exist in this case.

Now, let us suppose this system becomes more productive and it produces more outputs than what it uses as inputs, such as:

$$\begin{array}{c} 90p_i + 120p_c + 60p_w \le 180p_i \\ 50p_i + 125p_c + 150p_w \le 450p_c \\ 40p_i + 40p_c + 200p_w \le 480p_w \\ \hline 180p_i + 285p_c + 410p_w < 180p_i + 450p_c + 480p_w \end{array} \tag{ii}$$

Now the constraint of the aggregate equation of the subsistence system no longer holds, and therefore technically we do not have an equation system anymore. We have three independent inequalities with only two unknowns—the excess values of outputs must somehow be accounted for on the left hand side to turn it into a system of equations again. In this case, we do not know what exact ratio in which the three commodities must exchange; there can be several exchange ratios that can allow for this system to get back its original means of production to reproduce itself. It was at this stage that classical economists and Marx thought that they needed extra information from outside the input-output data and introduced the idea of market mechanics and rational behaviour on the part of the agents, which leads to adjustment of supplies with demands in such a way that the system comes to rest when each unit of capital or input receives equal returns. Thus on the basis of this extra information one can introduce one more unknown in the system as the rate of profits such that:

$$
\begin{array}{c}
(90p_i + 120p_c + 60p_w)(1+r) = 180p_i \\
(50p_i + 125p_c + 150p_w)(1+r) = 450p_c \\
(40p_i + 40p_c + 200p_w)(1+r) = 480p_w \\
\hline
(180p_i + 285p_c + 410p_w)(1+r) = 180p_i + 450p_c + 480p_w
\end{array} \qquad \text{(iii)}
$$

Now we can solve for the relative prices and the rate of profits simultaneously. Garegnani (1991) maintains that this has been Sraffa's main contribution to classical economics where he showed that equilibrium 'natural prices' and the uniform rate of profits are determined *simultaneously* and not *sequentially* by first determining 'natural prices' through a labour theory of value and then deriving the average or the uniform rate of profits from it as Marx tried to do.

It is my contention that Sraffa rejects this approach (see Sinha 2016 for details). The assumption of rational behaviour by the agents turns the system into a causal mechanism where supplies adjust to demands to bring about equal returns to factors and this requires knowledge of how changes in inputs relate to changes in outputs for every industry on the side of supply and consumers' subjectivities on the side of demand. Sraffa was of the opinion that the analyst does not have access to such data. He argues that instead of making any assumption about human behaviour and the technique of production one may stick to the data available after the 'harvest' without asking such questions as: why people did what they did or how people will behave if the system is not in 'equilibrium'? He succeeded in showing that there is enough information in this system to not only determine the unique set of prices and the rate of profits but also to establish, what he considered, the fundamental propositions of classical economics and Marx that can stand up as an alternative to the modern economics that is rooted in human subjectivity.

Let us rewrite the equation system (iii) without assuming that the industrial rates of profits are equal:

$$
\begin{array}{c}
(90p_i + 120p_c + 60p_w)(1+r_i) = 180p_i \\
(50p_i + 125p_c + 150p_w)(1+r_c) = 450p_c \\
(40p_i + 40p_c + 200p_w)(1+r_w) = 480p_w \\
\hline
(180p_i + 285p_c + 410p_w)(1+R) = 180p_i + 450p_c + 480p_w
\end{array} \qquad \text{(iv)}
$$

Where $r_j$'s represent the industrial rates of profits and $R$ stands for the weighted average rate of profits of the system as a whole. Objectively, the average rate of profits of this system is given by (165 t. coal + 70 qr. wheat)/(180 t. iron + 285 t. coal + 410 qr. wheat). This ratio is mathematically undefined because it is a ratio of disproportionate heterogeneous goods. It appears that the average rate of profits cannot be found without the knowledge of prices, which is supposed to homogenise these two collections of heterogeneous goods.

However, since $R$ is the average rate of profits of system (iv), each industry in the system should be able to receive a rate of profit equal to $R$.[1] In the case when each industry receives a rate of profit equal to the average rate, the system can be represented by:

$$\begin{array}{c}
(90p_i' + 120p_c' + 60p_w')(1+R) = 180p_i' \\
(50p_i' + 125p_c' + 150p_w')(1+R) = 450p_c' \\
(40p_i' + 40p_c' + 200p_w')(1+R) = 480p_w' \\
\hline
(180p_i' + 285p_c' + 410p_w')(1+R) = 180p_i' + 450p_c' + 480p_w'
\end{array} \quad \text{(iv')}$$

Where $p_j$'s are not equal to $p_j$s if $r_j$s are not equal to $R$.

We, however, know that multiplying any equation by a constant does not change the information set of the equation system in any way. So, let us multiply the equation for iron industry by 4/3 and the equation for coal industry by 4/5. This turns our equation system (iv') to:

$$\begin{array}{c}
(120p_i' + 160p_c' + 80p_w')(1+R) = 240p_i' \\
(40p_i' + 100p_c' + 120p_w')(1+R) = 360p_c' \\
(40p_i' + 40p_c' + 200p_w')(1+R) = 480p_w' \\
\hline
(200p_i' + 300p_c' + 400p_w')(1+R*) = 240p_i' + 360p_c' + 480p_w'
\end{array} \quad \text{(v)}$$

[1] During the early period of his theoretical breakthrough in the winter of 1927–28, Sraffa in a note wrote, 'But since we have (in the no surplus system) a spare equation, we can use it to determine, simultaneously with values, the ratio that total surplus bears to total initial stocks of the community' (D3/12/6: 17, quoted in Sinha 2016, p. 57). And went on to apply equal rate of profits in his system of equation with surplus to determine the ratio of total surplus to initial stock simultaneously with prices.

Sraffa called this system of equations the Standard system and demonstrated that there always exists a set of multipliers (such as 4/3, 4/5, 1) that will convert any given system of equations of basic goods to its Standard counterpart and that the Standard system is unique to all rescaled equations of the original set of equations. Now the objective average rate of profits of the Standard system, i.e., $R^*$, can be found out without the knowledge of prices. The ratio (*40 iron + 60 coal + 80 wheat*)/(*200 iron + 300 coal + 400 wheat*) must always be equal to 1/5 or 20% no matter what prices happen to be, as this ratio is a collection of heterogeneous goods collected in the same proportion. Since in this case, each industry receives a uniform rate of profits equal to $R$, $R^* = R$.

Now, let us go back to our original equation system (iv), if the industrial rates of profits are not equal then some will be greater than the average and some will be smaller than the average. Let us call $r_i = (R + e_i)$, $r_c = (R + e_c)$ and $r_w = (R + e_w)$. Thus we can write our equation system (iv) as:

$$\begin{array}{c} (90p_i + 120p_c + 60p_w)(1 + R + e_i) = 180p_i \\ (50p_i + 125p_c + 150p_w)(1 + R + e_c) = 450p_c \\ (40p_i + 40p_c + 200p_w)(1 + R + e_w) = 480p_w \\ \hline (180p_i + 285p_c + 410p_w)(1 + R) = 180p_i + 450p_c + 480p_w \end{array} \qquad \text{(vi)}$$

By definition $\{(90p_i + 120p_c + 60p_w)e_i + (50p_i + 125p_c + 150p_w)e_c + (40p_i + 40p_c + 200p_w)e_w\} = 0$. Without loss of generality, let us assume that $e_i > 0$ and $e_c$ and $e_w < 0$. Now, again rescale the equation system back to its Standard counterpart by multiplying iron-equation by 4/3 and coal-equation by 4/5. We obtain:

$$\begin{array}{c} (120p_i + 160p_c + 80p_w)(1 + R + e_i) = 240p_i \\ (40p_i + 100p_c + 120p_w)(1 + R + e_c) = 360p_c \\ (40p_i + 40p_c + 200p_w)(1 + R + e_w) = 480p_w \\ \hline (200p_i + 300p_c + 400p_w)(1 + R') = 240p_i + 360p_c + 480p_w \end{array} \qquad \text{(vii)}$$

We should expect the average rate of profits of equation system (vii), i.e., $R'$, to be greater than $R$ simply because we have increased the total weight of the iron industry in the system which has a higher rate of profits than the average $R$. However, from inspection we can see that $R' = R^* = 20\%$.

And since we have already established that $R^* = R$, it follows that all the $e$'s must be equal to zero. In other words, all the industrial rates of profit must be equal and equal to $R^*$, i.e., $r_i = r_c = r_w = R = R^*$. Now we can plug the value of $r_i = r_c = r_w = R = R^* = 20\%$ into equation system (iv) and solve for prices. This reduces the nature of our equation system to linear equation system just like it was for the subsistence system. Thus we do not need the market mechanics and rational human behaviour and the assumption of equilibrium to solve for prices in this case either—the required information to solve for the system of equations of basic goods could be found out by rearranging the data. In other words, the condition of a uniform industrial rate of profits is a structural property of the equation system. Prices yet again have no degree of freedom to be anywhere else than what the equations determine them to be. This explains why, while introducing the rate of profits in his system of equations on page 6 of his book, Sraffa (1960) asserted, without any explanation, that the rate of profits 'must be uniform for all industries'. It should be noted that the above argument will not be possible without the knowledge of the multipliers that converts the empirical system of equations to its Standard counterpart and that this Standard counterpart is unique to any given empirical system—the Standard system shows that the average rate of profits of any empirical system of basic goods is a purely physical property of the system.

Now let us introduce labour in the system explicitly and draw out a complete structural relation of any given system of production. We go back to Sraffa's original example of a three-commodity economy:

$$90\,\text{t.iron} + 120\,\text{t.coal} + 60\,\text{qr.wheat} + 3/16\,\text{labour} \rightarrow 180\,\text{t.iron}$$
$$50\,\text{t.iron} + 125\,\text{t.coal} + 150\,\text{qr.wheat} + 5/16\,\text{labour} \rightarrow 450\,\text{t.coal}$$
$$40\,\text{t.iron} + 40\,\text{t.coal} + 200\,\text{qr.wheat} + 8/16\,\text{labour} \rightarrow 480\,\text{qr.wheat}$$

This can be represented in price terms as:

$$\begin{array}{c}
(90p_i + 120p_c + 60p_w)(1+r) + 3/16w = 180p_i \\
(50p_i + 125p_c + 150p_w)(1+r) + 5/16w = 450p_c \\
(40p_i + 40p_c + 200p_w)(1+r) + 8/16w = 480p_w \\
\hline
(180p_i + 285p_c + 410p_w)(1+r) + w = 180p_i + 450p_c + 480p_w
\end{array} \qquad \text{(viii)}$$

Converting it to its Standard counterpart, we get:

$$\begin{array}{c}
(120p_i + 160p_c + 80p_w)(1+r) + 4/16w = 240p_i \\
(40p_i + 100p_c + 120p_w)(1+r) + 4/16w = 360p_c \\
(40p_i + 40p_c + 200p_w)(1+r) + 8/16w = 480p_w \\
\hline
(200p_i + 300p_c + 400p_w)(1+r) + w = 240p_i + 360p_c + 480p_w
\end{array} \quad \text{(ix)}$$

Now, let us normalise our Standard net output to one, i.e., put $(40p_i + 60p_c + 80p_w) = 1$, and call it the Standard commodity, which is our money or nominal commodity for measurement. If wages are given in terms of this money-commodity, i.e., as a proportion of the Standard net output, then we can derive the average rate of profits of this system for all the values of wages starting from zero to its maximum value $(40p_i + 60p_c + 80p_w)$, because the ratio of total profits to total capital remains in the Standard proportion and therefore can be determined without the knowledge of prices. This gives us a relationship between wages and profits, which is given by: $r = R^*(1 - w)$, where $w$ is given in terms of the Standard net output and $R^*$, which we have already derived from equation system (v), is the maximum rate of profits of the system—it is the ratio of net output to total capital or the productivity of the system. This relationship between the productivity of the system and the rate of profits and wages is derived on the basis of the objective data without any knowledge of prices. Thus it is the fundamental structural property of the equation system. Since the Standard system is only a rescaled system of the actual system of observation they are algebraically equivalent systems, and therefore, the mathematical properties of the two systems must be identical too. In other words, the relationship $r = R(1 - w)$ must also hold for the observed system, if the wages and prices in the observed system are measured by the Standard commodity as the chosen money-commodity. What we directly observe in the Standard system in terms of physical data must show up to be true in the empirical or observed system in terms of its calculations in prices:

> Such a relation is of interest only if it can be shown that its application is not limited to the imaginary Standard system but is capable of being extended to the actual economic system of observation. … But the actual system consists of the same basic equations as the Standard system, only in different proportions; so that, once the wage is given, the rate of profits is determined for both systems regardless of the proportions of the equations in either of them. Particular proportions, such as the Standard ones, may give transpar-

> ency to a system and render visible what was hidden, but they cannot alter its mathematical properties. ... The same rate of profits, which in the Standard system is obtained as a ratio between *quantities* of commodities, will in the actual system result from the ratio of aggregate *values*. (Sraffa 1960, pp. 22–23, original emphasis)

When we move wages from zero to their maximum value, we find that as wages and the rate of profits change, the set of prices change too. But these prices change only to ensure that for every given $w$, prices adjust in such a way that the structural property of the equation system, $r = R(1 - w)$, is satisfied throughout—i.e., prices play the role of accounting for the distribution of income, which is determined independently of prices. Notice that when all the income goes to wages then the value of the net Standard output is equal to the value of the net output of the observed system $\{(40p_i + 60p_c + 80p_w) = (165p_c + 70p_w)\}$, since in this case the prices would be proportional to labour-values and both the systems use the same technique and one unit of labour to produce their respective net outputs. For any other rate of wages the values of observed net output will not be equal to the value of the Standard net output. However, the ratios of net output to total capital will remain constant with respect to changes in prices throughout the variations of wages from zero to its maximum value, as $R^*$ is determined independently of prices.

In some sense, given wages, the average rate of profits determined by the Standard system gives us the multiplication factor for homogenising capital, which appears to us as heterogeneous collection of commodities. Sraffa's procedure of homogenisation of labour is dependent on the available objective data alone—it is simply a processing of proportions of total wage bills paid in various industries as proportions of total undifferentiated labour utilised in those industries:

> We suppose labour to be uniform in quality or, what amounts to the same thing, we assume any differences in quality to have previously reduced to equivalent differences in quantity so that each unit of labour receives the same wage. (Sraffa 1960, p. 10)

Similarly, homogenisation of capital requires us to measure capital in such a way that profits received by capital turns out to be equal for every unit of capital in the system—i.e., a return on the value of iron in each industry must be equal to a return on the value of coal and wheat when they are used as capital. This is possible only if returns on values of all commodities used as capital turn out to be equal to the average rate of profits of the system as a whole. The Standard average is the only average that can be distributed equally across the industries.

Clearly there is no room for 'market prices' and therefore classical gravitation mechanism in Sraffa's system. The notion of 'market prices' is based on the idea that when the system is out of equilibrium of demand and supply then prices are determined from outside the system of equations and when these prices are fed into the equations as givens then the industrial rates of profits get determined by the equations showing inequality in their rates as a result. But what goes unnoticed in this argument is that in this case the *system of simultaneous equations* gets extinguished—all industries become independent of each other since their rates of profits can now be determined by feeding the known 'market prices' into their equations independently of all other equations—we are back to looking at industries as vertically integrated silos. In a note written in 1943, Sraffa reinforces the idea of a system of equations and its relation to the aggregate equation by suggesting, 'Yet the aggregate is not a statistical result, but an algebraic one' (D3/12/36: 79, underlining in original, quoted in Sinha 2016, p. 134). And in a note written in 1956, Sraffa declares the idea of taking 'market prices' from outside the equations as 'ingrained notion of economists' and reiterates that the rate of profits must at any moment be equal throughout the changes in wages:

> If we now move in the opposite direction and, starting from the de facto level of wages, raise them until they absorb the whole net product of society and r = 0, we see that the distribution can be effected by a uniform rise of wages in all industries, through a readjustment of the prices of the products. The adaptation of the prices of commodities is what makes possible to reconcile the equality of the rate of surplus value and the equality of the rate of profits; [the equality of r holds at any moment, statically, the equality of both holds dynamically, in the sense that a change in either will change the other equally, till they both reach zero]. This relation is obscured by the ingrained notion of economists that prices of commodities are handed down from somewhere, so that they are independent of the way in which the proceeds are distributed. An illustration {above "consequence"} of this view is that two such opposite writers as v. H {von Hayek} and J.R. {J.R. Hicks} take it for granted that in a state in which the workers receive the whole product of industry this would inevitably result in the anomaly of higher

wages in industries using much capital. (Sraffa n.d., D3/12/54: 8, underlining in original, quoted in Sinha 2016, p. 150)

This may sound as curious or as bizarre as quantum entanglement. But this is a general property of a holistic system. Its structural properties do not allow for a trajectory of changes. Given Sraffa's equations, if wages are moved then prices must *jump* without loss of time in such a way that the relationship, $r = R(1 - w)$ is always maintained—'the equality of r holds at any moment'. In other words, what Sraffa shows is that the inputs and outputs data are facts and these facts are related in a particular relationship. This relationship always holds. Facts can change from time to time but the relationship must remain intact. This position is further reinforced in yet another note of 14.2.56 in relation to changes in $w$:

> The impulse towards price change is an internal one to each industry. It arises from its own internal conditions—not from those conditions compared with those of other industries. Hence the possibility of an invariable commodity. (Sraffa n.d., D3/12/59: 24, underlining in original, quoted in Sinha 2016, p. 148).

Here what is claimed is that with a change in $w$ the average rate of profits of the system must change by the formula $r = \mathrm{R}(1 - w)$. Now, if the direct-to-indirect labour ratios are not equal for all the industries in the system then some industries will have surplus and some deficit, which is solved by accompanying changes in the prices of commodities of those industries. In other words, given the structural property of the equation system that equal rate of profits must hold at any moment, a surplus or a deficit implies that industrial 'equations' no longer hold and the equations can be re-established only through changes in prices. There is no mechanism through which these prices change. It cannot be observed as a phenomenon taking place over a period of time. And it is this structural relation that reveals why the Standard commodity is not just any *numéraire*. It is because if there were an industry that would experience neither a surplus nor a deficit with a change in $w$ then its equation will remain intact and therefore it would be the ideal Standard by which the changes in prices, due to changes in $w$ or $r$, of all other commodities could be properly measured.

It is highly interesting that Sraffa clarifies this point by contrasting it with the classical case of *comparison* of differences in the rates of profits of industries and the accompanying mechanism of changes in prices due to changes in supply over a period of time. In the classical case, even if we assume constant returns to scale for all the industries as well as the condition that the use of total labour in the system remains constant all through the adjustment, a change in the commodity mix of the net output as a consequence of changes in *w* cannot be denied. It is because the change in prices in the classical case is not *structurally* related to changes in wages but is rather due to an external factor, i.e., due to the fact that the increase in quantities supplied of the commodities can only be sold at lower prices and the fall in the quantities supplied of commodities can only be sold at higher prices. In Sraffa's case, prices change but outputs remain constant.

The distance between Garegnani's interpretation of Sraffa's equations and Sraffa's own understanding of it comes out clearly when we juxtapose Garegnani's statements regarding his 'given outputs' with Sraffa's statements regarding his system of equations. As we have already quoted above, according to Garegnani, 'no economist had previously supposed the economy to ever actually *be* in equilibrium position, or more generally in a position of rest, except by fluke'; and further down in the same paper he writes, 'The "normal position" may be taken as a typical instance of Pareto's "ideal phenomena" in economics, centred as it is on Adam Smith's "central price", to which "the prices of all commodities are continually gravitating" (1776, I: 51) and therefore providing what Pareto calls here a "general or average fact"' (p. 95, n. 72). Contrast this with Sraffa's unambiguous statement that his equations refer to the 'actual economic system of observation' (Sraffa 1960, p. 22). Furthermore, as late as 1968, Sraffa, in a response to a letter from a German student named Soltwedel, wrote, 'As regards your own interpretation, I must say frankly that you have gone astray the moment you speak of "equilibrium" or of "elasticity of factor supply": all the quantities considered are what can be observed by taking a photograph, there are no rates of change, etc.' (C 294: 3, dated 1.3.68, quoted in Sinha 2016, p. 214). Clearly one can take a photograph only of something that exists—the metaphor of a photograph is meaningless for a situation which is ideal but non-existent. In fact, when Sraffa, in the opening sentence of the Preface to the book, stated, '[a]nyone accustomed to think in terms of the equilibrium of demand and supply may be inclined, on reading these pages, to suppose that the argument rests on a tacit assumption of constant returns in all industries. … In fact, however, no such assumption is made', he was not only referring to the absence of the constant returns to scale assumption but more importantly to an absence of the notion of 'equilibrium' in his book. This becomes clear from a look at the draft Preface to the book he wrote in April 1957, where the opening sentence began with the declaration: 'This is ~~no~~ <not proposed as a complete> system of equilibrium' (D3/12/46: 20, quoted in Sinha 2016, p. 151).

## 6 Where Did Garegnani Go Wrong?

Garegnani's understanding of the core of classical theory and Sraffa's book rests almost entirely on the second chapter of Sraffa's book with an added assumption of classical equilibrium as an obvious explanation for Sraffa's use of equal rate of profits in his equations in this chapter. This, in my opinion, was a serious mistake on Garegnani's part. The arguments of Chap. 2 were just a step towards a thesis that was not to establish that, given 'outputs', 'technique' and 'wages', the 'natural prices' and the rate of profits are determined simultaneously. What Sraffa establishes in his book is the thesis that 'the rate of profits is a non-price phenomenon' (D3/12/53: 32, 1955). This is what Sraffa believed was the central aspect of classical economics and Marx. Adam Smith had clearly stated that wages and the rate of profits are determined in the dynamic context of history, and for any point of time, they are given 'norms' and prices are determined by 'adding up' the given distributional variables. In other words, it is the distribution of income that determines prices—the rate of profits is not determined simultaneously with 'natural prices'. In Sraffa's interpretation (Sraffa 1951), Ricardo had started off with the proposition that 'it is the profits of the farmer that regulate the profits of all other trades'; implying that in agriculture both inputs and outputs can be treated as a single commodity 'corn'. In this case too, the rate of profits could be determined in physical terms independently of prices, and thus prices of all other commodities must adjust in such a way that all industries receive the same rate of profits. Marx's theory of surplus-value and his derivation of the rate of profits prior to the derivation of prices of production was also designed to show that profit is a non-price phenomenon. Sraffa's method of deriving the system's average rate of profits independently of prices and then applying them to production equations to derive the prices follows Marx's procedure almost step by step except that instead of deriving the production equations in terms of labour-values from the empirical input-output data Sraffa derives the equations of his Standard system from the same input-output data, which is a system of production of the weighted average commodity (i.e., the Standard commodity) of the empirical system. This thesis was so central to the book that, prior to its publication, Sraffa had thought of putting this idea in a couple of slogans or epigrams. One of the slogans was: 'A dividend could be declared before knowing what is the price of the company's product'. The other slogan was: 'The St. Syst {Standard System} provides tangible evidence of the rate of profits as a non-price phenomenon' (Sraffa n.d., H2/89, f. 56; quoted in Pasinetti 2001).

The establishment of the thesis that the rate of profits is a non-price phenomenon was crucially based on Sraffa's discovery of the 'commodity residue'. It was the discovery of the 'commodity residue' that gave rise to his 'Hypothesis' that the maximum rate of profits of the economic system must be its structural property and must remain fixed with respect to changes in prices due to changes in the rate of profits or wages. This is what gave rise to the development of the Standard system and the Standard commodity. The discovery of the 'commodity residue' was so central to Sraffa's book that he emphasised it in the very title of the book: *Production of Commodities by Means of Commodities.* In other words, there is no way out of the circle of commodities. Garegnani and most of the Sraffians simply failed to recognise its significance.

## References


Böhm-Bawerk, Eugen von. 1959[1884]. *Capital and Interest*, vols. 1–3, Illinois: Liberation Press.

Garegnani, P. 1959. *A Problem In The Theory Of Distribution From Ricardo To Wicksell*, Ph.D. Dissertation, University of Cambridge.

Garegnani, P. 1984. "Value and Distribution in the Classical Economists and Marx". *Oxford Economic Papers*, 36: 291–325.

Garegnani, P. 1987. 'Surplus Approach to Value and Distribution'. The New Palgrave. A Dictionary of Economics, edited by J. Eatwell, M. Milgate, and P. Newman, Vol. 4, pp. 560–74. London: Macmillan.

Garegnani, P. 1990. 'Sraffa: Classical versus Marginalist Analysis', in *Essays on Piero Sraffa critical perspectives on the revival of classical theory*, K. Bharadwaj and B. Schefold (eds), Delhi: Oxford University Press.

Garegnani, P. 1991. 'The Labour Theory of Value: "Detour or Technical Advance"', in G.A. caravale (ed.), *Marx and Modern Economic Analysis*, Vol. 1, Aldershot: Edward Elgar.

Garegnani, P. 2007. 'Professor Samuelson on Sraffa and the Classical Economics', *The European Journal of the History of Economic Thought*, 14(2): 181–242.

Garegnani, P. 2012. 'On the present state of the capital controversy', *Cambridge Journal of Economics*, 36(6): 1417–32.

Harrod, R.F. 1961. 'Review of *Production of Commodities by Means of Commodities. Prelude to a critique of economic theory*', *The Economic Journal*, LXXI, 783–787.

Hollander, S. 1992. *Classical Economics*, Toronto: University of Toronto Press.

Jevons, W.S. 1957[1871]. *The Theory of Political Economy*, New York: Kelly & Millman, Inc.

Kurz, H.D. (ed.). 2013. *The Theory of Value and Distribution in Economics Discussions between Pierangelo Garegnani and Paul Samuelson*, London: Routledge.

Marshall, A. 1949 [1890]. *Principles of Economics: An Introductory Volume*, eighth edition, London: Macmillan and Co. Limited.

Marx, K. 1977[1867]. *Capital*, Vol. I, New York: Vintage.

Marx, K. 1991[1894]. *Capital*, Vol. III, London: Penguin Classics.

Menger, Carl. 2007[1871]. *Principles of Economics*, Auburn: Ludwig von Mises Institute.

Pasinetti, L. 2001. 'Continuity and change in Sraffa's thought: an archival excurses', pp. 139–56 in Cozzi, T. and R. Marchionatti, (eds), *Piero Sraffa's Political Economy: A Centenary Estimate*, London: Routledge.

Ricardo, David. 1951[1821]. *On The Principles Of Political Economy And Taxation*, 3rd edition, Cambridge University Press.

Samuelson, P. 1978. 'The Canonical Classical Model of Political Economy', *Journal of Economic Literature*, 16: 1415–34.

Sinha, Ajit. 2016. *A Revolution in Economic Theory: The Economics of Piero Sraffa*, Cham: Palgrave Macmillan.

Smith, Adam. 1981[1776]. *An Inquiry Into The Nature And Causes Of The Wealth Of Nations Volume I*, Indianapolis: Library Fund.

Sraffa, Piero. 1925. 'Sulle relazioni tra costo e quantita prodotta', *Annali di Economia*, vol. 2: pp. 277–328. English Translation by John Eatwell and A, Roncaglia in *Italian Economic Papers*, vol. 3, ed. L.L. Pasinetti, Oxford: Oxford University Press.

Sraffa, Piero. 1951. 'Introduction', *Works and Correspondence of David Ricardo vol. I*, Cambridge, Cambridge University Press.

Sraffa, Piero. 1960. *Production of Commodities by Means of Commodities*, Cambridge, Cambridge University Press.

Sraffa, Piero. 1962. 'Production of commodities: A comment', *The Economic Journal*, 72(286): 477–79.

Sraffa, Piero. n.d. '*Sraffa Papers*', Wren Library, Trinity College, University of Cambridge, Cambridge.

Wicksell, Knut. 1934. *Lectures on Political Economy, vol. I: General Theory*, London: George Routledge and Sons, Ltd.

Comment

# Sraffa's Protoeconomics—A Comment on Ajit Sinha's 'Interpreting the Nature of Sraffa's Equations: A Critique of Garegnani's Interpretation'

*Bertram Schefold*

Sinha's chapter opens with an interesting question. What is new in the way of how Sraffa leads back to classical economics? According to Sinha, Sraffa disliked that Smith used subjectivist hypotheses like self-love. There are notes in the archive where Sraffa calls Smith the first of the vulgar economists. Smith's adding-up theory of value may be one of the reasons why Sraffa chose this denomination. Another could be that Sraffa strove to eliminate subjectivist foundations of economic thought from his analysis, and Sinha suggests that Smith's use of self-love constituted such a subjectivism. We know from Smith's *Theory of Moral Sentiments* that his psychology was more sophisticated than that, although it is true that, in the *Wealth of Nations*, not much remains of the more complex notions except this self-love. No economist would deny the importance of profit maximization as an explanation of capitalist behaviour. Smith's notion of self-love is broader than that, hence the ease with which neoclassical economists identify with Adam Smith in the belief that, where there is self-love, there is also the maximization of utility. Sraffa, according to Sinha, wished to eliminate such subjectivist notions. To be more precise, Sraffa did not deny the reality of profit maximization, but he sought to found his theory on another basis. It is true that profit maximization is not invoked explicitly in *Production of Commodities by Means of Commodities*, but there is the uniform rate of profits. Sraffa says, where he introduces this rate, that it "must" be uniform, without explaining the "must" (Sraffa 1960, p. 6). Garegnani does not hesitate to invoke competition, driven by the maximization of profits, to explain this "must", according to Sinha (2021). Ajit Sinha believes that here the pupil misunderstood his master. Sinha cannot seriously deny that Sraffa accepted the reality of profit maximization, but he is of the opinion that Sraffa did not make such an assumption. Hence he claims that Sraffa's theory of prices can be founded without it. How can we interpret such a claim?

We ask more generally how a new academic discipline can be founded or on which grounds a new theory can be erected. One difficulty of founding economic theory as a scientific discipline derives from the fact that we are part of the economy, our behaviour is shaped by our economic experiences and our language is full of economic terms. Their significance should be explained and not be taken for granted.

The conceptual framework of the new theory must be founded on something outside the subject under consideration; otherwise, one moves in a circle, because *explanans* and *explanandum* get mixed up.[2] The concepts thus introduced must refer to a universe of which we have an intuitive understanding. We have, in a Kantian tradition, to reflect on the conditions of cognition. Kant thought that the instruments for our intuitive apprehension of the empirical world were given to us, as far as space was concerned, by our vision of Euclidian space, and the conceptual apparatus for its description was given prior to experience in the form of mathematics. The foundations of physics therefore could not be anything. The movement of masses in mechanics could occur only according to laws compatible with Euclidian space, with our understanding of time and our consequent analysis of movement. Kant therefore undertook to derive the fundamentals of physics, for him the Newtonian axioms, from his philosophy in his *Metaphysische Anfangsgründe der Naturwissenschaft* (Kant 1787), and the result seemed convincing until it was understood that there were also other geometries beyond the Euclidian one, that the speed of light and

[2]This circle is on the contrary potentially virtuous, not vicious, according to the hermeneutical approach as an alternative to the Kantian position: The study of, for example, philosophical texts begins with a reading based on the ordinary understanding of the concepts used. These take on a deeper meaning for the reader as the concepts can be used to clarify their meanings mutually by repeated reading, by studying connexions between related texts, by contextual analysis and by communication with others engaged in same endeavour. This is an inductive approach to science, whereas Sinha takes *Production of Commodities by Means of Commodities* to be a piece of deductive reasoning.

communication had a specific limit and that elementary particles were quantized.[3] Kant had hoped thus to get to fundamental statements that would be true a priori and yet be more than tautologies (that would be "synthetic" and not only "analytical" truths). In this he failed, but the postulate that the instrument of the description should be given prior to the description has remained as a methodological principle for deductive reasoning.

But the postulate cannot be fulfilled completely. What could be the fundamental concepts outside economics which one might use to found economic theory? How shall we introduce concepts such as good, commodity, labour without a prior understanding of economic matters, basing ourselves exclusively on a non-economic discourse? The neoclassicals, at least in the Austrian variant (Mises 1940), reply that all action is economic and everything economic has to do with human action. The essential propositions of economics then follow from the understanding of action, by gradually filling up this deductive conception with intuitive content. The economic agents, in possession of their faculties and of goods, can produce for themselves or exchange. They exert their freedom to act in the act of exchange. While Weber wanted to understand and order different rationalities, there are not different rationalities but different degrees of understanding of what is going on in the Austrian approach. A rain dance is rational action, insofar as one tries to influence superhuman forces to make it rain, but one is in error regarding the causes, which make the weather change. Rationality then implies utility maximization along with other rules.

Marx, with his dialectical materialism, chose a different procedure. He took a number of elementary economic concepts such as labour and exchange and defined plausible relations from which he could deduce a theoretical framework, such that more complex notions could be derived. He assumed that there was labour, distinguished between abstract and concrete forms, and postulated that goods are exchanged according to their content of abstract labour—a proposition, which seemed roughly plausible, but of which Marx knew that it was only a provisional

[3] In what was in my recollection the most fascinating seminar during my time as a student, the physicist and philosopher Carl Friedrich von Weizsäcker took us through the *Metaphysische Anfangsgründe* and showed from paragraph to paragraph how Kant deduced the elements of Newtonian mechanics from his theory of cognition and failed to arrive at Einsteinian physics primarily because he took assumptions such as that of an infinite velocity in the transmission of information or Euclidian properties of space as obvious truths. A short introduction to this exercise is in von Weizsäcker (1971, pp. 189–192).

approximation to the true state of things in a competitive capitalist world. He got from a theory of value to a theory of exchange, money and production, from there to exploitation and accumulation. The drive of capital was to chase after surplus value, but it first had to be explained that profit was a form of surplus value, and before there could be talk of a maximization of the rate of profit, the rate of profit had to be defined. Competition was not any form of human rivalry, but it showed in the equalization of the rates of profit and the tendency for a general rate of profit to form. The uniform rate of profit was therefore not simply assumed as a result of competition, but, on the contrary, competition had first to be explained in terms of the process of the equalization of rates of profit, and these resulted from the drive to amass surplus value by exploiting labour.

As soon as this has been understood, one can see (cf. Schefold 2019), looking at the beginning of the Marxian construction, that it is really the opposite of what Austrian neoclassicism assumes: The value ratios do not result from the free intercourse of independent agents who may decide to swap or not to swap, but the exchange takes place between interdependent producers according to value relationships that are given to them. The material circumstances determine what they can do; Marx sticks to his materialist determinism even where he deals with entrepreneurial action. The capitalist is only the "character mask"; the process of surplus value creation is impersonal and anonymous.

Sinha is right that Sraffa has his own way of founding economics, at any rate that section of it which is concerned with value, price and distribution. I used to call this Sraffa's protoeconomics. He begins with his famous consideration of an "extremely simple society". Not the logic of action, not predisposed labour times, but the structure of the reproduction of commodities determines the price ratios. The author leads the reader from assumptions to conclusions. The homogeneity of the commodities is clearly simply assumed, and it is simply assumed that the structure of production is such that self-replacement is possible.

Various metaphors have been used to visualize how these magnitudes become known. Sinha prefers to speak of an observer. This is reminiscent of Einstein's thought experiments, and Sraffa, in taking up this particular metaphor, probably was influenced by contemporary discussions of modern physics, as Kurz and Salvadori (2005) argue. Thought experiments involve strong assumptions and abstractions. Sinha is in the danger of taking the metaphors too literally. Sraffa begins his protoeconomics with

$$\mathbf{Ap} = \mathbf{p} \text{ and } \mathbf{eA} = \mathbf{e}$$

where **e** is the summation vector, so that the second formula is the otherwise self-explanatory formalization of "self-replacement". If this is assumed and **A** is an indecomposable semi-positive input output matrix, the second equation shows that **e** is a Frobenius eigenvector of **A**, with an eigenvalue of 1, so that the right-hand side Frobenius eigenvector $\mathbf{p} > 0$ must also exist: The existence of prices is guaranteed. It is a foundation of prices, independent of a labour theory of value or of utility theory.

There follows the model of production with a surplus, where, instead of labour inputs, wage goods consumed by the labourers, whatever their social status, slaves, serfs or wage labourers, are treated on the same footing as the inputs to production. The formula summarizing this model is

$$(1+R)\mathbf{Ap} = \mathbf{p}$$

Here we have the uniform rate of profit $R$, of which Sraffa said that it "must" be uniform, without telling us why. Sinha sees two possibilities to explain the "must". There is his own attempt to provide a structural reason and Garegnani's reference to competition, which Sinha criticizes extensively. Sinha's procedure is curious. The structural reason, which he tries to give, is nowhere to be found in Sraffa's published writings. The opposition between Garegnani and Sraffa, which he tries to reconstruct, is a priori not plausible. Garegnani was without question Sraffa's most important pupil, his friend and support over many years to the end. Garegnani may have had his own agenda, but there was trust between them, and Garegnani was appointed Sraffa's literary executor. Finally, it does not occur to Sinha that a structuralist explanation of the "must", if one really exists, and Garegnani's interpretation do not necessarily exclude each other; they might be complementary on two different levels of abstraction. Garegnani's interpretation is indispensible, if one wants to get from protoeconomics to economics. For instance, the importance of competitive processes shows in classical theory in the treatment of market prices. If one moves from Sraffa's protoeconomic prices to the classical theory, whether one speaks of the Ricardian theory of rent or of the

Marxian theory of accumulation, whether one deals with Malthus and effective demand or with John Stuart Mill and the stationary state: One will always use natural prices to conceptualize different states of the economy in evolution, and one will know that market prices deviated from them, in ways that depend on competitive processes, because of harvest cycles, crisis and political disturbances. And what is the meaning of the natural prices, if the market prices do not somehow gravitate to them?

Sinha nowhere explains how he wants to get from protoeconomics to economics and lose his innocence, as it were. The diversity of his observations forces me also to make diverse remarks that cannot be ordered without some arbitrariness.

We can start from market prices, since Sinha begins with Garegnani's interpretation of the formation of market prices in Adam Smith. He speaks of demand and supply schedules in this context, as if Smith had a visualization of supply and demand as in neoclassical theory. Here Sinha seems to me to be uninformed about the state of discussion of supply and demand in the eighteenth century. In particular, reference should be made to James Steuart. In the eighteenth century, and still in Marx, demand and supply were seen as "forces". This visualization of the market process is fundamentally different from that expressed in terms of schedules. Such schedules are absent in Smith; they first appear in Rau and in Cournot. But this is another story which it would take longer to explain (Schefold 1997).

An example that we find in the earlier literature concerns the arrival of a ship in a colony (e.g. Steuart 1967 [1767, 1805], pp. 270–273). If people in the town can expect that other ships will follow soon, the merchant will sell his wares at more or less habitual prices, that is, prices that reflect normal costs, because the buyers will otherwise prefer to wait. If the arrival of other ships is more uncertain, moral suasion will play a role. The examples given by Adam Smith comprise the conditions in a beleaguered city, where prices can rise very high because of extreme scarcity. The gravitation of market prices to natural prices therefore is a matter of anecdotal circumstances. There is no theory of gravitation in Adam Smith, but he has an idea what it is and where it can lead, and we, of course, can construct modern models of gravitation.

In doing so, we have many possibilities. The natural price may be realized as an average over many years. This applies in particular to agricultural commodities. Or market prices can be close to natural prices, because charging mark-ups on direct costs such that normal profits are obtained in the long run is a good strategy for producers of industrial commodities.

The adaptation of prices can be formalized as a convergent process, if the quantities are rigidly given; an example is the following self-explanatory sequence which, if it starts from any prices, will converge to natural prices in terms of the wage rate:

$$\mathbf{p}_{t+1} = (1+r)\mathbf{A}\mathbf{p}_t + \mathbf{l}$$

There are more sophisticated gravitation models with cross-dual dynamics (the result, whether gravitation obtains, depends on assumptions). One can even regard intertemporal general equilibrium theory as the gravitation process of market prices to prices characterized by a uniform rate of profit, if conditions for the supply of unproduced factors on the one hand and demand conditions on the other are stationary, as I have discussed elsewhere. Sinha is right that Garegnani has made different attempts to formalize gravitation processes, which were not all equally convincing. My conclusion is that gravitation is an axiom of classical theory, not a result, and many and diverse models have been invented to justify this axiom.

His misconception of what gravitation is about leads Sinha to a number of unhelpful statements. It is true that Smith was not interested in the repercussions of gravitation in one market on gravitation in other markets. But that does not mean that Smith was generally unaware of the interdependence of markets; he was simply arguing, as one later would say, *ceteris paribus*. He was right to be cautious. If one talks about the interdependence of markets, one should not too easily postulate the law of one price. The same commodity can have different prices in different sectors in the real world. A detailed model of the gravitation process then becomes very complex; yet, the tendency remains theoretically and empirically plausible.

Why does Sraffa in *Production of Commodities by Means of Commodities* not need assumptions about returns to scale? One reason is this: Because the analysis of gravitation is not part of protoeconomics, although it may help to justify the assumptions. Sraffa hints however at some applications where constant returns to scale are the most convenient hypothesis for getting an intuition of what he discusses. An example is his rising supply schedules in the case of intensive rent.

I do not agree that the contrast between natural prices and market prices could not be introduced in the system without a surplus, as Sinha seems to believe, but there is no room to introduce such a special gravitation model here.

I hope we can agree that Sraffa's system is so interesting not the least because of its open character with regard to distribution, the composition of output and accumulation (Schefold 1987, Part Three). His theory is strong and rigorous, as far as the logic of the formation of natural prices is concerned, if distribution is given, but the influences on distribution may be varied and quite different, depending on whether one is dealing with an economy experiencing strong growth or one which is stagnant. The structure of the economy (the protoeconomics) does not suffice to explain how the economy will behave regarding changes of distribution, demand and employment. The question then is how the step from protoeconomics to economics is to be made. Sinha seems to be tempted to simplify too much, when he uses the image of the observer in this context. He believes that mere observation could reveal the protoeconomic structure of an actual economy. Can one not see what factories there are, how many workers there operate and how many consumption goods are being delivered?

But the observer of the real economy will not see homogeneous commodities, for competition regarding the quality of commodities will accompany competition regarding prices. The observer does not see one technique in use in any industry, but usually techniques that rival with each other and the dominant technique cannot easily be identified. Actual production is not based on single-product industries, but on a certain degree of joint production in most cases, and often it will be difficult to ascribe outputs and even inputs to specific processes. The observer, in trying to catch the underlying system, will not be able to avoid subjective interpretations, attempts to understand the intentions of producers and the desires of consumers. The bold simplicity of Sraffa's assumptions cannot be defended by the ingenuous postulate of the observer. I think it is one of the best and most original insights in *Production of Commodities by Means of Commodities* that Sraffa characterizes long period positions by an equality of the number of commodities produced and the number of processes used. There are not many economists out there who have understood what a long period position is, and the number of those who have understood this condition is even smaller. It is quite naive to believe that this condition could be verified in reality in a simple and straightforward manner. And yet I believe that phenomena, by which this law imposes itself, can be observed. I tried to show it using examples of energy production and consumption (Schefold 1985). It is a domain where the observer seems to be confronted with objective data: Household *a* uses gas, household *b* uses electricity for heating, country *A* uses nuclear power for electricity production and country $B$ uses wind. But which system will dominate? What is "socially necessary" here depends on a political process, which needs to be ***understood.***

We now turn to Sinha's assertion that the uniformity of the rate of profit can be described as a structural property of the system. In a quite formal sense this is trivial and a matter of definition. But if unequal rates of profit are admitted as a possibility, the question becomes whether a meaningful structural property can be found that imposes itself as a plausible assumption, that would imply uniformity ***and*** that would be in some sense more fundamental than the recourse to competition.

As we already indicated, Sraffa (1960, p. 6) introduces "Production with a Surplus" in the second chapter of his book (fourth paragraph) and says: "The distribution of the surplus must be determined through the same mechanism and at the same time as are the prices of commodities". And he continues: "We add the rate of profits (which must be uniform for all industries) as an unknown". What is the "mechanism", and what does "must" mean? According to the standard interpretation, rejected by Sinha and associated by him with Garegnani, the "mechanism" is competition.

It is clear, however, that there can be homogeneity of commodities, of prices and a structure of an economy producing a surplus such that unequal rates of profit result. Consider the following model of physiocracy, based on a suggestion by Jean Cartelier (1976): There is a physical surplus in both of two sectors. The artisans produce one unit of steel by means of 1/2 of steel and 1/4 of corn. Farmers produce one unit of corn by means of 1/4 of steel and 1/2 of corn, so that we get the following table:

| | *Steel* | *Corn* |
|---|---|---|
| Artisans | 1/2 | 1/4 |
| Farmers | 1/4 | 1/2 |

***Diagram 1: A Model of Physiocracy***

If we interpret this table as a Sraffa system, it is clear that we have standard proportions 1:1 and a maximum rate of profit of 1/3. Relative prices are also given 1:1. For the maximum rate of profit, we may write $R = 1/3$ and $1 + R = 4/3$. If $p$ is the price of steel in terms of corn, we get

$$(1+R)\left(\tfrac{1}{2}p+\tfrac{1}{4}\right)=p$$

$$(1+R)\left(\tfrac{1}{4}p+\tfrac{1}{2}\right)=1$$

But there is no full-scale capitalism if we interpret the table as a model of physiocracy. The wages of the workers are part of the means of production. The surplus is appropriated not by capitalists but by the monarchy in the form of a tax (***impôt unique***). We denote the price of steel in terms of corn by $u$ and the tax rate by $q$; the tax is to be levied in agriculture alone by virtue of the physiocratic doctrine: Agriculture is based on nature, nature is productive, the productive sector should be taxed. Hence we have the price system:

$$\tfrac{1}{2}u+\tfrac{1}{4}=u$$

$$(1+q)\left(\tfrac{1}{4}u+\tfrac{1}{2}\right)=1$$

Obviously, one obtains $u = 1/2$ and $1 + q = 8/5$, hence $q = 3/5$, so that $q–R = 4/15$.

This means that, with the tax rate $q = 3/5$, the entire surplus is levied in agriculture.

$q–R = 4/15$ means that the tax rate exceeds the rate of profit (if it is uniform) by 4/15. If physiocracy rules, the onus of surplus extraction is shifted to the farmers.

Historically, or, rather, in the history of ideas, one goes from the physiocratic system with the surplus levied in agriculture to the capitalist system with the uniform rate of profit, therefore from the model which we have introduced as the second to the first. Really, *institutions change*, if the surplus is now, other things being equal, appropriated by the capitalists. Marx would have said that they share the booty equally.

Instead of making competition and institutions responsible for the equalization of the rates of profit, Sinha has in his book a model with several sectors and profit differentials. He wants to argue that profit differentials must disappear thanks to a transformation into a standard system, but here we have, for the model of physiocracy, a standard system from the start, and the differentials persist. That the differentials become zero in Sinha is nothing but an assumption for which he provides a formal reasoning, but he has no economic argument to substantiate his formal procedure; he does not even state his assumption in clear and explicit form. I do not see why his formalism should be better than the simple assumption that the rate of profit is uniform; rather, I think it is confusing.

I do not deny that there is a deeper motivation. Sraffa, according to various notes in the archive quoted by Sinha, wanted to show that the rate of profit was not a price phenomenon. In a sense, this is obvious from the start, since the rate of profit is expressed in percentage terms, and the percentage can be announced before prices are made explicit. As it is a percentage that is given, it can be given independently of a numéraire. To visualize it, one may refer to the corn model and to the standard commodity as a physical analogue, as was once suggested by Eatwell (1975). The question then is how to express the wage, for the real wage depends on the standard of prices, and so Sraffa left theories of distribution behind which are based on a somehow given real wage. Instead, he advanced the hypothesis that it is the rate of profit, which is "determined from outside the system of production". He suggested that it might be determined in particular "by the level of the money rates of interest" (Sraffa 1960, p. 33). This has remained controversial. Distribution *does* depend on prices, if the theory of distribution must start from a real wage. Hence there is an ambiguity in the statement that distribution is not a price phenomenon, and I do not feel at ease when the attempt is made to remove this ambiguity by referring to the rate of profit instead. Is distribution primarily determined by the subsistence needs of workers, by effective demand, by the rates of interest or by marginal productivity? This is partly a theoretical and partly an empirical and historical question. To have laid the stress mainly on the rate of profit is for me not Sraffa's main achievement. By referring to the rate of profit, he posed a problem, but he did not solve it.

Sraffa's protoeconomics remain fascinating. In the end, the question is where we get when the step to economics is made.

## References

Cartelier, J.: *Surproduit et reproduction. La formation de l'économie politique classique*. Grenoble: Presses universitaires 1976.

Eatwell, J.: Mr. Sraffa's Standard commodity and the rate of exploitation. *The Quarterly Journal of Economics*. 89 (4), 1975, pp. 543–555.

Kant, I.: *Metaphysische Anfangsgründe der Naturwissenschaft*. Riga: J.F. Hartknoch, Zweyte Auflage 1787.

Kurz, Heinz D., and Neri Salvadori: Representing the Production and Circulation of Commodities in Material Terms: On Sraffa's Objectivism. *Review of Political Economy* 17 (3), 2005, pp. 413–441.

Mises, L.V.: *Nationalökonomie. Theorie des Handelns und Wirtschaftens*. Genf: Édition Union 1940.

Schefold, B.: Sraffa and applied economics: Joint production. *Political Economy. Studies in the Surplus Approach*, I, 1985, pp. 17–40.

Schefold, B.: *Mr. Sraffa on Joint Production and Other Essays*, London: Unwin & Hyman 1989 (now Andover: Routledge), XII, 378 pp.

Schefold, B.: "Einleitung zu Karl Heinrich Rau: Lehrbuch der politischen Ökonomie, Bände 1, 2, 3/1 und 3/2, *Historia Scientiarum* (Wirtschaftswissenschaften)". Ein Editions-Programm der Fritz Thyssen Stiftung zur Geschichte der Wissenschaften in Deutschland, ed. by Bernhard Fabian et al., Hildesheim: Olms, pp. 1–53.

Schefold, B.: The transformation of values into prices on the basis of random systems revisited. *Evolutionary and Institutional Economics Review*, 16 (2), 2019, pp. 261–302. Online DOI 10.1007/s40844-019-00140-6.

Sinha, A.: Interpreting the nature of Sraffa's equation: A Critique of Garegnani's interpretation. Paper prepared for this volume. 2021.

Sraffa, P.: *Production of Commodities by Means of Commodities*. Cambridge: University Press, 1960.

von Weizsäcker C. F.: *Die Einheit der Natur. Studien*. München: Hanser 1971.

Reply

# A Response to Professor Schefold

*Ajit Sinha*

Professor Schefold (2021) has done me honour by painting such a wide canvas while commenting on my chapter—its significance goes much beyond the narrow confines of my chapter. In my response, I will stick to his main criticisms of my thesis. It is noteworthy that Schefold does not find any flaw in my structuralist interpretation of or the proof of Sraffa's proposition regarding uniformity of the rate of profits in his system of equations. He, nevertheless, claims, without giving any reason, that it is 'trivial' and a matter of 'definition'. Now my argument that the uniformity of the rate of profits in Sraffa's system of equations of basic goods is a structural property of his *system* crucially depends upon or rather follows from Sraffa's earlier demonstration that for any given system of basic goods there exists a corresponding Standard system and that the Standard system is unique to the given system. Take, for example, my equation systems (vi) and (vii): if no Standard system existed or if there were more than one such Standard systems, then there will be no grounds for me to claim that $R'$ must be equal to $R$ and, therefore, there will be no way of establishing the proposition that all $r$'s must be equal. Thus my proof can be called 'trivial' if and only if one could characterise Sraffa's discovery of the Standard system and the demonstration of its existence and uniqueness for all systems of basic goods as 'trivial'—I suspect Professor Schefold will not go so far. As far as its being a matter of 'definition' is concerned, it depends on what one means by definition—all mathematical proofs are analytical in nature and to that extent a matter of 'definition'.

The root of the problem lies in the most crucial aspect of the *differentia specifica* of Sraffa's system of basic goods that gives rise to *commodity residue* and the classical idea of 'labour' as the sole primary factor of production along with *free* nature, which leaves no commodity residue. Schefold's comment does not even touch upon this matter—as a matter of fact, the concept of 'commodity residue' does not even once appear in his comment. Therefore, his criticism of my chapter remains peripheral and misses the point. As I have highlighted in my chapter (Sinha 2021), the existence of *commodity residue* in a system of production with basic goods does not conceptually allow for any industry to be completely vertically integrated. Thus it does not allow any industry to be *independent*—a consequence of this is that prices of all industries must be determined *simultaneously*—applying arbitrary prices (such as market prices) from outside the equation system effectively destroys the integrity of the interconnected nature of the *system*; it amounts to treating every industry as completely vertically integrated industries exchanging only their 'final goods'; in this context, the gravitation mechanism does not require the so-called *ceteris paribus* assumption, and Adam Smith, in my understanding, never made any such assumption as Schefold suggests.

Schefold (2021) presents a counter-example to my thesis that the rate of profits in Sraffa's system of equations *must* be uniform. He provides a simple example of a two-sector model with surplus production. Since his system is already in the Standard proportion, its average (and in this simple case also maximum) rate of profits can be calculated as a ratio of its net output to input, which turns out to be 1/3 and the prices turn out to be 1:1. He then introduces an institutional variable from outside the equation system—this institution is the State, which appropriates the total surplus (equal to ¼ steel + ¼ corn) by imposing taxes. Now, from the equations it is quite clear that the all-powerful State could directly tax both the sectors at the rate of 1/3 of their total outputs and appropriate the total surplus. In this case, the sectors would be reduced to their subsistence, leaving the prices intact. But Schefold suggests that the State wants to appropriate the total surplus by taxing only one sector, that is the agricultural sector. In that case, the State has to find out the terms of trade or the price ratio between the two sectors that would transfer all the surplus of the system to the agricultural sector alone. To solve this problem, the State solves the two equations given by: $q(1/4u + ½) = 1/4u + ¼$ and $(1 + q)(1/4u + ½) = 1$, where q is the rate of taxation and u is the price of steel in terms of corn. This gives us the value of $u = ½$ and $q = 3/5$.

Schefold then claims that 'here we have, for the model of physiocracy, a standard system from the start, and the differentials persist'. But all this is just a convoluted exercise of the case I have already taken care of in my chapter—as I have argued earlier, taking prices as 'market prices' from outside of the equation system and solving for unequal rates of profits amounts to breaking Sraffa's system of equations as each industry could solve for its rate of profits independently given the set of prices. In this case as well, the State must first decree the price ratio equal to 1:1/2 to ensure that all surplus is transferred to the agricultural sector. Imposition of these prices breaks the system in the same manner as imposition of the so-called market prices. Otherwise, 3/5 rate of taxation on agricultural sector with old prices given at 1:1 would contradict the two-equation system. My argument may be better illustrated by an analogy—a soap bubble has a natural shape, which is spherical, and the structural property of it can be described mathematically, but this does not mean that by putting outside pressure on it one cannot distort its shape—a State by putting taxes and subsidies on various commodities can always twist the prices; it can also destroy or break the economy by imposing taxes more than the surplus produced by the system just as we all know a sufficient pressure on a soap bubble breaks it too. Here I would also like to add that Schefold's characterisation of his example as a 'model of physiocracy' is rather problematic. In physiocracy, it is the rent income that is considered surplus and taxed. But rent is earned on land and is not determined as a percentage on the value of the means of production, which is the case in Schefold's example.

Now we come to Schefold's defence of classical gravitation of market prices to natural prices. On this account, he appears to be all over the place—I have not been able to find a central thread running through it except for an insistence that there must be natural prices, and market prices must gravitate to it. He first asserts that gravitation process is 'indispensable' if one wants to get from 'protoeconomics' to economics—this is argued in the context of evolution of an economy, where conceptualisation of the two states could be made only in terms of 'natural prices'—'And what is the meaning of the natural prices, if the market prices do not somehow gravitate to them?' (Schefold 2021). Then we are told that 'the gravitation of market prices to natural prices therefore is a matter of anecdotal circumstances.' So already we have based an 'indispensable' aspect of economic theory on something that is a matter of anecdotal circumstances! He further goes on to assert that Adam Smith had no theory of gravitation, but 'we, of course, can construct modern models.' He makes three suggestions in this regard. (i) In the case of agricultural goods, natural prices can be conceptualised as the average price of market prices over many years—but what if market prices turn out to be rising or falling over the years? It still will have an average but in what sense it could be considered the 'realized natural price'? And, of course, over the years techniques of some basic good could be changing as well—how does average market price take care of that? (ii) In the case of industrial goods, he suggests that one could conceive their pricing policies to be such that they use a mark-up on their cost equal to the normal rate of profits—in this case there is no gravitation mechanism as we are simply assuming that industries set their prices equal to their natural prices irrespective of whether their supplies are equal to their effectual demands or not. And (iii) if quantities are rigidly given then adaptation of market prices to natural prices can be formulised by the equation system given by: $\mathbf{P}_{t+1} = \mathbf{P}_t(1 + r) + \mathbf{l}$, where wage is taken as the numeraire—this, of course, is an alternative method of solving a simultaneous equation problem by assuming that the rate of profits must be equal. This simply cannot describe the gravitation mechanism since gravitation mechanism works precisely by adjusting quantities supplied to the given effectual demands—keeping supplies rigidly fixed takes away the very presupposition of the whole process. Interestingly, Schefold never asks: what if the gravitation process does not work? He finally goes on to claim that 'gravitation is an axiom of classical theory, and not a result.' But what good is an axiom that can be shown to be wrong, as it is shown in Dupertuis and Sinha (2009)? And even if it is an axiom of classical theory, the relevant question for us is thus: is it an axiom of Sraffa's theory? And the answer to that question is 'definitely not'. As Sraffa explicitly states that his equations represent 'the actual economic system of observation' (Sraffa 1960, p. 22) and also claims that 'the economic systems of reality are not self-replacing. They are in a constant state of transition and obsolescence, due to changes both in the types of (kind of) commodities produced and in methods of their production' (Sraffa n.d. 1955, D3/12/75, p. 16). In other words, his equation system is like a photograph of the existing economic system that is not in equilibrium or the centre of gravitation of any sort. Schefold criticises me for suggesting that Adam Smith's gravitation argument implicitly assumes a demand schedule (but not a function). He argues that '[h]ere Sinha seems to me to be uninformed about the state of discussion of supply and demand in the 18th century. In particular, reference should be made of James Steuart. In 18th century, and still in Marx, demand and supply were seen as "forces".' In this context, I will only like to point out that Sraffa in 1931 interpreted Marshall's demand and supply schedules as generating *forces* and clearly stated that he does not assume any forces—I have not found any evidence in Sraffa's archive that suggests that he repudiated his 1931 position on the question of 'not assuming any forces': 'The real point is that it is believed that Marshall's curves provide 'forces' which, in case the price falls below or above AB {equilibrium prices} by 'chance' will restore it to AB. Now I am not assuming any forces' (Sraffa n.d., D3/12/7: 65).

Schefold also criticises me for taking Sraffa's allusion to 'observation' too literally. He argues that Sraffa's equations assume homogeneous goods and a well-defined technique—both these aspects of his theory, according to Schefold, cannot be observed. As a matter of fact, there aren't two apples in the world that are identical or homogeneous. But this does not stop sellers and buyers of apples to buy and sell them in terms of either dozens or kilograms. If a kind of apples has a price then this very fact tells us that it has a homogeneous unit of measure as price is quoted in terms of a unit of measure. Whether all apples in a dozen or in a kilogram are physically and chemically identical or not is not the relevant question here—the observer simply observes that apples are socially measured as homogeneous commodity in certain socially accepted unit, and this information is available from the producers of apples. So where is the problem? Similarly, the spectre of a 'dominant technique' and how to establish it is a problem for Garegnani and his associates because they need to think in terms of supply adjustments for their gravitation mechanism but not for my interpretation of Sraffa's equations. Sraffa's equations are industrial equations, and they do not represent some sort of technique of production of a firm. If an economy uses coal, hydro, nuclear and wind technologies to produce electricity, then the observer will simply take down all the inputs used by various production units on the side of inputs and the total output of electricity on the side of output—this will not represent any particular technique. There is no reason for the data of inputs and outputs of Sraffa's equations to represent *a* technique for solving his equations as these data are *ex-post* data and do not require us to think in terms of adjustments of outputs.

Now we come to the fundamental distinction between Garegnani and his associates' interpretation of Sraffa's equations and Sraffa's own description of it. As I have pointed out in my chapter, Sraffa repeatedly claims that 'profit is a non-price phenomenon.' Schefold tries to underplay this significant theoretical claim by suggesting that it only refers to the absence of the unit of prices in the rate of profits, as it is a percentage. But this is not the case. Sraffa comes to this conclusion only after determining his fundamental equation, $r = R(1-w)$, where the measure of $w$ is in terms of the Standard commodity. This immediately determines the ***value*** of $r$ if we know $w$ since it is already established that $R$ is determined and remains fixed with respect to variations in $w$ or $r$. This shows that both $r$ and $w$ are determined ***independently*** of prices. Now the question is which one should be taken as 'given' from outside? Since wages in the real world are given in terms of 'money' and not in terms of the Standard commodity, the given 'money' wages could be translated into $w$ in the Standard commodity units only after knowing the prices but $r$, which is a percentage and is not given in terms of the Standard commodity, can always be taken as 'given' from outside independently of prices. This is the reason why Sraffa moves from taking wages as given to taking the rate of profits as given—but it crucially depends on the equation, $r = R(1-w)$, which shows that $r$ is a non-price phenomenon and therefore can be taken as given before the prices are determined. This, however, is not possible if you insert gravitation mechanism into Sraffa's system. The conceptual framework of the gravitation mechanism requires the rate of profits to be determined by prices. Schefold claims that 'Distribution *does* depend on prices, if the theory of distribution must start from a real wage'—this apparently is argued on the grounds that 'the real wage depends on the standard of prices.' But this is not true. If one takes real wages as the standard of prices then this problem will be taken care of or, for that matter, take Sraffa's equation system given in his book (Sraffa 1960, p. 6); in this case real wages are either zero or incorporated in the inputs data. The question is this: can we determine the equal rate of profits of this system of equations independently of prices without the help of the Standard system? And the answer is, no. Since the Standard system is not an integral element in the analysis of the system of equations in the gravitational framework, the rate of profits cannot be determined independently of the determination of prices—they ***must*** be ***simultaneously*** determined; and therefore, not only distribution turns out to be dependent on prices but also the rate of profits cannot be taken as given from outside. This reveals the fundamental dissonance between Sraffa's own interpretation or rendition of his theory and the interpretation given to it by Schefold or Garegnani and his associates.

Let me now take up the main concern of Professor Schefold. He characterises Sraffa's theory as a representative of 'protoeconomics' and argues that one needs to insert the idea of gravitation if one wants to get to economics from 'protoeconomics'. It seems to me that Schefold is largely in agreement with me or at least not in serious disagreement in so far as Sraffa's theory is characterised as a structure with well-defined relational properties among its variables, but he thinks that it only represents a 'protoeconomics'. For him, an economic theory must take into account changes in those variables based on 'causation'. For example, Sraffa's equation, $r = R(1-w)$, represents the relation in which the three variables must always be related given any input-output data of basic goods. In this case, empirically if we find that the value of $R = 20\%$ and the value of $r = 10\%$, then the value of $w$ must turn out to be ½ in terms of the Standard commodity of the system. But this does not mean that if for some reason $r$ changes to 5% then $w$ must necessarily rise to ¾ in the empirical or the real world. This is because a real change in $r$ could lead to choice of other techniques and many other changes in production decisions leading to a new set of data with a new $R$ and a new Standard commodity with a new $w$. All Sraffa's equation tells us is that the new set of data will again represent the same relationship between $R$, $w$ and $r$. Schefold seems to argue that an economic theory must be able to explain how a change in $r$ brings about the change in the structure itself. This is a problem of causation. In my understanding, Sraffa refrains from developing a theory based on causation because he thinks that a relation of causation is dependent on several contingent factors and can be discussed only in terms of particular cases—a general predictive theory based on causation is not possible—that's why the story of changes should be left to historians where hindsight is 20/20. This is a general problem with structuralist theories—they are good at explaining the relations of the structure and the conditions of its reproduction but do not have much to say when it comes to *predicting* changes in the structure itself—Marx tried to do it on a grand scale by introducing dialectics but could not get much beyond pointing to the internal contradictions in the conditions of the reproduction of the structure. Finally, if we need to develop even some tentative theories based on causation, we will have to introduce time dimension explicitly in it and I do not understand the insistence that all dynamic theory must have a well-defined equilibrium position as its reference point. If Sraffa is right in describing economic change as 'a constant state of transition and obsolescence, due to changes both in the types of (kind of) commodities produced and in methods of their production' (op. cit.), then the 'centre of gravitation' is of no help here and we must try to develop our dynamic theories without any reference to equilibrium. My interpretation of Sraffa's prices foots the bill as it makes the condition of equilibrium irrelevant.

It is a general refrain among many Sraffians that Garegnani's interpretation cannot be challenged given his close relationship with Sraffa. Schefold also repeats this trope: 'The opposition between Garegnani and Sraffa, which [Sinha] tries to reconstruct, is *a priori* not possible. Garegnani was without question Sraffa's most important pupil, his friend and support over many years to the end. Garegnani may have had his own agenda, but there was trust between them and Garegnani was appointed Sraffa's literary executor' (Schefold 2021). In this case, I would only like to point out that both James Mill and McCulloch were close friends of Ricardo (one in the role of a master and the other in the role of a pupil), but how many Sraffians would accept their interpretation of Ricardo over Sraffa's?—even though Sraffa was born 75 years after Ricardo's death and therefore did not even belong to Ricardo's intellectual era. The intellectual history of the world is full of such examples where close associates have been found to be wanting in understanding the works of geniuses they were closely associated with. I rest my case.

## References

Dupertuis, M.-S. and A. Sinha. 2009. 'A Sraffian critique of the classical notion of centre of gravitation', *Cambridge Journal of Economics*, 33(6): 1065–1087.

Schefold, B. 2021. 'A Comment on Ajit Sinha: Interpreting the nature of Sraffa's equations: A critique of Garegnani's interpretation', in *A Reflection on Sraffa's Revolution in Economic Theory*, (ed.) Ajit Sinha, London: Palgrave Macmillan.

Sinha, A. 2021. 'Interpreting the nature of Sraffa's equations: A critique of Garegnani's interpretation', in *A Reflection on Sraffa's Revolution in Economic Theory*, (ed.) Ajit Sinha, London: Palgrave Macmillan.

Sraffa, P. 1960. *Production of Commodities by Means of Commodities*, Cambridge: Cambridge University Press.

Sraffa, P. n.d. 'Sraffa Papers', Archive, Wren Library, Trinity College, University of Cambridge, Cambridge.

CHAPTER 2

# On Sraffa's Structuralism

*Roberto Scazzieri*

## 1 Introduction

Interdependence between processes within a self-replacing system is the hallmark of Sraffa's analytical representation of the economy as production of commodities by means of commodities (Sraffa 1960). This view highlights Sraffa's structuralism since a distinctive feature of the latter is to consider collections of elements as wholes 'subordinated to laws', in terms of which 'the structure *qua* whole or system is defined' (Piaget 1971 [1968], p. 7). In other words, a 'structure' is different from a simple 'aggregate' considered as a composite 'formed of elements that are independent of the complexes into which they enter' (Piaget 1971 [1968], p. 7). The aim of this essay is to investigate Sraffa's structuralism highlighting the central role in Sraffa's writings of the view of the economy as a system of elements whose interdependence assigns structure to the whole system by virtue of its own self-replacing condition, while *at the*

R. Scazzieri (✉)
Department of Economics, University of Bologna, Bologna, Italy

National Lincei Academy, Rome, Italy

Clare Hall and Gonville and Caius College, Cambridge, UK
e-mail: rs292@cam.ac.uk



A. Sinha (ed.), *A Reflection on Sraffa's Revolution in Economic Theory*, Palgrave Studies in the History of Economic Thought,
https://doi.org/10.1007/978-3-030-47206-1_2

*same time* being structured by it. This feature of Sraffa's economic structure is remarkably close to Jean Piaget's view that 'the character of structured wholes depends on their laws of composition', so that 'these laws must of their very nature be *structuring*', and thus must give rise to a 'constant duality or bipolarity, of always being simultaneously *structuring* and *structured*' (Piaget 1971 [1968], p. 10; author's emphasis). Similarly, in Sraffa's representation of the economy, interdependence is never likened to simple interaction between the autonomous elements of an aggregate. Instead, interdependence is seen as a relational pattern *identified* by the self-replacing condition. This condition is the 'law' that assigns structure to the collection of processes making up the economy as a system.

This essay discusses the roots and character of Sraffa's structuralism from his early theoretical writings through his reconstruction of Ricardo's theory of value up to *Production of Commodities by Means of Commodities* (Sraffa 1960). The essay is organized as follows. Section 2 examines Sraffa's early criticism of Marshall's demand and supply curves (Sraffa 1925, 1926) in the light of Sraffa's criticism of Philip Henry Wicksteed's substitution of 'functional curves' for the 'descriptive curves' associated with the classical economists' view of decreasing returns. Section 3 outlines a rational reconstruction of Sraffa's approach to the representation of the economy in terms of positional interdependence and contrasts this view with the approaches depending on functional interdependence. It is argued that positional interdependence allows consideration of structural transformation as change in the relative positions of elements (such as actors, productive sectors, socio-economic groups) within a given system. Positional interdependence is then discussed in the light of Sraffa's distinction between causality and causation, which points to the central role of sequential causal processes in determining the changes following any given source of change within a set of *feasible transformations* of relative positions of elements in the economy. Section 4 discusses the implications of Sraffa's structuralism for the construction of a political economy centred on the consideration of positional interdependence between units of analysis in the economy (such as industries, vertically integrated sectors, socio-economic groups). Section 5 brings the essay to close highlighting the character of Sraffa's *Production of Commodities by Means of Commodities* as the mature product of a structural heuristic open to a variety of structural configurations compatible with a given set of relatively invariant parameters.

## 2 'Descriptive Curves', Positionality and Structural Emergence

Sraffa's criticism of Marshall is based on a dual line of argument. On the one hand, Sraffa takes up Wicksteed's distinction between 'descriptive curves' and 'functional curves' (Wicksteed 1914) but twists Wicksteed's argument by making descriptive curves (that may be interpreted as the representation of *existing structures*) into building blocks of his theory of production and consumption relationships. Sraffa also highlights that structures themselves (such as the structures represented by descriptive curves) are located at a specific level of aggregation of economic units and are therefore likely to change as we move across the spectrum of intermediate levels of aggregation.

Concerning the distinction between 'descriptive curves' and 'functional curves', Wicksteed had written that:

> [A] diagram may easily be constructed in which different qualities of land are represented along the axis of X and their supposed relative fertilities to a fixed application of labour and capital along the axis of Y. The "marginal" land will occupy the extreme place to the right. This is not a functional curve; for the height of *y* does not depend upon the length of *x*, the units being expressly so placed on OX as to produce a declining *y*. It is applicable to land or to anything else of which typical units can be arranged in ascending or descending order of efficiency. But the same figure has been used as a functional curve in connection with the theory of rent. Take a given fixed area of land of a certain quality and consider what would be its yield if it were "dosed" with a certain quantity of labour and capital represented by a unit on the axis of X. Increase the doses till a further increment of labour and capital would not produce as large an increment in the yield of this land as it would if applied to some other piece of land of the same or different quality, or if turned to some non-agricultural business. The last increment actually applied is the "marginal" increment, and it measures the distributive share of a unit "dose" in the product. (Wicksteed 1914, pp. 17–18)

Wicksteed's conclusion is that:

> it is essential to point out that the descriptive and the functional curves just described both present the same appearance, both represent "rent" by a curvilinear surface, both use the term "margin," though in entirely different senses, as determining rent, and are both just applicable to anything else as to land, and (specifically) ignore the difference between "economic" and

> "commercial" rent, being just as applicable to one as to the other [...]. [A]t present the descriptive curve and "margin" have only been introduced to be dismissed. In the discussion of the functional curve, which we must now continue, I have used the term "marginal" in the sense of "differential" as applied throughout our whole investigation. It is not any peculiarity of the "marginal" increment that makes it yield less than the others. It does not. They all have exactly the same differential effect on the yield, as to which none is after or afore the other. The height of this differential or marginal yield is dependent not upon the nature of each several dose, but upon their aggregate number. (Wicksteed 1914, p. 18)

Sraffa takes up Wicksteed's distinction but *reverses* Wicksteed's emphasis on functional curves over descriptive curves highlighting that only descriptive curves allow the consideration of economic causes in the explanation of productivity curves:

> The characteristic of Ricardian theory which we have identified as fundamental, that is, attributing diminishing returns to an economic rather than a physical cause, has been very ably criticized by Wicksteed. He begins by dividing productivity curves into two categories: *descriptive curves* and *functional curves*. This distinction broadly coincides with the contrast between an *economic* law and a *physical* law of diminishing returns ...[T]he basis of the distinction is this: in the descriptive form the position in the sequence held by each dose is determined by the productivity of the dose; this productivity is therefore independent of the number of doses utilised. In the functional form, however, it is the place held by each dose which determines the productivity of that dose; this productivity is therefore strictly dependent on the number of doses utilised. In other words, in the first form it is assumed that all the doses considered are distinct from one another; and thus even if utilised under the same circumstances, they have different productivities. In the second form, the nature of all doses is assumed to be the same, but the doses have different productivities due to the differing circumstances of use. (Sraffa 1998 [1925], p. 336)

Distancing himself from Wicksteed's criticism of functional curves, Sraffa argues that the type of ordering presupposed by a functional curve is in fact groundless unless it is assumed that the identical doses are put to different uses, so that in fact one would be back to the descriptive form, even if applied to the different modes of use of identical doses of capital and labour rather than to the intrinsic characteristics of each plot of land:

> Any decreasing curve [that is, any decreasing productivity curve] with a general and not merely an accidental character, must be a "descriptive curve" … [I]f it is true that the doses are identical and yet give a different yield, this implies that they are put to different uses, and therefore the product of the marginal dose is dependent precisely on the nature of its use. Therefore, in the functional curve the productivity of the marginal dose does not depend directly on the aggregate number of doses, but rather because the previous doses having already been applied to the best uses, a less productive use is left for the last dose. *And the larger the number of doses the lower must we descend along the descending hierarchy of available uses.* (Sraffa 1998 [1925], p. 337; added emphasis)

Sraffa's reversal of Wicksteed's argument points to a fundamental characteristic of the structural point of view: positionality is central in determining the nature of any given structure, which may be described by the set of relative positions between its constituent elements and by the set of transformations of those relative positions that are feasible starting from the relative positions defining the existing structure at any given time. This point of view is made clear in Sraffa's remarks that follow the above passage:

> This hierarchy belongs to the genus of descriptive curves, for the uses have been placed "arbitrarily" in decreasing order, *and not because of any material necessity.* Thus the "functional curve" merely transfers the "difference in nature", and therefore the "arbitrary" ordering from the doses themselves to their modes of utilisation. But the relationship that links the number of doses to marginal productivity, that is the productivity curve, belongs, in both cases, to one genus alone [that is to say, the genus of descriptive curves]. (Sraffa 1998 [1925], p. 337; added emphasis)

Indeed, it is by switching from functional to descriptive curves that it becomes possible to enter the black box of productivity curves explaining the shape of those curves by human economic activity taking place under scale-constrained modes of utilization:

> Obviously … the arbitrariness is not, as Wicksteed seems to think, on the observer's part—who would arrange his pieces of land in decreasing order just as he would arrange a row of men in order of size.[1] But it is on the part of the producer himself, who, in effect, only uses his freedom to behave in the manner most rewarding to himself. (Sraffa 1998 [1925], pp. 337–338)

[1] Here Sraffa refers to a statement in Wicksteed's *Common Sense of Political Economy* (Wicksteed 1910, p. 539).

In short, positionality is central to the representation of any given structure and reflects the *intrinsic* properties of the different constituent elements of that structure, so that the relative positions of constituent elements along the sequence of utilization of the different modes are explained in terms of the intrinsic properties of those modes. In this case the relative positions of the different modes of utilization on the 'descending hierarchy of available uses' are associated with the degree of land use per unit of output for each mode of utilization. At the same time, more than one single sequence of modes of utilization are in general feasible given the degree of land use for each mode, and the scale constraints acting on it. This means that the actual 'hierarchy of available uses' (that is to say, the actual sequence of activation of the different modes) would also reflect the economic actors' 'freedom' to organize production activity 'in the manner most rewarding to [themselves]' (Sraffa 1998 [1925], p. 338).

The following scheme illustrates the mechanism at play when economic actors generate productivity curves under scale constraints due to limited availability of land, or of any other non produced means of production (see also Scazzieri 1981, 1982, 1993, pp. 136–153, 184–188):

1. There is a set *T* of different modes of use (technical practices), which are exogenously given and are associated with different land requirements per unit of output;
2. There is an upper bound on land available for cultivation (for each land type);
3. The upper bounds on land(s) available for cultivation generate upper bounds on the maximum level of activation for each mode of use;
4. Economic actors (producers) introduce a ranking between modes of use;
5. Scale of production *s* generates a set $T^*$ of modes of use whose elements are technical practices that (a) can deliver *s*; (b) satisfy the land constraints for each mode of use;
6. Economic actors (producers) adopt the 'most rewarding' mode of use of available land(s)

The above scheme shows the characteristic 'duality or bipolarity' of structured wholes, 'of always being simultaneously *structuring* and *structured*' (Piaget 1971 [1968], p. 10; author's emphasis). Production organization, which is associated with specific positions of the different modes of use relative to one another, reflects scale-dependent constraints and the principles guiding actors' actions. In this case, production organization is *structured* by those constraints and principles. At the same time, production organization at any given scale of production may constrain and orient production organization at different scales of production. For example, production organization at scale *s* may constrain and orient production organization at scale *s'* > *s* if the switch from *s* to *s'* makes certain modes of use unfeasible due to the land requirements of those modes and the constraints on land availability. In this case, production organization at the lower scale *s* is *structuring* production organization at the higher scale *s'*.

It is noteworthy that Sraffa considers the above relationship between different modes of use, scale-dependent constraints, and principles guiding actors' actions also relevant in explaining consumers' demand for different quantities of the same good:

> The same argument may be repeated for the case of diminishing utility (and therefore for the demand curves derived from it) which is a special case of diminishing productivity, when we consider utility as product, the commodities consumed as the variable factor of production, and the "sensitive organism" as the constant factor.[2] It is not any allegedly psycho-physical law which endows diminishing utility with generality, but the possibility of different doses of a commodity to satisfy different needs and the desire to utilise the first doses to satisfy the most urgent needs. (Sraffa 1998 [1925], p. 338)

Sraffa's early writings on economic theory highlight the interplay of 'objective conditions' setting the stage (that is, providing constraints and opportunities) for the emergence of structures, with human actions unfolding under those conditions and determining which causation processes would in fact make existing structures to evolve along one or another of the feasible paths open to them. This approach to the economic interdependencies making up any given structural arrangement (interdependencies that are both *structured* and *structuring* at the same time) has a counterpart in the *inherent plurality* of structural constraints and opportunities, which may be different depending on which type of aggregation is considered. This second strand of Sraffa's early structuralism is manifest in his investigation of the mechanism generating productivity curves under limited land availability. In this case, Sraffa highlights that a decreasing productivity curve *may or may not* arise depending on the level of aggregation of production processes using the same non-produced resource as an input:

[2] Here Sraffa refers to a statement in Wicksteed's *Common Sense of Political Economy* (Wicksteed 1910, p. 539).

> The cause of this decrease—the fact that one of the factors cannot be increased—operates only for the industry as a whole. The quantity of the factor available for the totality of the producers is constant, but the single producer can increase or decrease the quantity that he uses of it without appreciably influencing the price of the factor itself. In the case of agriculture, "land from the point of view of the individual cultivator is simply one form of capital".[3] It is therefore possible that, while the industry has increasing costs, the single cultivator might, up to a certain point, increase his production while lowering his own private cost of production, because he can take advantage of the economies of large scale production, and yet, without being forced to intensify the exploitation of the constant factor, can obtain for himself a large quantity of it at the expense of his competitors. (Sraffa 1998 [1925], pp. 342–343)

In the situation above, the level of aggregation at which the analysis is conducted (the level of the 'single producer') does not allow the detection of a decreasing productivity curve. However, the situation is different under institutional conditions making the switch-over of resources between different producers impossible:

> [A]lthough this is possible for each producer separately, it is not possible for the totality of producers, and therefore the sum of a series of individual curves of this kind is absurd, since each of them is valid only on condition that the production of the other individuals remains unchanged. In order to make it possible to add up the individual curves it is necessary to have recourse to a stratagem that moves the cause of the increase in cost from the conditions of the industry to the conditions of the single producer. This is achieved by supposing that the number of producers is fixed, and that *each of them, with the increase in his production, cannot increase the quantity used by him of the factor of which there exists a fixed quantity for industry as a whole*, so that the individual cost of production has to increase. In these conditions, the individuality of the "enterprise" is no longer characterised solely by the unit of management, that is, by the entrepreneur, but also by the presence of a unit of the "constant factor". (Sraffa 1998 [1925], p. 343; added emphasis)

More generally, decreasing (*or* increasing) productivity curves are more, or less, likely depending on the level of aggregation adopted:

[3] Here Sraffa cites from Marshall's *Principles of Economics* (Marshall 1890, p. 170).

> If every single industry is defined as the exclusive consumer of a given *factor of production* (for example, agriculture, the iron industry, etc.), a condition is at once assumed that implies a tendency towards increasing costs for the industry, since it is precisely the factor that is characteristic of the industry (cultivable land, iron mines, etc.) that, with the increase of production, generally remains constant. If, on the contrary, every industry is defined as the sole producer of a given *product*, and this is meant in a fairly restrictive sense, so that in general it can be thought that every industry uses only a small fraction of each factor of production ((negligible in comparison to the quantity used by all the other industries together), we thereby exclude from the industry the circumstance that generates increasing costs and make it more probably subject to the law of constant costs, or, in further specific conditions, to the law of decreasing costs. (Sraffa 1998 [1925], p. 357)

This argument entails that decreasing (or increasing) productivity curves are a possibility inherent in existing structures for any given technology in use, but that the specific condition allowing detection of a decreasing productivity curve (or, under different assumptions, of an increasing productivity curve) presupposes a particular type of aggregation of production processes. The structural condition allowing decreasing productivity curves may be expressed in terms of the matrix below (see also Scazzieri 1993, pp. 196–197):

$$\mathbf{Z} = \begin{pmatrix} z_{11} & z_{12}\cdots & 0 \\ 0 & z_{22}\cdots & 0 \\ 0 & 0\ldots & z_{nk} \end{pmatrix}$$

Matrix $\mathbf{Z}$ illustrates the resource constraints for a production technology in which $n$ non-produced resources are needed as inputs to produce $k$ different goods. Each element $z_{ij}$ of $\mathbf{Z}$ denotes the maximum quantity of non-produced resource $i$ that can be used to produce good $j$. Matrix $\mathbf{Z}$ shows a situation in which resource *2* is specific to the process producing good *2* and resource $n$ is specific to the process producing good $k$. This means that decreasing productivity curves may be detected for the 'industries' producing, respectively, good *2* and good $k$ in case of full utilization (at the level of the whole economy), respectively of resource *2* and resource $n$; on the other hand, a decreasing productivity curve should be excluded (for the reasons given in the passage quoted above) for the 'industry' producing good *1* in case of full utilization at the level of the whole economy of resource *1* (as this resource may in principle be shifted, say, between the 'industries' producing goods *1* and *2*). Matrix **Z** also shows that upper bounds on resource endowments are likely to hamper increasing productivity in the 'industries' producing, respectively, good *2* and good *k*, whereas they are unlikely to do so in the case of the 'industry' producing good *1* (unless it is institutionally impossible to shift resource *1* between the different 'industries' that need it as an input).

To sum up, Sraffa's early analysis shows a plurality of factors at play behind the emergence and dynamics of economic structures both in the production and the consumption spheres. On the one hand, structures presuppose the interplay of intrinsic characteristics of different modes of use of a specific input (either in production or in consumption) with the ranking of those modes of use by economic actors (with the proviso that this ranking may be different across different actors, or different socio-economic groups). On the other hand, structures are compatible with a *plurality* of possible configurations of the economy depending on which type of aggregation is considered. In either case, economic structures show a dual mechanism at play. For the intrinsic properties of the different modes of use determine a field of possibilities that is grounded in those properties and yet remains open to a variety of structural developments. Actions are required to turn this grounded virtuality into the emergence of specific structures.[4] There is thus a tension between determination and freedom, which is a distinctive mark of Sraffa's early contribution to economic theory.

## 3 Positional Interdependence and Functional Interdependence

As we have seen, positionality is central to Sraffa's early treatment of decreasing and increasing productivity curves: modes of use are to be identified in terms of their *intrinsic properties* (descriptive curves), and these properties are independent of the ranking that may be introduced between the different modes of use (which is what happens with functional curves). This approach to positionality suggests that interdependence itself takes two different meanings depending on whether 'positional interdependence' or 'functional interdependence' is considered.

[4] See also Cardinale and Scazzieri (2018) for an assessment of the relationship between structural virtuality, actions, and structural change.

Sraffa addresses the issue of interdependence in his 'Notes on interdependence and causality', which were drafted before 1928. There Sraffa writes:

> *Interdependence.* Muddled state of this notion. Marshall (App. on Ricardo) thinks that variables mutually determine each other. Pigou (*E.J.* 1902, p. 276) is also confused about interdependence. Interdependence is equivalent to causality as opposed to causation: it holds between categories; but between particulars only causality (one direction) can exist. (Sraffa, pre 1928)

Sraffa's remarks on interdependence should be read against what he had written in his 1925 paper when discussing Anne-Robert-Jacques Turgot's formulation of the law of variable proportions between productive factors. Turgot had written that:

> Seed sown on land that is naturally fertile, but totally unprepared, would be an advance almost entirely wasted. If the soil were tilled once, the produce would be greater; tilling it a second, a third time, might not merely increase the produce two or three times, but four times or ten times. The produce would thus be augmented in a much larger proportion than the advances increased, and this up to a certain point, at which the produce will be as great as possible compared with the advances. Past this point, if the advances are still increased, the produce will still increase, but less and less, until the fertility of the soil being exhausted, and art unable to add anything further, an addition to the advances will add nothing whatever to the produce. (Turgot 1768, as quoted in Sraffa 1998 [1925], pp. 327–328)

Turgot's argument describes a situation in which *identical* doses of capital and labour, when added to one another on the same plot of land, deliver a more than proportional increase of produce up to a certain point (the point of maximum 'marginal' productivity), beyond which additional doses of capital and labour would deliver a less than proportional increase of the produce. In this case, the position of each dose of capital and labour relatively to the other doses of capital and labour is of central importance: identical doses may deliver different results depending on how many doses are used on the same plot of land. No question arises as to the intrinsic differences between doses. This is an instance of what we may call *functional interdependence*, in which the marginal products from each dose 'mutually determine each other' (Sraffa, pre 1928). This type of interdependence is different from interdependence between *different* doses of capital and labour when those doses are located next to one another according to a descending (or ascending) order of efficiency. In this case, the doses are distinguished by intrinsic differences that are independent of the number of doses being employed. In other words, we are dealing here with what we may call ***positional interdependence***, in which the marginal products from doses of capital and labour ***would not*** 'mutually determine each other' (Sraffa, pre 1928) but instead reflect the ranking introduced by economic actors over the available doses of capital and labour by taking into account the intrinsic characteristics of each dose.[5]

The distinction between functional and positional interdependence is at the root of Wicksteed's distinction between functional and descriptive curves (see Sect. 2) and in turn implies a distinction between two different approaches to causal relationships. Functional interdependence points to mutual determination of variables (say, between *A* and *B*) and is by itself unable to identify actual causation from one variable to another (say, from *A* to *B*). In the case above, mutual determination highlights that the marginal product of dose *k* would be determined by the number of doses preceding it while also determining the marginal products of the doses added to it, but it is not concerned with the specific action leading a producer to add dose *k* to the doses already in use. On the other hand, positional interdependence points to an asymmetric relationship between variables and draws attention to the causal path leading from one condition to another (say, from a certain condition *x* for variable *A* to a certain condition *y* for variable *B*). In the example above, it draws attention to the reasons making a producer to use a specific dose of capital and labour (say, dose *k*) in a particular way, and to follow a particular ranking of modes of use in employing that dose (that is, adopting a particular mode of use only after other modes of use are no longer feasible, and before yet other modes of use would become feasible and/or profitable).

Sraffa is critical of interdependence circumscribed to the mutual determination of variables arguing that mutual determination leads to equational forms but that the form of equations (identification of parameters) is of central importance in causal analysis (Sraffa, pre 1928). He also maintains that causal analysis in the sense of the identification of causal paths (in his terminology, identification of 'causation' rather than 'causality') is

[5] Sraffa's analysis of the distinction between functional curves (functional interdependence) and descriptive curves (positional interdependence) is thus related to his criticism of 'marginism' (Marcuzzo and Rosselli 2011; Rosselli and Trabucchi 2018).

necessarily associated with the explanation of 'deviation from what is normal':

> Cause [is] required only when there is a deviation from what is normal, or uniform, or constant. That is to say, it is required only to explain change or difference. The habitual, normal or "natural" course of events does not require explanations; rather, it serves to explain "why" individuals behave in the way they behave—"everybody does it"' (Sraffa, pre 1928)

Positional interdependence highlights the relative positions of ***distinct*** economic units and determines which transformations of relative positions are feasible and which ones are not (as in a jigsaw puzzle). It allows the analysis of feasible causal paths by calling attention to the constraints and opportunities associated with a specific economic structure, which may be defined in terms of the *set* of possible transformations open to any given economy at any given time. It is in this way that the 'habitual, normal, or "natural" course of events' may be considered as a condition to explain why and in which way deviations 'from what is normal, or uniform, or constant' may take place. Sraffa's emphasis on descriptive over functional relationships entails a specific view of economic structure as a multi-layered set of relative positions. Each layer of relative positions may change subject to constraints and/or opportunities associated with the relative invariance of other sets of relative positions (in Sraffa's argument considered above, the relative invariance of available modes of use). This approach views change as ***displacement*** of relative positions within the set of possible transformations associated with any given layer of the economic structure. In other words, the analysis of change (as displacement of relative positions) entails a degree of structural invariance and change itself takes the form of a rearrangement within economic structure (***structural change***). This approach explains Sraffa's view of causal analysis as analysis of 'deviation from what is normal, or uniform, or constant' (see above). It also explains the view, subsequently expressed by Sraffa, that the best approach to a theory of economic change (economic dynamics) is to start from the analysis of stationary states.[6] The analysis of change as displacement of

[6] This view was expressed by Sraffa in private conversation with Professor Izumi Hishiyama. I am grateful to Professor Hishiyama for informing me about his exchanges with Sraffa on this particular issue. It may be interesting to recall in this connection Arthur Cecil Pigou's acknowledgement of Sraffa's comments and suggestions in the Preface to his *Economics of Stationary States* (Pigou 1935).

relative positions entails a view of change distinctly different from the approach that would follow from consideration of functional interdependence. This is because in the latter case change would be associated with the 'mutual determination' of economic units (such as the mutual determination of the marginal products of ***identical doses*** of 'capital and labour') and would not derive from displacement from one position to another along a hierarchy of independently identified states of the economy (such as the changes of technical practice along a descending or ascending ranking of techniques). In short, change would be defined as variation in the positions occupied by otherwise identical and interchangeable economic units, ***not*** as variation in the positions of different economic units relative to one another. In our example, change follows from whether a particular dose $k$ of 'capital and labour' shifts from, say, a position after dose $n$ to a position after dose $n + 1$ and does not involve a change in the relative positions of independently identified doses of 'capital and labour'. This means that the analysis of change does not presuppose a pre-existing structure since the elements of structure are not identified independently of their own position relative to one another. In our example, it is impossible to identify the set of feasible transformations constraining changes in the positions of the doses of 'capital and labour' due to the impossibility of assigning an identity to the different doses independently of the position they occupy relative to one another. The lack of a structurally invariant constellation of economic units makes it impossible to study change as transformation within the 'geometry' ordering those units. In short, ***positional interdependence*** presupposes that the identity of economic units is independent of the positions they occupy, thereby allowing the study of positional variation as structural change, whereas ***functional interdependence*** makes otherwise undifferentiated economic units to 'switch' from one ***pre-assigned*** position to another, thereby making it impossible to analyse positional variation as a change of structure. This approach makes functional interdependence conducive to the analysis of causality (which, according to Sraffa, holds 'between categories') rather than to the analysis of causation (which, according to Sraffa, holds 'between particulars'). For example, it would be impossible to identify causation between dose $k$ and dose $k + 1$ considering that they could be the same dose occupying two different positions in turn.

The two situations may be compared by considering the figures below:

Figure 2.1 shows a situation in which three different, and distinctly identified, economic units are subject to displacement due to structural

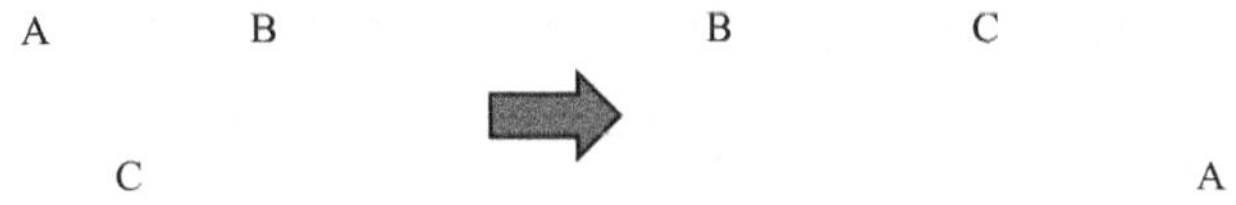


**Fig. 2.1** Positional interdependence and causation

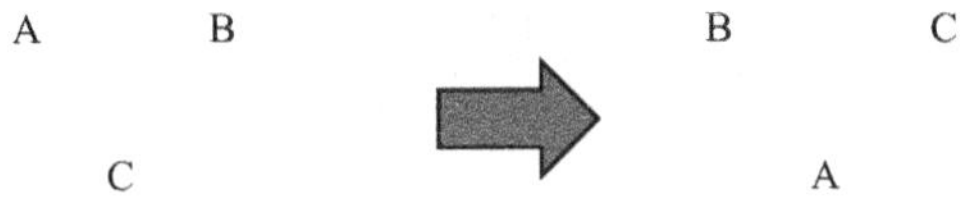


**Fig. 2.2** Functional interdependence and causality

change within the space in which they belong (a change in the set of available positions). Figure 2.2 shows a situation in which three economic units are subject to displacement due to their dislocation from one position to another in a space of given positions. In Sraffa's view, the situation in Fig. 2.1 allows the study of causal paths (causation) between particulars (say, between different modes of use of certain doses of 'capital and labour'), while the situation in Fig. 2.2 only allows the study of causality as a relationship between categories (say, between the virtual positions of a certain dose of 'capital and labour'), without highlighting causal links between 'particulars'.

Sraffa's assessment of Wicksteed's distinction between descriptive and functional curves, and his emphasis on the role of descriptive curves as a central device explaining productivity (and utility) curves highlights important features of Sraffa's early structuralism. It calls attention to causation between 'particulars' over causality between 'categories', and it does so by emphasizing causal analysis as identification of causal paths within a space of feasible transformations in the relative positions of economic units. Causation presupposes a hierarchical arrangement of states of the economy, which in turn highlights the relationship between causation and structural change as change in the relative positions of economic units.[7] This suggests that the reconstruction of causal paths may require

[7] There is a significant analogy between Sraffa's early structuralism and recent contributions to causal analysis as the investigation of causal paths under a plural causality framework (thus moving beyond a purely nomological approach to causality) (see Pearl 2009).

the substitution of 'recursive' for 'simultaneous' relationships between variables. In this case, causal paths may be reconstructed by considering systems of equations whose solution proceeds according to a sequence of steps:

> The first equation … can be immediately solved, thus obtaining the value of the unknown variable, while completely ignoring all the other relationships. We cannot, on the other hand, solve, for example, the last equation of the system … if not *after* having solved all the other equations of the system. In other words, it is possible to order the various relationships … in a precise and definite way. Whichever of the *n* equations we consider, it can be immediately ascertained that that equation can be solved only after all the preceding equations have been solved, while ignoring all the equations which follow. This means that there is, among the variables or unknowns of the system, *an asymmetrical relation which proceeds in a unique direction and not in the opposite direction*. … It is this notion of an asymmetrical relation among the variables that is here called "causality". The order in which the equations are placed—according to the chain of dependency which links them—will be called "causal order" or "causal chain". (Pasinetti 2019 [1965], p. 359; added emphasis)

The above point highlights that, in structural analysis, causation presupposes a degree of structural invariance that, in turn, guides causal paths along partially predetermined sequences of transformation.[8] Sraffa's early emphasis on causation as 'one direction' relationship is an important feature of the formative stage of his approach to economic structure and anticipates central further developments of his contribution to economic theory.

[8] The concept of 'relative structural invariance' is discussed in Landesmann and Scazzieri 1990. There it is argued that 'economic structure is generally described in such a way that certain elements of it are considered to be fixed while other elements are allowed to change' (Landesmann and Scazzieri 1990, p. 96). In view of this, 'the analysis of structural dynamics is associated with a general postulate of *relative invariance*' (Landesmann and Scazzieri 1990, p. 96). This postulate entails that 'any given economic system subject to an impulse or force is allowed to change its original state by following an adjustment path that belongs *to a limited set of feasible transformations*' (Landesmann and Scazzieri 1990, p. 96, added emphasis). The postulate of relative structural invariance highlights the role of 'structural permanence' in the analysis of change (Dagum 1969) and is closely associated with Herbert Simon's emphasis on 'intermediate stable forms' as a characteristic feature of evolving structures (Simon 1962).

## 4 Structural Analysis and Political Economy

Sraffa's subsequent research into classical economic theory and his endeavour to contribute to a 'critique of economic theory' (Sraffa 1960) may be interpreted as search of foundational principles for a theory of economic structure based on the condition of self-replacement as the 'law' in terms of which 'the structure *qua* whole or system is defined' (Piaget 1971 [1968], p. 7; see also de Vivo 2003). This approach may be seen at work in Sraffa's paper on Hayek's theory of money and capital (Sraffa 1932), in his development of an 'objective' theory of profits on Ricardian basis (Sraffa 1951), and in *Production of Commodities by Means of Commodities* (Sraffa 1960). In the Hayek paper, Sraffa highlights the structural character of debt-credit relationships by emphasizing the role of borrowing and lending in allowing interdependence of production activities in an economy based on division of labour:

> Loans are currently made in the present world in terms of every commodity for which there is a forward market. When a cotton spinner borrows a sum of money for three months and uses the proceeds to purchase spot, a quantity of raw cotton which he simultaneously sells three months forward, he is actually "borrowing cotton" for that period. The rate of interest which he pays, per hundred bales of cotton, is the number of bales that can be purchased with the following sum of money: the interest on the money required to buy spot 100 bales, plus the excess (or minus the deficiency) of the spot over the forward prices of the 100 bales. (Sraffa 1932, pp. 49–50)

The immediately following passage also highlights that the relationship between borrowing and lending may be affected by disequilibria between the quantities respectively demanded and supplied of the good in terms of which loans are made:

> In equilibrium the spot and forward price coincide, for cotton as for any other commodity; and all the "natural" or commodity rates are equal to one another, and to the money rate. But if, for any reason, the supply and the demand for a commodity are not in equilibrium (i.e. its market price exceeds or falls short of its cost of production), its spot and forward prices diverge, and the "natural" rate of interest on that commodity diverges from the "natural" rates on other commodities. (Sraffa 1932, p. 50)

The relationship between borrowing and lending is seen in terms of the structural requirements for the intertemporal coordination of productive activities (in the example above, raw cotton production vs. cotton spinning).[9] System-wise intertemporal coordination of productive activities brings about the equality between the spot and forward prices, as well as the equality of 'natural' rates of interest for all goods in the economy. Disruption of intertemporal coordination due to disequilibrium between demand and supply for the commodity in terms of which loans are expressed (in our case, cotton) makes its spot and forward prices to diverge, and makes the 'natural' rate of interest on that commodity to be different from the 'natural' rates of interest for other commodities. A systemic condition (intertemporal coordination of production activities) binds debts-credit relationships to one another and makes 'local' disequilibria (for example, an excess of the market price of cotton over its cost of production) to influence the whole credit economy inducing divergence between the 'natural' rates of interest for loans in terms of different goods.

Sraffa's interest in the interdependence of activities subject to a criterion of systemic coherence is also manifest in his 'Introduction' to David Ricardo's *On the Principles of Political Economy and Taxation* (Sraffa 1951). There Sraffa's attention switches to Ricardo's search of a uniform and invariable standard of measurement for the exchange value of commodities, which Sraffa describes as the search of a standard that would allow expressing commodity exchange values in terms of a magnitude that would in turn highlight an *immediate connection* between the exchange values of different commodities and the interdependence of all produced commodities within a system of division of labour. Sraffa sees the early steps of Ricardo's search in his *Essay on the Influence of a low Price of Corn on the Profits of Stock* (Ricardo 1951 [1815]). There Ricardo had argued that 'it is the profits of the farmer that regulate the profits of all other trades' (Ricardo 1951 [1815]), a statement whose 'rational foundation' Sraffa reconstructs as follows:

> in agriculture the same commodity, namely corn, forms both the capital (conceived as composed of the subsistence necessary for workers) and the product; so that the determination of profit by the difference between total product and capital advanced, and also the determination of the ratio of this

[9]On the 'material' debt-credit relationships in a system of interdependent production activities see also Cardinale and Scazzieri 2016.

> profit to the capital, is done directly between quantities of corn without any question of valuation. It is obvious that only one trade can be in the special position of not employing the products of other trades while all the others must employ *its* product as capital. It follows that if there is to be a uniform rate of profit in all trades it is the exchangeable values of the products of *other* trades relatively to their own capitals (i.e. relatively to corn) that must be adjusted so as to yield the same rate of profit as has been established in the growing of corn; since in the latter no value changes can alter the ratio of product to capital, both consisting of the same commodity. (Sraffa 1951, p. xxxi)

The above passage highlights the structural cohesion of a system of interdependent activities whose mutual fitting is achieved through the institutional assumption of a uniform rate of profit, which in turn reveals a hierarchy between activities (agricultural profits regulate profits in all other trades) under the additional assumption that in agriculture the determination of the rate of profits 'is done directly between quantities of corn without any question of valuation' (Sraffa 1951, p. xxxi). Sraffa acknowledges that the above 'rational foundation' is not explicitly stated in Ricardo's writings, but he also highlights that Ricardo came close to an explicit statement of that view in a letter to Malthus of June 1814: 'The rate of profits and of interest must depend on the proportion of production to the consumption necessary to such production' (Ricardo to Malthus, 26 June 1814, in Sraffa 1951, p. xxxii).

Sraffa's reconstruction of the evolution of Ricardo's' theory of value from the *Essay on Profits* through the three editions of the *Principles* highlights Sraffa's own interest in a theory of value grounded on a structural condition expressing interdependence of activities at the systemic level. The search for an invariable measure of value is not simply the search for an invariable measuring standard that would make relative exchange values of produced commodities independent of changes in the distribution of income. It is also the search for exchange ratios that would immediately follow from the technical conditions of production in an economy characterized by division of labour. From this point of view, Ricardo's labour theory of value may be seen as the expression of a structural approach: the *relative* quantities of labour embodied in the production of commodities express the mutual dependence of production processes in a system of 'divided labour': different producers (or groups of producers) make different commodities or different components of commodities. The labour theory of value expresses in a simple and fundamental form the reciprocal dependence of specialized producers, allowing the economy to be self-sustainable for any length of time.[10]

The switch to an objective theory of value (a labour theory of value) in Ricardo's *Principles* is explained by Sraffa as an attempt to measure 'the rate of profits in society' without assigning a special role to 'one special branch of production' (as is the case with agriculture in the *Essay on Profits*) (Sraffa 1951, p. xxxii). The labour theory of value enabled Ricardo 'to abandon the simplification that wages consist only of corn, which had been under frequent attack from Malthus, and to treat wages as composed of a variety of products (including manufactures), although food was still predominant among them' (Sraffa 1951, p. xxxii). As a result, it became possible to visualize the interdependence between economic activities by focussing on a synthetic measure of interdependence ('the rate of profits is society'), which could again be expressed in terms of physical quantities (quantities of labour rather than quantities of corn):

> It was now labour, instead of corn, that appeared on both sides of the account—in modern terms, both as input and output: as a result, the rate of profits was no longer determined by the ratio of the corn produced to the corn used up in production, but, instead, by the ratio of the total labour of the country to the labour required to produce the necessaries for that labour. (Sraffa 1951, p. xxxii)

The subsequent editions of Ricardo's *Principles* made him to acknowledge exceptions to a rigid formulation of the labour theory of value due to the different proportions between labour and instruments of production in producing different commodities, but did not make Ricardo to substantially alter his view that 'in fixing on the quantity of labour realised in commodities as the rule which governs their relative value we are in the right course.' (Ricardo, letter to McCulloch of 25 Jan. 1821, in Sraffa 1951,

[10] Self-sustainability may take different forms depending on technology in use. For example, self-sustainability for a pure labour economy (an economy of specialized producers making commodities by means of labour only without any need of capital goods) may be expressed in terms of a 'fundamental macroeconomic condition' relating the composition of the economy with effective demand for consumer goods (Pasinetti 1993). On the other hand, the self-sustainability for an economy in which commodities are produced by labour and capital goods has been expressed in terms of a self-replacement condition (the so-called 'viability condition') for intermediate goods (Hawkins and Simon 1949).

p. xl). In short, as Sraffa points out: 'no essential change was made in successive editions [of Ricardo's *Principles*] about the rule which determines value'[11] even if 'considerable alterations were made in connection with the choice of an invariable measure of value' (Sraffa 1951, p. xl). In this connection, Sraffa quotes passages from Ricardo's late essay on 'Absolute Value and Exchangeable Value' in which Ricardo distinguishes between determination of absolute value and measure of relative value, and highlights that the quantity of embodied labour should be considered as the governing principle of the former: 'I may be asked what I mean by the word value, and by what criterion I would judge whether a commodity had or had not changed its value. I answer, I know no other criterion of a thing being dear or cheap but by the sacrifices of labour made to obtain it.' (Ricardo, 'Absolute Value and Exchangeable Value', in Sraffa 1951, p. xlvi) and 'To me it appears a contradiction to say a thing has increased in natural value while it continues to be produced under precisely the same circumstances as before.' (Ricardo, 'Absolute Value and Exchangeable Value', in Sraffa 1951, p. xlvi).

In a letter to McCulloch, Ricardo had asked: '[i]s it not clear then that as soon as we are in possession of the knowledge of the circumstances which determine the value of commodities, we are enabled to say what is necessary to give us an invariable measure of value?' (Ricardo to McCulloch, 21 August 1823, in Sraffa 1951, p. xli). In that connection, Sraffa highlights that the core of Ricardo's search for an invariable standard of value is the search for 'the conditions which a commodity would have to satisfy in order to be invariable in value', which 'came close to identifying the problem of a measure with that of the law of value' (Sraffa 1951, p. xli). This view of Ricardo's theory of value is consistent with Sraffa's interpretation of Ricardo's *Essay on Profits* (see above) since the reference to the relative quantities of embodied labour (the hallmark of the labour theory of value) is meant to allow *both* the determination of 'the rate of profit in society as a whole' (Sraffa 1951, p. xxxii) and the expression of a material condition for the self-sustainability of a system of interdependent production activities when there is no physical equivalence between 'between total product and capital advanced' (Sraffa 1951, p. xxxi). In other words, the search for an invariable standard of value, which characterizes the evolution of Ricardo's theory of value, came to be seen by Sraffa as *equivalent*

[11] Sraffa acknowledges that this interpretation, while 'unusual', had been put forward by Huguette Biaujeaud in her *Essai sur la théorie ricardienne de la valeur* (Biaujeaud 1934).

to the search for the fundamental principles governing 'sustainability prices' in a system of division of labour based on intermediate products (production of commodities by means of commodities).[12] This point is of fundamental importance not only for the interpretation of Ricardo (in whose connection Sraffa made this claim) but also for the interpretation of Sraffa's own major theoretical contribution (Sraffa 1960).

The different strands characterizing Sraffa's structuralism come together as a coherent whole in *Production of Commodities by Means of Commodities* (*PCC*), Sraffa (1960): (1) *positional interdependence* is central to Sraffa's representation of economic structure (Sraffa's industries are interconnected by virtue of intrinsic features—intermediate input requirements—that are *prior* to the introduction of connectivity itself); (2) *self-replacement* in a system of interdependent activities—Sraffa's industries—is the fundamental 'law' determining the identity of the economy as a self-contained system; (3) *hierarchical causal paths* are embedded in the system of interdependencies defining the economy (there is an asymmetrical relation between variables so that the price-distribution system is of the recursive type); (4) a *systemic constant* provides physical 'unity' to the system (this systemic constant is the maximum rate of profit, or rate of surplus ($R$), in the standard system) while also allowing for a plurality of alternative constellations of distributive variables, and thus for a plurality of price systems compatible with the self-replacement of the means of production used in the economy.

The components of Sraffa's structuralism are at work in Sraffa's analysis of the determination of reproduction prices under different assumptions concerning technology in use and the distribution of value added (surplus) between different categories of income (wages, profits, and rents). Positional interdependence is visible in all models of the economy considered by Sraffa in *PCC*: the interdependence between the corn industry and the iron industry reflects the technological structure of each industry and posits on that basis a structurally determined set of exchange values allowing self-replacement. Other features of positional interdependence become visible in the last chapters of *PCC*, such as the hierarchical relationship between basic and non-basic commodities and the treatment of decreasing productive efficiency. In the latter case, positionality highlights a feature of productive structure that goes back to Sraffa's

[12] The analytics of sustainability pricing in a multisectoral production framework has recently been investigated by Carlo D'Adda (D'Adda 2019).

earlier work on the relation between cost and quantity produced (Sraffa 1998 [1925]). This is a property of what we may call the 'natural' ranking of techniques (the 'modes of use' discussed in the 1925 article), which must be independent of activity levels (the scale of production) to allow producers to generate efficiency ranking reflecting their economic judgement and disposition (see also Scazzieri 1981, 1982). Self-replacement is the most visible hallmark of Sraffa's circular economies and it is central to his different models of the economy. However, the later chapters of *PCC* introduce a variety of qualifications: non basic commodities are not essential to self-replacement; joint production makes self-replacement open to a variety of different solutions for any given technology; the utilization of land or other non-produced means of production makes circularity to depend on resource availability and constrains self-replacement of any given circular economy (associated with a specific technology in use) within the range of activity levels compatible with existing resources (Quadrio Curzio 1986, 1996; Quadrio Curzio and Pellizzari 1999, 2018; Scazzieri et al. 2015). Hierarchical causal paths are associated with Sraffa's approach to interdependence as a *positional characteristic* of the relationships between economic units, such as the 'industries' of the *PCC* models of the economy. This entails that each industry is *by its own internal structure* connected with the other industries of any given economic system, so that the relationship between industries is not one of mutual determination (suggesting 'causality' of the interdependence type) but one leading to 'one direction' causal paths (suggesting 'causation') whenever 'there is a deviation from what is normal, or uniform, or constant' (Sraffa, pre 1928). The type of causal analysis that may be associated with Sraffa's approach in *PCC* is one that would highlight relationships 'between particulars', rather than relationships 'between categories' (Sraffa, pre 1928). This means that causal analysis would not be an issue in the case of interdependence between industries under a given structure|, while causal analysis would be required to explain the switch over from one configuration of 'particulars' to another (for example to explain the switch from one technical structure to another). This situation is typical of sequential paths of structural change, in which the relationship between variables is asymmetrical (it proceeds, say, from *A* to *B* but not also from *B* to *A*) (see Pasinetti 2019 [1965]; see also Hicks 1979). Sraffa's view of causal investigation is grafted upon the view of the economic system as a *coherent structure* of relationships, in which 'causality' is a virtual condition (a relationship between 'categories') to be distinguished from 'causation' as a realized path involving change from one state of the economy to another (a relationship between 'particulars'). The distinction between 'causality' as a set of feasible transformations and 'causation' as a specific relationship between 'particular' states of the economy highlights the central role of systemic constants in Sraffa's view of economic structure. This characteristic of *PCC* is most clearly in view in Sraffa's construction and exploration of the 'Standard System' as a central heuristic device for the detection of structural (that is to say, positional) properties of actual economic systems. As we have seen, in his discussion of Ricardo's *Essay on Profits* Sraffa had argued that, under the assumption that in agriculture the same commodity ('corn') forms both the capital and the product, 'the determination of profit by the difference between total product and capital advanced, and also the determination of the ratio of this profit to the capital, is done directly between quantities of corn without any question of valuation' (Sraffa 1951, p. xxxi). In *PCC* this insight is used in the investigation of the structural properties of a multisectoral economy in which the same physical homogeneity between the 'capital' and the 'product' is obtained by extracting from the 'actual economic system… a miniature system' such that 'the various commodities are represented among its aggregate means of production in the same proportions as they are among its products' (Sraffa 1960, p. 21). In this type of economic system (which Sraffa calls 'Standard system'), 'the various commodities are produced in the same proportions as they enter the aggregate means of production', which implies that 'the rate by which the quantity produced exceeds the quantity used up in production *is the same for each of them*' (Sraffa 1960, p. 23; added emphasis). In this situation, '[t]he rate that applies to the individual commodities is naturally also the rate by which the total product of the Standard system exceeds its aggregate means of production, or the ratio of the net product to the means of production of the system. This ratio we shall call the *Standard ratio*' (Sraffa 1960, p. 21). A remarkable consequence of this property is that in the Standard system:

> The possibility of speaking of a ratio between two collections of miscellaneous commodities without need of reducing them to the common measure of price arises of course from the circumstance that both collections are made up in the same proportions—from their being in fact quantities of the same composite commodity. The result would therefore not be affected by multiplying the individual component commodities by their prices. The

> ratio of the values of the two aggregates would inevitably be always equal to the ratio of the quantities of their several components. Nor, once the commodities had been multiplied by their prices, would the ratio be disturbed if those individual prices were to vary in all sorts of divergent ways. Thus in the Standard system the ratio of the net product to the means of production would remain the same whatever variations occurred in the division of the net product between wages and profits and whatever the consequent price changes. (Sraffa 1960, p. 21)

The above reasoning leads directly to the discovery of a linear inverse relation between wages as a proportion of the net product of the Standard system and the rate of profits:

> What has just been said of the ratio of the net product to the means of production in the Standard system applies equally if we replace the net product by any fraction of it: the ratio of such fraction to the means of production will remain unaffected by any variation of price … The rate of profits in the Standard system thus appears as a ratio between quantities of commodities *irrespective of their prices*. To restate the position in general terms, as far as the Standard system is concerned, we may say that, if $R$ is the Standard ratio or Maximum rate of profits and $w$ the proportion of the net product that goes to wages, the rate of profits is $r = R\ (1-w)$. (Sraffa 1960, pp. 24–25; added emphasis)

A particular heuristic device (the Standard system) allows discovery of a structural property (the linear relation between wages and the rate of profits) that is also at the root of the distribution of the net product in actual economies provided we make such distribution independent of the movement of relative prices:

> [T]he actual system consists of the same basic equations as the Standard system, only in different proportions; so that, once the wage is given, the rate of profits is determined for both systems regardless of the proportions of the equations in either of them. Particular proportions, such as the Standard ones, *may give transparency to a system and render visible what was hidden, but they cannot alter its mathematical properties.* (Sraffa 1960, p. 26, added emphasis)

As a result:

> The straight-line relation between the wage and the rate of profits will … hold in all cases, provided only that the wage is expressed in terms of the Standard product. The same rate of profits, which in the Standard system is obtained as a ratio between quantities of commodities, will in the actual system result from the ratio of aggregate values. (Sraffa 1960, p. 26)

The above set of propositions makes clear the instrumental character of the Standard system in Sraffa's investigation, and in particular the role of the 'mathematical properties' of the Standard system in bringing to light objective structural properties of actual economic systems independently of the actual proportions between industries and of the specific institutional arrangements governing the formation of prices in those systems. Sraffa had acknowledged that: 'the economic systems of reality are not self-replacing. They are in a constant state of transition and obsolescence, due to changes both in the types of (kind of) commodities produced and in methods of their production' (Sraffa, D3/12/75:16, as quoted in Sinha 2016, p. 131). However, the re-proportioning device, which Sraffa introduces at the close of Chap. 1 of *PCC* ('Production for Subsistence'), allows the detection of properties that (under the clause of relative structural invariance) may be considered as given in spite of changes in the activity levels and relative proportions of the different industries. Under relative structural invariance of technology in use 'the Standard system is nothing but an equivalent system to the "original or actual" system' (Sinha 2016, p. 133). In this way, Sraffa fulfilled the aim of building '[a] theory which would be akin to geometry', in which he did not allow 'any changes in the parameters of his equations because he reasoned that causal explanations can come into play only if there is *change*' (Sinha 2016, p. 179).

Sraffa's 'geometrical' approach to the construction of economic theory is a distinctive mark of his structuralist perspective. Sraffa's economic theorizing, from his early discussion of productivity curves to *PCC*, highlights the role of a relatively invariant constellation of relative positions as the central benchmark against which to assess the contingent features of the economy and the causal paths linking one state of the economy to another. In short, Sraffa's economic structure is a space of constants constraining the actual constellations of relative positions within a set of feasible transformations. This is the generating mechanism of 'modes of use' along descending or ascending productivity curves (Sraffa 1998 [1925],

1926), as well as the mechanism generating debt-credit relations in a production economy (Sraffa 1932), and the mechanism determining self-replacement prices on the basis of an exogenously given distribution of the net product (Sraffa 1960). This approach to economic structure makes structure inherently dynamic, as any given structure is compatible with manifold alternative constellations of relative positions (say, alternative constellations of relative proportions between industries, or alternative constellations of distributive variables). The inherent openness of Sraffa's structuralism makes it compatible with the analysis of manifold criteria by which relative positions may be arranged within any given structure, and with manifold criteria by which institutions may determine the way in which socio-economic groups take part in the distribution of the net product subject to the objective constraints of existing structure.

Alternative representations of structure are discussed in Appendix A ('On "Sub-Systems"') of *PCC*. There Sraffa argues that a given 'system of industries' in a self-replacing state, provided each industry is assumed to produce a different commodity, 'can be subdivided into as many parts as there are commodities in its net product, in such a way that each part forms a smaller self-replacing system the net product of which consists of only one kind of commodity. These parts we shall call "Sub-systems"' (Sraffa 1960, p. 89). This approach 'involves subdividing each of the industries of the original system (namely, the means of production, the labour and the product of each) into parts of such size as will ensure self-replacement for each sub-system' (Sraffa 1960, p. 89). Any given economic structure (that is, any given system of economic interdependencies) may thus be represented in two different although mutually compatible ways. In one case (Sraffa's original 'system of industries'), the representation of structure highlights the horizontal relationship between industries delivering commodities as intermediate products to one another; in the other case, the representation of structure highlights the vertical relationship between 'commodities in the net product' and the bundles of produced means of production and quantities of labour needed to produce those commodities. Both horizontal and vertical representations reflect the same economic structure but highlight different features of the economy and different possible causal paths from one state of the economy to another. Horizontal structures call attention to the mutual dependencies between different industries in a system of intermediate productions that are mutually constraining one another (in the extreme case in which all commodities are intermediate inputs for all commodities, no production is possible unless the other productions also take place). On the other hand, vertical structures highlight dependencies between commodities in the final demand vector (consumption goods or investment goods) and the specific miniature systems needed for each component of that vector. In this case, the system of interdependencies is partitioned and, in principle, the production of certain components of the final demand vector is possible even if the production of other components of that vector does not take place. Causal paths may be different in the two cases, since a horizontal structure makes the whole set of interdependencies at, say, time $t$ to affect the whole set of interdependencies at, say, time $t + 1$. On the other hand, vertical structures allow more specific causal paths linking each component of final demand with a distinct subsystem of means of production and labour quantities.

The openness of Sraffa's structural analysis allows not only a plurality of structural representations but also a *hierarchy* of constraints for any given type of representation. An instance of the latter feature may be found in the way in which a given structural constraint (corresponding to a given technology in use) may be compatible with different arrangements for the distribution of the net product between socio-economic groups. As we have seen, any given technology may be associated with a 'physical' straight-line relation between wages and the rate of profits provided a suitable unit of measurement is adopted. However, this 'physical' rate of profits is a systemic magnitude expressing the economic system's potential for allocating the whole net product (or a part of it) to a destination other than the remuneration of wages. This fraction of the net product (which coincides with the whole net product in the case of the maximum rate of profits) may be allocated differently depending on the institutional arrangement under consideration. This may be explained by the fact that the fundamental structure of the economic system (as expressed by the 'physical' straight-line relation between wages and the rate of profits) allows different approaches to the determination of self-replacement prices depending on which features of that structure are considered (see also Zambelli 2018).

A horizontal representation of structure generates the following price equation (**p** denotes the vector of prices, **a** [$n$] the vector of labour coefficients, **A** the matrix of intermediate-product technical coefficients, $\pi$ the rate and profits, and $w$ the unit wage):

$$\mathbf{p} = \mathbf{a}[\mathrm{n}]w + \mathbf{pA} + \mathbf{pA}\pi = \mathbf{a}[\mathrm{n}]w + \mathbf{pA}(1+\pi) \tag{2.1}$$

The above equation makes self-replacement prices to reflect the relative conditions of production in different industries (as expressed by their different proportions of labour to means of production) as well as the specific institutional setting of a competitive capitalist economy (as express by the uniform unit wage and the uniform rate of profits).

A vertical representation of structure generates the following price equation ($\mathbf{p}$ denotes the vector of prices, $\mathbf{v}$ the vector of vertically integrated labour coefficients, $\mathbf{H}$ the matrix of units of vertically integrated productive capacity, and $\pi$ the rate of profits, while the unit wage is chosen as *numéraire*) (Pasinetti 1980 [1973], pp. 16–24):

$$\mathbf{p} = \mathbf{v} + \mathbf{pH}\pi, \tag{2.2}$$

Equation (2.2) presupposes the vertical integration of production processes discussed by Sraffa in the Appendix 'On Sub-Systems' of *PCC* (see above). If vertical integration is repeated a sufficient number of times, Eq. (2.2) can be written as follows (Pasinetti 1980 [1973], p. 31):

$$\mathbf{p} = \mathbf{v} + \mathbf{v}_k\pi + \mathbf{v}_{k2}\pi^2 + \ldots + \mathbf{v}_{ks}\pi^s + \mathbf{p}_{ks}\mathbf{H}^{\pi s+1} \tag{2.3}$$

Equation (2.3) allows one 'to resolve the price of every commodity into a sum of profit-weighted quantities of labour' (Pasinetti 1980 [1973], p. 31) considering that the $(s + 1)$ th order physical residual 'can be made as small as may suit one's purpose by making $s$ as great as is necessary' (Pasinetti 1980 [1973], p. 30).

The comparison of Eqs. (2.1) and (2.3) shows that different representations of structure may highlight different causal paths explaining changes of self-replacement prices compatible with the same structural parameters. Equations (2.2) and (2.3) also show the way in which different institutional contexts may exert an influence on self-replacement prices (Scazzieri 2012). Equation (2.1) presupposes an institutional set-up consistent with a uniform unit wage and a uniform rate of profits. Equation (2.3) opens the way to the consideration of a possible cleavage between an institutional context assigning increasing weights to 'remote' quantities of labour (possibly the context most closely resembling a conventional and stationary wage-profit economy) and an institutional context assigning decreasing weights to 'remote' quantities of labour (possibly the context of a dynamic economy assigning increasingly lower weights to remote quantities of labour and providing no structural justification for the reward of 'abstinence').

Sraffa's structuralism highlights a multi-layered visualization of economic structure, the plurality of causal paths allowed by different structural representations under any given set of structural constants, and the plurality of institutional arrangements that may be compatible with any given representation of economic structure.

## 5 Concluding Remarks

This chapter has examined Sraffa's contribution to economic theory as an exploration into structural heuristic. The analysis has been carried out in a sequence of steps. First, the fundamental characteristics of the structural approach have been outlined by highlighting the conception of structure as a set of relative positions within a relatively invariant space of feasible positional changes (that is, a relatively invariant set of feasible transformations). Second, the chapter has examined Sraffa's contributions to economic theory, from his early criticism of Marshall's productivity curves to *Production of Commodities by Means of Commodities (PCC)*, as the gradual unfolding of a structural heuristic based on classical economic theory but increasingly distinct from the original formulations of that theory. In that connection, the chapter has highlighted the role of central elements of structural heuristic, such as the 'geometric' emphasis on positional interdependence in lieu of the 'analytic' emphasis on functional interdependence, which is apparent already in Sraffa's early discussion of decreasing and increasing productivity curves. The chapter has then examined the subsequent development of Sraffa's structuralism where relative positioning is gradually associated with emphasis on structural constants as defining parameters in a space of relative invariances. Finally, the chapter has addressed *PCC* as the mature product of a structural heuristic in which relative positions and structural constants come together as complementary features of a type of structuralism open to a variety of structural configurations and to a corresponding plurality of causal paths compatible with any given set of relatively invariant parameters.

## References


Biaujcaud, H. (1934) *Essai sur la théorie ricardienne de la valeur,* Paris, Librairie du Recueil Sirey.

Cardinale, I. and Scazzieri, R. (2016) 'Structural Liquidity: The Money-Industry Nexus', *Structural Change and Economic Dynamics,* 39 (December), pp. 46–53.

Cardinale, I. and Scazzieri, R. (2018) 'Explaining Structural Change', *Structural Change and Economic Dynamics,* online 13 December 2018, https://doi.org/10.1016/j.strueco.2018.12.002.

D'Adda, C. (2019) 'Note on Multi-Sectoral Growth and Sustainability Pricing in a Leontief-Type Model with Technical Progress', *Italian Economic Journal,* online 04 July 2019, https://link.springer.com/article/10.1007%2Fs40797-019-00102-2.

Dagum, C. (1969) 'Structural Permanence: its Role in the Analysis of Structural Dualisms and Dependences and for Prediction and Decision Purposes', *Zeitschrift für die gesamte Staatswissenschaft / Journal of Institutional and Theoretical Economics,* 125, (2, April), pp. 211–235.

de Vivo, G. (2003) 'Sraffa's Path to *Production of Commodities by Means of Commodities*', *Contributions to Political Economy,* 22(1), pp. 1–25.

Hawkins, D. and Simon, H. A. (1949) 'Note: Some Conditions of Macroeconomic Stability', *Econometrica,* 17 (3/4, July–October), pp. 245–48.

Hicks, J. (1979) *Causality in Economics,* Oxford, Basil Blackwell.

Landesmann, M. and Scazzieri, R. (1990) 'Specification of Structure and Economic Dynamics', in M. Baranzini and R. Scazzieri (eds.), *The Economic Theory of Structure and Change,* Cambridge, Cambridge University Press, pp. 95–121.

Marcuzzo, M.C. and Rosselli, A. (2011) 'Sraffa and his Arguments against "Marginism"', *Cambridge Journal of Economics,* 35 (1, January), pp. 219–231.

Marshall, A. (1890) *Principles of Economics,* London, Macmillan.

Pasinetti, L.L. (1980 [1973]) 'The Notion of Vertical Integration in Economic Analysis', in L.L. Pasinetti (ed.), *Essays on the Theory of Joint Production,* London, Macmillan, pp. 16–43.

Pasinetti, L.L. (1993) *Structural Economic Dynamics. A Theory of the Economic Consequences of Human Learning,* Cambridge, Cambridge University Press.

Pasinetti, L.L. (2019 [1965]) 'Causality and Interdependence in Econometric Analysis and in Economic Theory', *Structural Change and Economic Dynamics,* 49, pp. 357–363.

Pearl, J. (2009) *Causality: Models, Reasoning and Inference,* second edition, Cambridge, Cambridge University Press.

Piaget, J. (1971 [1968]) *Structuralism,* London, Routledge and Kegan Paul.

Pigou, A.C. (1935) *The Economics of Stationary States,* London, Macmillan.

Quadrio Curzio, A. (1986) 'Technological Scarcity: An Essay on Production and Structural Change', in M. Baranzini and R. Scazzieri (eds.), *Foundations of Economics. Structures of Inquiry and Economic Theory*, Oxford e New York, Basil Blackwell, pp. 311–38.

Quadrio Curzio, A. (1996) 'Production and Efficiency with Global Technologies', in M. Landesmann and R. Scazzieri (eds), *Production and Economic Dynamics*, Cambridge, Cambridge University Press, pp. 105–126.

Quadrio Curzio, A. and Pellizzari, F. (1999) *Rent, Resources, Technologies*, Berlin, Springer.

Quadrio Curzio, A. and Pellizzari, F. (2018) 'Political Economy of Resources, Technologies, and Rent', in I. Cardinale and R. Scazzieri (eds), *The Palgrave Handbook of Political Economy*, London, Palgrave Macmillan, pp. 657–704.

Ricardo, D. (1951 [1815]) *Essay on the Influence of a low Price of Corn on the Profits of Stock*, in *The Works and Correspondence of David Ricardo* (edited by Piero Sraffa with the collaboration of M. H. Dobb), vol. IV: *Pamphlets and Papers 1815–1823*, pp. 1–41.

Rosselli, A. and Trabucchi, P. (2018) 'Sraffa, the "Marginal" Method and Change', *Structural Change and Economic Dynamics*, available online 10 October 2018, https://doi.org/10.1016/j.strueco.2018.09.006.

Scazzieri, R. (1981) *Efficienza produttiva e livelli di attività*, Bologna, Il Mulino.

Scazzieri, R. (1982) 'Scale and Efficiency in Models of Production', in M. Baranzini (ed.), *Advances in Economic Theory*, Oxford, Basil Blackwell, pp. 19–42.

Scazzieri, R. (1993) *A Theory of Production. Tasks, Processes and Technical Practices*, Oxford, Clarendon Press.

Scazzieri, R. (2012) 'The Political Economy of *Production of Commodities by Means of Commodities*: A Comment on Pasinetti and Sraffa', *Cambridge Journal of Economics*, 36 (6, November), pp. 1315–1322.

Scazzieri, R., Baranzini, M. and Rotondi, C. (2015), 'Resources, Scarcities and Rents: Technological Interdependence and the Dynamics of Socio-Economic Structures', in M. Baranzini, C. Rotondi and R. Scazzieri (eds.), *Resources, Production and Structural Dynamics*, Cambridge, Cambridge University Press, pp. 427–484.

Simon, H. A. (1962) 'The Architecture of Complexity', *Proceedings of the American Philosophical Society*, 106, No. 6. (December 12), pp. 467–482.

Sinha, A. (2016) *A Revolution in Economic Theory: The Economics of Piero Sraffa*, Basel, Springer.

Sraffa, P. (1925) 'Sulle relazioni fra costo e quantità prodotta', *Annali di economia*, 2 (1), pp. 277–328.

Sraffa, P. (1926) 'The Laws of Returns under Competitive Conditions', *The Economic Journal*, 36 (December), pp. 535–50.

Sraffa, P. (pre 1928) 'Notes on interdependence and causality', *Sraffa papers*, Trinity College Archive, Cambridge.

Sraffa, P. (1932) 'Dr. Hayek on Money and Capital', *The Economic Journal*, 42 (March), pp. 42–53.
Sraffa, P. (1951) 'Introduction', in *The Works and Correspondence of David Ricardo* (edited by Piero Sraffa with the collaboration of M. H. Dobb), vol. I: *On the Principles of Political Economy and Taxation*, Cambridge, Cambridge University Press for the Royal Economic Society, pp. xiii–lxii.
Sraffa, P. (1960) *Production of Commodities by Means of Commodities. Prelude to a Critique of Economic Theory*, Cambridge, Cambridge University Press.
Sraffa, P. (1998 [1925]) 'On the Relations between Cost and Quantity Produced', in L. Pasinetti, ed., *Italian Economic Papers*, vol. III, Oxford, Oxford University Press and Bologna, Il Mulino, pp. 323–63.
Turgot, A.-R.-J. (1768) 'Observations sur le Mémoire de M. de Saint-Péravy en faveur de l'impôt indirect', in *Oeuvres de Turgot*, Paris, 1844.
Wicksteed, P.H. (1910) *The Common Sense of Political Economy: including a study of the human basis of economic law*, London Macmillan.
Wicksteed, P.H. (1914) 'The Scope and Method of Political Economy in the Light of the "Marginal" Theory of Value and of Distribution', *The Economic Journal*, 24 (March), pp. 1–23.
Zambelli, S. (2018) 'Production of Commodities by Means of Commodities and Non-Uniform Rates of Profits', *Metroeconomica*, 69(4), pp. 791–819.

Comment

# A Comment on Scazzieri's 'On Sraffa's Structuralism'

*Ajit Sinha*

In this chapter, Professor Scazzieri (2021) presents a highly original interpretation of the development of Sraffa's economic analysis as well as the nature of his early critique of the marginalist analysis of the neoclassical economics. Scazzieri begins with the young Sraffa's (1998 [1925]) criticism of Wicksteed's (1914) distinction between the neoclassical notion of marginal productivity and Ricardo's diminishing returns on land. Wicksteed argues that Ricardo's representation of diminishing returns on land in his theory of extensive rent rests on the arrangement of land in terms of their natural fertility in declining order, and the argument is that the extension of cultivation chooses the highest fertility of land available in the descending order. This, according to Wicksteed, is a descriptive relationship of productivity of a dose of capital and labour on land—it is the difference in the natural productivity of land and their arrangement that explains the declining outcome of the successive application of the homogeneous dose of capital and labour. He juxtaposes this arrangement of the Ricardian explanation of *extensive* rent with the other scenario of *intensive* rent when equal doses of capital and labour are applied one after another on the same plot of land. In this case, the diminishing productivity of every subsequent dose of capital and labour, according to Wicksteed, represents a functional relation between doses of capital and labour and the output of land—this is supposed to be a technical relation that results from the order in which the homogeneous dose of capital and labour is applied. In this case, the relationship can be inverted—one can fix the number of doses of capital and labour and successively apply equal and homogeneous units of land on them and find that the outcome would be the same when one arrives at the stage when the aggregate of all the land reaches its original plot size and the number of doses of capital and labour applied in the original case were equal to the number fixed in the latter case.

Sraffa's criticism of Wicksteed is that both his scenarios are descriptive and not functional. To understand Sraffa's point, let us suppose I have ten acres of homogeneous land, and a dose of labour and capital consists of one man, one horse and ten kilograms of seed. Now if I cultivate all my land by spreading my first dose of labour and capital on my ten acres of land, then the total output I reap is one quintal; if I increase the dose to two, the total output I reap is three quintals; when I add my dose to three, then the total output I reap is six quintals; when I add another dose, the total output I reap is eight quintals; when I add another dose, the total output I reap is nine quintals, and so on. Here we can see that as we go on adding the doses the marginal productivity of the successive doses rises from one to two and then three but eventually begins to decline from three to two to one. Sraffa argues that if I were a rational farmer then I would not have followed this scheme of production. Instead of spreading the first dose of labour and capital on the whole plot of land, I, as a rational farmer, would find the optimal combination of land that goes with one dose of labour and capital and apply my first dose only to that much of land. I will parcel all my land into such optimal size and apply every subsequent dose on them one by one, reaping equal and optimum returns on all the doses till all my land is exhausted by this most productive technique. After which, I will again find the second best, since the first best is exhausted, combination of land size to go along successive doses of labour and capital till that exhausts the whole land and so on. Yet again, we find that successive techniques are arranged in a descending order of productivity and it is a result of a rational arrangement by the farmer. Thus, this too is a descriptive and not a functional relation.

Here, Scazzieri teases out Sraffa's embryonic structuralism. Following Piaget's description of structural representations as collections of elements as wholes subordinated to laws, in terms of which the structure *qua* whole or system is defined (Piaget 1971 [1968], p. 7), Scazzieri argues that Sraffa's claim that productivity curves are descriptive and not functional highlights the positionality of the relation in which the techniques are arranged, which is subordinated to the 'law' of rational behaviour or as Sraffa (1998 [1925]) puts it, 'freedom to behave in the manner most rewarding to [the farmer]'. Further, Scazzieri analyses Sraffa's (1998 [1925]) critique of Marshall's supply or cost curve for an industry derived from aggregating the supply curves of all the firms in an industry. Sraffa argues that though a firm within an industry can always increase its output without having any impact on the price of the constant factor, thus experiencing constant or even decreasing cost, this cannot be true for the industry as a whole if the factor is fixed. Therefore, the idea of the whole as an aggregation of individual parts collapses—the behaviour of the whole is constrained by a structure; however, the individual elements have more freedom of movement. Similarly, whether an industry must experience increasing cost or decreasing cost or constant cost largely depends on the level of aggregation used for classifying an 'industry'; for example, if an industry is defined as 'agriculture' then there is a good chance that this will show increasing cost as land for agriculture is more or less fixed, but if an industry is classified as 'tomato industry' then there is a good chance that it experiences no fixed factor and would most likely show either constant or decreasing cost. Yet again, the whole thing depends upon a broader structure of classification (see Sinha 2016, pp. 32–39 for my reading of Sraffa's 1925 paper).

Moving on from 1925 to 1927–28, Scazzieri takes up a cryptic note of Sraffa where he criticizes Marshall and Pigou for confusing causation for causality. Sraffa argues that a relationship of causality is a relation of determination that holds only between two categories such as the relationship between bodies and their shadows—bodies determine shadows and movement of bodies determine movement of shadows, and existence of shadows imply existence of bodies. These are categorical relations, and they always exist between categories. However, if a mushroom grows in the shadow of a tree then such a relationship is a relationship of causation—it is a relationship between particulars—this particular shadow causes this particular mushroom to grow but shadows do not necessarily cause mushrooms to grow, that is they don't always go together, nor do existence of mushrooms imply shadows of trees. This sets the stage for understanding the difference between the natures of change within a structural framework from change in a usual causal framework of causation. Take, for example, Sraffa's most consequential equation from *PCMC*, $r = R(1-w)$, where $R$ is the maximum rate of profits and $r$ and $w$ are the rates of profits and wages given in terms of the Standard commodity. In this relationship, $R$ remains constant when either $r$ or $w$ changes—it is the constancy of $R$ with respect to changes in $r$ and $w$ that defines a structure given by an input-output system of production. In this case, the relationship between $r$ and $w$ exists in the same sense as the relationship between a tree and its shadow—for every given value of $w$, a determinate value of $r$ exists and vice versa. Furthermore, for every given value of $w$ and $r$, there exists a set of unique prices of commodities—changes in $w$ and $r$ yet again determine changes in the set of prices as movement of bodies determine movement of shadows. Now contrast this with the idea that a change in wages leads to changes in consumers' demand which in its turn leads to changes in production and thus the set of input-output data, which generates different $R$ and $r$ along with a different set of prices. Here the relationship between $w$ and $r$ and the set of prices is a relationship of causation and not causality. It is a particular relationship—given a socio-economic setup, a change in $w$ leads to several links of changes in a causal manner, but they are not *determinations*—they cannot claim to hold in all circumstances whenever $w$ changes by that amount. However, once the causation has produced a new set of input-output data, we once again will have the same structural relationship, given by $r = R(1-w)$, holding with a new value of $R$.

Scazzieri detects a shift in Sraffa's use of the law that subordinates the structure since his critique of Hayek in 1932 (see Sraffa 1932). Instead of using 'rational behaviour' on the part of the economic agents as the law of structural determination, Sraffa from 1932 onwards moved to the 'condition of reproduction' of the structure itself as the law underpinning the structure. In this context, Scazzieri considers the possibility of unequal rates of profits given wages, 'This fraction of the net product (which coincides with the whole net product in the case of the maximum rate of profits) may be allocated differently depending on the institutional arrangement under consideration. This may be explained by the fact that the fundamental structure of the economic system (as expressed by the 'physical' straight-line relation between wages and the rate of profits) allows different approaches to the determination of self-replacement prices depending on which features of that structure are considered.' It is true that outside institutions such as a State through its tax and subsidy policy could affect prices in such a way that the rates of profits may not remain equal. But in this context one cannot do the exercise of varying wages or the average rate of profits from zero to its maximum value. That is why any such outside institutions cannot be brought in the analysis at this stage—they can only be brought in in particular concrete cases to explain the existence of unequal rates of profits. This, however, does not mean that one can also introduce the idea of market mechanics of demand and supply to explain unequal rates of profits because it introduces an alternative theory of price determination and given these 'market' prices from outside the equation system one can always solve for industrial rates of profits for every industry independently—in other words, the idea that reproduction prices could be taken from outside the equation system destroys the integrity of Sraffa's 'system of interconnected basic goods'.

## References

Piaget, J. (1971 [1968]) *Structuralism*, London, Routledge and Kegan Paul.

Scazzieri, R. 2021. On Sraffa's Structuralism', in *A Reflection on Sraffa's Revolution in Economic Theory* (ed. Ajit Sinha), Cham: Palgrave Macmillan.

Sinha, A. (2016) *A Revolution in Economic Theory: The Economics of Piero Sraffa*, Cham: Palgrave Macmillan.

Sinha, A. (2021). 'Interpreting the nature of Sraffa's equations: A Critique of Garegnani's interpretation' in *A Reflection on Sraffa's Revolution in Economic Theory* (ed. Ajit Sinha), Cham: Palgrave Macmillan.

Sraffa, P. (1925) 'Sulle relazioni fra costo e quantità prodotta', *Annali di economia*, 2 (1), pp. 277–328.

Sraffa, P. (1932) 'Dr. Hayek on Money and Capital', *The Economic Journal*, 42 (March), pp. 42–53.

Sraffa, P. (1960) *Production of Commodities by Means of Commodities. Prelude to a Critique of Economic Theory*, Cambridge, Cambridge University Press.

Sraffa, P. (1998 [1925]) 'On the Relations between Cost and Quantity Produced', in L. Pasinetti, ed., *Italian Economic Papers*, vol. III, Oxford, Oxford University Press and Bologna, Il Mulino, pp. 323–63.

Wicksteed, P.H. (1914) 'The Scope and Method of Political Economy in the Light of the "Marginal" Theory of Value and of Distribution', *The Economic Journal*, 24 (March), pp. 1–23.

Reply

# A Response to Ajit Sinha's 'A Comment on Scazzieri's "On Sraffa's Structuralism"'

*Roberto Scazzieri*

As Professor Ajit Sinha points out in his insightful comment, the main strategy of my chapter is to take on board Sraffa's early criticism of Marshallian economics and to follow the analytical thread leading from that early criticism to Sraffa's own theoretical contribution in *Production of Commodities by Means of Commodities* (*PCC*). The aim of the chapter is *analytical*, rather than exegetical or historical, and its core idea is that there is a deep methodological thread linking Sraffa's early criticism of Marshallian economics to the conceptual framework of *PCC*, this thread being what I call in the chapter Sraffa's 'structuralism', that is, the belief that the fundamental configuration of the economic system should be understood primarily in terms of a system of (relatively invariant) relative positions rather than in terms of the manifold systems of interactions that may develop out of the structure built on those relative positions. My chapter argues that Sraffa's 'structuralism' (as described above) provides an analytical bridge between Sraffa's criticism of Wicksteed's functional curves in the treatment of diminishing returns (where relative positions of doses of labour and capital are varying and independent of 'deep' structural characteristics of those doses) and Sraffa's investigation of the reproduction condition of a system of industries associated with given relative positions of industries vis-à-vis one another. In short, my view is that Sraffa's 'structuralism' is always at play through Sraffa's lifework, that such 'structuralism' aims at identifying the relatively invariant constellation of relative positions behind the manifold contingent interactions that any given structure allows, and more generally that Sraffa's 'structuralism' provides an important example of the interplay between grounded virtuality (virtuality rooted in a relatively invariant set of relations) and historical particulars.[1]

In view of the above, I completely agree with the full first paragraph of Professor Sinha's Comment. As to the second paragraph, it deals with a part of Sraffa's 1925 paper that is not central to my own argument, that is, Sraffa's criticism of the law of variable proportions as originally formulated in A.R.J. Turgot's 'Observations sur M. de Saint Peravy', in which Turgot attempts to provide a type of coordination between increasing and decreasing returns. In my view, Sraffa's argument there, while interesting in itself, is in fact a corollary of his own criticism of the formulation of the law of diminishing returns in terms of functional curves (see also Scazzieri 1982). In view of this, I would still associate 'Turgot's curve' (as drawn by Sraffa) to functional rather than to descriptive curves.

I fully agree with the third paragraph of Sinha's Comment. I am grateful for Professor Sinha's remark that Sraffa's argument makes 'the idea of the whole as an aggregation of individual parts' to collapse (Sinha 2021, p. 84). This point has deep theoretical relevance as it shows that 'the behaviour of the whole is constrained by a structure' even if 'the individual elements have more freedom of movement' (Sinha 2021, p. 2). Indeed, this is a point that highlights an important feature of Sraffa's 'structuralism' at the interface between grounded virtuality and historical particulars (see above).

The fourth section of Professor Sinha's comment takes up my discussion of Sraffa's distinction between causality and causation and highlights the link between that distinction and Sraffa's approach to the analysis of change. Here Sinha calls attention to 'Sraffa's most consequential equation from *PCC*, $r = R\,(1-w)$' (Sinha 2021) and rightly points out that that equation is related to Sraffa's distinction between categories (rooted in the system structure) and particulars (showing the freedom of movement allowed within a given structure).

Section five of Sinha's Comment examines the relationship between the structural properties associated with Sraffa's 'most consequential equation' $r = R\,(1-w)$ and the various system configurations possible under that equation. In fact, the conceptual framework of Sraffa's 'structuralism' suggests that the uniqueness of $r$ stemming from equation $r = R\,(1-w)$ is a fundamental structural property (a relationship between categories) while the empirical (or historical) plurality of $r_i$'s ($i = 1,\ldots, k$) might be a contingent property (a relationship between particulars). I do not deal with this issue in my chapter, but I think it would be an interesting problem to investigate.

## References

Cardinale, I. and Scazzieri, R. (2018) 'Explaining Structural Change', *Structural Change and Economic Dynamics*, December, DOI: https://doi.org/10.1016/j.strueco.2018.12.002.

Scazzieri, R. (1982) 'Scale and Efficiency in Models of Production', in M. Baranzini (ed.), *Advances in Economic Theory*, Oxford and New York, Basil Blackwell, pp. 19–42.

Sinha, A. (2021) 'A Comment on Scazzieri's "On Sraffa's Structuralism"', this volume.

CHAPTER 3

# On Sraffa's Challenge to Causality in Economics

*Maria Cristina Marcuzzo and Annalisa Rosselli*

## 1 Foreword

The opening of Sraffa's Archives[1] has given us the opportunity to unveil what lies behind his published works, offering clues to interpret them. Among the contributions found in Sraffa's unpublished papers are the arguments he developed over the years on the role of causality in economics.

In the last few years several scholars have tackled the issue which, however, remains to be untangled and fully understood. This more recent

[1] References to the Sraffa Papers, which are kept at Trinity College Library, Cambridge, and mostly available online, will be given hereafter following the catalogue classification. In a few cases, we completed Sraffa's abbreviations in order to make the text more readable.

M. C. Marcuzzo (✉)
Sapienza, Università di Roma, Rome, Italy
e-mail: cristina.marcuzzo@uniroma1.it

A. Rosselli
Università di Roma "Torvergata", Rome, Italy
e-mail: annalisa.rosselli@uniroma2.it



A. Sinha (ed.), *A Reflection on Sraffa's Revolution in Economic Theory*, Palgrave Studies in the History of Economic Thought,
https://doi.org/10.1007/978-3-030-47206-1_3

literature converges on the interpretation—suggested by Sraffa's himself (see below)—that his entire research project is a struggle to escape from "mechanical", i.e. causal theory, and to develop a "geometrical" representation of the economic system. While the geometrical theory refers to an instant in time and concerns logical relations, the mechanical theory refers to processes that happen in real time, in which causality is involved. The question remains open, however, as to the reasons *why* Sraffa embarked on such a project; they certainly are complex and possibly related to his beliefs that the requirements for causal explanations—like those met with in the natural sciences—are too stringent to be applicable in economics or because economic theory has often invoked causes that are of a metaphysical (and therefore ideological) nature. Some have stressed the similarity with Wittgenstein's criticism of causality (Davis 2012; Arena 2015), others the influence of his early interest in quantum mechanics (Sinha 2016), physics, chemistry and philosophy of science (Kurz and Salvadori 2005, 2018), and yet others the influence of Gramsci (Ginzburg 2013; Davis 2021). Possibly many factors concur to fill out the background picture, which still needs to be fully discovered. In this paper, we review this recent literature to raise a few points that may lead to further discussion.

## 2 Change and Difference

In the Preface to *Production of Commodities by Means of Commodities*, the reader is warned that the book: "is concerned exclusively with such properties of an economic system as do not depend on changes in the scale of production or in the proportions of 'factors'". In this respect the work is presented as an alternative to the "marginal approach" which "requires *attention to be focused on change*, for without change either in the scale of an industry or in the 'proportions of the factors of productions' there can be neither marginal product nor marginal cost" (Sraffa 1960: v, emphasis added).

In fact, in neoclassical economics change in the quantities produced (or consumed) is necessary to find the marginal cost (or utility) upon which supply-and-demand curves are built.

Krishna Bharadwaj was perhaps the first to expand upon Sraffa's 1960 brief reference to the "requisite kind of change" (ibidem), by arguing that the role of change in supply-and-demand theories is to allow convergence towards equilibrium and it must therefore be conceived as following a well-specified path.[2] Thus the direction of change required (i.e. the assumption that demand and supply are well-behaved functions) imposes "constraints which operate against an effective handling of problems of change" (Bharadwaj 1986: 3). On the contrary, changes observed in the real world are descriptions of events that may occur outside the strictures of the predictive relation between price and quantity described by supply-and-demand functions, which are necessary for the stability of equilibrium. She writes:

> Thus, a statement about one position (at equilibrium) is tied up with a theory *specifying in exact terms* the behaviour of the system under 'change' away from that position, even though these variations may be considered 'hypothetical' or theoretically confined to the infinitesimally small (or to points 'in the neighbourhood'). [...] The prices of production in the classical theory, in contrast, do not posit and therefore do not depend for their stipulation upon the postulate of actual or 'potential' change. (Bharadwaj 1986: 39)

And she adds that the classical system is more general in scope and versatile in dealing with historic-specific factors because

> it does not commit itself through its theoretical structure to any rigid form and direction of change; [on the contrary, neoclassical theory] sought to explain a single observed situation in terms of potential changes—and as brought about by the balancing of marginal quantities operating through the principle of substitution. In order to be consistent with these explanations, the changes had to be in a direction and of the type postulated by theory. (ibid.: 63–64)

The question of how change is conceptualised in economics appeared quite early in the evolution of Sraffa's thinking.[3] In winter 1927–28, while preparing for the lectures he was to give at Cambridge, he came to the

[2] See Marcuzzo (2014), from which the paragraph draws.

[3] See Sraffa's comment in Summer 1927 "[...] the main stream of modern economic thought proceeds to analyse the ways in which change takes place, without being hindered by the fact that little is known of the ultimate causes of change" (D3/12/3/14). In fact, here Sraffa is making a distinction between the notions of 'ultimate cause' and 'mechanical cause'. He is identifying the difference between the Classical and the Modern by associating the Classical with the notion of 'ultimate cause' and the Modern with the 'mechanical cause'. We are indebted to the Editor of this volume for suggesting clarification of this distinction.

important conclusion that classical political economy and neoclassical economics rested on alternative theories of value and distribution; only the latter explains distribution and relative prices by means of the equilibrium of the two opposing sets of forces, demand for and supply of goods and factors of production; the classical political economy determines the distribution of the surplus, i.e. of product other than wages.

Another point that Sraffa made in this context was that in economics there is also the need to distinguish between situations in which change is necessary to construct the elements of the theory and those in which it is not required while analysis is conducted in terms of differences that co-exist in the same instant.

This led Sraffa to raise the related point of the distinction between "change" and "difference" in economic analysis. In one of the manuscripts of the period 1927–28, possibly where mention of the "Difference (simultaneous) versus Change (succession in time)" first occurs, Sraffa wrote:

> The general confusion in all theories of value [….] must be explained by the failure to distinguish between two entirely distinct types of questions and the universal attempt of solving them both by one single theory.

The two questions are:

1. what determines the [difference in the?] values at which various commodities are exchanged in a given market on a given instant?
2. what determines the changes in the values of commodities at different times? (e.g. of one commodity)." (D3/ 12/ 7/115; emphasis in the original).

And he continues:

> The first problem gives rise to a geometrical theory, the second to a mechanical one. The first is so much timeless that it cannot even be called statical. It does not represent an ideal stationary state in which it is assumed that no change takes place: but it represents a situation at one instant of time, that is to say something indistinguishable from a real state of things in such a short period of time that no visible movement takes place. Its object is, as it were, the photograph of a market place […] Marshall's theory of value, with its increasing and diminishing costs and marg[inal] utility, scissors, pillars

> and forces, can only be understood as an attempt to solve the first question in terms of the second. (D3/12/7/117)

And in the same set of manuscripts, while working on his price equations, reflecting on the possibility of letting the reader assume constant returns to scale, Sraffa wrote:

> [...] these equations cannot possibly answer as to how or why prices change. They only explain why, at a given moment (?), prices of different things bear to one another the proportions which they do. They explain variation (difference) between individual commodities at one time, not variations of one commodity at different times.
>
> No system of equations, whether it considers variable returns or not, could tell this if time does not enter as a variable. (D3/12/ 7/85-6)

The same point, but related to the theory of distribution, is reiterated more than 30 years later in one of the many extant drafts of the Preface to *Production of Commodities*:

> The fundamental difference is that the extensive (different qualities of land) is truly a purely timeless, or geometrical representation: all the different lands exist simultaneously/at one instant, they and their products can be ascertained, distinguished and measured at one instant, without changing anything in the present arrangements.
>
> On the contrary, the intensive (successive doses of c[apital] and l[abour] on a piece of land) dim[inishing] ret[urns] do not exist at any one instant [...] We can only find these dim[inishing] ret[urns] by change, or movement: that is to say, we require *time*. (D3/12/13/23.2; emphasis in the original).[4]

In conclusion Sraffa's point is the distinction between two types of change, those which involve the passage of time and those which do not. See the annotation by Sraffa dated October 1929:

[4] In a contemporaneous manuscript dated 1–3 January 1958 and inscribed as "Margins and margins", which he might have considered including in his book, Sraffa made the same point, returning to his earlier distinction between difference and change (D3/12/46/50). Finally, Sraffa reiterated the same point as late as 17 August 1965, when he was still trying to carry out his project for a comprehensive critique of marginal theory (D3/12/42/5). On these points, see Marcuzzo and Rosselli (2011).

> [The following] notion of time is important: it really substitutes 'instantaneous photographs' as opposed to ordinary time. It is only a part of ordinary time, it has only some of its connotations: it includes events, also different events, but no change of events. It enables us to compare two simultaneous, but not instantaneous, events—just as if they were 'things'. (D3/12/13/1.3 emphasis added)

In a different context, in the Introduction to Ricardo's *Principles*, we find Sraffa explicitly referring to "the two points of view of difference and of change" (Sraffa 1951: xlix) to distinguish between real and apparent change. The distinction lies behind the question of how to measure "the magnitude of aggregate of commodities" in order to study the distribution of surplus: there *seems* to be a change in the quantity of output to be distributed whenever there is a change in its value due to a change either in wages or in profits. However, this is an *apparent* change,[5] since the conditions of production of the commodities and the quantities produced remain the same. It is the kind of change studied in *Production of Commodities* when analysing the effects on prices of a change in distribution, not to be confused with the effects deriving from the introduction of the time factor.

## 3 Determinism

"Real" changes imply a process which occurs in time. Under what circumstances can we consider the effect of a change that implies comparison of the same object at two different instants, equivalent to the observation of two distinct objects at the same time? There are two possibilities and Sraffa seems to have considered both. The first is that we assume that the time interval is so short that the assumption of *ceteris paribus* holds because of lack of time for modifications of the existing situation other than that in

[5] Finally, the distinction between difference and change is employed by Sraffa to distinguish two different aspects of the effects on the relative value of two commodities of different proportions or durabilities of capital employed in their production. Sraffa writes: "[the first aspect] is that of occasioning a *difference* in the relative values of two commodities which are produced by equal quantities of labour. Second, that of the effect which a rise of wages has in producing a *change* in their relative value" (Sraffa 1951: xlvii). This testifies to the importance Sraffa attached to the distinction between change and difference in economics.

the independent variable.[6] The second is that once the initial position is known, any other position is known with absolute certainty, in the same way in which, studying the motion of a physical body, the knowledge of its initial position and of the forces applied to it allows us to determine any successive position. As Sraffa wonders, discussing marginal productivity "are the potential, hypothetical returns which would be obtained by additional doses *part of the existing situation*? (D3/12/49.5 recto; emphasis added).

To appreciate the relevance of this question fully, let us recall that in 1929 Sraffa had written that solely to be included in the "data" of the theory were "things that actually happen". Marginal magnitudes would appear to be excluded from this condition by definition, being called into existence by thought experiments. But now Sraffa seems to be considering the possibility that marginal magnitudes might be conceived as "part of the existing situation" as if they were "things that actually happen".[7] The question Sraffa poses is, then, a crucial one. For if the answer were to be affirmative, then marginal magnitudes could after all be taken as a basis for economic theory to determine price and distribution. We hear in this question an echo of Sraffa's reflections on the developments in physics in the early decades of the twentieth century that challenged the determinism of Newtonian mechanics, which had been the source of inspiration for Marshall. We find in his notes several passages copied from an essay of 1931 by Erwin Schroedinger with the title "Indeterminism in Physics". In one of these passages Sraffa summarises the "so-called problem of causality" in physics (D1/91/66 recto), quoting from Schroedinger: "Given any physical system, is it possible, at any rate in theory, to make an exact prediction of its future behaviour, provided that its nature and condition at *one* given point of time are exactly known?" And Sraffa seems interested in the conclusions that identical initial conditions do not lead to identical results, according to the most recent experiments in physics of his time.

[6] Referring to Russell's interpretation of differential calculus in physics, Sraffa seems to be thinking of differential calculus as a model because it considers "time in which effects follow causes, but so closely that there is no room either for dispersion or for entering of foreign influences: it does this by differentiation (making time so short as actually to leave no room for change in circumstances: the cause and effect are perfectly contiguous—nothing happens in between" (quoted in Martins 2013: 44).

[7] We interpret the existing situation as also including alternative techniques as long as they are comparable because, even if not used, they "exist" at the same time. We are grateful to the Editor of this volume for raising this point.

But under precisely what conditions would any future development be included into the initial condition?[8] To answer this question, in what might be a draft of the Preface to *Production of Commodities*, Sraffa makes reference to a passage in Wicksteed's *Alphabet of Economic Science* on the existence of a labour supply curve. Wicksteed's passage runs as follows:

> But even if he [the labourer] cannot tell what amount of work he will be willing to do under the varying circumstances, obviously *there is* a given amount which, as a matter of fact, he would be willing to do under any given circumstances. Thus the curve *really exists* whether he is able to trace it or not. (Wicksteed 1888: 55)

Variation in the labour supply "under varying circumstances" is clearly a process of change. The point is, can the effects of this change be included within the "existing situation"? "Wicksteed", Sraffa observes, "with great and indeed reckless consistency, thinks that they do." (D3/12/49.5 recto). In fact, according to Wicksteed such change appears to be determined in its entirety by the initial position as if it were known a priori. To this Sraffa replies:[9] "Wicksteed considers that the path to be followed when one of the quantities is changed is prescribed a priori like the rails prescribe the path of a tramcar". (D3/12/46.43b recto). And yet again, in another draft:

> This sounds like a declaration of faith in universal determinism ('not a bus, not a bus, but a tram'): however Wicksteed considers that when one of the quantities is changed, the dependent variable will follow a path which is prescribed a priori although no rails or other circumstances capable of directing its motion are visible. But of course, saying that it will follow 'one' path at a time does not prove that it will follow one particular prescribed path rather than another. (D3/12/46.43b recto)

[8] The following paragraphs draw on Rosselli and Trabucchi (2019).

[9] For the vivid contrast of the image of a tram vs. that of a bus, we have preferred to quote this version, which Sraffa crossed out, to the one he kept in the manuscript which runs as follows:

> This is nothing less than a declaration of faith in universal determinism, for nothing less can support the belief in the actual existence of a prescribed path which must inevitably be followed, whether by the consumer or by the producer, such as is described by the demand- and supply-curves: for no observation, however minute, of the existing situation (in our case, of the existing methods of production) can bring out the path along which they must move in any given circumstances. (D3/12/46.43b recto)

In the light of the distinction between difference and change we can appreciate the sense of Sraffa's forthright reference to a "faith in universal determinism" as the only conceivable foundation for the "marginist" method (as Sraffa used to call it). Variations in the labour supply "under varying circumstances"—and the same obviously applies to variations in the cost of production of a commodity attending variations in the quantity produced or, again, to variations in output due to variations in the proportions in which different means of production are used—are all processes of change. As such they should be open to a wide range of outcomes. However, this would preclude them from serving as a basis for an explanation of value and distribution. Or, as Sraffa notes analysing the change in output as a result of the increase in the use of a factor of production: "The 'change' or historical method would in practice involve trial and error but the theory assumes that 'the best' or at any rate prescribed method is known a priori" (D3/12/49.6 recto).

Thus the only way to support the marginist method would appear to lie in postulating a certain determinism: in the form, for example, of a marked degree of regularity in the relation between cost and quantity produced—which, we may note, is precisely what would happen if the product obtained with *n* units of a factor and the product obtained with *n + 1* units were two *different* magnitudes existing side by side.

In fact, Sraffa speaks of faith in such regularity, because no actual experience can confirm it. Here it is worth noting that the role of the experiments that should lead to definition of the marginal magnitudes is, in fact, completely different from the part experiments normally have to play in the physical sciences. Unlike physics, experiments for marginalist theory do not consist in application of the conceptual tools of the theory to investigate certain properties of the economic systems under study. Rather, the very creation of these tools is made to depend upon one and only one possible result of the experiments. If the experiments were to yield different outcomes, we would not be dealing with a different theory, i.e. a theory yielding different results, for there would no longer be any theory from which results could be derived.

This criticism of "marginism" (again, Sraffa's wording), focussed on the regularity of the outcome of changes that must be assumed in order to build supply-and-demand curves, has more general implications for the study of causal relations in economics.

## 4 Continuity and Structural Change

Let us now look at the other possibility that we can take into account to consider the effect of a change, which implies comparison of the same object at two different instants, equivalent to the observation of two distinct objects at the same time. Could we consider variations so small that the assumption that nothing else changes is plausible? Sraffa seems to argue that in economics it is not possible to resort to infinitesimal calculus. First, variations are highly unlikely to come about in a continuous manner but rather occur in discrete form. Moreover, change hardly ever takes the form of variations in magnitudes that leave the overall structure unchanged.

On various occasions over the years Sraffa objected to Marshall's case of an "alert" railway manager who deals with the increasing number of passengers by altering the composition and size of carriages in a train, "constantly weighing the net product in saving of time and of annoyance to passengers, that will accrue from the aid of a second guard on an important train and considering whether it will be worth its costs" (Marshall 1920: 427).

On 29 March 1963, Sraffa again wrote:

> This suggests that his [the alert manager's] main task is to sack a porter here, add a coach to a train there, or shorten a platform elsewhere. The idea is that the process of change can be reduced to a continuous process, like shortening platforms: 'a penny is the basis of a million', and so a process of shortening, adding [,] sacking in detail is the route from one position to another. In Marshall's view the 'alert' Dr. B…. never needs to take bird's eye view of his enterprise. (D3/12/42.12)

This awareness that in economics changes in one variable affect the whole structure connects with Sraffa's repeated observation that the marginal productivity is always measured in the case of two factors, which is the only case in which changing the quantity of one factor and changing the proportions in which factors combine amount to the same thing. However, when we have more than two factors, can we still maintain that the proportion between the unchanged factors remains constant in the presence of a change in the quantity of one factor?

We have several passages showing that Sraffa was critical of the assumption that one could deal with change in real time as if the *ceteris paribus* assumption held.

> M[arshall] is constantly on the defensive against objections to continuity based on facts [...] This is not the basis on which this [my argument] is based. That can be granted in detail, wherever it is possible. It is against the logical possibility of the type of continuity assumed. That type is only possible with two factors. (D3/12/42.12)

And again:

> Where marginism goes astray is in (falsely) assuming [...] that it has general applicability whereas in fact it only applies exceptionally (in cases where partial change is feasible, there is independence, the whole is not affected). (D3/12/42.9)

However, even if supply-and-demand functions (for goods and factors) were continuous so that marginal calculus were applicable, there would still be one obstacle to adopting the methods of mechanics in economic analysis. Sraffa wrote:

> [...] in mechanics if the experiment is repeated in similar circumstances (say, on the elasticity of a metal) the same results will be obtained. But with supply and demand, even if the external circumstances were the same, the result would be different because man learns from experience, or at any rate is changed by it, forms and transforms habits, etc. (D3/12/42/11recto)

In conclusion, differential calculus and demand and supply functions are not tools applicable to economic analysis borrowing the method of classical physics—i.e. mechanical causality does not work in economics. Does this mean that Sraffa gave up the possibility of establishing causal relationships to explain change in economics? In the next section we look at what the literature had to say on the matter.

## 5 Interpreting Mechanical Vs. Geometrical Causality

On Sraffa's challenge to the notion of mechanical causality in economics there have recently appeared a fair number of interesting contributions which, although not differing greatly, nevertheless highlight different aspects of the matter. In common they offer a defence of Sraffa's theory against the criticism of lacking generality (for example, valid only by assuming constant returns to scale) or of being narrow in scope (for example, it allegedly describes an equilibrium situation without specifying how the system reaches it). Moreover, investigation into Sraffa's research project has brought to light the fact that his theory of prices and distribution is not subject to the limitations that constrain the marginalist approach which determines prices through the balancing of the forces of supply and demand. This is the view that Sraffa expressed at the beginning of his work on *Production of Commodities* in the famous parable of the "Man from the moon" (D3/12/7.87) and reiterated in 1942:

> The problem is that of ascertaining the conditions of equilibrium of a system of prices and the rate of profits, independently of the study of the forces which may bring about such a state of equilibrium. Since the solution of the second problem carries with it a solution of the first, that is the course usually adopted in modern theory. The first problem however is susceptible of a more general treatment, independent of the particular forces assumed for the second; and in view of the unsatisfactory character of the latter, there is advantage in maintaining its independence. (D3/12/15.2)

Ginzburg (2013) provides an interesting interpretation of this point, highlighting the distinction between causal representations, which answer the question "why" and non-causal representations which answer the question "how".

The distinction is between representations which investigate the causes of phenomena (and are therefore presumably able to predict them) and merely descriptive representations, like "snapshots",[10] which capture the system as it actually is at a given instant in time. Ginzburg claims that this "metaphor, if taken literally as a photograph of a real economic system, is misleading. It can be considered a snapshot only in the sense that all the phenomena taken into account relate to the same period" (Ginzburg 2013: 113). Ginzburg argues that non-causal representations are not realistic descriptions of a situation, but require a mode of abstraction: for example, the methods of production under consideration are those prevailing in the system and not those actually adopted under specific circumstances. Non-causal representations capture meaningful features of the

[10] The first to employ the "snapshot" metaphor was Roncaglia in his 1975 Italian book, translated into English as Roncaglia (1978). For a more recent restatement of his interpretation of Sraffa's approach, see Roncaglia (2009). On the evidence of Sraffa's reference to the snapshot, see Kurz and Salvadori (2018).

object of analysis and the properties of the structure without the need to bring in specifications of their developments over time, which remain open to a variety of possibilities.

The idea that equilibrium prices can be defined independently of the forces which guarantee that equilibrium is reached does not imply that equilibrium is independent of other factors which call for a different kind of analysis. The argument that social analysis, and economic theory in particular, must address questions which require different levels of abstraction and generality had been put forward by Garegnani in several contributions (1984, 1990a, b) as well as various other scholars who adopted his approach.

This view—which is known as the "core" interpretation of Sraffa's theory—has found support in a number of studies by Kurz and Salvadori (2004, 2005, 2018). In particular, they have introduced two elements which clarify Sraffa's research project through a careful reconstruction of the complex work underlying the final text of *Production of Commodities*. The first element is the influence on Sraffa of his perusal of a great many books on the natural sciences and how this led to a search for "objectivism" which, at an early stage[11] consisted in an attempt to construct a theory of exchange values whose elements would be measurable quantities, thus excluding any reference to motivations. The economic system could then be represented as production of commodities by commodities, since any output would be the result of the destruction of all the inputs used up in its production.

The second element highlighted by Kurz and Salvadori and, with his alternative interpretation, by Sinha (2016)—and recently by Davis (2020)—is Sraffa's important note on "Surplus Product" written in August 1931. It marks the abandonment of the "natural science point of view" which links one effect to one cause.

Sraffa realised that the distinction between surplus and necessities brought him up against an apparently insoluble dilemma: either all the

[11] Sraffa had endorsed Francis Bacon's principle of "efficient causes" as against "final causes". He wrote in the December 1927–December 1928 period: "'Efficient causes' are facts of the past that act on the present: 'final causes' are facts of the future that act on the present. The existence of the latter is at best dubious and they are better called 'illusions'. The classical P[olitical] E[conomy] dealt only with the first sort of causes, i.e. of 'material things' that have existed in the past. Modern economics deals with the second class, i.e. hopes for the future, such as utility, abstinence, disutility, etc.; these things, it must be noticed, refer only to the foreseeing of future acts." (D3/12/10/ 61.1recto).

surplus is necessities, i.e. all that is produced is necessary for the reiteration of the production and the surplus vanishes, or the components of the surplus, like profits and rents, must be represented as inducements and then the distinction between objective and subjective disappears. This led Sraffa to conclude that "The surplus may be the effect of the outside causes; and the effects of the distribution of the surplus may lie outside." (D3/12/7/161.5recto).

In the search for the influences which might have led Sraffa along the path to abandoning the notion of mechanical causality in economics, it has been argued in the literature that the exchanges between Sraffa and Wittgenstein from 1929 to 1946 provide evidence that Sraffa had an influence on the preparation of Wittgenstein's *Philosophical Investigations* and in turn Wittgenstein had an influence on Sraffa's *Production of Commodities.* Arena (2015) in particular has argued that the letters between them show that both Sraffa and Wittgenstein criticised various notions of causality, and specifically, the notion of mechanical causality.

Wittgenstein abandoned the idea that there is only one type of causality that we could discover by observation and experiments, interpreting "causality as one grammar rule amongst others" (Arena 2015: 1097). He also criticises the notion of causes based on individual motivations,

> such as beliefs, intentions or desires, arguing that if agent motivations do exist, they do not necessarily produce and therefore explain their actions: motivations (even social ones) are not causes. (ibid.)

Accordingly, Arena finds a clear parallelism with what Sraffa had been arguing since his early period:

> Economists who do not take this objective test as the standard of what is the cause of an event, are always driven back to trace the 'ultimate causes, causae causantes, etc.' to the wants, desires, aversions, decisions, volitions and intentions (or inducements and rewards) of individuals. In fact, if we do not use an objective standard and rely simply upon an unconscious 'feeling that this must be the real cause' … we are bound to base the conclusions on our own individual experience, from which it appears that we do what we want, or what we like, etc., and this seems the only convincing final conclusion". (D 1/9/6, pre-1928[12])

[12] According to Nerio Naldi in a private communication, the papers in D1/9—in the catalogue dated pre-1928—in fact belong to the 1928–1931 period, as revealed by the annotation "next lecture" on one sheet.

Wittgenstein introduced a new type of representation, that can be defined as *surveyable* or *perspicuous representation*, which aims at bringing out "mutual links and connections in the data and corresponds to the synthesis or the synopsis of the grammatical rules of a form of life, but without making any assumptions about their evolution over time" (Arena 2015: 1099).

So Arena also agrees that Sraffa must have come to the conclusion that a "snapshot" was best suited to escape from causality in providing a representation of the economic system.

Davis (2012) also re-assessed the nature of the influence on and by Wittgenstein, but in Davis (2020) he adds a further element to the present discussion, i.e. the distinction between closed and open systems. A "closed system" is isolated from the environment and is subject to principles which are invariant relative to it; on the contrary the functioning of an open system is influenced by the environment.

Davis argues that Sraffa confined the objectivist, sufficient cause principle to the field of commodity prices, but also supposed that this "closed system" was operated upon as a whole by causes that lay outside of it. Davis contends that Sraffa both maintained and modified his objectivism by allowing for different types of causal factors associated respectively with production and distribution. In conclusion, Sraffa is said to allow for two types of causal forces, one associated with the natural world, the other with the social world, and these two types of causal forces are assumed to be interacting "in a complex manner" (Davis 2012: 9).

The distinction between an open and a closed system seems to find corroboration in the following passage dating to August 1931:

> Thus there must be a leak at one end or the other: the 'closed system' is in communication with the world.
>
> When we have defined our 'economic field', there are still outside causes which operate in it; and its effects go beyond the boundary. This must happen in any concrete case. (D3/12/7/161.5recto)

Sinha (2016) places at the centre of his interpretation Sraffa's insistence on the difference between two types of explanations, meant to answer two types of questions:

> why ‘x’ has increased in price from $2 to $3 […] why ‘x’ is sold at $3. The first question needs an answer in terms of a discovery of a cause that explains the change… the second question […] requires an answer in terms of a relation (a logical relation), that is, why ‘x’ must relate to $ or other commodities ‘y’, ‘z’ and so on in a precise quantitative association at one moment. The first case is mechanical in nature whereas the second is geometrical. In the second case, the problem of causation does not arise because time is absent from the problem—hence the explanation must be a description of what exists”. (Sinha 2016: xii)

Sinha has clarified the core point of the discussion on the nature of Sraffa’s objections to the notion of causality in economics, or at least the notion of causality borrowed from classical mechanics, namely the nexus time-change-cause. Observed phenomena could be traced back to a given cause only if we could observe the “forces” that produced that effect, but this presupposes going back in time to produce a change in the existing situation. This, however, does not mean that we cannot study situations following a change, for instance in a given variable (wage or profit), although we can only do so as long as these changes have no time dimension, and are merely a logical derivation.

## 6 The Boundaries of the “Economic Field”

As in the case of objectivism, which has been interpreted as if Sraffa were a crude positivist, unable to see the importance of subjective magnitudes, like expectations, which although unobservable play an important role in economics, his rejection of mechanical causality should not be interpreted in a restrictive way. Obviously, Sraffa set out to understand economic reality and therefore how economic phenomena could best be explained; his concern seems to have been primarily to avoid the employment of tools or concepts that appeared from the outset to be of limited generality or dubious existence (utility), or indeed ideologically vitiated (abstinence). The pursuit of a scientific method to apply to economics required the utmost care and attention to avoid metaphysics, ideology and ad hoc assumptions.

The way Sraffa went about it was a very drastic delimitation of the “economic field”, as he called it, leaving out “causes” from the investigation. It is argued by many interpreters, as we have seen, that Sraffa makes a distinction between the “economic field”—the plane to which only a geometrical theory can be applied, i.e. satisfying the criterion of using only logical and timeless relationships, without dealing with "ultimate causes"—and the outside realm of circumstances and historically given context, as was clear to him from the very beginning of his research programme:

> Clearly, we must reduce all the data to things that actually happen, excluding inexistent possibilities. Only such things are measurable, and can enter the theory as 'knowns', or 'constants'; and, in reality, only really happening things can be real causes and determine effects. (D3/ 12/ 13/ 1.2)

What this implies is that institutions, technology, social norms and levels of activity cannot be approached with the same level of abstraction as within the "economic field". However, this leaves open the question of the nature of the relationships prevailing outside. The question arises if we can take them as causal forces even though we cannot represent them in a functional deterministic form.

Any attempt to assess how Sraffa would have answered this question takes us onto slippery ground, because direct evidence is lacking. Seeking out Sraffa's motivations and objectives—e.g. demolition of the marginal productivity theory, rehabilitation of classical political economy—and/or searching for influences by the contemporary scientific discussion (quantum physics) or personal interactions with particular individuals (Wittgenstein, Gramsci) may well prove illuminating, but hardly conclusive on this point.

What is clear is that we are left with alternative options to continue Sraffa's research programme and accept his legacy. We could attempt to expand the "economic field" beyond the boundaries marked by Sraffa, introducing further relationships of the wage-profit type, namely those that have the property of bringing in an "if and only if" clause. Another approach could be to enlarge the realm of the "economic field", applying price equations to problems that imply changes, for instance, in the quantities produced, but introducing assumptions (such as constant returns to scale) that restrict the level of generality Sraffa aimed at.

In any case, we have to continue looking into the causes at work outside the economic field and build a bridge between the core and the outside world. This might lead us to accept that the level of abstraction and therefore of generality will be lower, gaining in relevance what is lost in rigour.

Only by keeping the uses we can make of Sraffa's system of equations separate from the reasons which may have led Sraffa to travel the road of the geometrical representation and, above all, of what is really meant by it, can we make any further headway, without being bogged down by disputes on Sraffa's motivations or being unfaithful to his text.

What we are left with is a powerful warning against invoking spurious causes to explain economic phenomena; we are also left with means to prevent us from mistaking changes for differences and above all a fence against false claims of rigour and generality.

## References

Arena R. (2015), Order, process and morphology: Sraffa and Wittgenstein, *Cambridge Journal of Economics* 39: 1087–1108.

Bharadwaj K. (1986), *Classical Political Economy and the Rise to Dominance of Supply and Demand Theories* (2nd ed), Calcutta: Orient Longmans.

Davis J. (2012), The change in Sraffa's philosophical thinking, *Cambridge Journal of Economics* 36: 1341–56.

Davis J. (2021), 'Openness' as a methodological principle of Sraffa's economic thinking; this volume.

Garegnani P. (1984), Value and distribution in the classical economists and Marx, *Oxford Economic Papers* 36: 291–325.

Garegnani P. (1990a), On a supposed obstacle to the tendency of market prices towards natural prices, *Political Economy: Studies in the Surplus Approach* 6: 329-59.

Garegnani P. (1990b), Sraffa's Production of Commodities: classical versus marginalist analysis in K. Bharadwaj and B. Schefold (eds), *Essays on Piero Sraffa*, London: Unwin Hyman.

Ginzburg A. (2013), Sraffa, Sen and Non-Causal Representations in Economic Analysis. In E.S. Levrero, A. Palumbo, A. Stirati (eds), *Sraffa and the Reconstruction of Economic Theory, Volume III: Sraffa's Legacy: Interpretations and Historical Perspectives,* Basingstoke: Palgrave Macmillan.

Kurz H. and Salvadori N. (2004), Man from the Moon: on Sraffa's objectivism, *Économies et Sociétés* 35: 1545–57.

Kurz H. and Salvadori N. (2005), Representing the production and circulation of commodities in material terms: on Sraffa's objectivism, *Review of Political Economy* 17: 69–97.

Kurz H. and Salvadori N. (2018), On The "Photograph" Interpretation of Piero Sraffa's Production Equations: A View From The Sraffa Archive in J. Kregel, M. Corsi and C. D'Ippoliti (eds), *Classical Economics Today*, London and New York: Anthem Press.

Marcuzzo, M.C. (2014), On alternative notions of change and choice. Krishna Bharadwaj's legacy, *Cambridge Journal of Economics* 38: 49–62.

Marcuzzo, M.C. and Rosselli, A. (2011), Sraffa and his arguments against marginism, *Cambridge Journal of Economics* 35: 219–31.
Marshall A. 1920 [1890], *Principles of Economics*. London: Macmillan.
Martins N. (2013), *The Cambridge Revival of Political Economy*, London: Routledge.
Roncaglia A. (1975), *Sraffa e la teoria dei prezzi*, Bari: Laterza.
Roncaglia A. (1978), *Sraffa and the Theory of Prices*, Chichester: Wiley.
Roncaglia A. (2009), *Piero Sraffa and his life, thought and cultural heritage*. London: Routledge.
Rosselli A. and Trabucchi P. (2019), Sraffa, the 'marginal' method and change, *Structural Change and Economic Dynamics* 51: 334–340.
Sinha A. (2016), *A Revolution in Economic Theory: The Economics of Piero Sraffa*, London: Palgrave Macmillan.
Sraffa P. (1951), Introduction in D. Ricardo, *Works and Correspondence*, edited by P. Sraffa with the collaboration of M. Dobb, vol. I, Cambridge: Cambridge University Press.
Sraffa P. (1960). *Production of Commodities by Means of Commodities*, Cambridge: Cambridge University Press.
Wicksteed P. (1888), *The Alphabet of Economic Science*, London: Macmillan.

Comment

# A Comment on 'On Sraffa's Challenge to Causality in Economics'

*Ajit Sinha*

Cristina Marcuzzo and Annalisa Rosselli make a significant contribution to our understanding of the nature of Sraffa's critique of causation in economic theory. Their foremost concern in this respect is with the relationship between the notion of causation and the 'marginal method' of analysis adopted by the neoclassical, as opposed to the classical, economics. As is well known, Sraffa, in the preface to his book, announced that 'although [the propositions of his book] do not enter into any discussion of the marginal theory of value and distribution, they have nevertheless been designed to serve as the basis for a critique of that theory' (Sraffa 1960, p. vi). Hence their concern is central to our understanding of Sraffa's book. As far as the relation of the notion of causation with the 'marginal theory of value and distribution' is concerned, it would be better to separate the criticisms that might apply to the 'marginal method' in general from the criticisms that might only apply to Marshall's partial equilibrium 'method' with its necessary *ceteris paribus* clause. Since the problems with Marshall's partial equilibrium method is now well accepted by the orthodoxy in economics, I leave them out in this brief comment. By 'marginal method' Sraffa means a method of analysis that bases its understanding of the nature and behaviour of a system by analysing the results of a small or a slight change in one variable of the system.

A functional relation between two variables (say, price of a commodity and quantity supplied of it) is of the nature: 'if an independent variable $x$ takes a specific value then the dependent variable $y$ must also take a particular specific value.' In this case, changes in $y$ is related to changes in $x$ in a determinate manner. In pure mathematics, a functional relation is simply a mapping between two sets. This does not require any notion of time as changes in the elements of a set do not relate to changes in any entity in the real world. But when we talk about *change* in a variable existing in the real world, then we necessarily bring in *time* into the picture as *change* in one entity requires time. Sraffa's argument seems to be that the propositions of neoclassical economics such as 'marginal product' of a factor of production or 'marginal utility' of a commodity that constitute their fundamental supply and demand functions apparently refer to changes in real entities. Hence time is implicitly present in those functional relations in terms of a *causal relation* between $x$ and $y$. At this stage of analysis, Marcuzzo and Rosselli register two specific criticisms of such neoclassical functions offered by Sraffa: the question is how does one know the precise relation in which $y$ is supposed to be related to $x$.

First, Sraffa refers to Wicksteed (1888) as an authority on the proposed use of the 'marginal method' in economics. Wicksteed evidently acknowledges that the precise relation may not be known to either the analysts or the agents themselves but still asserts that a precise relationship in reality nevertheless must exist. In other words, on the assurance or the supposition that a precise relation exists in reality, a theoretician may go on to build a theory by simply postulating such relationship without worrying about the exact empirics of it. Sraffa characterises Wicksteed's assumption as 'faith in universal determinism'. This, of course, is a criticism launched from the vantage point of the quantum physics' criticism of deterministic physics and its championing of indeterminism—the real entities, as Sraffa puts it, are more like buses rather than trams that move on predetermined lines. A precise prediction of the movements of a bus cannot be made. Marcuzzo and Rosselli provide some of Sraffa's excerpts from Schrödinger as evidence to this respect; in addition to this, there are also several books, most notably by Eddington and Max Plank, in Sraffa's library with his annotations that further attest to this conclusion (see Sinha 2016 for more details).

Second, Sraffa goes on to argue that even if one could produce changes in $y$ by changing $x$ in an experimental setting, since the changes, in the final analysis, relate to human subjects and since humans do learn and change with experience, a repeated experimentation on the same subject will most likely not reproduce the same results as every repeated experiment necessarily changes the prior givens of the experiment—yet again violating the postulate of determinism assumed by the causal functional relations. The point to note in this respect is that once we take into account 'human experience' as a variable that changes the human subject, we have introduced unidirectional arrow of time into the picture—now taking back the experimental stimulus will no longer bring the agent back to the original position; however, the mathematical functional relation implies unfettered movement back and forth on the curve. Thus with human subjects one can only draw a descriptive historical curve between the two variables that moves only in one direction and not a functional curve.

Now, let us suppose that we take these functional relations as purely mathematical equations and try to find a solution to the simultaneous equations problem. A solution or multiple solutions, if they exist, of this system of equations only give us the specific values of all the different variables that satisfy all the equations simultaneously. In neoclassical economics we interpret such solutions as 'equilibrium' of the system. 'Equilibrium', however, typically refers to balancing of opposing forces, which is not an aspect of the mathematical 'solution'. Where do these 'forces' come from then? They come from two interrelated sources: (1) from some hypothesis about human psychology such as consumers' desire to maximise their levels of satisfaction and producers' desire to maximise their profits, and (2) the functional relations that determine the shape of the function such as the relationships between prices and quantity demanded and quantity supplied are such that when the real variables are not found at the 'solution' points, then the shapes of the functional relations depicted by the equations ensure that the agents' desires will relentlessly drive the variables to their solution points. Here *change* explicitly brings in real time on the axis of which the system must move. As a matter of fact, the popular version of the neoclassical theory contends that this depicts the actual functioning of the markets. This is the context of Sraffa's reference to Marshall's 'scissors, pillars and forces', where the determination of an exchange relation of a commodity with respect to some other commodity at a given point of time is explained through changes in its own price over a period of time. Marcuzzo and Rosselli's reference to Bharadwaj (1986) also relates to this situation where Bharadwaj relates the hypotheses regarding shapes of the utility and production functions based on 'notional marginal changes' to the condition of the stability of the equilibrium prices.

After which Marcuzzo and Rosselli quote Sraffa from his note of early 1942, where he states that 'The problem is that of ascertaining the conditions of equilibrium of a system of prices and the rate of profits, independently of the study of the forces which may bring about such a state of equilibrium' (Sraffa n.d., D3/12/15.2). They then leave the matter at that. This quotation, however, has been used in the past by other scholars to argue that Sraffa's own position was to keep the notion of 'equilibrium prices' and only jettison the functional relations of the neoclassical economics—this is the Garegnani-led orthodox Sraffian interpretation of Sraffa's final position, which I contest. The orthodox position is that Sraffa replaced the neoclassical adjustment mechanism of prices to the equilibrium position with Adam Smith's or classical gravitational mechanism that explains the movement of 'market prices' to 'natural prices'.

In my chapter to this volume as well as in Sinha (2015), I have argued that the classical gravitation mechanism necessarily, though implicitly, assumes constant returns to scale—an assumption Sraffa emphatically denied making for his equations in the preface to his book (Sraffa 1960, p. v). Furthermore, there has been no proof of the proposition that the classical centre of gravitation does indeed exist, and without such a proof it would have been foolhardy for Sraffa to build his entire 'prelude to a critique of economic theory' on a mere fairy-tale (see Dupertuis and Sinha (2009) for an argument that the mathematical probability of the existence of classical centre of gravitation for a system of three or more basic goods is equal to zero). Moreover, Sraffa's consistent and repeated use of the metaphor of a 'photograph' for his system of equations that Marcuzzo and Rosselli themselves have highlighted clearly suggests that his reference is to an empirical system and not to the classical centre of gravitation, as the centre of gravitation is considered to be not an empirical but an ideal situation by the classical economists themselves, and clearly the idea of a photograph of something ideal and non-existent is quite absurd. Besides, it should be kept in mind that once Sraffa hit upon his physical equations in the winter of 1927–28, he realised that he would need to somehow *justify* imposing a uniform rate of profits on his surplus-equations system to ensure a solution of the system. In this context, Sraffa rejected the classical gravitation mechanism on the grounds that it necessarily brings in some hypothesis regarding human subjectivity into the picture—something that he wanted to completely keep out of his system. Though he had convinced himself that by imposing a uniform rate of profits for all industries he could solve for the average rate of profits of the system, the justification of a uniform rate of profits still remained a problem for him since the solution for prices required that there cannot be more than one unknown for all the industrial rates of profits in the system. In this context it should also be noted that when Sraffa got down to work on this problem in 1942, he found Marx's approach to deriving the average rate of profits quite attractive. He interpreted Marx's average rate of profits derived from the labour-value magnitudes as the rate of profit that must hold for the industry made up of the average organic composition of capital in both the value and prices of production regimes. But once he realised that the organic composition of capital of all the industries would themselves change bringing along a change in the average of the system as labour-values are transformed to prices of production, he abandoned this approach and went on to search for a more robust way of finding an average commodity. This resulted in the discovery of the Standard system and the Standard commodity in 1943–44 period (see Sinha 2016 for the details of the story). It was the Standard system and the Standard commodity as the measuring rod for prices and wages that clearly established that the average rate of profit is a physical property of the system and can be discovered independently of prices—it is a consequence of this fact that prices must always be associated with a uniform rate of profits in the system. Hence the condition of a uniform rate of profits is no longer tied to the notion of 'equilibrium' of the system—it turns out to be a structural property of the system itself. Thus Sraffa rejected not only neoclassical functional relations but also the notion of equilibrium along with it. It should also be noted that in Sraffa's vast archive one is hard-pressed to find any serious deliberation on the merits of classical gravitation mechanism—his general disinterest in this aspect of classical economics goes to show that it would be foolhardy for us to search for Sraffa's dividing line between neoclassical economics and the classical economics by suggesting that one relies on functional relations to achieve the equilibrium prices and the other does not.

Thus, against the neoclassical system, Sraffa counterposes his system as the 'classical standpoint' by not admitting either the functional relations or the notion of equilibrium in it. Though the neoclassical economics determines total production, prices of commodities and income distribution *simultaneously* by the same mechanism based on demand and supply functions, Sraffa strongly maintained that the classical theory takes income distribution as given or determined independently of prices, and subsequently it determines prices on the basis of the given technique of production in use and the given distribution of income. Thus Marcuzzo and Rosselli put their finger on the right spot by highlighting the problem of determination of 'surplus product', which is distributed among the competing recipients, as the central problem of classical economics. They discuss an interesting and important note written by Sraffa in August 1931 on 'surplus product'. Sraffa poses the problem of surplus in economic theory as paradoxical, if it is posed within the deterministic framework of the principle of sufficient reason. If every effect must have a sufficient cause then in the field of economics all outputs must have an equivalent cost, and so once the cost of surplus product is discovered the surplus must disappear as the 'surplus' is a surplus only because it appears to have no cost associated with it. In this context, they mention the interpretation given to it by Kurz and Salvadori (2005). Kurz and Salvadori argue that the paradox of how to determine the 'surplus' in a completely objective and scientific framework led Sraffa to abandon his objective scientific approach. This position seems to stem from their desire to open some room for subjectivism for the working of a causal mechanism in Sraffa's system, as the classical gravitation mechanism requires some hypothesis about human agents' behaviour to work itself out, and it is their position, along with Garegnani, that the classical gravitation is an integral part of Sraffa's theory. My detailed reading of this note is presented in my book (Sinha 2016, pp. 82–89) where I disagree with Kurz and Salvadori and point out that Sraffa always held the quantum physics' indeterminism where a precise position of a particle can be established only in relation to the observer—it does not exist independently of the observer. In a similar manner, Sraffa argued that 'surplus' in economics exists only in relation to the position the theoretician adopts. He himself adopted the 'classless human standpoint' from which the necessities are only the inputs used in production, and all the outputs over and above these necessary inputs are taken as 'surplus'. Thus Sraffa never abandoned his scientific outlook and never allowed subjective motives to enter his theory—he held the modern scientific approach of quantum physics and did not maintain the classical scientific position of Newton's universal determinism.

## References

Bharadwaj, K. 1986 (2nd ed). *Classical Political Economy and the Rise to Dominance of Supply and Demand Theories*, Calcutta: Orient Longmans.

Dupertuis, M.-S. and A. Sinha. 2009. 'A Sraffian critique of the classical notion of centre of gravitation', *Cambridge Journal of Economics*, 33(6): 1065–87.

Kurz, H. and N. Salvadori. 2005. Representing the production and circulation of commodities in material terms: on Sraffa's objectivism, *Review of Political Economy* 17: 69–97.

Sinha, A. 2015. 'A reflection on the Samuelson-Garegnani debate', *Economic Thought*, 4(2): 48–67.
Sinha, A. 2016. *A Revolution in Economic Theory: The Economics of Piero Sraffa*, Cham: Palgrave Macmillan.
Sraffa, P. 1960. *Production of Commodities by Means of Commodities: Prelude to a Critique of Economic Theory*, Cambridge, UK: Cambridge University Press.
Sraffa, P. n.d. 'Sraffa's unpublished papers', Cambridge, Wren Library, Trinity College, University of Cambridge.
Wicksteed, P. 1888. *The Alphabet of Economic Science*, London: Macmillan.

CHAPTER 4

# The Meaning of Sraffa Prices

*Goddanti Omkarnath*

## 1 Introduction

The precise theoretical content of the celebrated Sraffa system (Sraffa 1960) is hard to grasp. In a system which has both quantities and a rule of distribution given, there could be neither a theory of outputs nor a substantive theory of distribution. One may have recourse, then, to a theory of value. Indeed, Sraffa system is most commonly identified with a thorough re-statement of classical theory of exchangeable value or prices.[1] Within classical theory, the question of value of particular commodities arose as a by-product of a central interest in a theory of distribution.[2] To discuss the same question, Sraffa chooses the particular setting of given output levels and methods of production which allows him (1) to criticize marginal theory from the standpoint of its parent notion of 'marginal change', and (2) to solve some of Ricardo's puzzles that partly led to the decline of classical political economy after him.

[1] Roncaglia (1978), an early interpretative text, is titled *Sraffa and the theory of prices*.

[2] '...in the background of every theory of value is a theory of distribution' Sraffa (D3/4/3:4).

Goddanti Omkarnath (✉)
University of Hyderabad, Hyderabad, Telangana, India



A. Sinha (ed.), *A Reflection on Sraffa's Revolution in Economic Theory*, Palgrave Studies in the History of Economic Thought,
https://doi.org/10.1007/978-3-030-47206-1_4

It is well known that a major riddle that engaged Ricardo to his last days was inter se changes in distributive shares in the context of accumulation. In trying to understand the impact on the rate of profits of an increase in wages, he had to contend with the intervening effects of price changes. He understood the problem very well,[3] but lacked the analytical tool necessary to separate distribution theory from value theory. Now when Sraffa succeeds in determining the rate of profits independently of prices (Sraffa, ibid., Ch. IV), he solves Ricardo's riddle. It is then pertinent to ask what might remain of the classical theory of value after this solution. In one way or the other, the theoretical status and role of prices must undergo deep change. The present chapter is an attempt to understand this change.

It will be argued that a *theory* of prices stands dissolved in Sraffa while his *prices* come to represent minimal conditions of a viable economic system. They become an apparatus of logic and arise strictly from the necessity of a given system of production to replace itself. Their conventional accounting function in the measurement of distribution of product and other aggregates is only derivative of this logical function. If our case that there is no theory of prices—whether 'causal or 'non-causal'—in Sraffa holds, then we will run the risk of concluding that his entire system is bereft of theory. However, we show that his system as a whole represents the theory of economic *structure*, which forms the basis for a theory of accumulation and *change*. It is, of course, in questions of the latter kind that classical economists were primarily interested.

The chapter is organized as follows. As a preliminary to the main thesis, we begin by asking whether Sraffa system meets Ronald Meek's requirement of a theory of value in political economy. We shall argue that there is no central 'cause' or 'chain of causes' of prices in the system (Sect. 2). In Sect. 3, we present the main thesis in three steps. We explore Sraffa's crucial assumption of a 'self-replacing' economic system to find the first link of Sraffa prices to system's viability. This assumption is shown to be an adaptation of Adam Smith's concept of 'natural states' (Subsection 3.1). We then examine the critical role of Sraffa's Standard commodity in liberating the classical theory of distribution from the confounding effects of prices. Sraffa reduces the *circular* relation between distribution and prices,

[3] He wrote to McCulloch on 13 June 1820: '...the great questions of rent, wages and profits must be explained by the proportions in which the whole produce is divided between landlords, capitalists and labourers, and which are not essentially connected with the doctrine of value' (Ricardo 1951–1973, vol. VIII).

which vexed Ricardo, to a *simultaneous* relation between them. By this act, Sraffa eliminates any vestigial remains of a classical *theory* of prices. This paves the way to the conviction that Sraffa prices represent conditions of logical consistency between production and distribution of social surplus (Subsection 3.2). We then look at the asymmetry as between conditions of production and conditions of exchange in Sraffa. It will be argued that in his system it is the conditions of production which must regulate the conditions of exchange, and not the other way. In particular, the rate of profits in Sraffa system (even when it is given from outside the system as a 'ratio') must accord with the general rate of profits determined in the Standard system. The prices that correspond to this rate of profits assume the form of minimal conditions of a self-replacing and therefore viable economic system (Subsection 3.3). After concluding that there is no theory of prices in Sraffa, we shall argue, in Sect. 4, that the system as a whole represents a theory of economic *structure* which forms the basis for the theory of accumulation and *change* in historical time. Sect. 5 is a concluding statement.

## 2 A Requirement of the Theory of Value

Meek (1977, p. 151) lays down the requirement of a causal theory of value in political economy. According to him, there should be an independent 'determining constant' from which a 'one-directional *catena* of causes' should follow. It is appropriate that we find out if Sraffa system satisfies this. It is well known that Sraffa rejected, early on, the idea of an 'ultimate cause' or 'force', such as *labour* or *utility*, explaining values of individual commodities. He considered them to be 'metaphysical'.[4] If Meek's 'determining constant' is labour, Sraffa's critique digs deeper into the very measure of labour adopted by classical economists. According to him, Petty and the Physiocrats measured 'labour' correctly in terms of *commodities* that are required in the sustenance of workers, e.g. 'food'. Smith, Ricardo and Marx, however, 'corrupted' this measure and regarded labour as 'a quantity measured in hours or kilowatts of human energy, and thus commensurated to value'. This was their 'fatal error' (Sraffa, E2/56). Further, Sraffa's demonstration of 'reduction to dated quantities of labour' (Sraffa 1960, Chap. VI) is as much a critique of the labour theory of value

[4] Although he later softened his position on the labour theory of value, because it allowed Marx to place a finite upper limit to the rate of profits (Sraffa, ibid., p. 94).

as it is of the Austrian theory of capital. We may conclude that Sraffa presupposes no independent 'determining constant' for his system of values.

It may be asked if a 'chain of causes' of prices is otherwise discernible. Most interpreters who believe in a causal theory of prices in Sraffa implicitly identify two independent causes: (1) methods of production, and (2) an exogenously given rule of distribution. The obvious difficulty with such a position is that both these are part of the data of Sraffa system and as such cannot be viewed as 'causes.' Cause-effect relations can be gauged only in a system that admits of change, but Sraffa system *ex hypothesi* rules out change, either in the scale of production or in methods of production. Quite apart from this problem, there is a question as to the supposed *independence* of the two 'causes.' Insofar as the maximum rate of profits as well as the inverse relation between wages and the rate of profits are consistently embedded in the given conditions of production in Sraffa, the argument about independence of causes cannot be sustained. It may be argued, however, that the actual or ruling rate of profits used for 'closure' of the equation system is given from outside the system of production. But to say this is to confuse the theory of prices with the theory of distribution. Even so, an exogenously given rate of profits can be deemed to have been adopted by the system of production only when it helps generate such a set of prices as to yield a *level* of profits and capital consistent with that rate. But in this process of price formation, as many causes of prices must simultaneously come into operation as there are basic commodities. In the interdependent structure that the Standard system represents, no specific 'chain' of causation will be identified. We may conclude that Sraffa prices have no foundation either in an independent 'determining constant' or in any 'one-directional' chain of causation. In other words, there seems to be no basis for the supposition that Sraffa prices represent a causal *theory* of prices.

However, it may yet be suggested that the aforesaid analysis still leaves open the door for a *non-causal* theory of prices in Sraffa. Scholars such as Sen (1989) and Sinha (2016) have attributed to Sraffa such non-causal theories of prices. They do not find, quite rightly from our standpoint, a central cause or chain of causes of prices in Sraffa equations. However, the implication or purpose of such non-causally determined prices is not clear in these accounts. For Sen, Sraffa prices have an important descriptive value for public discourse on political and social issues. For Sinha these prices have a purely 'functional' role in consistently accounting for income distribution. While these uses of prices are important in themselves, we shall argue that the main function of Sraffa prices is of maintaining the social reproduction of the economy. They signify a viable economic system.

## 3 Viability: The Sole Foundation of Value in Sraffa

In this section, we present our main thesis that Sraffa prices do not constitute a *theory* of prices, causal or non-causal, but that their major purpose and justification is to serve as indicators of a viable economic system. We begin by explaining the significance of Sraffa's basic assumption of 'self-replacing' economic system (3.1). This is followed by an examination of the implication for prices of Sraffa's derivation of a price-independent rate of profits (3.2). Finally, we go into the primacy which Sraffa accords to conditions of production over conditions of exchange. This helps us relate our analysis of prices back to the condition of self-replacement and viability (3.3).

### *3.1 The 'Self-Replacing' Economic System and Viability*

The key to understanding the meaning of Sraffa prices seems to lie in the one assumption about which he is insistent throughout. It is the assumption of a 'self-replacing' economic system. This assumption is maintained consistently across the four commodity production models he specifies in Part I of his classic.[5] Perhaps because the 'assumption' refers to a primitive and self-evident truth of social life, namely that the material basis for the survival of society must always exist, Sraffa does not go on to justify it. A close reading of the text suggests that his objective is not price-determination per se by some ingenious means. This way of posing the problem only reifies the belief that there is not only a definite theory of prices there but also that such a theory is the primary objective of Sraffa. The question he asks (even for his very basic model of the system without surplus) is rather as follows: Which set of prices will ensure that the commodities required as means of production and in support of labour will be replaced for the

[5] Specific references to 'self-replacement' are: (1) pp. 4–5, model of a subsistence economy; (2) p. 6, model of an economy with surplus but which accrues entirely to capitalists; (3) p. 11, the model of an economy with surplus but which is shared between workers and capitalists (numeraire—national income); and (4) pp. 26–7, same model as (3) but with Standard commodity as the numeraire. See also, p. 89, Appendix A (On Sub-Systems).

next cycle of production? Or which set of prices will render the economic system viable at the observed scale of output and methods of production? Although Sraffa uses the terms 'reproduction' and 'replacement' interchangeably, it is important that we distinguish carefully between them. Following Bharadwaj (1989, p. 246), we define 'self-reproduction' as a characteristic feature of a subsistence system which tends to exactly reproduce itself through time. The concept of 'self-replacement' is apt for the economic system with surplus because it refers only to the system's viability and 'does not indicate how the system would change in the next period' (Bharadwaj, ibid.). In other words, Sraffa's frame of analysis allows only for the class distribution, and hence prices and valuation, of given social surplus, leaving open the question of uses to which that surplus may be put.

The assumption of self-replacement is critical for Sraffa system, indeed for surplus-based theories in general, because it is the foundation upon which the entire analysis of the system of production, conceived as a circular process, stands. The concept itself has origins in pre-classical literature and that of the Scottish Historical School. According to the 'pre-analytic vision' of this literature, in any given historical stage of their development national economic systems display certain 'uniformities and regularities' which were observable (Meek 1977, p. 19). This vision was later developed by Adam Smith as the concept of the 'natural state' of the economy, relative to each historical stage of development of society or 'mode of subsistence' (Bharadwaj 1989, p. 23). Smith's 'natural prices', corresponding to the discoverable regularities and tendencies of the economic system, were indicators that the whole economic process could repeat itself. Sraffa's concept of self-replacement is an adaptation of the natural state in its clipped-down version to fit into the bare and essential form of 'equations' in which he represents the economic system.

There is a second dimension to Sraffa's adaptation of Smith. Given the incipient stage of capitalism of Smith's times, marked by low rates of accumulation and technical progress, the natural state of the economy was a rather simple and self-evident historical truth. Natural prices themselves were transparent and more or less amenable to direct observation. They did not need to be extracted by means of any considerable theoretical effort or logical operation. Sraffa needed to adapt the concept to modern conditions of industrial capitalism marked by rapid rates of accumulation and technical progress, and therefore by *continuous change*.

Sraffa's adaptation is best illustrated if we place his economic system on a growth path, itself uniquely located in the historical time-space continuum. The analyst, whenever she has reason to believe that the system has undergone structural change following, say, changes in methods of production in one or more basic industries, or a permanent shift in income distribution, can begin to observe it.[6] She can remove layer after layer from an actually existing economic system, markets, traders, transport, money, government, taxes and so on till she hits at the skeletal form of Sraffa equation system comprising only physical quantities. She can then solve the equation system the Sraffa way for prices which, 'if adopted' (by the market), would guarantee replacement of the means of production system without impairing production on the same scale and with same methods in the next cycle. What needs noting is that the test of viability is conducted through the medium of prices which turn out to be *logical* in character. It is neither necessary nor possible under modern conditions to expect history to throw up Smith-type natural state of the economic system comprising a constellation of economic variables which remain stable over very long periods of time such that prices become more or less directly observable.

### 3.2 Standard Commodity and the Wedge Between Distribution and Prices

The argument about the logical character of Sraffa prices will gain in strength if we extend the discourse to his model of the economic system with surplus and Standard commodity as the *numeraire*. We will see that in this model, which establishes the price-independent rate of profits, the identity and power of a classical *theory* of prices all but disappear. All too often, the Standard commodity is seen as an ingenious *numeraire* commodity with the desirable property that Ricardo was eagerly but vainly seeking out, namely a measure of value whose own value is invariant with respect to changes in distribution. Sraffa's focus on the mathematics of constructing the Standard commodity, proving its uniqueness, and above all calling it 'an auxiliary construction', have only buttressed this view. Such a view, however, tends to discount the significance of Standard commodity for theory, which is to liberate the theory of distribution from the apron strings of value.

[6] No presumption regarding equality or otherwise between outputs and demand at the time point of observation is warranted, because viability of an economic system need not, and in realty does not, imply full employment.

Ricardo's greatest theoretical puzzle was the circularity between prices and distribution which frustrated his attempt at an unambiguous measure of the rate of profits. He faced the puzzle in his analysis of change in the rate of profits consequent upon a rise in wages and the intervening price changes (see Sraffa 1951). It has been argued that Ricardo was in fact interested in shielding inter se changes in distributive shares from intervening price changes, rather than in any *theory* of prices (see Cassels 1935).[7] Sraffa succeeds in liberating the theory of distribution from the problem of value in the general case of multiple basic commodities.[8] The separation of the problem of distribution from that of prices in his system is obvious from the fact that with Standard commodity as the measure of wages, both the maximum rate of profits and the inverse relation between the rate of profits and wages are determined within the physical quantity system without the need for prices. This drives a wedge between distribution and prices. Sraffa reduces Ricardo's problem of circularity between distribution and prices to one of *simultaneous* determination of profits (given wages) and prices. Prices come unstuck in Sraffa. In his preferred commodity production model with the rate of profits given exogenously as a pure ratio, Sraffa demonstrates that wages and prices, both measured by Standard commodity, are simultaneously fixed. In this sense, prices become merely an accounting device for measuring aggregates. In the present instance they fix the level of profits and total capital, such that the rate of profits in the given system of production conforms to the exogenously given rate of profits as a pure ratio from the financial system. But prices acquire a new, logical function. They emerge as conditions of logical consistency between the production and distribution of surplus. Sraffa prices have no central cause or a chain of causes behind them. Neither do they determine anything in the system. A *theory* of prices is thus neutralized and dissolved in his system.

Perhaps another implication of Standard commodity for political economy must be noted. The classical theory of value, as indeed others that followed it, had traditionally served as a cover in some sense for class conflict over distributive shares (see Sect. 1). Depending on how the theory of value determines the cause(s) of value, one or the other pattern of

[7] Sraffa's neutral response to Samuelson's query on whether Ricardo believed in labour theory of value—'he did and he didn't'—may have some bearing on this point (Samuelson 2000).

[8] Ricardo achieved this in his model with corn as the only basic commodity.

distribution of net product could be rationalized. Sraffa effectively relieves the value problem of this normative burden and transfers it to the theory of distribution. His system opens out for us only two pillars of economic theory: a historical theory of quantities and a historical and 'moral' theory of distribution. Prices are logical strings joining the two pillars.

### *3.3 Sraffa Prices and Conditions of Exchange*

Relative prices are numerical relations among commodities. They are ratios which effect multilateral exchanges between diverse commodities. Besides, the 'surplus' approach to economic theory entails that profits, while they are generated in the sphere of production, must be realized in exchange. Both these considerations make it imperative that we consider how Sraffa prices relate to conditions of exchange. Any relation Sraffa prices may have with conditions of exchange must be through the rate of profits. The view held by Adam Smith was that a uniform rate of profit—the natural rate of profit—emerged through 'competition of capitals', namely through mobility of real capital across industries in search of higher profits. However, this view is inconsistent with the Sraffa system because such movements of capital would entail *changes* in output and/or methods of production which are the givens in his system. Such capital movements, being changes in historical time, form an integral part of the theory of accumulation.

It was left to Ricardo to show consistently that profits are part of the surplus originating in, and constrained by, the conditions of production. Even if competition set up the tendency towards the uniformity of the *rate* of profit, it could not determine the *level* of profit which depended (inversely) upon wages. However, Ricardo considered the inter-relation between distributive variables in the dynamic context of accumulation. Sraffa integrated into a consistent framework Ricardo's analyses of rent, profits, wages and prices, after removing some of his puzzles (see Subsection 3.2 above). In particular, Sraffa system scrupulously avoids any conflation of given economic system with economic change. His analysis of 'the effects which changes in the division of the product between profits and wages may have on the values of individual commodities' (Sraffa 1960, Chaps. III & IV) may not be construed as historical changes in distribution. Those 'changes' are meant to show (1) that alternative patterns of distribution are conceivable for a given system of production (because the maximum rate of profits in the Standard system is invariant with respect to changes in distribution), and (2) that for the given system of production, a new set of prices would be necessary for its viability if a different pattern of distribution were to be prevalent.

Sraffa prices are meant to be logical conditions ensuring self-replacement of the production system and therefore must be internal to the system. So must be the case with the rate of profits. Such a rate of profits could be none other than the general rate of profits of the Standard system which holds for the actual system (viz. the system including non-basics) as well. This will appear to be contrary to the fact that Sraffa takes the rate of profits to be exogenously given as a 'ratio' independent of prices and governed 'by the level of the money rates of interest' (Sraffa 1960, p. 33). However, this 'ratio' given before prices will acquire substance only when it enters the Standard system and generate prices such that the resultant levels of profits and capital, as value magnitudes, vindicate this ratio. In other words, the rate of profits, emerging as the *value* of surplus distributed over the *value* of capital, is a mechanism internal to the system of production.

That Sraffa does not allow the flux of exchange conditions to affect the integrity of the given system of production is also evident at another level. While his system of production is defined for a unit period, no similar unit period is defined for circulation of commodities. They are assumed to be sold and bought at a single time point. 'Commodities are produced by separate industries and are exchanged for one another at a market held after the harvest' (Sraffa, ibid., p. 3). If this is held for the subsistence system, 'the supposition of an annual cycle of production with an annual market' is retained for the system with surplus (Sraffa, ibid., p. 10). An essential feature of his system is that it is the conditions of production that must determine the conditions of exchange and not the other way. Thus, he speaks of his prices 'which if adopted restore the initial position' of quantities (Sraffa, ibid., p. 4). This reiterates the primacy conventionally accorded in political economy to production over exchange. Prices, as logical conditions of consistency between production and distribution of surplus, are determined internally and independently of conditions of exchange. It is these 'necessary' or 'natural' prices or 'prices of production' (Sraffa, ibid., p. 9) which must persistently underlie the 'market' prices at which actual trades take place. There is no presumption, nor is it a requirement of theory, that the latter should ever 'converge' to the former. It perhaps remains to be added that viability of the production system does not presuppose that sales must equal outputs, much less that such equality must hold at the level of full employment.

## 4 A Theory of Economic Structure

Our conclusion is that Sraffa prices do not constitute a theory of prices. Together with our opening statement (see Sect. 1) that there can be neither a theory of quantities nor a theory of distribution because they are the data of the system, this pushes us into an awkward corner of believing that the whole of Sraffa system is bereft of theory. However, we shall presently argue that the system represents a definite theory of economic *structure*, the necessary foundation for a theory of economic *change*.

Once prices indicating viability of the economic system are determined, the system 'reveals' its essential structure. All aggregates are consistently measured, e.g. net product, wages, profits, capital, as also their ratios, in particular, the rate of profits. The structure of outputs and technology and a solution to the problem of choice of technique become evident. We know the extent of unemployment of labour and industrial capacity. Capital consumption is correctly accounted for. Not only are all the concepts precisely defined but the corresponding measures are consistent with the concepts. In particular, the rate of profits, the rate which no economic theory of capitalism can do without, is precisely defined and consistently measured. Critical knowledge needed for policy-making is available. In other words, the structure of the economic system under observation is revealed, but there is no presumption about the extent or direction of change in the system. Sraffa system provides us with a nucleus of precise concepts and consistent measures to study quantity and distributional dynamics. It liberates us from dependence on individual and subjective motives in the study of the economy. Above all else, it does not lead us astray into the blind alley of dichotomies plaguing neo-classical theory, namely the dichotomies of micro versus macro theories, of short-run versus long-run theories, and of statics versus dynamics.

A high point of Sraffa's theory of economic structure is the precise concept and consistent measure of the rate of profits, which is also its main connecting link with the theory of accumulation. It is striking how even major systems of heterodox economic theory flounder on this simple test. Marx, for example, has very rich concepts of capital and profits, but seems to lose a determinate and consistent measure of the rate of profits to his 'transformation' of embodied labour values into prices of production. In Keynes, both the concept and measure of the rate of profits seem to have been sacrificed at the altar of uncertainty. It is hard to imagine how any economic theory of capitalism could survive this lacuna and still flourish.

## 5 Conclusion

We shall not summarize the discussion. In arguing that Sraffa prices do not constitute a theory of prices, and that his prices signify conditions of system's viability, we have mainly relied on the text of his own classic. Piero Sraffa ranks among the finest of economic thinkers. But did he, 'from whom nothing is hid' (Keynes 1933, p. 138), choose to hide something from his readers?

## References

Bharadwaj, Krishna (1989) *Themes in Value and Distribution—Classical Theory Reappraised*, Oxford: Oxford University Press.

Cassels, John M. (1935) 'A Re-interpretation of Ricardo on Value', *Quarterly Journal of Economics*, May. Reprinted in Ingrid Rima (ed.) (1970), *Readings in the History of Economic Theory*, New York: Hold, Rinehart and Winston.

Keynes, J. M. (1933) *Essays in Biography*, London: Macmillan.

Meek, R. L. (1977) *Smith, Marx and After—Ten Essays in the Development of Economic Thought*, London: Chapman and Hall.

Ricardo, David (1951–1973) *The Works and Correspondence*, P. Sraffa (ed.), Cambridge: Cambridge University Press. Vol. VIII, *Letters, 1819–June 1821*.

Roncaglia, A. (1978) *Sraffa and the Theory of Prices*, Chichester: John Wiley & Sons.

Samuelson, P. A. (2000) 'Sraffa's Hits and Misses', in H. D. Kurz (ed.) *Critical Essays on Piero Sraffa's Legacy in Economics*, Cambridge: Cambridge University Press.

Sen, Amartya (1989) 'Economic Methodology: Heterogeneity and Relevance', *Social Research*, 56 (Summer).

Sinha, Ajit (2016) *A Revolution in Economic Theory—The Economics of Piero Sraffa*, Palgrave Macmillan.

Sinha, Ajit (2018) Theories of Value from Adam Smith to Piero Sraffa, Routledge, Second edition with an 'Afterword'.

Sraffa, P. (1951). 'Introduction', in P. Sraffa (ed.), *Works and Correspondence of David Ricardo*, vol. 1, Cambridge: Cambridge University Press.

Sraffa, P. (ed.) (1955) Works and Correspondence of David Ricardo, vol. VIII, Letters, 1819-June 1821, Cambridge: Cambridge University Press.

Sraffa, P. (1960) *Production of Commodities by Means of Commodities*, Cambridge: Cambridge University Press.

Comment

# A Comment on 'The Meaning of Sraffa Prices'

*Ajit Sinha*

Some years ago, Louvre, Paris had organized a comprehensive exhibition of Leonardo da Vinci's works, which contained many of his working sketches and diaries. What struck me, while I was observing them, was that Leonardo was consistently sketching and re-sketching a series of things in motion—how they change with time—he would sketch a human face ageing over years and things of that nature. I felt he was constantly struggling to capture a moment of an incessantly changing dynamic subject. It appears to me that Sraffa was trying to do something of a similar sort. In a note of 1940s, Sraffa writes, 'the economic systems of reality are not self-replacing. They are in a constant state of transition and obsolescence, due to changes both in the types of (kind of) commodities produced and in methods of their production…' (Sraffa n.d., 1955, D3/12/75, p. 16). This, in the light of his remarks that, '[t]he question asked of the theory of value is the following: Given (from experience) the prices of all commodities at one moment find a set of conditions that will make these prices appear to be necessary…' (Sraffa n.d., D3/12/9: 65) and his repeated use of the metaphor of a 'photograph of a market place' for his system of equations clearly points in that direction.

Professor Omkarnath (2021), in this remarkable paper, has brilliantly articulated this aspect of Sraffa's work. He argues that we would look in vain if we look for a causal theory of value or prices in Sraffa's book. Though in the classical theory we find a theory of value, this disappears in Sraffa's theory in his opinion. The classical theory of value, particularly in Ricardo, was a by-product of his concern for a theory of distribution in a dynamic context. On the assumption of given real wage, the question was how to understand the movements or changes in rent and the rate of profits as population rises and the agriculture is forced to either expand on less fertile land or use the same land more intensely, leading to a fall in returns to capital and labour. This, from the perspective of Ricardo's theory of rent, where marginal land or the marginal dose of labour and capital pays no rent, should unambiguously raise the rents on all intra-marginal lands both by virtue of more physical surplus left for them as well as the price of agricultural goods (i.e., of those physical surplus) rising due to the rise in their 'cost of production' relative to industrial goods. Now, if the wages remain constant then the rate of profits must fall. But shouldn't the wages rise in value terms? Since the real physical wage basket is assumed to remain fixed then due to rise in the value of food, as explained above, the 'money' wages or the value of wages must rise. Thus in value terms it seems that a rise in wages must be related to a fall in the rate of profits even if the marginal land and therefore the rent remains fixed. But could this proposition be established? The proposition would be unassailable if one could establish that changes in the rate of profits or wages have no impact on prices or values, i.e., if prices remain constant when wages rise then obviously the total value divided between wages and profits remain constant and as wages rise the total profits must fall and given the assumption that this does not change prices, the value of total capital invested remains constant and therefore the rate of profits must fall. But here is the snag. Ricardo realized that in a general case, given his condition that competition among capital must lead to equalization of the rate of profits across industries, it could not be assumed or at least established that changes in the rate of profits will not have any impact on prices. Once this was recognized, Ricardo did not know how to analyse the knock-on effect of changes in prices on the rate of profits back again and so on and so forth.

Sraffa (1951) interprets Ricardo's problem in terms of constancy of the measure of the total net product produced on the marginal land. If this net output has to be divided between wages and profits then changes in its distribution should not have any effect on the measure of the total—that is, changes in the size of the slices should not change the size of the proverbial pie. But that is what Ricardo noticed happening when he changes the rate of profits or 'money' wages and allows the prices to adjust so that all the industrial rates of profits become equal. Ricardo, according to Sraffa, thought that this was happening because he had no option then to arbitrarily take a commodity as the Standard of measure for prices and wages and since changes in the rate of profits affected all commodities, including the commodity chosen as the numeraire, the size of the net output could not remain the same after the change in the rate of profits. Thus, in Sraffa's view, this led Ricardo to think in terms of an 'average commodity'—a commodity that is produced by such a technical mix of direct and indirect labour that changes in the rate of profits do not affect this commodity. In that case, Ricardo thought that against such a commodity, if used as the Standard of measure for prices and wages, changes in prices of all other commodities would cancel each other out and leave the measure of total net output constant with respect to changes in the rate of profits. Now, the question was how to find the condition of production of such an 'average commodity'—this is where, according to Sraffa, the problem of finding a measure of value gets entwined with the true theory of value itself.[9] Omkarnath raises the question that after the discovery of the Standard commodity by Sraffa in his book (Sraffa 1960), which apparently solves Ricardo's problem of the 'average commodity', what remains of the classical theory of value?[10]

Now, let us suppose that Ricardo could find a commodity against which the value of total net output remains constant with respect to changes in wages and profits. What would this accomplish? Well, as we have argued above, it would establish that a rise in wages would lead to a fall in the rate of profits and vice-versa. However, since changes in wages bring about changes in prices of all other commodities, it will also bring about changes in the value of capital. Thus the determination of the rate of profits remains tied to prices in this case. But Sraffa (1951) also argued that before embarking on his book in 1815, Ricardo in 1814–1815 period had held the view that 'it is the profits of the farmer that regulate the profits of all other trade'. Sraffa had interpreted this statement of Ricardo as a statement of what has now come to be known as 'corn-profit model'. In Sraffa's interpretation, Ricardo was suggesting that in agriculture both the inputs and outputs could be conceived as being one and the same commodity, 'corn'—corn is used as seed and payment for wages and output also comes out as corn. In this case, both the net output and capital are given in the

[9] See Hollander (1979) and Sinha (2018) for alternative interpretations of Ricardo on this issue.

[10] Strictly speaking, the Standard commodity does not solve Ricardo's problem as it was presented by Sraffa. What it establishes is that the ratio of the value of net output to total capital, i.e., the maximum rate of profits, and not the value of the net output itself, remains constant with respect to changes in the rates of profits and wages. See Sinha (2016).

same unit and thus their ratio can be calculated directly without any recourse to their prices. This ratio gives us the rate of profits independently of prices and since the rate of profits must be equalized across the sectors and industries the prices of all commodities must be so adjusted that the corn-ratio rate of profits prevails in all industries. In this case, the rate of profits is determined independently of prices but prices are dependent on it.

Now the question is, which one of the two propositions Sraffa establishes in his book? Both Professor Omkarnath and I would agree that it is the second proposition and not the first one that Sraffa establishes through the means of the Standard commodity. The single commodity 'corn' gives way to a more general system of interconnected industries of 'basic goods'. Sraffa's most consequential equation: $r = R(1 - w)$, where $w$ is given in terms of the Standard commodity, achieves exactly that. The only difference between Omkarnath and I on this issue appears to be that he thinks that even in this case the rate of profits, $r$, is determined simultaneously with prices. But there is no need for this anymore. One of the fundamental proposition that Sraffa wanted to establish, and which he considered to be a fundamental proposition of the classical economics as well, was that *profit is a non-price phenomenon*. Thus the rate of profits must come unstuck from prices—the set of prices is the *shadow* of the real structure given by the equation, $r = R(1 - w)$, they only project the body in their own dimension.

The property, $r = R(1 - w)$, is the relations of the aggregate variables of the system, captured directly, i.e., independently of prices, in the aggregate equation of the Standard system, which represents the weighted 'average commodity' of the real system. When the average is expanded into its individual components as separate industries, we see that they are all inter-related and have no existence independently of each other. How are these industries related to each other? This relationship is given by the ratio of exchanges that must hold between the commodities of these industries—prices represent those relations. If I understand correctly, Omkarnath argues that given the values of the aggregate variables, the exchange ratios come out as a logical deduction or a mathematical derivation. Here, there is no causal theory of prices nor do prices play any causal role as 'market prices' do in classical economics. Omkarnath relates these 'logical prices' with Sraffa's condition of self-replacement. By which I understand that even though the dynamic real system may not be in a self-replacing state, these prices represent the conditions that ensures that the system could go on spinning on the same axis. Omkarnath also speculates that Sraffa's structural analysis of a system at one moment may lend itself for a fruitful dynamic analysis through the variable of changes in the rate of profits (also see the last chapter of Sinha 2016 on this score).

## References

Hollander, S. (1979). *The Economics of David Ricardo*, Toronto: University of Toronto Press.

Omkarnath, G. (2021). 'The Meaning of Sraffa Prices', in A. Sinha (ed.), *A Reflection on Sraffa's Revolution in Economic Theory*, London: Palgrave Macmillan.

Sinha, Ajit. (2016). *A Revolution in Economic Theory—The Economics of Piero Sraffa*, Cham: Palgrave Macmillan.

Sraffa, P. (1951). 'Introduction', in P. Sraffa (ed.), *Works and Correspondence of David Ricardo*, vol. 1, Cambridge: Cambridge University Press.

Sraffa, P. (1960) *Production of Commodities by Means of Commodities*, Cambridge: Cambridge University Press.

Sraffa, P. (n.d.). 'Sraffa Papers', Wren Library, Trinity College, Cambridge University, Cambridge.

CHAPTER 5

# 'Openness' as a Methodological Principle of Sraffa's Economic Thinking

*John B. Davis*

## 1 Introduction: The Impact of Sraffa's Thinking

Piero Sraffa's *Production of Commodities by Means of Commodities* (1960) was path-breaking as a contribution to political economy and penetrating as a critique of the orthodox economics of the twentieth century. As Ajit Sinha recently put it, the book produced a 'revolution in economic theory' (Sinha 2016; cf. Martins 2019), the impact and significance of which continues to be investigated. At the same time, despite the depth and far-reaching implications of Sraffa's critique of orthodox economics it was ignored by the great majority of economists, and this not only complicates our understanding of its impact on economic theory, but also creates a

I am grateful for comments on an earlier version of this chapter from Cristina Marcuzzo, Geoffrey Harcourt, Nuno Martins, and Ajit Sinha. All errors remain mine alone.

J. B. Davis (✉)
Marquette University, Milwaukee, WI, USA

University of Amsterdam, Amsterdam, The Netherlands
e-mail: john.davis@marquette.edu



A. Sinha (ed.), *A Reflection on Sraffa's Revolution in Economic Theory*, Palgrave Studies in the History of Economic Thought,
https://doi.org/10.1007/978-3-030-47206-1_5

paradox regarding our interpretation of 'the' history of economics. For Sraffa, 'the' history of economics dates back at least to Adam Smith and David Ricardo as founders of a subject specifically understood as political economy. Yet economics today is no longer identified as political economy by most people in the field, but is conventionally said to be a science independent of history, politics, and social values, and thus makes little reference to how the social organization of the economy was a distinctive characteristic of the thinking of Smith, Ricardo, Marx, and others in the history of political economy.

We might explain this paradox, and Sraffa's contemporary influence, as an instance of 'scientific amnesia'—a phenomenon whereby the leading edge of a science is so focused on particular problems at hand that the past commitments and achievements of the field are neglected and forgotten. One can argue that all sciences, as their subjects of investigation are further and further extended, undertake increasingly specialized types of research, and that this produces a fragmentation in science whereby researchers no longer assess a science's overall goals and its history of development and only focus on very narrow issues. I have recently argued that this sort of development characterizes contemporary economics—an 'all trees, no forest' type scenario (Davis 2019). One possible consequence, then, of this sort of development might be a rise in pluralism in science. Pluralism of course has different meanings. In its more normative sense, it involves a culture in a science that embraces and promotes diverse research strategies. In its more descriptive sense, it involves how the practice and social organization of a science happens to foster diverse research strategies—irrespective of the attitudes of researchers toward diversity in research. It is this latter state of affairs, I believe, that seems to characterize current economics, especially since there is little evidence of a culture of pluralism in today's economics. The advance of specialization in economics, made possible by the modern growth of the field in the resources it commands and the number of researchers it involves, has generated a diversity in new research strategies that exceeds what was the case only quite recently up to the end of the first half of the twentieth century and arguably into the first postwar decades of the second half of the twentieth century.

This, then, tells us something about interpreting the impact and significance of Sraffa's thinking. The publication of the *Production of Commodities* and the debates immediately thereafter over its meaning and implications occurred at the end of a long period in the history of economics prior to the recent emergence of a *de facto* greater diversity of research strategies in the field. During this period neoclassical economics possessed a dominant presence in the field, and diversity in research was relatively modest if not actively discouraged. However, as Beatrice Cherrier has shown in her study of the evolution of the JEL (*Journal of Economic Literature*) code used to classify different economic approaches and areas of investigation in economics, over the last half century—that is, since the publication of Sraffa's book—there has been a significant increase in new categories and new sub-categories of research in economics associated with new fields of research, new techniques and methods of research, and new domains of application (Cherrier 2017).

This suggests, then, that our evaluation of the impact of Sraffa's book up to now may have been, in effect, backward-looking in that views of its reception have been framed by the history of economics of the first half of the twentieth century, a period when neoclassical economics was commonly identified with the field. If we instead adopt a forward-looking perspective, and frame the book's impact in terms of how economics has developed since 1960, we arguably get a different view. The earlier frame is one in which orthodox economics had largely replaced the political economy arguments of Smith, Ricardo, and Sraffa. In contrast, the later frame is one in which political economy arguments have re-emerged in a range of new research venues, especially, for example, in the re-invigorated recent debate over economic inequality (Chetty et al. 2017, 2018; Milanovic 2016; Piketty and Saez 2003; Saez and Zucman 2014), which again makes relationships between classes and social groups central to economics in the manner of classical political economy. Indeed, the whole issue of globalization, much discussed today, raises fundamental questions about how economies are organized and grow, essentially just what concerned Smith, Ricardo, and Marx as political economists. From this perspective, Sraffa's contributions could well have an increasing impact and significance in the future—for reasons I will discuss in this paper.

However, left unaddressed in this picture, and potentially counteracting it, is the intellectual inertia of the past, and the role that economists' past attitudes and beliefs about the nature of economics might play in economics' future evolution. Needless to say, it is difficult to summarize the attitudes and beliefs of a large number of individuals, so claims about what economists' attitudes and beliefs are should be treated with caution. At the same time, one ought not deny that there exist collective views about economics and what it investigates on the part of great numbers of economists. If we term these views an ideology of economics, they represent assumptions that are likely slow to change but which might nonetheless also evolve over time as research strategies in economics diversify and what economists investigate changes. My conjecture, then, is that the emergence of a *de facto* greater diversity of research strategies in the field associated with increasing specialization in research is likely to act contrary to the field's past dominant conception of economics, if with a lag, and lead us increasingly back to a political economic thinking in the tradition of Smith, Ricardo, Marx, and Sraffa. In short, as the world has become economically more complicated in the future, and as economics research becomes increasingly diversified, a natural ideological posture for economists is to adopt is to be 'open' to thinking that confronts the changing ways in which the economies of the world are organized and function.

If this is so, it is worth examining what an increasingly 'open' ideological conception of economics might involve. This chapter seeks to develop a conception of what this might involve by following Sraffa's early lead in formulating a distinction between 'open' and 'closed.' Sraffa's *Production of Commodities* is not generally associated with this issue, and much of the research it generated is concerned with complex technical issues surrounding the nature of commodity values and production. Yet Sraffa's views regarding the relationship between production and distribution—views that departed in fundamental ways from neoclassical economics—raise important issues regarding how the economy may be open to history and change, and thus provide grounds for saying what an 'open' ideological conception of economics might involve.

This chapter traces Sraffa's views on this back to an early turning point in his intellectual development *en route* to the *Production of Commodities* that he commented upon in his unpublished 1931 'Surplus Product' manuscript (Sraffa 1931). I revisit this document and argue that it involves an important contribution to the philosophy of science of economics in regard to how understanding the relationship between economics and the economy depends on working with a methodological conception of openness. While there is a great deal of careful scholarship on the *Production of Commodities*, less extensively investigated is Sraffa's philosophy of science thinking, and the role it plays in his economic thinking. This chapter is intended as a contribution to this side of his thought.

To introduce this discussion, Sect. 2 reviews Sraffa's economic contribution and critique of orthodox neoclassical economics, especially as they concern the relationship between production and distribution. It then places the *Production of Commodities* in its historical context, and discusses the puzzling nature of its reception in economics, treating this as methodological problem concerning how sciences evaluate and respond to internal critique.

Section 3 discusses Sraffa's 1931 manuscript. Here, building on arguments I previously developed (Davis 2012, 2018), I argue that Sraffa used an open-closed distinction to explain the relationship between production and distribution, and engaged in a kind of philosophy of science reasoning that has come to be associated with systems theory.

Section 4 connects Sraffa's philosophy of science, methodological reasoning to his Turin, Italy origins and connection to Antonio Gramsci. It argues that Gramsci in his development of the concept of hegemony was concerned with how capitalism organized different domains of social activity, and then argues that there are important parallels between this and Sraffa's thinking regarding the open-closed distinction.

Section 5 closes the chapter by briefly commenting upon openness as a principle of explanation, and returning to the issue of economics' ideological conception of the relationship between economics and the economy. If how the changing of the world's economies promotes a *de facto* greater diversity in economic thinking resonant of the political economic views of Smith, Ricardo, and Marx, then Sraffa's thinking, with its focus on the relation between distribution and production, may enjoy new appreciation.

## 2 Sraffa's Entry Point and Challenge to Neoclassical Economics

To begin, let me summarize the domain of research stimulated by Sraffa's important book. I rely on Tony Aspromourgos' extensive review of the scholarly literature on Sraffa's thinking (through 2004) that distinguishes five separate research programs that he sees Sraffa as having been responsible for having generated: (a) the nature and significance of long-period equilibria in the theory of distribution and production prices; (b) the Cambridge Growth Equation and closures of the distribution system via the accumulation rate; (c) closures of the distribution system via the monetary determination of the rate of interest; (d) production prices, effective demand, and long-period adjustment—the Sraffa–Keynes synthesis; (e) the critique of marginalism (Aspromourgos 2004, Table 1). As

Aspromourgos shows, research on Sraffa's economic thinking is quite advanced and far-reaching. But what is it that unites all these different research programs?

I take as an entry point, as many others have also argued, Sraffa's profound investigation of the relationship between production and distribution, and how it contrasts in fundamental ways with marginalist neoclassical economics. In the latter, individual economic agents' optimization behavior generates demand and supply functions for all goods and factors of production under market-clearing competitive conditions, such that production and distribution are simultaneously determined—a framework fully explainable in terms of a self-contained economic logic of scarcity. In contrast, Sraffa rejected the factor substitution production analysis this involved, and developed an analysis of commodity prices in production systems in which, given the methods of production and the level of output produced, one distributive variable is fixed exogenously (for example, the general rate of profit), or is set outside of the analysis, thereby determining all other distributive variables (in particular, the wage rate). This effectively 'opens' the 'closed' neoclassical system of production and distribution in which all economic relationships are endogenous to one another and in which individual economic agents determine prices independently of any historical reference or context. For Sraffa, whatever the rate of profit is, or alternatively whatever the wage rate is, is determined by historical conditions, not by a self-contained economic logic independent of history, so that economic analysis always needs to be integrated with an analysis of the social forces determining those rates.

Sraffa's rejection of marginalist thinking—the foundation of the neoclassical analysis of prices—has made his own analysis of commodity prices a central focus of Sraffa scholarship. That explanation depends on input-output ratios and production coefficients tied to methods of production and technologies in use when there are many capital goods, or capital is heterogeneous. Neoclassical economists, following in the lead of J.B. Clark (1891), had come to rely on the idea that capital goods could be represented as a single quantity of capital and the economy as a whole could be represented as a single aggregate production function. This led them to claim that scarcity relationships governed the economy, and in the case of capital that there accordingly exists an inverse monotonic relation between the quantity of capital and the rate of interest. Sraffa's analysis showing that such phenomena as capital-reversing would occur when a lower capital/labor ratio is associated with a lower interest rate was contrary to neoclassical scarcity reasoning and the idea that a demand for capital curve must always be downward sloping.

This result led to a lengthy, complex debate—the so-called 'capital theory controversy' (cf. Cohen and Harcourt 2003)—in which leading neoclassical economists from Cambridge in the US attempted to defend the Clark vision and set aside Sraffa's thinking, while leading economists from Cambridge in the UK and Italy defended Sraffa's conclusions. The result of this debate was that the neoclassical substitution analysis was shown to fail as an interpretation of the economy as a whole, and that Sraffa's explanation of commodity prices was agreed to provide a more accurate and more sophisticated account of the nature of production. Yet, as is well known, most neoclassical economists subsequently ignored this outcome and proceeded as if the 'capital theory controversy' debate had never occurred.

This is puzzling from a history of science perspective since a widely shared belief is that scientific progress prevails in the long run, meaning that discredited theories and unsupportable explanations are ultimately abandoned. It might be thought that when the issues debated are highly complex, as they were in this case, that this outcome can be delayed, on the grounds that it would then take considerable time for important scientific results to be acknowledged and understood. Yet this should not have been the case in regard to the 'capital theory controversy' since the most influential neoclassical economist of the time, Paul Samuelson, who was centrally involved in the debate, explicitly and clearly stated at its end that the neoclassical aggregate production function analysis could not be maintained (Samuelson 1966). This seems to imply, then, that the conventional view of science—that progress ultimately prevails—is somehow incorrect. This is both worrisome and confusing. It is worrisome because it suggests humanity's best efforts to understand the world may sometimes be fruitless; it is confusing because it challenges our whole conception of science as rational dialogue.

I believe, however, that we can retain the general view that science is a progressive human endeavor if we modify the view that science is essentially a rational dialogue. Let me explain this in connection with the evolution of philosophy of science thinking in the postwar period. Though that thinking is largely the product of reflection on the history and character of natural and physical science, where historians of science generally believe there is strong support for a progress view, the argument has also been made that progress operates in the social and human sciences as well.

Influential in this regard is Karl Popper's defense of falsificationist inquiry and rationality throughout science (Popper 1963). In his view, though practitioners invested in past research are often guilty of 'immunizing' it from confounding evidence and competing theories, in the long run these efforts are likely to succumb to rational inquiry.

Popper's arguments, however, encountered a problem in the form of the Duhem-Quine thesis, which was that any particular scientific proposition subject to testing and apparent falsification, which was an implication of a theory with its many supporting assumptions and confirming evidence, could still be retained if that theory and the assumptions from which it is derived are sufficiently modified and reconfigured (cf. Boumans and Davis 2016, pp. 90–92). In effect, questionable propositions in theories that fail to be empirically supported can be insulated from critical evaluation by creative re-elaboration of the theories in which they are embedded. Imre Lakatos, in his methodology of scientific research programs approach, accordingly revised Popper's views to account for this, arguing that scientific theories typically possessed 'hard cores' of basic principles which scientists often aimed to preserve under all circumstances (Lakatos 1978). But Lakatos did not explain what motivated scientists' defenses of theories' 'hard cores.'

Consider, then, what happened in the case of the 'capital controversy' critique of neoclassical substitution analysis. While it was agreed that the aggregate production function view had been falsified as Popper would have emphasized, nonetheless, in Duhem-Quine terms, neoclassical theory after the debate largely came to be seen as represented by the Walrasian multi-market general equilibrium framework to which it was argued this critique did not apply (Hahn 1982). In a general equilibrium framework there are many capital goods, not a single aggregate capital. At the same time, the static character of the Walrasian framework meant that such problems as capital-reversing (which depend on temporal analysis) do not arise, so the critique of scarcity reasoning lost its immediate target. What, then, was the 'hard core' of neoclassical theory, as Lakatos would have put it, and what, moreover, motivated its defense?

There are different possible answers to these questions, but I emphasize one specific to Sraffa's political economic concern with history and the relationship between production and distribution. As we saw, central to this is his rejection of the marginalist idea that all economic activities are endogenous to one another—that is, that economic analysis is closed, and involves a fully self-contained scarcity logic sufficient for explaining any historical context. Whether economics should be thought to be closed in this way or should be thought to be historically open as Sraffa believed, then, concerns a fundamental matter, namely how one understands the relationship between economics and the economy. Neither view, it seems fair to say, can ultimately be empirically verified or falsified. Indeed, what evidence would ever clearly tell us which view should be favored? In addition, proponents of each view, committed to their respective beliefs, are likely to always be able to reframe their theories in a Duhem-Quine manner to accommodate or set aside evidence inconsistent with them. Thus, I characterize both views, following Lakatos, as 'hard core' beliefs, or alternatively as ideological conceptions that underlie economists' philosophy of science thinking.

Yet if economists' 'hard core' or ideological beliefs are not subject to empirical verification or falsification, might something else affect them? Looking at the change in economics especially in the more recent postwar period, I argued above that economics is becoming more diverse as a consequence of the growth in numbers of economists and the field's increasing rate of specialization, and that this is occurring in a globalizing world seen as exhibiting greater inequality. I suggested that this change may produce a more political economic type thinking in economics, especially compared with economics' recent past, as the range of types of issues investigated expands together with a rising awareness of the social dimensions of economies in the world today. That is, economics may be becoming *de facto* more pluralistic despite the discipline's past orientation. If this is indeed the case, then, economists' 'hard core' views and ideological conceptions as pertain to the relationship between economics and the economy may change as well. What this change could then promote would be a more open view of economics emphasizing how the relationship between production and distribution varies over time and across societies, rather than the more 'closed' understanding of that relationship which makes them endogenous to one another in a way that fits all societies in an ahistorical and identical way.

To explore further what an open ideological conception of economics might involve, the next section, discusses Sraffa's distinction between 'open' and 'closed' in his 1931 'Surplus Product' manuscript. Sraffa sought to restore political economic thinking in economics. As we will see, not only did he intend to develop a powerful economic analysis of commodity values alternative to that given by marginalism, but also sought to build that analysis upon secure philosophy of science foundations. Those foundations, I will argue, depend on his open-closed distinction.

## 3 Sraffa's 1931 Manuscript and the Open-Closed Distinction

In his 1931 "Surplus Product" manuscript, Sraffa distinguished clearly between what he called the 'economic field'—later the subject of his *Production of Commodities* analysis of commodity values—and distribution particularly as pertained to the surplus product, thus suggesting that they constituted two different types of economic systems. But why should we say these two domains constitute different 'systems' rather than, say, just different aspects of the economy, as in marginalist neoclassical economics where everything is endogenous within one system? The answer for Sraffa, I believe, derived in important ways from his philosophy of science thinking about the scientific principle of causality, and how maintaining this principle created a problem for explaining surplus product.

For Sraffa, then, causality operated in a clear way in the 'economic field' following established thinking in natural science. At the same time, he saw that how causality operated in the 'economic field' could not be readily transferred to an explanation of distribution and the surplus product. Indeed, as he expressed it:

> If one attempts to take an entirely objectivist point of view, the very conception of a surplus melts away. For if we take this natural science point of view, we must start by assuming that for every effect there must be a sufficient cause, that the causes are identical with their effects, and that there can be nothing in the effect which was not in the causes; in our case, there can be no product for which there has not been an equivalent cost, and all costs (=expenses) must be necessary to produce it (Sraffa, D3/12/7: 161: 3).

The "objectivist point of view" he refers to contrasts with neoclassicism's subjectivist point of view, and holds that how the natural world works is independent of human valuation. For Sraffa, it is embodied in "the natural science" principle that "for every effect there must be a sufficient cause," a corollary of which is that "there can be nothing in the effect which was not in the causes." For economics to be scientific, then, this principle needs to be applied to the analysis of production in terms of the idea that the costs incurred in producing goods constitute their causes, where these costs or "expenses" were only the commodities consumed in the production of commodities.

Thus, clearly, Sraffa aimed to build his analysis on sound philosophy of science thinking. Yet paradoxically this 'natural science point of view' led him to conclude that "the very conception of a surplus melts away" since a surplus product amounts to an excess of value over costs or "expenses" incurred in producing goods, thus violating the corollary that "there can be nothing in the effect which was not in the causes." Indeed, Sraffa believed that the existence of the surplus in capitalism depended on capitalists' willingness to withdraw circulating capital from production in order to engage in luxury consumption. Their motivation for doing so was clearly subjective and consequently at odds with a natural science point of view. Thus when it came to the surplus product, as Heinz Kurz and Neri Salvadori have put it, "as soon as it is explained, [and] a cause is found for it ... [it] ceases to be a surplus," given the cost of production analysis of commodity values Sraffa sought to develop (Kurz and Salvadori 2008, p. 267; also cf. Sinha 2016, pp. 82–89).

Sraffa, then, resolved this problem by first assuming that production and distribution involved two different types of causal systems, not just two different types of economic activities, which therefore could not be explained with identical forms of reasoning. Production was still to be explained in an objectivist way with a natural science account of commodity values based on costs, but social forces operated in a different causal way in connection with distribution. To reinforce this conception, Sraffa accordingly characterized commodity production and the 'economic field' as a 'closed system.' But he also asserted that this 'closed system' interacted with "outside causes" that somehow operate both upon it and within it at the same time. As he put it, this 'closed system' somehow had a "leak" by which it communicated with the world permitting these "outside causes" to affect the determination of commodity values.

> There must be a leak at one end or the other: the 'closed system' is in communication with the world. When we have defined our 'economic field', there are still outside causes which operate in it; and its effects go beyond the boundary (Sraffa, D3/12/7: 161: 5).

One might argue that since "outside causes" also affect commodity values that it makes no sense to say that the 'economic field' should be modeled as a 'closed system.' However, Sraffa recognized that subsistence economies producing no surplus represented production in its most basic form, implying that later stages of historical development which produced a surplus only modified the operation of the 'economic field.' We should say, then, that the 'economic field' is merely a relatively independent 'closed system' in which commodity values are dominated by, but not exclusively determined by, their costs of production. This characterization is reinforced by the fact that different historical regimes would be expected to affect the 'economic field' differently, implying that over time the 'economic field' was independent of its different historical manifestations and only modified in its functioning by social forces.

Thus, the position Sraffa adopted for explaining commodity values when a surplus is generated involves two types of causal systems whose different principles affect how production and distribution interact and work together. The *Production of Commodities* later showed how distributive variables mathematically enter into the determination of commodity values, but does not interpret this in system terms or raise philosophy of science issues regarding causality. Perhaps Sraffa believed a philosophy of science treatment of the matter would have made his analysis in the book unnecessarily more complicated, and could have distracted from its main conclusions when he wanted his analysis to be as clear and powerful as possible. Yet the issues he addressed in his 1931 manuscript remain.

On one level, then, we might explore further Sraffa's conception of how two types of causality combine to produce one causal system. However, I put aside this investigation because Sraffa's thinking about science does not lead in this direction. As Sinha notes, Sraffa "never adopted a deterministic scientific point of view," but was rather "committed to the quantum physics point of view of indeterminacy" (Sinha 2016, p. 88), so the whole idea that two types of causality might combine in some single way is foreign to his thinking.

On a second level, however, we can say something more about how Sraffa thought in systems terms, since he explicitly speaks of systems in open and closed terms. In addition, there also existed at the time Sraffa wrote an incipient literature on systems thinking, particularly as developed by Ludwig von Bertalanffy (1968). Though there is no evidence archival or otherwise that I am aware of that Sraffa knew or drew on Bertalanffy's work, at the same time, as I have argued (Davis 2018), there is an important overlap in their thinking in regard to what open and closed systems are. I thus draw on this overlap to say more regarding the nature of the open-closed distinction in Sraffa's philosophy of science thinking.

Bertalanffy (1901–1972) was an Austrian biologist who in the 1920s and 1930s was one of the early developers of general systems theory, formulated in terms of an open-closed systems distinction comparing how the physical sciences differ from the natural or life sciences (cf. Davidson 1983). On the one hand, he associated the former with physics and the second law of thermodynamics (the entropy law), and argued that the physical sciences are closed systems in the sense that their principles work in isolation from their particular environments. Alternatively, physical science principles work in the same way in all environments no matter what the time or place. On the other hand, he argued that living systems, especially those seen as undergoing growth and change such as studied in biology, are open in that their activity must always be explained in terms of their interaction with their environments, since the environments they occupy influence how they function. Thus, whereas closed systems always operate in the same way, open systems always work differently according to the environments they occupy.

We can recast Bertalanffy's distinction between open and closed sciences as a distinction between open systems that are dependent on time and place and closed systems that are in an important sense independent of time and place. For Sraffa, then, whereas the distribution of the surplus in terms of wages and profits is settled in an open socio-historical way, production relationships are rather closed in a technically determined way.[1]

In economics, this is the difference between a type of analysis that is fully self-contained and applies to any and all historical contexts, and a type of analysis that always needs to be integrated with its historical context. For Sraffa, neoclassical economics is a closed system because all relationships are endogenous to one another, and the scarcity reasoning which this allows can be applied to any environment, circumstance, or application no matter what the time, place, or setting. In contrast, his political economic distinction between production and distribution as two different domains of economic activity essentially treated the former as a closed system (in that it still obeyed the principles of natural science) which was nonetheless still open to the world, because it was influenced by social forces that reflected historical context.

Bertalanffy did not explain how open and closed systems communicated. He recognized that the natural or life sciences drew on principles

[1] Thus, in regard to Sraffa's later relationship, $r = R(1 - w)$ – the variables are determined independently of prices of production. I thank Ajit Sinha for this point.

from the physical sciences, but his goal as a biologist was to make the case that the natural or life sciences operated on a different basis from the physical sciences despite their drawing on physical science. Sraffa, however, was not in a position to avoid this issue of the relationship between open and closed, since economies such as capitalism function as wholes that are not compartmentalized in the way that the sciences are. At the same time, the neoclassical way of dealing with two types of domains was not acceptable to him because the subjectivism on which it is based was incompatible with a natural science account of production. How, then, did he understand the open-closed distinction?

What his remarks in his 1931 manuscript then specifically tell us is that the 'economic field' as a (relatively) 'closed system' is acted upon by "outside causes" but these "outside causes" are not referred to as a different type of system or appear to constitute a system in the manner that the 'economic field' does. In contrast to Bertalanffy, who was concerned with the differences and relationships between two types of science systems, Sraffa was rather concerned with how a (relatively) 'closed system' might be embedded in a larger world that lacks the characteristics of a system. That is, the world itself is, in effect, 'unsystematic' and constitutes a wider process that influences a (relatively) 'closed system' such as production. This suggests that the nature of that influence is not easily determined. Were we investigating the interaction of two different kinds of systems, we might explain their interaction in terms of how their respective sets of principles combine, as did Bertalanffy when he allowed that physical science principles affect how living systems grow and change. That strategy, however, was not one that Sraffa undertook. Rather, his use of the open-closed distinction was not an open systems-closed systems one, but rather one that embedded closed systems in an open world which operated on an entirely different basis.

If an open world is 'unsystematic' in the sense that it is not governed by deterministic principles in the way that a 'closed system' can be, this does not imply that it involves an entirely chaotic, random set of processes. As a political economist, Sraffa thought in historical terms, where this meant that economies developed in an 'arrow of history' sort of way whereby the past influences the future. Let us put this as he did in terms of how distribution affects production. Determining the profit (or wage) rate contributes to the determination of commodity values, which, given distributive shares, influences distribution, so how what is outside the relatively closed system of production not only acts upon it but also feeds back indirectly upon itself in regard to the effects on distribution. Seen as a repeated, reflexive process, this can have cumulative effects on how economies change and develop. Yet that distribution still belongs to an open world governed by social forces tells us that the specific direction that the interaction of production and distribution takes is not determined. All that is determined is that the future will be different from the past—that is, that the economy is an open historical process.

Sraffa, then, differed from Bertalanffy in how he used open-closed thinking. Whereas the latter was motivated to explain and emphasize the distinctiveness of the natural or life sciences, Sraffa was motivated by the fact that economics had abandoned its identity as political economy. He consequently appropriated and reconceived the open-closed distinction as a means of developing the scientific foundations he thought necessary for saying that economics must be political economy. Those foundations allowed him to distinguish different types of causal principles operating in production and distribution, and this made a philosophy of science analysis of causality central to the explanation of the economy as an historical process. This also meant he avoided the subjectivism on which neoclassical thinking depended, and can be seen as thinking—as reported by Ludwig Wittgenstein—as thinking in an 'anthropological way' (Monk 1990).

From the perspective of how philosophy of science thinking evolved from Popper to Lakatos, Sraffa may have regarded neoclassicism's scarcity logic as equivalent to a type of 'hard core' or ideological conception unlikely to be successfully contested in arguments about evidence. He certainly saw that scarcity reasoning maintained a powerful hold on most economists' thinking, and perhaps reasoned that he needed to make the case for a political economic approach on solid philosophy of science grounds. That endeavor was complicated by the challenges he encountered in applying causal reasoning to economies that produce a surplus, but his use of the open-closed distinction provided a means of addressing them in a way that deepened his thinking about the role of history in political economy. In the following section, I briefly compare this use of the open-closed distinction to reasoning to how Gramsci can be interpreted as reasoning in a similar way about how capitalism organized different domains of social activity.

## 4 Gramsci and Sraffa on the Open-Closed Distinction

Sraffa and Gramsci were both from Turin, Italy, and knew each other well (Naldi 2000). Sraffa came to know Gramsci in 1919, and their friendship continued until the latter's death in 1937. While evidence regarding how they influenced each other's thinking philosophically is lacking and much debated, how Gramsci reasoned about his central concept of hegemony can be argued to exhibit clear parallels to Sraffa's open-closed distinction. Gramsci developed the concept, then, to broaden the concept of state power in class society (Gramsci 1971; cf. Bates 1975). Traditionally the power of the state had only been associated with control over the machinery of government such as the police and the courts, but Gramsci argued that class power was also exercised through various non-state institutions such as the Church and the press. This meant that power manifested itself in various ways in that it took on different forms in connection with how it was exercised in these relatively independent social domains and institutions.

We can characterize these relatively independent social domains and institutions such as the Church and the press as closed systems since they can be distinguished from one another according to their own sets of rules and principles. Or rather, we should characterize them as relatively closed systems, in the sense employed by Sraffa, since for Gramsci how they operate is influenced by power relationships that lie outside of them acting upon them and affecting their internal functioning. At the same time, since Gramsci believed that class power pervades society, it lacks the characteristics of a system itself in the way that institutions function as systems. Moreover for Gramsci, class power was the product of class conflict that involved an historical process that was open-ended in nature. Thus interpreted, Gramsci's thinking about hegemonic power made similar use of the open-closed distinction to the way that Sraffa employed it. For both, the world is an open historical process in which social forces get differentially embedded in relatively closed systems of activity.

Gramsci did not frame his thinking in a philosophy of science manner in terms of different kinds of causality and how they might interact, as I have argued Sraffa did. But similar to Sraffa, he did adopt ideas suggesting he believed broad, historical social forces interacted with the functioning of multiple different social domains and institutions in a two-way street manner or as a repeated, reflexive process. Sraffa focused on the interaction between social forces and only one relatively closed system, production, given his primary concern was the economy. This allowed him to emphasize the primacy of the relationship between production and distribution, but he avoided making any claims about what sort of balance existed between them, or in Marshallian terms, whether this interaction produced some sort of equilibrium state of affairs.

Gramsci, however, with his focus on power, did label that balance using the concept of equilibrium, though in a critical way. In commenting on Louis Bonaparte's conventional exercise of power by means of the military in nineteenth century France—Caesarism—he asserted that in this we saw a circumstance in which "the existing social form had not yet exhausted its possibilities for development," whereas "[i]n the modern world, the equilibrium with catastrophic prospects occurs … between forces whose opposition is historically incurable" (Gramsci 1971, p. 222). Gramsci's concept of a 'catastrophic equilibrium' is the idea of unstable equilibrium reflecting a combination of opposed, conflicting forces rather than the neoclassical idea of a harmony of forces. For Gramsci, this reflected his view that class societies were always in conflict. We may also, however, explain this in terms of the open-closed distinction. The idea of a relatively closed system, whether it be production or a social institution, is of something whose activities are primarily governed by principles specific to its nature as a particular type of system. Yet that such systems are affected by forces from outside them means that their internal principles of operation are altered, or more strongly we might say disrupted, since as only relatively closed they are prevented from operating as they would in the absence of those outside forces.

Open-closed reasoning, then, gives us a way of looking at an economic system that transforms it from its neoclassical, ahistorical meaning to a meaning that treats as temporary states of affairs appearing to be in balance yet as transient and impermanent. This constitutes such a departure from its meaning in neoclassical economics that one might conclude it would be better to give up the concept altogether. For Sraffa, however, the concept was part of the language of economics and political economy, and proceeding explicitly in this way risked making his work inaccessible. Gramsci's 'catastrophic equilibrium' idea also employed the concept but criticized it at the same time. We cannot say whether he was influenced by his contact with Sraffa to use the idea in this way or that this was Sraffa's view of the concept. But they both appear to have used the open-closed distinction which Sraffa developed in his 1931 deliberations over how to reason in causal terms about the relationship between production and distribution, and this provides grounds for understanding equilibrium as a transient, impermanent state of affairs.

## 5 Concluding Comments: Openness as a Principle of Explanation

This chapter's focus was how Sraffa thought about the open-closed distinction, but its larger theme is the concept of openness as a principle of explanation in economic thinking. Taking Sraffa's entry point as the relationship between production and distribution, the economic approach he rejected—that was responsible in his view for the demise of political economy and for placing neoclassical economics in its place—was one that enshrined a fully closed conception of the economy in which all relationships are endogenous to one another. Thus his goal was to open up economic analysis in order to rehabilitate political economy, which I argue required that he identify openness as a systematic principle of explanation.

My argument in this regard is from the perspective of the present looking backward, but Sraffa's pathway began almost a century ago from within the Marshallian system in his quest to make history matter. He did not fully foresee how he needed to proceed, but adopted an "objectivist point of view" rooted in the science thinking of his time he believed offered a way forward. To his surprise and dismay, he discovered, as he reported in his "Surplus Product" manuscript, that his goal of rehabilitating political economy was undermined by that same "objectivist point of view." This, however, impelled him to creatively re-think how causal forces could take different forms and interacted in ways that he then outlined in open-closed terms.

Yet the destination of his work, the *Production of Commodities*, did not make this underlying thinking clear, and the Duhem-Quine general equilibrium response to the debate the book generated concealed its critique of the closed production-distribution neoclassical system in the 'hard core' neoclassical view that scarcity logic explains all economies, whatever the time or place. I have treated 'hard core' views in economics, then, as ideological conceptions that concern the relationship between the field and the economic world it is intended to explain. Sadly, evidence and reasoned debate seem to have little effect on what ideological conceptions people hold. But if Sraffa's political economic vision is correct, and if history does change how we see the world, if slowly, then the change in the world's economies today may well make defense of the idea that the economy can be explained as a closed system increasingly difficult. Then perhaps Sraffa's impact on the history of economics will be seen in a brighter light and the field again identified as political economy as thought by Smith, Ricardo, and Marx.

## References


Aspromourgos, Tony (2004) "Sraffian Research Programmes and Unorthodox Economics," *Review of Political Economy* 16 (2): 179–206.

Bates, Thomas (1975) "Gramsci and the Theory of Hegemony," *Journal of the History of Ideas* 36 (2): 351–366.

Bertalanffy, L. von (1968) *General Systems Theory: Foundations, Development, Applications*, New York: George Braziller.

Boumans, Marcel, and John Davis (2016) *Economic Methodology: Understanding Economics as a Science*, Second edition, with contributions from Mark Blaug, Harro Maas, and Andrej Svorenčík, London: Palgrave Macmillan.

Cherrier, Béatrice (2017) "Classifying Economics: A History of the *JEL* Codes," *Journal of Economic Literature* 55 (2): 545–579.

Chetty, Raj, David Grusky, Maximilian Hell, Nathaniel Hendren, Robert Manduca, and Jimmy Narang (2017) "The Fading American Dream: Trends in Absolute Income Mobility since 1940," *Science* 356 (6336): 398–406.

Chetty, Raj, Nathaniel Hendren, Maggie Jones, and Sonya Porter (2018) "Race and Economic Opportunity in the United States: An Intergenerational Perspective," National Bureau of Economic Research Working Paper No. 24441 (March).

Clark, John Bates (1891) "Distribution as Determined by a Law of Rent," *Quarterly Journal of Economics* 5 (3): 289–318.

Cohen, Avi J., and Geoffrey C. Harcourt (2003) "Retrospectives: Whatever Happened to the Cambridge Capital Theory Controversies?" *Journal of Economic Perspectives* 17 (1): 199–214.

Davidson, Mark (1983) *Uncommon Sense: The Life and Thought of Ludwig von Bertalanffy (1901–1972), Father of General Systems Theory*, Los Angeles: Tarcher.

Davis, John (2012) "The Change in Sraffa's Philosophical Thinking," *Cambridge Journal of Economics* 36 (6): 1341–1356.

Davis, John (2018) "Sraffa on the Open vs. 'Closed Systems' Distinction and Causality," *Research in the History of Economic Thought and Methodology* 35B: 153–170.

Davis, John (2019) "Specialization, Fragmentation, and Pluralism in Economics," *European Journal of the History of Economic Thought* 26 (2): 271–293.

Gramsci, A. (1971) *Selections from the Prison Notebooks of Antonio Gramsci*, Q. Hoare and G. Nowell Smith, eds., New York: International.
Hahn, Frank (1982) "The Neo-Ricardians," *Cambridge Journal of Economics* 6: 353–374.
Kurz, Heinz and Neri Salvadori (2008) "Representing the Production and Circulation of Commodities: On Sraffa's Objectivism," in H. Kurz, L. Pasinetti, and N. Salvadori (eds.), *Piero Sraffa: The Man and the Scholar*, London: Routledge.
Lakatos, Imre (1978) *The Methodology of Scientific Research Programmes: Philosophical Papers*, vol. 1, Cambridge: Cambridge University Press.
Martins, Nuno (2019) "The Sraffian *Methodenstreit* and the Revolution in Economic Theory," *Cambridge Journal of Economics* 43 (2): 507–525.
Milanovic, Branko (2016) *Global Inequality: A New Approach for the Age of Globalization*, Cambridge, MA: Harvard University Press.
Monk, Ray (1990) *Ludwig Wittgenstein: The Duty of Genius*, London: Jonathan Cape.
Naldi, Nerio (2000) "Piero Sraffa and Antonio Gramsci: Their Friendship between 1919 and 1927," *European Journal of the History of Economic Thought* 7 (1): 79–114.
Piketty, Thomas, and Emmanuel Saez (2003) "Income Inequality in the United States, 1913–1998," *The Quarterly Journal of Economics* 118 (1): 1–39.
Popper, Karl (1963) *Conjectures and Refutations: The Growth of Scientific Knowledge*, London: Routledge.
Saez, Emmanuel, and Gabriel Zucman (2014) "Wealth Inequality in the United States since 1913: Evidence from Capitalized Income Tax Data," *Quarterly Journal of Economics* 131 (2): 519–578.
Samuelson, Paul (1966) "A Summing Up," *Quarterly Journal of Economics* 80 (4): 568–583.
Sinha, Ajit (2016) *A Revolution in Economic Theory: The Economics of Piero Sraffa*, London: Palgrave Macmillan.
Sraffa, Piero (1931) "Surplus Product," D3/12/7 161: 1–5, August, Wren Library, Trinity College, Cambridge.
Sraffa, Piero (1960) *Production of Commodities by Means of Commodities: Prelude to a Critique of Economic Theory*, Cambridge: Cambridge University Press.

Comment

# A Comment on "'Openness' as a Methodological Principle of Sraffa's Economic Thinking'

*Guglielmo Chiodi*

John Davis has written a paper on an issue not so much fashionable in regard to Sraffa, although of a primary importance to better understand his thought and particularly his critique of the dominant neoclassical economic theory—to which the Italian economist has mainly devoted practically all his research work over the years.

The issue concerns the open-closed distinction, on which at different times and in other disciplines has also been the object of much reflection and discussion.

The starting point of Davis' insightful paper is the recognition that despite Sraffa's path-breaking and penetrating critique of neoclassical theory, its impact in changing (and possibly displacing) the dominant economic thought has been so far quite feeble—if not almost non-existent.

Davis, however, seems to show too much optimism on the possibility that a turning point in the contemporary economic thought has already made its appearance. By taking a forward-looking perspective since the publication of Sraffa's book and up to date—he argues—'political economy arguments have re-emerged in a range of new research venues, especially, for example, in the re-invigorated recent debate over economic inequality [...]which again makes relationships between classes and social groups central to economics in the manner of classical political economy'. I agree with Davis only on the appearance of new research venues in the political economy debates over the last decades, while I still continue to be quite skeptical that a radical change has ever taken place, to the effect of making classical political economy a genuine rival of the neoclassical economic paradigm. The latter, which is characterized by the market-centered way of thinking, continues to be dominating and all-pervasive, so deeply rooted as it is into the general economic culture, well supported by the still alive tool kit of highly questionable concepts like 'production function', 'capital', 'efficiency', 'optimality', 'maximization', 'equilibrium'.

The still prevailing neoclassical economic thought can practically be seen by looking at some important international institutions like, for example, the International Monetary Fund, The World Bank, the European Economic Commission, which have given in the recent past, and are still giving up to now, 'guidelines' to countries and corporate firms as to the 'best' policies to be pursued. In recent years, the prescriptions of most of these policies have obliged several countries in the world to adopt austerity measures, which almost infallibly have led them into deep long-lasting depression and persistent unemployment. Despite evidence, Alesina et al. (2019) still continue to think austerity as a useful measure for a sound macroeconomic governance, in sharp contrast, for example, to Pasinetti (1998) and Nuti (2013), who have instead brought to the fore some foolish and perverse consequences of it.

The 2007–2008 financial crisis, which started in the US and that soon afterwards hit most European countries, had at its theoretical background the dominant neoclassical paradigm, whose specific roots might easily be found in very well-known theorems. The Modigliani-Miller theorem (1958, 1961), despite its highly restrictive and peculiar assumptions, has since then been taken as one of the most important reference points of any practical operation in the financial markets all over the world. Pasinetti (2012, pp. 1445–1446), has singled out several consequences which could spring out from the actions and policies inspired by that theorem. Among these '[a] wild rush toward short term profits and a rapid distribution of dividends', 'a decreasing preoccupation over unemployment', a pathological 'increase in the process of financialisation of our economies'. Much responsibility heavily lies also in the hypothesis of *efficient markets*—originally put forward by Fama (1965) within the Stock Market context—which emphasizes what is supposed to be the most important feature of a market-centered theory, *viz.* the *perfect functionality* of market mechanism. In the simple words by Fama (1970, p. 383), 'A market in which prices always "fully reflect" all available information is called "efficient"'.

It should also be mentioned the 'disastrous' Black-Scholes formula for determining the value of derivatives. It is worth noticing that Scholes and

Merton have been the winners of the Sveriges Riksbank Prize in Economic Sciences in Memory of Alfred Nobel in 1997, whose works—according to the Swedish Academy of Sciences—'generated new financial instruments and [have] facilitated more effective risk management in society'. Quite ironically, less than one year later, in 1998, the Long-Term Capital Management (a hedge fund having both Scholes and Merton as principal shareholders) collapsed—and since it was 'too big to fail', the Federal Reserve had to bail it out.

Davis' brilliant paper very clearly puts forward one of the most characterizing features of Sraffa's work, devoted 'to restore political economic thinking in economics'—a very much needed breakthrough nowadays. Crucial, in this respect, is the open-closed distinction, to which Davis devotes with extreme care and passion much of his attention. It seems to me that its recognition ultimately proves the unavoidable necessity of introducing the historical and political environment of the circumstances, thus making the economic discipline not only *logical consistent*, but above all absolutely *sensible*.

It may be added that the open-closed distinction makes it evident the fact that the economic discipline cannot make any 'progress', in the way the latter is intended in other genuine sciences, like mathematics. Economics—as Hicks (1983, p. 4) once put it—'is more like art or philosophy than science', and thus it can only undertake *change*, and it can only change *spasmodically*. It proves also that no 'hard core' can exist, for the core of an economic paradigm is not unique and it too can change, and only in a magmatic unpredictable way.

## References

Alesina, A., C. Favero, and F. Giavazzi (2019) *Austerity. When It Works and When It Doesn't*, Princeton and Oxford: Princeton University Press.

Fama, E. (1965) "Random Walks in Stock Market Prices," *Financial Analysts Journal* 21: 55–59.

Fama, E. (1970) "Efficient Capital Markets: A Review of Theory and Empirical Work," *Journal of Finance* 25 (May): 383–417.

Hicks, J. R. (1983) *Classics and Moderns. Collected Essays on Economic Theory*, vol. III, Oxford: Basil Blackwell.

Miller, M. H., and F. Modigliani (1961) "Dividend Policy, Growth and the Valuation of Shares," *The Journal of Business* 34 (October): 411–433.

Modigliani, F., and M. H. Miller (1958) "The Cost of Capital, Corporation Finance and the Theory of Investment," *American Economic Review* 48 (June): 261–297.

Nuti, D. M. (2013) "Perverse Fiscal Consolidation," https://dmarionuti.blogspot.com

Pasinetti, L. L. (1998) "The Myth (or Folly) of the 3% Deficit/GDP Maastricht 'Parameter'," *Cambridge Journal of Economics* 22: 103–116.

Pasinetti, L. L. (2012) "A Few Counter-Factual Hypotheses on the Current Economic Crisis," *Cambridge Journal of Economics* 36: 1433–1453.

CHAPTER 6

# Sraffa and Manara: The Mystery of the Last Article of Piero Sraffa

*Yoann Verger*

## 1 Introduction

Piero Sraffa was a very peculiar kind of a scholar. On the one hand, his influence on history of economic thought is immense, as it is acknowledged by one of his most famous critics, Paul A. Samuelson:

> Did any scholar have so great an impact on economic science as Piero Sraffa did in so few writings? One doubts it. And there cannot be many scholars in any field whose greatest works were published exclusively in their second half century of life. Piero Sraffa was much respected and much loved. With each passing year, economists perceive new grounds for admiring his genius. (2008)

On the other hand, he is one of the few modern economists for whom we can say that his most important publications are almost all of his publications. For example, out of the total of 18 "scientific" publications by

Y. Verger (✉)
UMR ECOSYS, INRA, AgroParisTech, Université Paris-Saclay, Thiverval-Grignon, France
e-mail: yoann.verger@uvsq.fr



A. Sinha (ed.), *A Reflection on Sraffa's Revolution in Economic Theory*, Palgrave Studies in the History of Economic Thought,
https://doi.org/10.1007/978-3-030-47206-1_6

my count (either books or articles published in scientific journals), Beaud and Dostaler (1997) list 15 of his publications as "main publications". Hence, Sraffa was not trying to publish much, and he did only when a really important problem was concerning him (he does not even write more than a few unpublished notes about his talks with Wittgenstein!).

In his unpublished papers,[1] two files are titled as unpublished draft articles: a draft from 1923, on the economics of fascism (D3/3), and a draft written around 1969, titled: "Produzione congiunta di merci a mezzo di merci. Riposta a C F Manara" (D3/14). The latter is the one that interests us here.

This unpublished article was supposed to be an answer to Manara's article of 1968, and, if published, it would have been the last published article of Sraffa. As Sraffa was really rarely publishing something, it means that Manara was asking a serious question to Sraffa. Then two questions arise immediately: why was it so important for Sraffa to answer Manara? And why he did not publish his draft?

In Sect. 2, I present Sraffa's draft. In Sect. 3, I answer the first question, I show how this draft helps us understand why it was so important for Sraffa to respond to Manara. In Sect. 4, I answer the second question, why Sraffa finally decided not to publish this draft. In Sect. 5, I present Dupertuis and Sinha's (2009) solution to Manara's problem and show how it would have helped Sraffa to answer Manara satisfactorily.

## 2 Sraffa's Draft

I will now present the file containing the drafts of Sraffa's answer, and then describe its content.

### *2.1 Content of the File D3/14*

The file D3/14 gathers 115 notes (the notes refer sometimes only to one recto page, or to a recto-verso page, or sometimes to a group of pages), which we can order in this way:

[1] They have been published online by Giancarlo de Vivo and Murray Milgate with the collaboration of the Wren archivist Jonathan Smith. They can be found at the following URL: https://janus.lib.cam.ac.uk/db/node.xsp?id=EAD%2FGBR%2F0016%2FSRAFFA.

- D3/14/2–D3/14/6: first typewritten and annotated draft.
- D3/14/7–D3/14/11: second typewritten and annotated draft (the typewritten draft is identical to the first, but the handwritten additions are slightly different).
- D3/14/13–D3/14/20: handwritten draft.
- D3/14/22–D3/14/23: letter from Pasinetti to Sraffa, dated 4th February 1969, where Pasinetti writes that he met Manara in Milan, and presents Manara's problem to Sraffa.
- D3/14/24: letter from Pasinetti to Sraffa, dated 11th December 1969, where Pasinetti writes that he quickly read the note made by Sraffa to answer Manara's problem, and makes a comment about it.
- D3/14/25 (pp. 1–18): printed version of Manara's (1968) article.
- D3/14/26–D3/14/115: handwritten comments and drafts about Manara's problem.

My guess, after reading all the notes, is that the first typewritten drafts (D3/A4/2–D3/14/6 and D3/14/7–D3/14/11) are the final version, and that all other handwritten comments and drafts are earlier to it. This last version was probably written after 11 December 1969, i.e., after the second letter from Pasinetti to Sraffa. In this letter Pasinetti tells Sraffa that he should not use the term "incompatibilità" (incompatibility, which we can find in the handwritten notes, for instance in D3/14/28), but rather the terms "assoluta inferiorità" (absolute inferiority), which convey no mathematical signification while having an economic meaning. In the typewritten drafts, we find these terms, and no mention of the term "incompatibilità".

### 2.2 *Transcription of the Presumed Last Version*[2]

I present below the integral transcription of this presumed final version. The translation (as well as the translation of the other notes presented in this article) was made by a qualified translator, Sylvia Summaria (see the Appendix for the original text). I use [ ] to signal footnotes and handwritten insertions in the margin or above or below the line, and I use {} to indicate my additions:

[2] The file is available at the following URL: http://trin-sites-pub.trin.cam.ac.uk/manuscripts/Sraffa_D3_14/manuscript.php?fullpage=1;startingpage=1.

## JOINT PRODUCTION OF COMMODITIES BY MEANS OF COMMODITIES. IN REPLY TO C.F. MANARA

1—On an article issued by this journal (1) Carlo Felice Manara provided a mathematical analysis of one of the chapters of my book Production of commodities by means of commodities (2), and I am grateful for this.

{*the next sentences are crossed out*}

Since the beginning of his essay, Manara refuses to take any responsibility for the economic content both of the hypothesis and of the examples suggested (3). This would have been an honorable intention if form and content were easy to separate. But it immediately becomes clear that the real core of his criticism derives from his mistake in economic terms.

At first I thought that this issue was patent for any economist, so that I didn't even need to rectify. Nevertheless, I recently realized that the aforementioned article misled many young scholars, hence a reply became imperative.

{*end of the crossed out section*}

2—Before analyzing the main topic, something needs to be clarified. Manara focuses on a chapter whose title is "The Standard system with joint products" (4)

[(1) "Il modello di Piero Sraffa per la produzione congiunta di merci a mezzo di merci", L'industria, 1968, Pag. 3–18.

(2) Edizione Einaudi, 1960; re-printed in 1969 with same layout.

~~(3) Manara, pag. 4.~~

(4) Production of commodities by means of commodities, pag. 59–70. We can construct the Standard system by finding a set of multipliers which, applied to the equations representing the several industries (or methods of production), will result in the quantity of each commodity in the aggregate means of production of the system bearing to the quantity of the same commodity in the aggregate product a ratio which is equal for all commodities. The "Standard product" is the aggregate of the products of the Standard system[, with the same ratios]]

His main argument is: "It doesn't appear that Sraffa has experienced the slightest doubt concerning the possibility of imagining the existence of a Standard product, even though this possibility is not generally verified but must be postulated by means of a suitable hypothesis on the matrices" (pag. 12).

After a two pages inquiry, he states this relevant hypothesis: "the algebraic equation in the unknown $r$ ($r$ is the maximum rate of profit) has at least one real and positive root (5)" (pag. 14).

When writing this, Manara seems to realize that this condition had already been presented in my book, and he tries to get out of trouble with a footnote that it's worth to be entirely quoted:

The terms 'real number' and 'real solution' are used here (by Manara) in the precise technical sense of mathematics and not in the rather vague sense adopted by Sraffa pg. 56. In this section, indeed, so far as I can understand, Sraffa uses the expression 'real solutions' to mean perhaps, 'solutions that have economic meaning' or 'solutions that have a correspondence in reality'. (Manara, footnote pag. 13)

[(5) {*the next sentence is crossed out*} While the rest of this answer to Manara will focus on the need for a real root, the negativity issue will require just a few words. {*end of the crossed out section*} The $r$ non negativity condition had already been presented at the beginning of the book (Sraffa, footnote p. 5); positivity is not necessary for we can have $r = 0$.]

Since Manara didn't consider useful to quote the text of which he gives such an uneasy interpretation, I will: "subject only to the general requirement of the resulting equations being mutually independent and having at least one system of real solutions". (Sraffa, pag. 56). I have no further comments to make.

{*the next sentences are crossed out*}

3—So far Manara, despite his polemical tone, does nothing but confirm what was stated in the chapter he criticizes. One can even ask why once he realized it, he did not erase this redundant passage of the article.

There must be something more. This must be [the belief] that the requirement for a real solution (it doesn't matter if the author asserted it before the critic) is so restrictive that the construction of a Standard product is generally inapplicable to a normal economic system. As a matter of fact, Manara resorts to a numerical example to present the non-applicability of the Standard product. Such a simple and consistent example, once verified would be decisive to confirm this belief.

The example is based on two industries (two methods of production) each of them producing jointly, using the same means of production and the same commodities but in different ratios. Data are presented through two matrices, the former ($A$) for means of production and the latter ($B$) for products:

{*end of the crossed section out*}

$$A = \begin{bmatrix} I & I.I \\ I.I & I \end{bmatrix}$$

$$B = \begin{bmatrix} I.09 & I.144 \\ I.144 & 0.99 \end{bmatrix}$$

Manara proves that the resulting equations have not any real solutions, so the construction of the Standard product would not be possible (pag. 13).

{*the next sentences are crossed out*}

Let us now consider these data from an economic standpoint, comparing the two industries' means of production to the corresponding products. Industries are I and II, commodities are <u>c</u> and <u>d</u>:

<u>Industry I</u>

employs 1 <u>c</u> + 1.1 <u>d</u> produces 1.09 <u>c</u> + 1.144 <u>d</u>

<u>Industry II</u>

employs 1.1 <u>c</u> + 1 <u>d</u> produces 1.144 <u>c</u> + 0.99 <u>d</u>

Put in this way, it's suddenly obvious that the example given by Manara is economically impossible. Industry II is indeed absolutely inferior compared to industry I. In fact, for both the commodity <u>c</u> and <u>d</u>, industry II has an inferior ratio between produced quantity and employed quantity. Namely:

| | IndustryI | | IndustryII |
|---|---|---|---|
| for comodity <u>c</u> | $\frac{1.09}{1}$ | > | $\frac{1.144}{1.1}$ |
| for comodity <u>d</u> | $\frac{1.144}{1.1}$ | > | $\frac{0.99}{1}$ |

{*the table is also crossed out*}

This absolute inferiority prevents the construction of the Standard product.

One can finally add that this inferiority has other prominent consequences. Manara indeed ignored that the same reasons that in his case prevent the construction of the Standard product, would equally prevent the existence of a uniform profit rate in the several industries as well as the existence of a uniform price for each unit of the same commodity.(6)

{*end of the crossed out section*}

[(6) At the end of his article (pag. 18), Manara asks for an explanation. In his opinion: "It seems that he (Sraffa) has not thought it necessary to interpret negative quantities of labour absorbed by industries". The explanation is to be found in the same passage that Manara, to say it in his own words, "didn't seem to have thought it necessary" to read, namely pag. 60–61 of the book [in question]. It is indeed written that: "since no meaning can be attached to the 'negative industries' which such multipliers entail, it becomes impossible to visualize the Standard system as a conceivable rearrangement of the actual process. We must therefore in the case of joint-products be content with the system of abstract equations, transformed by

> appropriate multipliers, without thinking of it as having a bodily existence". As regards the Standard product (which is the only raison d'être of the Standard system), it consists only of commodities and doesn't include labour quantity.]. (Sraffa's unpublished papers, D3/14: 2–6)

## 3 Why Was It So Important for Sraffa to Answer Manara?

Manara's paper is about the impossibility to find a real Standard system from some real systems of production, and about the possibility that negative prices may appear when the wage rate is evolving from zero to one. The answer of Sraffa is focusing on the first problem: the impossibility to construct a Standard system. Hence the reason why it was important for Sraffa to prove that Manara was wrong is related to his proof that one can correctly understand the distribution of income among workers, landowners, and capitalists independently of the complications of demand theory.

### *3.1 Content of Manara's Article*

Manara's article was first published in Italian in 1968, and then republished in English translation in 1980 (Manara 1980, in Pasinetti 1980). This article was widely read and cited (we find in Google Scholar 34 citations for the 1980 article and 46 citations for the original one—05/04/2019) and it was also cited in two influential reviews of the Sraffian literature: in Pasinetti's (2012) article headed "Piero Sraffa and the future of economics" and in Samuelson's (2008) article on "Sraffian economics" published in *The New Palgrave Dictionary of Economics*. So what is its content, and what is at stake concerning its results?

Let us introduce Manara's article by one of the most competent scholars in the field of Sraffian economics, and especially concerning the questions raised by Manara (1968), Bertram Schefold:

> While the first part of Sraffa's 'Production of Commodities by Means of Commodities' received quite considerable attention when the book was first published, the second passed almost unnoticed for some time. The third part on 'Switching of Techniques' stood at the centre of the most interesting controversy on economic theory of the 1960s, but the discussion was again focused on single-product industries. The first exception to this pattern was Manara's article on 'Il modello di Piero Sraffa per la produzione

congiunta di merci a mezzo di merci' in the Italian journal L'industria, which appeared in 1968. The author, apart from giving an elegant mathematical formulation for the distinction between basics and non-basics in the case of joint production,[3] pointed to two of the difficulties of joint-production systems: namely the likelihood of negative prices being associated with some labour vectors at some rates of profit in almost any joint-production system and the nonexistence of the maximum rate of profit and the Standard commodity in certain cases. It was not said whether these were to be considered normal or exceptional. Manara, a mathematician, left the tasks of analysis and interpretation of the 'anomalies' to the economists. Manara failed to note that Sraffa had simply and explicitly assumed that positive prices and a real Standard commodity exist (*PCMC*, sec. 50)—but then Sraffa had failed to indicate under what conditions these assumptions would be fulfilled. (1989, p. 44)

Manara (1968) says that these assumptions would be fulfilled under the condition that, for any system of joint production of commodities by interdependent industries (i.e., with only basic commodities) where **A** is the matrix of inputs per industry and **B** the matrix of outputs per industry (with commodities in columns and industries in rows), a real solution to the following equation can be found for the scalar $R$:

$$\det\left[\mathbf{B}-(1+R)\mathbf{A}\right]=0$$

If a real solution for $R$ cannot be found, then it means that, from this system of joint production, no physical rate of surplus and no Standard

[3] In single production (i.e. each industry only produces one commodity, and each commodity is only produced by one industry), basic commodities are the commodities that are directly or indirectly needed for the production of all commodities of the system. A system with only basic commodities can be said to be a system of interdependent industries: each industry needs all the other ones to ensure its production.

In joint production (one commodity can be produced by different industries, and industries can produce several commodities), basic and non-basic commodities cannot be separated so easily. For instance, a commodity that enters all processes of production can be considered non basic in some cases. Sraffa (1960, § 60) gives a formal (mathematical) definition of the distinction between basic and non-basic commodities that applies for both single and joint production, but he also indicates that, even in the latter case, an economic sense can be given to the distinction: if a tax is imposed on a commodity, and that all prices are impacted by this change then that means that the commodity is basic (1960, § 65).

commodity can be constructed. Furthermore this root must be simple, otherwise the Standard system will not be unique.

In his article, Manara provides two numerical examples, and he shows that, for the second example, a real solution cannot be found for *R*. In the first numerical example, he shows that for some labour vectors, negative prices can appear when the wage rate is different from zero.

### 3.2 *Comments on Sraffa's Draft*

Sraffa's draft can be separated in three parts. First, "before analyzing the main topic", Sraffa answers Manara's claim that he did not take the precaution to state the conditions under which the Standard system exists. He recalls that, in his book, he precisely states that (i) equations must be independent, (ii) a system of real solution must exist, and (iii) that the rate of profits must be non-negative.

Secondly, Sraffa focuses on what is for him the "main topic", i.e., the second numerical example, where no Standard system can be constructed, because no real solution exists for the rate of profits. In this part, Sraffa dismisses Manara's example because it has no economic meaning: in empirical reality, this system could not appear because the second industry is absolutely inferior to the first one and would not be adopted.

Finally, he answers in the last footnote the point made by Manara that the negative multipliers that may appear during the construction of the Standard system in the case of joint production have no economic meaning. He recalls the passage in his book where he states that, in joint production, the Standard system may only be considered as a theoretical construct whose only goal is to construct the Standard product, and then to be able to separate the question of distribution with the question of prices.

Two important remarks can be made. First he does not consider the first numerical example given by Manara, and the remark made by Manara that, in some joint production cases, and for some labour vectors, negative prices will appear when the wage rate is different from zero. In my point of view, this means that Sraffa did not consider this point as important enough to require an answer. What was really important for Sraffa was to have a clear answer to the question of the conditions that are necessary to ensure that a Standard system can be constructed.

Secondly, the main topic for Sraffa is to understand if "the requirement for a real solution {…} is so restrictive that the construction of a Standard product is generally inapplicable to a normal economic system". The strategy used by Sraffa to answer this is to show that in an economically viable system, there will be a real solution for the uniform rate of profits. He then proceeds to define what an economically viable system is. So his strategy is to prove that the requirement for a real solution has an economic meaning.

While we discuss the results of this strategy in the next section, we shall, in the present section, focus on the reason why the impossibility to construct a Standard system was an important matter for Sraffa.

### 3.3 *Consequences of Manara's Findings about Sraffa's Theory*

Why the Standard system is so important for Sraffa? In the last sentence of the draft, he provides the answer: "As regards the Standard product (which is the only raison d'être of the Standard system) …" So it is important for Sraffa that there is a Standard system for every system of production because, if this is true, a Standard product can always be constructed. And why is the Standard product important? Because it is the only numeraire whose value will not change when distribution of surplus changes.

Sraffa (1960, Chap. 4) constructs his imaginary Standard system by applying to the initial system of interdependent industries a set of multipliers, so that the commodities used as inputs are in the same proportions as the commodities produced as outputs, bringing out a Standard surplus and a rate of physical surplus $R$. Sraffa shows that the rate of physical surplus $R$ is also the maximum uniform rate of profits of the initial system. Hence, the construction of an imaginary Standard system helps Sraffa to show that, in a system of interdependent industries, the maximum rate of profits is independent of prices (it is a "physical" property of the system).

Furthermore, by defining a Standard commodity as a composite commodity where the commodities are in the same proportion as in the Standard surplus, i.e., the physical net output of the Standard system, and by using this Standard commodity as the numeraire, Sraffa shows that distribution is also independent of prices (i.e., it can be known prior to the knowledge of prices, once, for instance, the uniform rate of profits is given). Then prices appear to be dependent on the techniques of production and on the distribution of the surplus, the size of the surplus appears to be dependent on the technique of production (which set the maximum uniform rate of profits), whereas distribution of the surplus appears to be decided outside the system: distribution neither depend on prices, nor on techniques of production, nor on demand.

This provides a theory of distribution radically opposed to the neoclassical theory of distribution, which amounts to calculating distribution as optimal prices determined by marginal productivity of factors and preferences of consumers. It re-establishes the perspective that classical economists would have liked to have on this phenomenon, i.e., that distribution depends on external factors, such as political struggle, institutional rules, ethical arguments, etc.[4]

What are the consequences of Manara's findings? The fact that a Standard system cannot be constructed reveals that in some case, distribution may depend on the techniques of production, or on prices, or on demand. For some cases of joint production, it is not possible to calculate a physical rate of surplus and a Standard commodity and, hence, to disconnect distribution theory from price theory. This is precisely the conclusion drawn by Samuelson from Manara's paper:

> Sraffa's models, we have by now seen, tellingly reject the following Ricardian stereotypes: ... {o}ne can correctly understand the distribution of income among workers, landowners, and capitalists independently of the complications of demand theory (consumers' demand functions, marginal utility, revealed preferences, etc.). (2008)

We can find in the drafts some arguments that support this interpretation. First, more than hundred pages of notes in the D3/14 file show that Sraffa spent a lot of time on the subject, trying to express his answer in the best possible way. Second, we find in the file some notes saying that if the Manara's article was right, it would shake the whole economic theory starting with Adam Smith, for example in the following note:

> One might conclude that the [possibility to construct] a Standard product [is under the same restrictive conditions] as any other general theory in an interdependent industries system.
>
> If this is true, M. would have destroyed my humble hut and as a new Samson, he would have torn down the entire palace of modern economics. ~~che Adam Smith in~~

[4] Sraffa also provides direct argument against the neoclassical theory of distribution, such as (i) the impossibility to measure 'capital' independently of the variables that that theory aims to determine; (ii) the possible instability of the equilibrium solution, due to the phenomenon of reverse capital deepening.

> If in the example given by Manara we modify the last element from 0.99d to 1.09d the example becomes possible and already constructed as a Standard system, where the maximum rate of profit can be calculated depending on one or the other commodity, = 2.234-2.1 2.1 = 0.133 2.1 or 6.6 … %. (Sraffa's unpublished papers, D3/14: 61)

Or again in this note:

> If the example [given by M.] were verified, together with my humble building all the milestones of traditional economics would be swept away. (Sraffa's unpublished papers, D3/14: 63)

## 4 Why Sraffa Did Not Publish It?

In this section I try to formulate a plausible conjecture explaining why Sraffa decided not to publish the above draft. Sraffa says that Manara is wrong because his example shows "absolute inferiority" of one process regarding the other one. So his method to dismiss Manara was to find an economic significance of the fact that, sometimes, a Standard system cannot be constructed, rather than argue on the mathematics of Manara. But he perhaps did not find his answer convincing enough, both from the mathematical as well as the economical point of views.

### 4.1 *On the Mathematical Point of View*

We have a first hint that he was not satisfied by his answer in some notes in the file D3/14, for instance here:

> Introduction
>
> 1 S. has no doubt
> 2 proves that it's not the case
> 3 determines that it is "rational"
> 4 that is exactly what I had said
> 5 does he realize it? Funny passage: I quote again without adding any comment
>
> Denied Generality: false.
> but the example would be interesting.

> While it is not proven (I never said that it is always verified but that it is generally verified) Manara fails: his mistake does not prove anything and does not affect generality: on the contrary, the assumption is strengthened.
> M.'s failure strengthens the assumption that in those cases economically possible it is general. But it would be interesting if someone more patient and luckier than M. found it.
> Please note that the example given by M., not only was excluded from the real solutions condition that I had explicitly stated: it would have also been excluded from the elementary economic condition of convenience. It would be interesting if M. or someone more patient or luckier, found an example of a basic products system economically possible, that doesn't have any real solution for the Standard commodity. (Sraffa's unpublished papers, D3/14: 72–74)

Sraffa thinks that his explanation does not prove in a general way that Manara is wrong, and that if an example were to be found where no industry shows "absolute inferiority" and no Standard system can be constructed, then it would be a bad hit for his theory.

Besides Pasinetti, it is possible that Sraffa talked about this matter with Schefold, who was in Cambridge at that time.[5] But I directly asked Bertram Schefold about it and he replied to me that Sraffa never told him about his draft; however, he did work on Manara's paper, possibly after a discussion with Sraffa:

> Dear Colleague,
>
> Thank you for your letter of July 6. {…}
>
> According to your investigation, Pasinetti presented the Manara problem to Sraffa on February 4, 1969. You also state that Sraffa showed his note on Manara to Pasinetti on December 11. This fits in with my recollection. I distinctly remember how Pasinetti confronted me with Manara's article—I think in Sraffa's presence—, and I believe that it was early in spring 1970.
>
> {…}
>
> I suppose that I was confronted with the Manara example early in 1970, because Sraffa wanted to test his idea on some younger researcher. He did the same, when he had written PCMC, and showed the draft to Amartya

[5] "My PhD thesis 'Mr Sraffa On Joint Production' … was written in 1969/70 when I was first a Visitor to the Faculty of Economics, then an Advanced Student at King's College, Cambridge" (1989, p. vii).

Sen prior to publication, as you may know. In the course of the year 1970, I would discuss these matters further with Sraffa. He never told me that he had been trying to write an article about the matter. So he could simply abandon his attempt when he had understood that his main argument was not convincing. (personal communication, July 11th, 2018)

In his PhD thesis, Schefold uses exactly the same argument previously exposed by Sraffa in his draft to explain that, in a case of "absolute inferiority" of one industry (Schefold uses the term "imbalance"), as in the second numerical example of Manara, no Standard system can be constructed:

Consider (it makes sense for $n = 2$) systems with $\mathbf{A} > \mathbf{O}$.

$$g_i^j = \frac{b_i^j}{a_i^j}(i, j = 1,2)$$

is the rate of reproduction of good $j$ in process $i$.

We define: a system involving two goods and two processes is imbalanced if one process is better than the other for both goods in that

$$\left(g_i^1,g_i^2\right) > \left(g_j^1,g_j^2\right) \text{ for } (i,j) = (1,2) \text{ or } (2,1)$$

The example proposed by Professor Manara is inefficient in this sense.

*Theorem*

If the inefficiency above (imbalance) is ruled out, a positive Standard ratio $R$ {...} exist{*s*}. (1989, p. 146)

But Schefold also goes a bit further, by defining the condition where, in joint production systems where the number of industries is more than two, a physical rate of surplus and a Standard commodity can be constructed (1989, pp. 96–99). This condition is that, for all $a_j^i$ and $b_j^i$, we should observe (possibly after having made the right substitutions of rows and columns in matrices $\mathbf{A}$ and $\mathbf{B}$);

$$0 \le \frac{b_i^j}{a_i^j} < 1 + R < \frac{b_j^j}{a_j^j} \le \infty$$

Schefold, by assuming the above condition, can then also avoid cases such as the one presented in the first numerical example of Manara. He also shows that, if there are more than two joint production industries, the simple argument of "absolute inferiority" does not work anymore, and the above requirement should be used instead. Hence, he could construct cases of joint production where there is no "absolute inferiority" while also being a case where no Standard system can be constructed. The argument developed by Sraffa in his draft is refuted.

### 4.2 On the Economic Point of View

Schefold concludes his development on this above condition by saying:

> The main insight obtained in this section is the one derived from the von Neumann model: systems without a semi-positive Standard commodity {…} suffer from specific inefficiencies revealed at $r = R$. (1989, p. 98)

Hence, as Sraffa, he explains the fact that in some cases of joint-production, a Standard system may not be constructed because of the fact that some industries are less efficient than others.

Unfortunately for Sraffa's theory, this above condition is rather strong ("slightly more stringent", as it is acknowledged by Schefold 1989, p. 97): for instance an industry which does not use one of its outputs as input should be ruled out, and so independently of the input-output coefficients of the other industries. Hence, this condition seems to convey no real economic significance, as it eliminates a large number of normal economic systems.

This is why the demonstration by Schefold of the requirement for the existence of a Standard system when the number of industries are more than two is problematic: first, it is more a mathematical requirement than an economic one; second, the requirement for a real solution is so restrictive that the construction of a Standard product is inapplicable for a large number of normal economic systems.

It is then possible that Sraffa, discovering this development from Schefold, finally decided not to publish his draft. The part where he exposes that it is because of the "absolute inferiority" that Manara's example does not work is totally crossed out in the draft above. This can be explained by the fact that he found out from Schefold's work that the case where no Standard system can be constructed does not necessarily require the condition of "absolute inferiority". However, as we will see in the next section, it is possible to find an economic justification, based on the notion of mutual dependence of industries, to dismiss such cases. I believe this justification would have been preferred by Sraffa, rather than a purely mathematical one which seems to engender a restriction too important on the number of economic systems that can have a Standard system (on the objective approach of Sraffa, see Sinha 2012).

## 5 Dupertuis and Sinha's Solution to Manara's Problem

Sraffa found it really important to answer Manara on the precise point that in some cases a Standard system cannot be constructed. So the Standard system was really important for Sraffa. And he did not find his answer on the ground of "absolute inferiority" convincing enough.

Actually, Dupertuis and Sinha (2009) have found a good way to answer Manara both on the mathematical as well as the economical levels: they show that the concept of "basic goods" as defined by Sraffa for the case of joint-production is not adequate, and that a focus on the notion of "interdependence" helps to reconstruct the equivalent of basic goods for joint production systems. Then they provide a method to find in any real systems the "core" with only "basic interdependent goods" from which a Standard system can always be constructed.

But in order to understand why they provide a solution to Manara's problem, we must first go back to the goal of Sraffa and the way he achieves it in his book, *Production of Commodities by Means of Commodities* (1960).

### 5.1 To Prove the Existence of a Maximum Rate of Profits

Sraffa's goal was to undermine the dominant economic theory. This goal was achieved in Sraffa (1960) by showing that the rate of surplus has an objective signification, a "physical" meaning, so that it is not necessary to enter into the "complications of demand theory (consumers' demand functions, marginal utility, revealed preferences, etc.)" to define a surplus and analyse its distribution.

If the system is reproducing itself, the surplus cannot exceed a certain amount, and then a maximum rate of surplus can be defined, which is equal to the maximum rate of profits. It is easy to calculate it without any notion of prices if we consider a unique process (this is what Sraffa does in the introduction to *The Works and Correspondence of David Ricardo*, 1951–1973, with the example of a corn industry). But Sraffa shows that it is also possible to calculate it in the case of multiple commodities, if the processes of production are interdependent.

Interdependence is the key notion, or rather "mutual dependence":

> It is desirable at this stage to explain why the ratios which satisfy the conditions of production have been called 'values' or 'prices' rather than, as might be thought more appropriate, 'costs of production'.
>
> The latter description would be adequate so far as *non*-basic products were concerned, since, as it follows from what we have seen in the preceding section, their exchange ratio is merely a reflection of what must be paid for means of production, labour and profits in order to produce them—there is no mutual dependence.
>
> But for a basic product there is another aspect to be considered. Its exchange-ratio depends as much on the *use* that is made of it in the production of other basic commodities as on the extent to which those commodities enter its own production. (1960, § 7)

Interdependence can be defined in this way: two industries, or processes of production, are interdependent, if each industry needs, directly or indirectly, the output of the other to achieve its reproduction.[6] When two industries are interdependent, it is possible to calculate the unique maximum rate of profits without knowing the prices, thanks to the Standard system. This is so because the rate of profits in each industry depends on the rate of profits decided in the other industry: all rates of profits are interdependent.

Outside interdependence, relation between industries falls into two categories. First, two industries can be independent. Then a unique maximum rate of profits cannot be defined because there is no reason why the maximum rate of profits should be the same in the two industries. Second, two industries can be in a relation of dependence (one industry needs the output of the other to produce—let us call this industry the dependent industry—while the other does not). Then in order to calculate the rate of

[6] In the same way that the maximum rate of profits can be defined in terms of reproduction, as the maximum rate of surplus that allows the reproduction of the system, interdependence can be defined as the relation of mutual dependence between industries so that they are able to perform reproduction.

profits of the dependent industry, one needs to know the price of the input coming from the other industry and therefore the notion of surplus is not anymore independent from the prices.

This is why Sraffa, in his book, provides a method to eliminate these two cases: first he assumes that in the system under consideration, there is at least one basic good, so that the first case disappears. Second, he provides a method to eliminate the non-basic goods during the construction of the Standard system. In this way, the second case is also eliminated. However, while this works perfectly fine for single production, it does not do the job for joint production. The two numerical examples of Manara show that Sraffa's distinction between basic and non-basic goods and his method to eliminate non-basic goods are not sufficient to eliminate all cases where there is no dependence. This is why a maximum rate of profits cannot be constructed in Manara's second example, and why negative prices appear as soon as the wage rate is different from zero in Manara's first example.

### 5.2 *The New Method of Dupertuis and Sinha*

Dupertuis and Sinha (2009) show that there is no interdependence between all industries in Manara's examples. Remember the second numerical example (the processes are in column):

$$A = \begin{bmatrix} 1 & 1.1 \\ 1.1 & 1 \end{bmatrix}$$

$$B = \begin{bmatrix} 1.09 & 1.144 \\ 1.144 & 0.99 \end{bmatrix}$$

We can see clearly that the first process does not need the second to reproduce itself: it is self-reproducing (while the second process is not, as only 0.99 of the second commodity is produced whereas one unit is required as input). We can then limit the analysis at the first process, the second being a "non-basic" process.

And here is the first numerical example of Manara (processes are again in column):

$$A = \begin{bmatrix} 1 & 2 & 1 \\ 2 & 1 & 3 \\ 1 & 2 & 2 \end{bmatrix}$$

$$B = \begin{bmatrix} 2.9 & 1.2 & 1.9 \\ 1.2 & 2.9 & 3.9 \\ 0.1 & 1.2 & 3.9 \end{bmatrix}$$

Here, it is the third process which is self-reproducing. Thus there is no need to take into account the two others to search for rate of physical surplus of this system.

The method provided by Sraffa to eliminate cases of dependence and independence between industries is not sufficient for joint production. Therefore, Dupertuis and Sinha (2009) provide a new method to eliminate these cases, even in the most complicated cases of joint production. They begin by defining a system as "the exchanges the various sectors or processes must make among themselves to be able to reproduce themselves" (2009). Then they apply a mathematical procedure, called "transformation of goods and processes", in order to highlight what they call "superfluous" goods and processes and eliminate them.

In this way, they can reduce any given system of production to the set of interdependent processes that constitutes its heart (what they call an "atomic system"), and that gives the key to the maximum rate of surplus:

> In brief, the procedure of transforming goods amounts to repackaging the goods such that we get one full column on the input side with only zero entries. Once we get that we suppress that column and go on to see if further repackaging would lead to finding another column of inputs with only zero entries. If we find that, we suppress that column as well. We continue this procedure till we find that we can no longer get a column of inputs with only zero entries. Then we apply the procedure of transforming the processes. In this case we multiply all the outputs entries by the same factor. This does not change the equality of rates of profit. As we decrease the factor, it may turn out that a group of processes is globally consuming more of each good than it is producing. If it is the case, we suppress those processes. We go on suppressing such unproductive processes till we find that no such processes are left. After that we again go back to see if further transformation of goods is possible and subsequently for the transformation of pro-

cesses till we arrive at a combination of goods and processes where no more transformation is possible. This system will always be square and we call it an *atomic system*. (2009)

By proving that from all atomic systems a Standard system can be constructed, they finally show that in every system of production, if the industries are not independent (i.e., there is at least one basic good), a Standard system can be constructed, and thus a maximum rate of profits can be calculated without knowing the prices.

## 6 Conclusion

Focusing on the notion of interdependence, Dupertuis and Sinha (2009) show that in Manara's examples, a Standard system cannot be constructed because all the industries are not interdependent. Once the dependent industries are eliminated, a Standard system can be constructed and a maximum rate of profits can be found without knowing the prices. Thus an economic reason to dismiss Manara's argument that in some cases of joint production it is not possible to construct a Standard system is found.

This means that Sraffa's theory is applicable to all system of production, even in the most intricate cases of joint production, as soon as the interdependent industries that form the heart of the system of production are identified. Thus distribution can be known prior to prices, and this theory appears to be a radical contestant of the neoclassical distribution theory. What would have been history if Sraffa had found this way to dismiss Manara's argument? Surely it would have been much harder for neoclassical theorists such as Samuelson to argue that Sraffa's theory is just a special case of the neoclassical theory.

Of course, it may be that my conjecture is wrong, and that Sraffa, after some attempts to respond, "became convinced that it was not necessary",[7]

[7] Responding to an email in which I asked him why Sraffa did not publish the above draft, Pasinetti responded:

Dear Mr. Verger,

Thank you for contacting me.

After almost 60 years, I do not recall why Piero Sraffa decided not to publish a reply to Manara's 1968 article.

My presumption is that in the end he became convinced that it was not necessary.

My best wishes for your research work.

Sincerely yours,

Luigi Pasinetti. (personal communication, October 5th, 2017)

i.e., that the problem was not important. But I do not think so. Looking at the notes in the D3/14 file, one can see the urgent need of Sraffa to try to save his life's work. And one can only wonder how broken was his heart when he finally crossed out the last version of what would have been his last article.

## Appendix

Here is the transcription of the drafts and notes that have been translated in this article.

Sraffa's unpublished papers, D3/14: 2–6:

PRODUZIONE CONG[U]INTA DE MERCI A MEZZO DI MERCI. RIPOSTA A C.F. MANARA

1—In un articolo pubblicato su questa rivista (1) Carlo Felice Manara ha presentato un'analisi in termini matematici di un capitolo del mio libro <u>Produzione di merci a mezzo di merci</u> (2) e di questo gli sono grato.

[a capo] {*the next sentences are crossed*} Fin dall'inizio del suo scritto il Manara ha voluto esimersi da ogni responsabilità per il contenuto economico delle ipotesi o esempi che egli propone (3). Questa sarebbe stata una felice ispirazione se forma e contenuto potessero così facilmente essere separati. Ma subito diventa chiaro che la critica che egli fa è appunto basata su di un suo sbaglio di natura economica.

[a capo] Mi era sembrato che la cosa dovesse essere evidente ad ogni economista così da rendere superflua una rettifica. Viceversa, ho constatato recentemente che l'articolo in questione ha messo fuori strada alcuni giovani studiosi; e di qui è sorta la necessità di una riposta. {*end of the crossed section*}

2—Prima di venire alla questione essenziale c'è un punto da chiarire. Il capit~~a~~[o]~~le~~[o] al quale il Manara dedica la sua analisi è intitolato: "Il sistema tipo con prodotti congiunti".(4)

[(1) "Il modello di Piero Sraffa per la produzione congiunta di merci a mezzo di merci", L'industria, 1968, Pag. 3–18.

(2) Edizione Einaudi, 1960; ristampa 1969 con la stessa paginatura.

~~(3) Manara, pag. 4.~~

(4) <u>Produzione di merci a mezzo di merc[i]</u>, pagg. 59–70. Il "systema tipo è una costruzione che si [o]ttiene applicando alle equazioni che rappresentano le varie industrie (o metodi di produzione) moltiplicatori tali che l'in~~i~~sieme dei prodotti risulti composto delle stesse merci che compongono l'insieme dei mezzi di produzione imiegati; e che le propozioni delle merci

fra i prodotti risultino uguali alle loro propozioni fra i mezzi produzione. Il "prodotto tipo" è l'insieme dei prodotti del sistema tipo[, e in quelle proporzioni.]

L'obbiezione principale che egli fa è questa: "non par[/]eche" Sraffa sia stato sfiorato dal minimo dubbio sulla possibilità di immaginare l'esistenza di un prodotto tipo, mentre invece tale possibilità non si verifica in generale ma deve essere postulata mediante una opportuna ipotesi sulla costituzione della matrici" (pag. 12).

[di seguito] La necessaria ipotesi che il Manara arriva ad enunciare dopo due altre pagine di elaborazione è la seguente: "la equazione algebrica nella incognita $r$ [$r$ é il saggio massimo del profitto] ammette almeno una radice reale e positiva (5)" (pag. 14).

[a capo] A questo punto il Manara deve essersi accorto che tale condizione era già stata posta nel mio libro e cerca di cavarsi ~~di~~ [d']imp[a]ccio con una nota che val la pena di riprodurre integralmente:

[a capo] "Il termine 'numero reale' e 'soluzione reale' è usato qui [cioè dal Manara] nel senso tecnico preciso della matematica e non nel senso un po' vago in cui è usato in Sraffa pag. 56; in questa ultima paginal, invero, per quanto è dato di poter capire, l'Autore usa l'espressione 'soluzion[i] reali' per dire, forse, 'soluzioni che abbiano significato economico' ovvero 'che abbiano una rispondenza nella realà.['"] (Manara pag. 13 nota).

[(5) {*the next sentence is crossed*} Mentre l'esigenza di una radice reale occuperà il resto di questa riposta, la questione della negatività o no richiede solo poche parole. Ifatti {*end of the crossed section*} [L]a condizione della non-negatività di $r$ era stata messa per tutta la trattazione sin dal principio del libro (Sraffa, p. 6 nota); la positività non è necessaria poichè può essere $r = 0$.]

[a capo] Poichè il Manra non ha rituneto utile di riportare il testo d[el] [quale] egli d[à] questa travagliata interpretazione, lo riporterò io: "salvo solo la condizione generale che le equazioni risultanti siano fra di loro indipendenti e che abbiano almeno un sistem di soluzioni reali". (Sraffa, pag. 56). Non occorre altro commento.

3—{*the next sentences are crossed*} Fin qui il Manara, nonostante il suo tono polemico, in sostanza non fa altro che confermare quanto era stato detto nell'opera che egli critica; e ci si potrebbe chiedere perchè una volta accortosene non abbia ritirato questa parte, divuneta a dir poco superflua, del sua articolo.

[a capo] Ma evidentemente c'è qualcosa di più. E questo qualcosa à la [convinzione che la] esigenza di una soluzione reale (e qui non importa se forumlata dall'autore prima che dal critico) sia talmente restritiva da rendere la costruzione del prodotto tipo generalmente inapplicabile ad un sistema economic normale. Infatti il[/]Manarar per illustrare la inaplicabilità del

prodotto tipo presentta un esempio numerico così semplice e calzante che, se valido, sarebbe decisivo per confermare tale convinzione.

[a capo] L'esempio è basato su due industrie (cioè due metodi di produzione) ciascuna delle quali produce congiuntamente[,]ed usa come mezzi di produzione, le stesse due merci [ma] in proporzioni diverse. I dati sono presentati in forma di due matrici, la prima ($A$) per i mezzi di produzione e la seconda ($B$) per i prodotti: {*end of the crossed section*}

[in tutti i casi per I leggi 1]

$$A = \begin{bmatrix} I & I.I \\ I.I & I \end{bmatrix}$$

$$B = \begin{bmatrix} I.09 & I.144 \\ I.144 & 0.99 \end{bmatrix}$$

Il Manara dimostra che le risultanti equazioni non hanno soluzioni reali, e quindi non consentono la costruzione del prodotto tipo (pag. 13).

[a capo] {*the next sentences are crossed*} Ma guardiamo ora e questi stessi dati dal punto di vista economico, confrontando cioè i mezzi di produziono delle due industrie con i rispettivi prodotti. Chiamando le industrie I e II e le merci <u>*c*</u> e <u>*d*</u> abbiamo:

[centrare la tabella] Industria I

[centrare il segno + ] impiega 1 <u>*c*</u> +1.1 <u>*d*</u> produce 1.09 <u>*c*</u> +1.144 <u>*d*</u> [spaziare ingualmente a destra e a sinistra di "produce]

Industria II

[centrare il segno + ] impiega 1.1 <u>*c*</u> +1 <u>*d*</u> produce 1.144 <u>*c*</u> +0.99 <u>*d*</u> [spaziare ingualmente a destra e a sinistra di "produce]

Messa la cosa in questi termini salta subito agli occhi che [l'essempio] del Manara è economicamente impossibile. Infatti la industria II è assolutamente inferiore all'industria I, poichè sia per la merce <u>*c*</u> che per la merce <u>*d*</u> l'industria II dà un rapporto fra quantità prodotta e quantità impiegata inferiore a quello dell'industria I. E precisamente:

[centrare la tabella]

| | | IndustriaI | | | IndustriaII |
|---|---|---|---|---|---|
| [menospazioaiduelatidi >] | perlamerce$\underline{c}$ | $\frac{1.09}{1}$ | > | [menospazio] | $\frac{1.144}{1.1}$ |
| [menospazioaiduelatidi >] | perlamerce$\underline{d}$ | $\frac{1.144}{1.1}$ | > | [menospazio] | $\frac{0.99}{1}$ |

{*the table is also crossed*}

E' questa assoluta inferiorità di un'industria rispetto all[ʻ] altra che rende impossibilie in questo caso di costruire il prodotto tipo.

[a capo] Si può aggiungere, per finire, che questa stessa assoluta inferiorità ha ben altre conseguenze. Infatti, cosa che è stata ignorato dal Manara, le stesse ragioni che rendono nel suo caso impossibilie la costruzione del prodotto tipo renderebbero impossibile anche l'esistenza di un saggio del profitto uniforme nelle varie industrie e di un prezzo uniforme per tutte le unità di una stessa merce.(6) {*end of the crossed section*}

[(6) Alla fine del suo articolo (pag. 18), il Manara chiede una spiegazione poichè, egli dice, lo Sraffa "non pare si sia preoccupato di dare una interpretazione a quantità di laoro negative assorbite da industrie". La spegazione si trova in un passo che il Manara, per usare la sua propia frase, "non pare si sia preoccupato" di leggere, e precisamente a pag. 60–61 del libro [in questione] dove si dice: "non potendosi dare un significato quasiasi alle 'industrie negative' che tali moltipicatori negativi porterebbero con se', diventa impossiblie rappresentarsi il sistema tipo come un concepibilie riproporzionamento delle industrie reali. Dobbiamo quindi, nel caso di prodotti conguiunti, accontentarci del sistema di equazioni astratte, trasformato da opportuni moltiplicatori, senza tentare di immaginare che esso abbia une esistenza concreta". Quanto al prodotto tipo (che è la sola ragion d'eessere del sistema tipo), esso è composto esclusivamente di merci e non comprende quantità di lavoro.]

Sraffa's unpublished papers, D3/14: 61:

~~Si può concludere che la generalità della costruzione di un prodotto tipo, e con non e diversa da quella che~~ [non è soggesta a condizioni più restrittive di quella] di qual~~unque~~[siasi] altra teoria generale di un sistema di industrie interdipendenti[.] ~~la capace di mantenersi~~

Se questo fosse vero il M. non solo avrebbe ~~demolito~~ [schiacciato] la mia povera [~~modesta~~] capanne ma, nuovo Sansone, avrebbe fatto crollare tutto i édificio ~~colonnato~~dell' economia moderna. ~~che Adam Smith in~~

Se nell'esempio del Manara cambiano l'ultimo membro da 0.99d a 1.09d il caso diventa possibile: e il sistema si trove già formato in sistema tipo, con il saggio massimo del profitto che può essere calcolato in termin della une o dell'altro merce, = 2.234-2.1 2.1 = 0.133 2.1 o 6.6 … %.

Sraffa's unpublished papers, D3/14: 63:

Se l'esempio fosse valido con la mia poverite costruzione crollenbbero anche i pilaobri della economia tradizionali

Sraffa's unpublished papers, D3/14: 72–74:

Introduz—magistrale

1. S. Non è sfiorato da dubbio
2. prova che no
3. conclude che "razionale"
4. avevo detto proprio quello
5. se ne avvede? nota spiritosa cito di nuvo scuza commenti

generalità negata:
non vero.
Ma esempio serabbe interessante
Mentre non è provato (no ha mai detto che sia scupre ma solo in generale) M. fallisce: il suo errore non prova niente, ~~sua rafforza la~~ così che la generalità novi è scossa: caso mai la presunzione è rafforzata
Il fallimento di M. rinforze la presuzione che nei casi economicamente possiblii ~~non c'è~~ è generale Ma sarebbe interessanta se qualcuno che maggio pazienza e fortuna del M. lo trovarse
Si noti che
~~L'esempio del M., non solo era stato esclusivo dall condiz. di soluzioni reali che avevo fatti explicitamente; ma sarebbe stato esclusi dalla condizione prelim~~[economica] elementare ~~che il sistema~~ [della

convenienza].~~che si tratti di un sistema capace vitale capace di esistere o di montenersi in vita.~~ Sarebbe interessante se il manara o ~~qualcuno~~ [più paziente] ~~meno~~ meno sfortunato di lui trovarse un esempio di un sistema di prodotti base [in sè] economicamente ~~accettabile~~ [possibile], che non avesse soluzione raz. per~~il sistema~~ la merce tipo.

## References


Abraham-Frois, G., Berrebi, E.: *Théorie de la valeur, des prix et de l'accumulation*. Paris: Economica 1976.

Beaud, Michel and Dostaler, Gilles, Economic Thought Since Keynes, A History and Dictionary of Major Economists (London: Routledge, 1997).

Manara, C. F., "Sraffa's Model for the Joint Production of Commodities by Means of Commodities", Macmillan (1980), pp. 1–15.

Pasinetti, L. L., ed., Essays on the theory of joint production (The Macmillan Press LTD, 1980).

Pasinetti, Luigi L., "Piero Sraffa and the future of economics", Cambridge Journal of Economics 36 (2012), pp. 1303–1314.

Ricardo, David, The Works and Correspondence of David Ricardo (Cambridge: Cambridge University Press, 1951–1973).

Samuelson, Paul A., "Sraffian economics", in Durlauf, Steven N. and Blume, Lawrence E., ed., The New Palgrave Dictionary of Economics (Basingstoke: Palgrave Macmillan, 2008).

Sinha, Ajit, "Listen to Sraffa's silences: an new interpretation of Sraffa's Production of Commodities", Cambridge Journal of Economics 36 (2012), pp. 1323–1339.

Sinha, Ajit.: A Revolution in Economic Theory. The Economics of Piero Sraffa, London: Palgrave Macmillan, 2016.

Sraffa, Piero, Production of Commodities by Means of Commodities (Bombay: K. K. Vora, Vora & Co, 1960).

Comment

# Why Sraffa was Right not to Publish his Last Article. A Rectification of Yoann Verger

*Bertram Schefold*

Yoann Verger (2021) needs to be congratulated on the discovery of Sraffa's unpublished paper on Manara. This draft is very interesting, since it concerns the beginnings of one of the major debates, namely on the viability of Sraffa's theory of joint production—a debate which has died down without all questions having been resolved. Verger's initiative might help to rekindle it. The matter is of special concern for me, since it provided the starting point for my academic career.

Manara's paper (Manara 1968) provided the first attempt to deal with Sraffa's theory of joint production in mathematical terms. He first gave conditions for joint production systems to result in semi-positive prices, but the conditions he proposed were little more than a restatement of the requirement that semi-positive prices had to result at some positive rate of profit. The formulation of economically meaningful conditions that would secure this outcome came later. Manara understood that such conditions could be derived in the single product case, using the well-known theories by Perron and Frobenius. They ensured essentially meaningful solutions, as soon as the system was capable of self-reproduction and contained a basic commodity, but the definition of basic commodities could not easily be extended to joint production, and the possibility of self-reproduction did not suffice to exclude negative prices. The last part of Manara's article showed how one of the formal definitions of basic commodities—the one which Sraffa had given in a footnote—could be used to isolate the basic commodities in this system by rearranging it and by applying multipliers. The non-basic processes could be separated from the basic system in such a way that a tax on basics would affect the prices of basics and non-basics, while a tax on non-basics would not affect the prices of basics. This was clarifying, but not really innovative.

Manara made the main point of his paper in its middle part by showing that a maximum rate of profit need not exist in a joint production system, in that the corresponding characteristic equation did not necessarily have a real, let alone positive, solution. This was surprising from the economic point of view, since Manara's joint production system did have a definite physical surplus, and one had been accustomed to think, knowing the properties of single product systems, that the maximum rate of profit was the appropriate economic measure of the magnitude of the surplus. If there was no maximum rate of profit, there was no standard commodity either, and, with this, the specificity of Sraffa's approach in the attempt to revive classical economics seemed to have gone. This explains why Sraffa felt so much concerned about Manara's objection.

In his paper, Yoann Verger draws attention to the fact that there is a draft of a paper in the Sraffa archive (Sraffa Archive D3/14, listed by Verger), in which the author, now past his 70th year, tried to answer Manara's challenge. This was especially remarkable since Sraffa had published his book almost ten years earlier. A number of reviews and papers had appeared about it, but in most cases he would not answer his critics. He must have felt—and indeed he stated it in his draft—that this time a central point of his theory was at stake.

Verger's paper provides the chronology of the events and explains the details of what is to be found in the archive. It is clear from Verger's reproduction of Sraffa's draft that, while he was writing it, Sraffa was convinced that Manara's example was economically not legitimate. One of the processes seemed technically to be superior to the other so that they could not coexist for economic reasons. But Sraffa did not publish the paper in the end, although it seemed nearly complete. Verger speculates that Sraffa abandoned his draft when he read in my "Mr Sraffa on Joint Production" (Schefold 1971) that the question of technical superiority could be analysed by means of rates of reproduction of the commodities, as Sraffa himself had done in his draft, but the sufficient condition for guaranteeing the existence of a standard commodity on the basis of rates of reproduction was so restrictive that Sraffa, according to Verger's conjecture, decided not to pursue the approach further. Verger believes that Sraffa's approach can be rescued, however, if one extends the analysis of basics and non-basics in Sraffa systems according to the proposal by Dupertuis and Sinha (2009).

If non-basics are eliminated and, moreover, pathologies such as processes which are technically superior to all other processes, are excluded from a Sraffa system, it can be reduced to a so-called "atomic" core of processes and commodities that are all mutually interdependent, and then the standard commodity exists—at least according to the paper by Dupertuis and Sinha.

Verger thus has made a valiant attempt to rescue Sraffa's theory with as small a modification as possible, but it fails. The proof of the existence of the "atomic" core with the required properties can be falsified by an example. Moreover, the "atomic" core is not necessarily unique. Verger does not even manage to explain clearly why Manara's example was, despite Sraffa's attempt to criticise it, economically significant. To get to the truth in this matter, one must start the story elsewhere.

1. I convinced Sraffa and Pasinetti that Manara's example could be economically legitimate, provided it was interpreted in the right context, that is, taking appropriate costs of labour into account. This is what caused Sraffa to abandon his attempt; it will be shown below.
2. It is true that the conditions, which ensure the existence of a standard commodity are restrictive, if there is joint production and if the only operation to be performed on the system consists in the elimination of non-basics. The conditions on rates of reproduction referred to above has this restrictive character. But there is a by now vast literature on the theory of joint production, mainly written in the1980s, showing that, by means of truncation and dual truncation, one can arrive at square Sraffa systems, from which negative prices, overproduced goods and inefficient processes have been eliminated so that, if only the system is capable of producing a surplus, one can, by gradually raising the rate of profit from zero, follow the wage curve of the system, which is then an envelope of individual wage curves of individual truncations succeeding each other as minimizing the cost of production, until eventually the wage is zero and a maximum rate of profit is reached. The only important restriction is that the truncations themselves have to be regular. Regularity, first defined in Schefold (1971, p. 20), is not a restrictive property, since it is generic.
3. However, it is a question whether the standard commodity thus found still serves the purposes it serves in the case of single product systems.
4. There is no room here to analyse fully whether the approach by Dupertuis and Sinha goes beyond the more familiar analysis of trun-

**Table 6.1** Manara's example, with labour made explicit

| | Pikes | Carps | Labour | | Pikes | Carps |
|---|---|---|---|---|---|---|
| Breeding | 1 | 1.1 | $x$ | → | 1.09 | 1.144 |
| Wild | 1.1 | 1 | $1 - x$ | → | 1.144 | 0.99 |

cations of regular systems; they achieve rather less, it seems to me. At any rate, I here draw attention to the theory of intensive rent. An example was given already in 1971 for which the standard commodity does not exist, and yet the basic processes *are* interdependent; we shall return to it at the end of this paper, after having discussed the other objections.

We first take up Manara's example. It is represented in Table 6.1. As far as the numbers are concerned, it is as in the original publication, except that, like Verger, we preserve Manara's ordering of the processes, but exhibit the industries on the rows and the commodities on the columns.

However, I here present an interpretation of the given numbers, which renders the economic meaning more transparent. There is fish production, of pikes and carps. It takes place in an artificial tank (first process) and in the "wild", in a lake (second process). The fish are periodically harvested. The input matrix shows with which quantities the reproduction begins, after the captured fish have been taken out, the output matrix shows the stocks immediately prior to the harvest. In the "wild", the pikes eat carps uncontrolled; they often catch the young so that the reproduction of the carps is slowed down, and taking the fishing activity into account, it is even necessary to replenish the stock of carps periodically by one per cent (compare the second input to the second output in the second process). In the tank, the pikes are kept apart and fed by means of labour. Hence they multiply more slowly and the carps more rapidly. Additional labour is required, apart from the fishing activity. The rates of reproduction thus become plausible. The system functions economically, because two processes are needed *to adapt output to demand*. The net outputs for the two kinds of fish differ somewhat in Manara's example, taken at unit activity levels, but net outputs could easily be changed within a certain range that is limited, as always with joint production (net output of pikes is 0.134 and of carps 0.034 at unit activity levels). The example is also economically meaningful *from the point of view of cost*. In terms of reproduction of *both* kinds of fish, breeding is more efficient than letting the fish grow in the "wild", but breeding requires more work, and it is easy to see, with appropriate values chosen for $x$, that the prices of both goods will be positive at some moderate level of the rate of profit. We choose $x = 0.9$.

It really was Manara's fault to forget labour in his analysis of the example. One might try to defend the omission by arguing that labour could be implicit in the means of production as in Sraffa's second model (production with a surplus). But that would be a mistake: the purpose of the standard commodity is to provide a measure for variations of distribution between wages and profits. Sraffa, when discussing Manara's example, could be content with leaving Manara in his error if he just wanted to show that there was a snag in Manara's argument. But, as reported by Verger, I discussed the example, which had been presented to me by Pasinetti, with Pasinetti and Sraffa in the spring of 1970. Although the first process is superior in terms of rates of reproduction, it had become clear to me that Manara's example was economically viable, if labour was introduced in appropriate relative amounts and received a positive wage. I had not yet begun to think about existence theorems for the standard commodity, based on assumptions about rates of reproduction; this I began to do only sometime after I had been confronted with Manara's challenge. I only argued in the discussion that the standard commodity was there to explain changes of relative prices as caused by changes in distribution between wages and profits. Manara's example, complemented by a labour vector, with more labour allocated to the first process, was economically viable; relative prices then would move with the rate of profit. That would have to be discussed. Was it a problem for Sraffa's theory if the standard commodity for this kind of analysis was not there? If an appropriate labour vector complemented it, one could not object to Manara's example. Sraffa and Pasinetti took the point, but they never really told me what that meant for them. In particular, as already stated, I was not told that Sraffa had made a prior attempt to reply to Manara in a paper to be published.

Of course, neither they nor I knew that I would go to work on joint production for many years. I was just one of the students on whom they could try their arguments. Amartya Sen once recounted that he had been used as a test person on a much grander scale. He was, as a student at Trinity College, one of the few persons, to whom Sraffa showed the manuscript of *Production of Commodities by Means of Commodities* before publication. Actually, Amartya Sen was asked to sit at a table in Sraffa's rooms, Sraffa would bring the manuscript and watch how Amartya Sen would read it, during several days. Amartya Sen told me that he thought he understood what he was reading under such circumstances. I am still impressed by that.

It is not easy to grasp the movement of relative prices intuitively in a model such as that of Table 6.1, where there are four rates of reproduction and two labour inputs. A surplus is being produced, but a falling wage need not result in a rise of a rate of profit up to a maximum, where wages are zero; it may just as well happen in such models that one of the prices turns negative. Abraham-Frois and Edmond Berrebi (1976, p. 119) have called *Théorème de Schefold* the proposition (Schefold 1971, p. 33) that, for any joint production system with positive prices at a given rate of profit, either "a maximum rate of profit and a standard commodity exist and/or prices in any standard do not remain positive for all positive $r$". The theorem meant that even if no maximum rate of profits existed, the rate of profits could not rise indefinitely, because negative prices would come in.

What could negative prices mean? It was Joan Robinson who would provide the relevant example in her lectures with reference to her experience in India. The lower the real wage, she said, the longer lorries would be kept on the street, because repair work was cheap. This sounded like neoclassical theory, but substitution works in some contexts, and the effect could be reproduced in Sraffa systems with fixed capital, with lorries as machines, and with, as they grow older each year, the older machine being a joint product of transportation. Sraffa had used subsystems to explain negative values. He showed that more could be produced of some commodity, if a negative labour value was involved, by reducing the amount of labour employed (assuming constant returns to scale). The paradoxical character of the consideration disappears with fixed capital. If a machine needs an increasing amount of repairs, as it grows older, it pays to scrap it and to replace it by a new machine. The old machine will have had a negative labour value, for the new machine will allow to save labour (less repair work) and the old machine, since it is not used any more in production, will appear as a newly produced output. Since there is no demand for it, it will receive a zero price.

The generalisation of the idea that a negative labour value indicates the possibility to increase output and to reduce labour input by scrapping a commodity is known under the heading "truncation". Assuming constant returns and making the golden rule assumption, it turns out that, at the rate of profit where the price of some commodity turns negative, a process can be eliminated, saving labour and producing as much as before, but overproducing the commodity, which thus receives a zero price and ceases to be a commodity, as the old lorry in the example.

The rise of the rate of profit in joint production systems thus is not necessarily limited by a maximum rate of profit, with a standard commodity associated with it, but it can also be limited by a price turning negative, and it can rise further, if there is profit maximisation, only after truncation so that the remaining commodities have positive prices. Sraffa's attempt to refute Manara's example by pointing to the inferiority of one of the processes without taking wage costs into account was formally not mistaken, because Manara had made such assumptions. But Sraffa's attempted answer would not be effective, for as soon as somebody asked what the standard commodity was for, the answer had to be: to analyse effects of changes of distribution on prices, but where was such change in a model without labour? And there is another point on which Sraffa was only formally right. In the draft, as here published by Verger, Sraffa points to the fact that he had made the assumption of a positive real root of the characteristic equation explicitly, in order to have the standard commodity, but that he had done it at the advice of Besicovic; the problem to justify the assumption economically remained. In later conversations, Sraffa was annoyed that the standard commodity did not necessarily exist in the general case, but he had accepted the fact. It did exist in the fixed capital case; he was happy and visibly content when I could confirm that he had described the composition of the standard commodity correctly, regarding the proportions in which machines of different ages appear in it, when it is a matter of finding the standard commodity for pure fixed capital systems.

Of course, one can find a "true" maximum rate of profit by applying truncations to a given joint production system, which will then be "square", with a number of commodities produced equal to the number of positive prices and to the number of processes activated. This came out in my paper "On Counting Equations" of 1978, after an important suggestion made by Steedman, under the assumption of an equality of the rate of profit and the rate of balanced growth $g$ (golden rule), and this was extended to cost minimizing systems (where $r$ and $g$ could be different) in papers by Lippi, Salvadori and myself. The classical assumption is $r > g = 0$. Of course, one needs to make regularity assumptions. For me, this research was essentially closed with the second edition of my thesis (Schefold

1989). The procedure may be illustrated by means of the model with pikes and carps. As long as the rate of profit is low enough and the wage cost of raising fish in the basin is high, both processes will be operated in a stationary state at low rates of profit. But if the rate of profit rises and wages fall, harvesting fish in the wild will be become unprofitable. Carps will tend to be overproduced, their price drops to zero. If the workers in the breeding industry receive a sufficiently low wage, harvesting in the wild thus will be abandoned, carps are overproduced and a free good, and the rate of profit can rise up to a level determined by the rate of reproduction of pikes. The standard commodity then consists of pikes only.

The modern reader is invited to program the Manara example on his or her laptop to confirm this insight. One finds, with a labour input to tank production of 9/10 and one of 1/10 to lake production, that prices in terms of labour commanded are at first positive. The price of pikes rises monotonically up to a distinct sharp maximum near r = 11%. The price of carps in terms of labour commanded falls and becomes zero near r = 3.3%. Hence the movement depends on the composition of the wage. We measure it in terms of the surplus produced in the stationary state at unit activity levels. The wage rate then is equal to unity at $r = 0$ and falls monotonically. Truncation obviously must take place at $r = 3.3\%$, for the system consisting of both processes and both kind of fish ceases to be viable, as soon as the price of carps has turned negative. Only the tank process will be used at high rates of profit, when wages are so low that their cost does not compensate for the relative inefficiency of the production of fish in the lake. Truncation means that the surplus of pikes will then be produced by activating only the first process, carps will be overproduced and the wage rate becomes linear. It falls off and becomes zero at what may now be called the true maximum rate of profit at r = 9%. The second process, lake production, now is not used, because labour has become cheap. This means that lake production is not profitable at the wage resulting from the truncation, and this remains true up to the rate of profit, where the wage rate is zero.

The analysis can be rendered exact, using the chapter on the analysis of prices in steady states in Schefold (1989, pp. 114–123); it discusses a similar numerical example in detail. Here it should suffice to say, using the concepts introduced in that chapter, that the wage curve in the stationary state is given by two cost-minimizing systems successively. The first consists of both processes and both commodities, the second only of the first commodity and the first process. This truncation is q-feasible in the stationary state (the second commodity is overproduced) and $p$-feasible for all rates of profit from the point onwards, where the price of the second commodity turns zero (the second process is unprofitable). To complete the analysis, one must look also at the other three truncations that are formally possible and show that they are economically not relevant, because they are not $q$-feasible or not $p$-feasible.

So the standard commodity consists of pikes only. But what is the use of such a standard commodity? The standard commodity shall simplify the wage curve for a given technique and render the movement of relative prices transparent. It also is a physical analogue (Eatwell 1975) of a corn model and helps to analyse distribution as if we were in a one commodity world. The analogy has its limits: one needs prices to define a given real wage. The standard commodity finally is most useful in relation to what I call the Ricardian exercise (Ricardo 1966 [1951], p. 35) for single product systems: If the rate of profit rises, the wage rate must fall, whatever the composition of the numéraire, so that the cost of production, including normal profits, rises more in a capital intensive industry than in a labour intensive industry. For, to the extent that profits rise, the price must rise, but, to the extent that wages fall, labour costs fall. Hence one expects the prices of capital-intensive goods to rise relative to the prices of labour-intensive goods, the cost of which, in the given numéraire, will fall. Sraffa, in presenting this Ricardian exercise, has a reservation to make: If the prices of the capital goods used in the production of the capital intensive good fall strongly during the process, the conclusion cannot be drawn, hence one must look for the capital composition of the means of production of the goods currently used in production, and so further on backwards, and unaffected are only baskets of a composition of outputs which is the same as that of the inputs. The ideal numéraire, the standard commodity, therefore is that commodity, for which the causes for changes of relative prices are absent. This is the invariability of the standard of value, for every numéraire is otherwise invariant by definition.

This entire derivation, which can be explained well only in a full paper (Schefold 1986), has always been presented for single product systems only. The intensity of capital $\mathbf{a}_i\mathbf{p}/l_i$ (where $\mathbf{a}i$ is the vector of inputs, $\mathbf{p}$ the vector of Sraffa prices, expressed in some provisional numéraire, and $l_i$ the amount of labour used in the industry under consideration, $i$) explains the influence of distribution on the price of the output, if we are dealing with single products, but what, if the changing costs have to be ascribed to different outputs? It is true that one can reduce joint production systems to vertically integrated systems by the procedure explained in Schefold (1971, p. 30)—the term "vertical integration" was introduced later—but the standard commodity loses much intuitive appeal.

Now back to our example: What is the use of a standard commodity consisting only of pikes, if the task is to analyse the movement of relative prices (here the relative price of pikes ***and*** carps) at low rates of profit, i.e. in the system consisting of both processes? This is the problem of truncation again. A joint production system cannot easily be regarded as one technique, as in the case of a single product system, since the possibility of truncation is inherent. If we have, to use another example, a pure fixed capital system with one machine which grows ten years old at the maximum rate of profit, lowering this rate of profit and having a positive wage then may entail that the machine grows only five years old. This is like having a different technique. What does the standard commodity, containing ten old machines, mean at that level of distribution, where there are positive prices only for five machines? We know that the standard commodity is suitable as a standard of value, if there is one given technique. If there are many techniques to be compared, because there is a choice to be made, a numéraire must be common to all the techniques. To take the standard commodity of one of the techniques as the numéraire is a possibility, but it is not very useful. A large system of joint production is, because of the possibility of many truncations, like a spectrum of techniques, and that is why the standard commodity here is not of much help. It is probably agreed that the standard commodity must be unique, given a system. But what if a system, which is economically viable at one rate of profit, yields another truncation as a solution, as soon as the rate of profit is somewhat changed in a joint production example? If each is associated with a standard commodity, which is the relevant one?

All this seems to be ignored in the paper by Dupertuis and Sinha. They make much, on the other hand, of the idea that commodities could come as packets of inputs which stand in the same proportion in all lines of production. But this means that two columns of inputs of the input matrix are proportional. Hence the system would not be regular (it is one of the assumptions of regularity that the determinant of the matrix does not vanish).

Pasinetti wrote to Sraffa on 11 December 1969 (Sraffa Archive D3/14/24) and advised him for his "nota sul M" rather to speak of "assoluta inferiorità (absolute inferiority) than of "incompatibilità". The notion chosen by Dupertuis and Sinha in order to exclude such cases is "interdependence". By means of linear combinations of rows and columns according to certain rules, they want to reduce the system to its "atomic" core where all processes are interdependent. Then, they assert, a standard commodity must exist.

They overlooked that I gave an example in my thesis where such interdependence results from the construction of the basic system, and yet no standard commodity exists. It seems worthwhile to reproduce this example fully, taken from Schefold (1971, p. 88). Corn is produced on one kind of land by means of two different techniques (intensive rent) by means of corn and ploughs, and ploughs are produced by means of ploughs and corn in a single product process (Table 6.2).

Land is eliminated and the basic system is obtained by doubling the coefficients of the second process and deducting the result from the third, changing the sign. One obtains the basic system shown in Table 6.3.

It is not obvious that a surplus is being produced according to Table 6.3, but we can reassure ourselves by looking at Table 6.2. We can see that, at unit activity levels, the surplus of ploughs is equal to 1/10 and the surplus of corn equal to 2/5. Moreover, it must be assumed that the land is used up, hence that three units of land are available in total. The economy is therefore viable from the point of view of quantities. It is also viable from the point of view of prices, if the labour inputs stand in the following proportion: 2 : 25 : 1. The proportions cannot be changed much; otherwise, the rent turns negative. Again, the reader is invited to program the system on a laptop. The solution laboriously calculated by hand in Schefold (1971) will be confirmed.

The characteristic equation, following from Table 6.3, is $(11/25)(1 + r)^2 - (1 + r) + 1 = 0$.

**Table 6.2** Intensive rent, no standard commodity

| Ploughs | Corn | Land | | Ploughs | Corn |
|---|---|---|---|---|---|
| → | 1 | 0 | → | 1 | 0 |
| 1/10 | 2/5 | 1 | → | 0 | 1 |
| 2/5 | 1/5 | 2 | → | 0 | 1 |

**Table 6.3** The basic system, pertaining to the system of Table 6.2

| Ploughs | Corn | | Ploughs | Corn |
|---|---|---|---|---|
| → | 1 | → | 1 | 0 |
| -1/5 | 3/5 | → | 0 | 1 |

**Table 6.4** Intensive rent, no standard commodity

| Ploughs | Corn | Land | | Ploughs | Corn |
|---|---|---|---|---|---|
| → | 1 | 0 | → | 1 | 0 |
| 1/10 | 2/5 | 1 | → | 0 | 1 |
| 2/5 | 1/5 | 2 | → | 0 | 1 |

The discriminant is negative: 1 - 44/25 = - 19/25, hence there is no maximum rate of profit nor a real standard commodity. But it is clear from Table 6.3 that there is a maximum of interdependence, in that the basic system is a single product system: each process produces one of the commodities that are necessary for reproduction. There are two anomalies in the basic system: one input is negative (ploughs in corn production) and the sum of the corn inputs is greater than the corn output at unit activity levels. This is the consequence of the introduction of negative multipliers, which Sraffa defended in his book. In fact, what matters for reproduction is the existence of a surplus in the original system, according to Table 6.2.

Hence Verger seems to err in asserting that the problem of the missing standard commodity could be overcome by reducing the system to a state in which there is full interdependence of the processes. The mistake must be traced back to the original paper by Dupertuis and Sinha.

Sraffa himself said in one of the drafts of his attempted reply to Manara (Sraffa Archive, D/14: pp. 72–74): "Please note that the example given by M. not only was excluded from the real solutions condition that I had explicitly stated: it would have also been excluded from the elementary economic condition of convenience. It would be interesting if M. or someone more patient or luckier, found an example of a basic product system economically possible, that doesn't have any real solution for the standard commodity". Such an example has here been given, after reducing the processes using land to the corresponding basic system.

Yoann Verger has done us a favour by publishing Sraffa's draft and by clarifying some of the circumstances surrounding its origin. But his paper would be better if he had concentrated on the history of economic thought and avoided presenting an analytical solution, which makes Sraffa's mistake worse.

## References

Abraham-Frois, G., Berrebi, E.: *Théorie de la valeur, des prix et de l'accumulation*. Paris: Economica 1976.

Dupertuis, M.-S., Sinha, A.: Existence of the Standard system in the multiple production case: a solution to the Manara problem. *Metroeconomica* 60(3), 2009, pp. 432–454.

Eatwell, J.: Mr. Sraffa's standard commodity and the rate of exploitation. *The Quarterly Journal of Economics,* 89(4), 1975, pp. 543–555.

Manara, C. F., "II modello di Sraffa per la produzione congiunta di merci a mezzo di merci", L'industria 1 (1968), pp. 3–18.

Ricardo, David: *On the Principles of Political Economy and Taxation*. The Works and Correspondence of David Ricardo, ed. by Piero Sraffa, vol. I. Cambridge: University Press 1966 [1951].

Schefold, B.: *Mr. Sraffa on Joint Production*, in *Piero Sraffas Theorie der Kuppelproduktion, des fixen Kapitals und der Rente*. Basel: Private print 1971, S. 1–89. https://www.wiwi.uni-frankfurt.de/fileadmin/user_upload/dateien_abteilungen/abt_ewf/Economic_Theory/publications/Theorie_der_Kuppelproduktion_Bertram_Schefold_1971.pdf.

Schefold, B.: The standard commodity as a tool of economic analysis. *Zeitschrift für die gesamte Staatswissenschaft* 142 (3), 1986, pp. 603–622.

Schefold, B.: *Mr Sraffa on Joint Production and Other Essays*, London: Unwin Hyman (now Routledge) 1989.

Verger, Y.: Sraffa and Manara: The mystery of the last article of Piero Sraffa. Paper prepared for this volume, 2021.

Reply

# A Response to Schefold's 'Why Sraffa was Right not to Publish his Last Article'

*Ajit Sinha and Yoann Verger*

Since Professor Schefold's comment on Verger (2021) as well as Verger (2021) deal with Dupertuis and Sinha (2009) in a significant way, we thought it proper to write this rejoinder in co-authorship. We thank Professor Schefold (2021) for offering his critical comments on Verger (2021). We are, however, not convinced by his criticisms. It seems to us that our basic difference regarding Sraffa's concern over Manara's criticism relates to the question of the importance and the role of the Standard commodity in Sraffa's economic analysis. The crux of Manara's critique of Sraffa is that a Standard commodity does not exist in some cases of joint production. Now, if it could be shown that such examples as Manara's could be ruled out by economic arguments such as Manara-type examples necessarily involve inefficient or inferior techniques in use then the problem could be taken care of at that level. But if it could be shown that Manara-type examples could also be constructed in economically viable scenarios then Sraffa's reaction turns out to be quite revealing:

> One might conclude that the [possibility to construct] a Standard product [is under the same restrictive conditions] as any other general theory in ***an interdependent industries system***. If this is true, M. would have destroyed my humble hut and as a new Samson, he would have torn down the entire palace of modern economics. ~~che Adam Smith in~~ … If the examples [given by M.] were verified, together with my humble building all the milestones of traditional economics would be swept away. (Sraffa ND, D3/14: 61 and D3/14: 63, emphasis added)

This shows that the existence of a Standard system and a Standard commodity was at the core of Sraffa's analysis and definitely not a peripheral matter that could be treated as a minor inconvenience.

Schefold boldly claims that 'I convinced Sraffa and Pasinetti that Manara's example could be economically legitimate, provided it was interpreted in the right context, that is, taking appropriate costs of labour into account. *This is what caused Sraffa to abandon his attempt*' (Schefold 2021, emphasis added). This is not convincing as no evidence to this effect is provided in the paper and particularly in the light of his own admission that '*Sraffa and Pasinetti took the point, but they never really told me what that meant for them*'. (ibid., emphasis in original). So how could he be so convinced that his argument caused Sraffa to abandon his attempt? As a matter of fact, yet again this reveals the fundamental nature of the difference in our understanding of Sraffa's analysis. In Sraffa's analysis the rate of profits reaches its finite maximum at zero wages when labour is an input in production. But the equation system would remain exactly the same if, instead of wages put equal to zero, one puts the labour inputs equal to zero. Thus the solution of the system of equations for the maximum rate of profits must remain exactly the same even if labour is removed from the system as an input. That is why we do not find any note of complaint on Sraffa's part against Manara for not introducing labour in his examples.

Schefold argues that in the general cases of joint production with old machines as joint product a rise or fall in the rate of profits would change the economic calculation of when to scrap the old machine. In other words, the life time of a machine changes with changes in the rate of profits. In such cases, he argues that it makes no sense of having a Standard system determined for a machine of life time 10 years when at certain rate of profits it is profitable to scrap the machine after 5 years. Now, strictly speaking this is a case of change in technique with respect to changes in the rate of profits or wages. This is not particularly a problem of joint-production but it also exists in single-production system with choice of technique as is discussed by Sraffa in the chapter on 'switch in methods of production', the last chapter of his book (Sraffa 1960). This reveals that Schefold's method of truncation misses the point. As Sraffa shows in the chapter on 'switch in methods of production', a change in the rate of profits may lead to switch in technique, which amounts to change in systems with two different Standard systems and Standard commodities: 'There is, however, one statement of general validity that can be made in this connection. But for this purpose it is convenient to transfer our attention from two methods of producing the commodity in question to the two corresponding economic systems' (Sraffa 1960, p. 84). What is important for Sraffa is that for every system there exists a Standard system and a Standard commodity even though changes in the rate of profits may lead to switches from one system to another, which is what Schefold's truncation plots out. The existence of a Standard system is the evidence that the economy is an interconnected system of basic goods. If a Standard system does not exist then, for Sraffa, that implies that no 'basic good' exists. Thus the very foundation of his analysis evaporates, as early in his book Sraffa wrote '[w]e shall assume throughout that any system contains at least one basic good' (Sraffa 1960, p. 8). And this is why Manara's examples could be devastating for his whole enterprise.

Schefold claims that 'the purpose of the standard commodity is to provide a measure for variations of distribution between wages and profits'. It is not clear what Schefold means by this as in his own examples such measure is provided in other numeraire as he admits that in those examples Standard commodities do not exist. As a matter of fact, as Sinha (2016) has argued, the role of the Standard commodity was to show that the Maximum rate of profits of the system remains constant with respect to changes in the rate of profits. As in August 1942, Sraffa wrote: '[w]hat is demanded of a model is that it should show a constant (*constant with respect to variations of r*) ratio between quantity of capital & quantity of product. If this can be constructed, and proved to be general, a number of important "consequences" follow' (Sraffa ND, D3/12/16: 14, emphasis added, quoted in Sinha 2016, p. 115). It was to prove this proposition that Sraffa developed the Standard system and the Standard commodity. Thus the proposition that a maximum rate of profits exists and that it is independent of the variations in the rate of profits was the central proposition of his book.

In the end, Schefold provides an example that purports to prove Dupertuis and Sinha (2009) wrong. His example is reproduced below:

Schefold claims that Table 6.5 is an example of joint-production with only basic goods, which does not have a real Standard commodity. But clearly this example suffers from the same problem as Manara's. In this case, the second process is self-reproducing and does not need the first one. Negative inputs are equivalent to positive outputs. As Sraffa himself had clarified, '[n]ow it is not exact that wool does not "enter in the cost"

**Table 6.5** The basic system, pertaining to the system of Table 6.4

| Ploughs | Corn | | Ploughs | Corn |
|---|---|---|---|---|
| 2/5 | 1 | → | 1 | 0 |
| -1/5 | 3/5 | → | 0 | 1 |

of sheep—it enters negatively: in effect to produce a sheep, some wool must be produced while it grows up, as a by-product, inevitably: this must be sold & its proceeds deducted from the cost of the sheep—thus a negative "cost"' (Sraffa ND, D3/12/28-1).

The merit of Dupertuis and Sinha's (2009) approach is that they highlight the nature of Sraffa's concept of 'basic goods' as goods that are interconnected in their production technique such that the removal of any one of them would amount to cessation of the whole system. In the case of single-product system the interconnection of commodities as inputs are clear cut but in the joint production cases it may so happen that inputs relations of the structure of production treat some combination of commodities as one composite commodity. If such composite commodities are not properly identified as one commodity then some self-reproducing sub-system would hide in the production system of apparently all 'basic goods'. Dupertuis and Sinha (2009) provide a method to identify such composite goods and self-reproducing sub-systems and remove them from their 'atomic' system, which paves the way for a real Standard system and the Standard commodity to be identified.

In the case of rent, because two processes using land are producing the same output, two self-reproducing subsystems are necessarily present. What should be the method to choose between these two self-reproducing subsystems? We must choose the one that gives the smallest positive maximal rate of profits, because this is the only maximal rate of profit that ensures that solution of prices is all positive. In order to do this, we must first remove land by setting the rent equal to zero, in the same way that labour is removed by setting wages to zero. By doing this, the analysis of the interdependence between industries is not corrupted by any institutional set-up.

Pasinetti has also highlighted the pre-institutional nature of interdependence: 'There are economic relationships that are so fundamental to an industrial society that they can be defined independently of the institutional-political-juridical set-up that a society has adopted. Think of the structural interdependence linking together the industrial sectors of an economic system… These relationships can be stated in terms of objective efficiency, or as they have been called: in "natural" terms. They remain the same in any institutional set-up, whether it is a market economy or a centrally planned economy' (Pasinetti 2019, p. 362).

## References

Dupertuis, M.-S., Sinha, A. 2009. 'Existence of the Standard system in the multiple production case: a solution to the Manara problem'. *Metroeconomica* 60(3), pp. 432–454.

Pasinetti, L. L. 2019. 'Causality and interdependence in econometric analysis and in economic theory'. *Structural Change and Economic Dynamics*, Volume 49, June, pp. 357–363.

Schefold, B. 2021. 'Why Sraffa was right not to publish his last article' in A. Sinha (ed.), *A Reflection on Sraffa's Revolution in Economic Theory*, London: Palgrave Macmillan.

Verger, Y. 2021. 'Sraffa and Manara: The mystery of the last article of Piero Sraffa', in A. Sinha (ed.), *A Reflection on Sraffa's Revolution in Economic Theory*, London: Palgrave Macmillan.

CHAPTER 7

# Sraffa and Wittgenstein

*Ajit Sinha*

## 1 The Friendship Between Sraffa and Wittgenstein

Ludwig Wittgenstein had come back to Cambridge in 1929. Soon after he met Sraffa, and they struck up a friendship—a relationship that was intellectually intense and at times highly stressful to both. Sraffa's first entry in his appointment diary for a meeting with Wittgenstein is listed at 4:30 pm on 17 February (Sunday), 1929 and the last one on 19 March 1946, although among the extant letters we have the last letter from Wittgenstein to Sraffa dated 24 October 1950—six months before Wittgenstein died. So the relationship continued till the end of Wittgenstein's life. During this period, they met regularly, at times interrupted by their sojourns abroad or for other reasons such as Sraffa's internment for three months from 4 July to 9 October 1940 at the Isle of Man for being a citizen of an enemy country or Wittgenstein's voluntary work during the war as a hospital aid at Guy's Hospital (first as a porter and then as a pharmacist's

This chapter is a slightly edited version of the fourth chapter of my book, *A Revolution in Economic Theory: The Economics of Piero Sraffa*, Palgrave Macmillan. Here it is reprinted with the permission of the publisher.

A. Sinha (✉)
Thapar School of Liberal Arts and Sciences, Patiala, India



A. Sinha (ed.), *A Reflection on Sraffa's Revolution in Economic Theory*, Palgrave Studies in the History of Economic Thought,
https://doi.org/10.1007/978-3-030-47206-1_7

assistant) during 1941–1942 for more than six months. At times these discussions were aided by written notes and letters.

The intellectual relationship between Sraffa and Wittgenstein remained extremely intense and difficult throughout. On 19 January 1934, Wittgenstein wrote to Sraffa that '... Now you know that the conversations which we had in recent times (meaning roughly the last 6 or 9 months) were always a very great strain for me; and I'm sure for you too. This in itself wouldn't matter but I think it is clear that, for the present at any rate, we have given each other all that we can give. I have learnt an enormous amount from you in the conversations we had during the past 2 or 3 years; but I can't say that I have learnt much from you in our last conversations. Not that I've learnt all you can teach! But I have learnt most of what at present can be assimilated by me. That's why our last conversations haven't been profitable. Now this, I think, is no ground for never meeting again; but it is the reason why, until I feel more powerful, I avoid having a conversation with you.' Interestingly however, there was at least one conversation between 19th of January and the 31st of January, as we have a letter to Sraffa from Wittgenstein of 31 January 1934 where Wittgenstein refers to our 'last conversation'. We will come back to the intellectual content of this letter later. On the topic of the nature of their relationship, however, we find Wittgenstein yet again complaining, 'I think that your fault in a discussion is this: YOU ARE NOT HELPFUL! I am like a man inviting you to tea to my room; but my room is hardly furnished, one has to sit on boxes and the teacups stand on the floor and the cups have no handles, etc etc. I hustle about fetching anything I can think of to make it possible that we should have tea together. *You* stand about with a sulky face; say that you cann't {sic} sit down on a box, and cann't hold a cup without a handle, and generally make things difficult.— At least that's how it seems to me.' Again in a note to Sraffa written on 21 February 1934, Wittgenstein wrote, 'I think that writing down my arguments might possibly be the only way of making it worthwhile expressing them at all.—For correct or incorrect I believe they are in *most* cases worth hearing and properly considering & the mere fact that I know they are wasted on you when I just *say* them fills me with a kind of despair while I'm stating them. It's like trying *hard* to fill a barrel which has no bottom.' Sraffa, on his part, felt that Wittgenstein had a tendency to saunter from one point to the other, apparently disconnected. He responded by saying that 'I am much too slow for that, and cannot find the hidden connection. Also I cannot be content with hints or allusions (or things that cannot be laid down black or white), I must have it all thrashed out. But there is no use laying down conditions to which anyhow we could not stick; if in practice it works out in a way that is intolerable to either of us, then we shall give it up.' By 1941 the relationship had deteriorated considerably. In a letter dated 8 January 1941, Wittgenstein wrote to Sraffa that a 'deep discussion' had become impossible with him because he had 'gone *soft*'—'Its symptom is that you are unable now to stand decently *strong contradiction*, the contradiction of someone, I mean, who mistrust your reasoning … you used to take contradiction as medicine; and by contradiction I don't mean the expression of polite disagreement but a *challenge*! You did not always take it graciously (but who would?) but you used not when contradicted to kick with forelegs and hindlegs like some animals.' Finally, in another letter to Sraffa dated 23 August 1949, Wittgenstein wrote: 'In order to understand why it's impossible, or almost impossible, for certain people to understand each other, one has to think *not* of the few occasions on which they meet, but of the differences of their *whole lives*; and there can be nothing more different than your interest and mine, and your movements of thought and mine. Only by a real tour de force it was possible for us to talk to each other years ago when we were younger. And if I may compare you to a mine in which I worked to get some precious ore, I must say that my labour was *extremely* hard; though also that what I got out of it was well worth the labour. But later, when we no longer could give each other anything (which does *not* mean that each of us had got all the other had), it was natural that only an almost complete lack of understanding should remain; and, at least on my part, for a long time a wish that an understanding should again be possible.'[1]

Such were the exchanges that Wittgenstein refers to in 1945 in the 'Preface' to his *Philosophical Investigations* when he writes:

> For since beginning to occupy myself with philosophy again, sixteen years ago, I have been forced to recognize grave mistakes in what I wrote in the first book. I was helped to realize these mistakes—to a degree which I myself am hardly able to estimate—by the criticism which my ideas encountered from Frank Ramsey, with whom I discussed them in innumerable conversa-

[1] All the known extant exchanges between Wittgenstein and Sraffa are now published (except for one letter by W to S written in 1935, which was bought by an anonymous buyer at Berlin auction house Stargart on 22 March 2006) in McGuinness 2008. I follow McGuinness's practice of maintaining the authors' idiosyncratic spellings and *Italics* for single underline, small capitals for double underlines and normal capitals for thrice underline.

> tions during the last two years of his life. Even more than to this—always certain and forcible—criticism I am indebted to that which a teacher of this university, Mr. P. Sraffa, for many years unceasingly practised on my thoughts. I am indebted to *this* stimulus for the most consequential ideas of this book. (p. viii)

Unfortunately not much of intellectual value of these exchanges have survived, except for a few letters and notes for discussions; however, the importance of them cannot be underestimated. Here we have one of the greatest philosophers of his time crediting Sraffa for making him realize that he had made grave mistakes in his earlier classic, the *Tractatus Logico-Philosophicus*, and that his criticisms provided the 'stimulus' for the 'most consequential ideas' of his second classic, the *Philosophical Investigations*. In his 'Biographical Sketch' of Wittgenstein, Von Wright reports that '[Wittgenstein] said that his discussion with Sraffa finally made him feel like a tree from which all branches had been cut' (1955, p. 539).

One way to speculate about the nature of Sraffa's influence on the later Wittgenstein's philosophy would be to look for the decisive shifts in Wittgenstein's philosophy from the *Tractatus* to the *Investigations*. This would give us some indication of what might be the 'grave' mistakes of the *Tractatus* that Sraffa could be pointing to. We should, however, keep in mind that these discussions were on a variety of topics (e.g. on 18.2.1931, Wittgenstein wrote to Sraffa, 'I want to talk with you about vivisection, I think it is closely related to the things we are talking about') and perhaps never on the *Tractatus* directly, though Sraffa had read *Tractatus* and had made several annotations in the early part of the book. We should also keep in mind that at one point in 1935, precisely on 13.7.1935, Wittgenstein had come to the conclusion that Sraffa was exposing some fundamental problems with his thinking but he was not sure whether Sraffa's finger was pointing to the right spot; as he wrote to Sraffa: 'I said to you today that I would write to you sometime, but in fact I think that it wouldn't be quite natural for me to write to you; unless one day I see clearly that you were right in the essential points of our conversations, or, if I should ever see clearly that you were wrong. All I can now say is this: that there is *something* fundamentally wrong with me.—if a layman with healthy eyes looks at a bad portrait he sees that it is bad and will often tell you straightaway *what* is wrong in it, in his opinion; he will insist, say, that the nose is too long. The painter can then safely take it from him that the portrait is bad; but he would as a rule be ill advised to shorten the nose.

For it is one thing to see that a picture is bad, and an entirely different thing to see where the fault lies.—Thus, when you look into me you see that here is something wrong—and I agree—but whether in showing what is wrong you point to the right spot is very doubtful,—and perhaps not really relevant, as it is *my* job anyway to do something to put it right. Let us hope that this shall really be done and that one day I shall feel all-right {sic}'. Keeping such caveats in mind, we now turn to Wittgenstein's two books to locate Sraffa's possible pointers.

## 2 The Philosophy of *Tractatus*

In the *Tractatus Logico-Philosophicus* Wittgenstein took the position that the philosophical questions or rather confusions arise because of the lack of understanding of what language *can do*. His task was to draw the limit of language and he thought that the limit of language could be drawn by an investigation of the foundations of logic, as logic covers all necessary truths.[2] The fundamental question he poses to himself is: *Where from* a proposition gets its meaning or sense? His answer to this question boils down to this: Any proposition can be analysed or broken down into *elementary* propositions. An elementary proposition represents one 'fact'. All elementary propositions are *logically independent* of each other. An elementary proposition is made of a string of words. Words are names of simple 'objects' or things, whose form is to combine with other simple objects in a determinate order.[3] The order in which simple objects can combine with other simple objects is the structure of the 'state of affairs'. The structure of the state of affairs exhausts all the logical possibilities of the possible state of affairs. A fact is a realization of a state of affair out of all possible state of affairs. Thus the structure of a fact must be the structure of the state of affairs. Now, for an elementary proposition to represent a fact, it must mirror the structure of the fact that it depicts,

[2] See David Pears (1969, 1985) for the development of this thesis.

[3] The simple "objects" are not necessarily the things to which proper names are attached in our day-to-day life. For example, a statement that 'Excalibur, a proper name, has a sharp blade' makes sense even if the Excalibur is broken into pieces. Thus in this case the sentence 'Excalibur has a sharp blade' makes sense even when there is a word in it to which nothing corresponds. Thus, for this sentence to have sense, Excalibur must disappear when it is analysed and its place is taken by the words that name simples. Thus, simple "objects" are the logical necessity of Wittgenstein's theory, even though he is unable to give concrete example of his "objects" or "things". (See Wittgenstein 1953).

similar to a pictorial representation of facts. As McGuinness (2002) explains, 'A picture which represents that a is to the left of b by putting 'a' to the left of 'b' has as its form of depiction 'that x is to the left of y' and is a spatial picture, and has spatial form in common with reality. On the other hand, a picture which represents that a loves b by putting 'a' to the left of 'b' has its form of depiction 'that x stands in relation to y' and is a logical picture, having logical form common with reality. It will be clear that a spatial form of depiction is one kind of, and thus presupposes, a logical form of depiction' (p. 72). However, as a picture cannot depict its pictorial form but can only display it, the logical form cannot be said in language through propositions—it can only be shown. Thus on the basis of the above analysis, we can conclude that a factual proposition derives its meaning or sense from reality which it depicts and the *essence* of reality is the *logical form*, a form that language must share with reality to get its sense. The criteria for distinguishing sense from nonsense are clear. Any proposition that entails logical contradictions or fails to give meaning to a sign in the proposition is non-sense.[4] The *logical independence* of one possible state of affair from another implies that no causal connection can be drawn between two successive states of affairs in reality, as Wittgenstein states: 'superstition is nothing but belief in the causal network' (Wittgenstein 1974 [1922], 5.1361, p. 47). Yet, it appears, there is some kind of a notion of causation hidden behind the Tractarian explanation of meaning. The words of the elementary propositions are supposed to invoke or bring forth images of real 'things', and their combination in a possible 'logical form' gives 'sense' to the proposition. Thus it is a mental process of remembering or picturing the real that seems to give or *cause* the meaning of the words and the proposition.

## 3 The Nature of the *Break*

Wittgenstein, in the *Philosophical Investigations*, begins with a subtle attack on the idea that 'Every word has a meaning. This meaning is correlated with the word. It is the object for which the word stands' ($2^{e}$). With careful construction of extremely simple languages, which he calls

[4] Thus the *Tractatus* argues that language has severe limitations. Only factual propositions can be sensible as only factual propositions can have 'things' correlated with words. More important aspects of life such as ethics, aesthetics, mystic and spiritual fall in the realm of silence, "what we cannot speak about we must pass over in silence." (p. 89).

'language games', he shows that the words do not get their 'meaning' by reminding us of some mental images of things. He argues that the meaning of a word is its *use*. As Oswald Hanfling (1989) writes, 'The meaning will consist in the work the word is doing in a given context, and not in a corresponding entity' (p. 49). Such contexts are like games with their rules known to the players. The meaning of a word is neither attached to the things it names (a word is like a tool that can be used for many purposes) nor derived from the subjectivity of its user. The rules of the game are *objective* and are known to all the players who use words in a particular language game. Thus learning a language is akin to getting *training* in how to play a game—'"language game" is meant to bring into prominence the fact that *speaking* of language is part of an activity, or of a form of life' (Wittgenstein 1953, p. 11[e]).

As mentioned above (f.n. 3), in the *Tractatus* Wittgenstein is unable to give examples of 'objects'. It is taken as a logical necessity of the theory.[5] In *Philosophical Investigations*, however, he points out that the *Tractarian* dictum that 'a word has no meaning if nothing corresponds to it' uses the word 'meaning' illicitly if it is used to signify the thing that 'corresponds' to that word. The first casualty of this is the disintegration of the concept of simple 'objects' that was the basic building block of his atomistic theory of meaning. After discussing several examples, he concludes that:

> We use the word "composite" (and therefore the word "simple") in an enormous number of different and differently related ways. (Is the colour of a square on a chessboard simple, or does it consist of pure white and pure yellow? And is white simple, or does it consist of the colours of the rainbow?—Is this length of 2 cm. simple, or does it consist of two parts, each 1 cm long? But why not of one bit 3 cm. long, and one bit 1 cm. long measured in the opposite direction?)
>
> To the *Philosophical* question: "Is the visual image of this tree composite, and what are its component parts?" the correct answer is: "That depends on what you understand by 'composite'." (And that is of course not an answer but a rejection of the question.). (pp. 22[c]–23[c])

[5] "To Norman Malcolm, who in later years asked what he [Wittgenstein] would have regarded as an example of an object, Wittgenstein replied that his thought at the time of the *Tractatus* had been 'that he was a *logician*; and that it was not his business, as a logician, to try to decide whether this thing or that was a simple thing or a complex thing, that being a purely *empirical* matter!" (Hanfling 1989, 13).

Along with the collapse of the atomistic theory of meaning of the *Tractatus*, the *essentialism* of his previous theory also falls by the wayside, as Wittgenstein recognizes that in general the meaning of words cannot be strictly defined. It varies from one context of its use to another and these contexts, which he calls language games, are not reducible to any common denominator such as 'logical form' but rather only resemble one another like faces of individuals in a family. Thus the *essence* of language does not lie outside of the language in the so-called *real*—that is, the meaning of words or propositions cannot look for support outside of language. A proposition is not a *pictorial representation* of the real world or facts. Thus Wittgenstein's theory of meaning makes a decisive and dramatic shift from an *abstract causal* theory based on *logical necessity* to an *anthropocentric description* of language.

It is important to note that Wittgenstein himself credited Sraffa for this crucial move in his philosophy. Monk (1990) reports that once Wittgenstein told his student Rush Rhees that 'the most important thing he gained from talking to Sraffa was an "anthropological" way of looking at philosophical problems' (260–261). This point needs some elaboration. As we have seen, the logical analysis of *Tractatus* is a search for the *hidden* essence of language and the world. It provides us with a logical analysis (or an abstract theory) that brings to us the hidden essence of language, which is its logical form. Whereas, in the *Investigations*, we are told to "look and see" (31[e]) how language is actually used. As Pears (1969) writes, "... the source of the mysterious character of language is no longer its deep essence: it is everywhere, and it is on the surface" (p. 32). To quote Wittgenstein:

> We may not advance any kind of theory... We must do away with *explanation*, and description alone must take its place... The problems are solved, not by giving new information, but by arranging what we have always known. (1953, p. 47[e])

The point of the 'anthropological way' is an appeal to move away from search for *essence*, i.e. from questions raised in the general context to the questions raised in the context of particular cases. Philosophical questions such as 'what is meaning?', 'what is knowledge?', etc. cut the words 'meaning', 'knowledge', etc. from the stream of life. Whether these questions are sensible depends upon concrete circumstances in which such questions could be asked:

> When philosophers use a word—'knowledge', 'being', 'object', 'I', 'proposition', 'name'—and try to grasp the *essence* of the thing one must always ask

oneself; is the word ever actually used in this way in the language-game which is its original home?—

What *we* do is to bring words back from their metaphysical to their everyday use. (Wittgenstein 1953, p. 48[e])

But once such questions are placed in concrete circumstances where they could legitimately arise, the philosophical problems dissolve:

If we are using the word "know" as it is normally used (and how else are we to use it!), then other people very often know when I am in pain. (Wittgenstein 1953, p. 89[e])

David Pears (1969, 1985) has argued that one of the fundamental concerns of both the *Tractatus* and the *Investigations* was to draw a line between sense and non-sense. In the *Tractatus* Wittgenstein drew an outer limit of language, which pushed all other propositions except the factual propositions to the other side of the limit. In the *Investigations*, however, there is no single outer limit of language but several internal limits are drawn. The 'language games' represent those internal limits. A word may have sense within a particular language game but an attempt to drag it from one language game to another produces non-sense. For example, words such as God or Soul may have well understood meanings in a religious language game but produce non-sense in a scientific language game. Similarly, the general philosophical problem of looking for a substance corresponding to a substantive such as Hume's problem of 'sensations' is caused by forgetting the internal boundaries of the language games. In other words, philosophical problems arise when we forget what language *does*. As Wittgenstein once remarked, "philosophical problems arise when language *goes on holiday*" (1953, p. 19[e]). In a nutshell, we find Wittgenstein moving from the problem of *what language can do* to the description of *what language does*.

## 4 The Nature of Sraffa's Influence on Wittgenstein

Now let us come back to the few extant notes and letters of Sraffa and Wittgenstein that relate to the intellectual exchanges between them. Most of them are from the year 1934, notably the year of the lectures contained in the *Brown Book* of Wittgenstein. We, however, begin with a note on

'Language' written by Sraffa in early 1932, presumably for a future discussion with Wittgenstein and most likely a part of an ongoing discussion:

> If the rule of language can be constructed only by observation, there can never be any nonsense said. This identifies the cause & the meaning of a word.
>
> The language of birds, as well as the language of metaphysians can be interpreted consistently in this way.
>
> It is only a matter of finding the occasion on which they say a thing, just as one finds the occasion on which they sneeze.
>
> And if nonsense is "a mere noise" it certainly must happen, as sneeze, when there is cause: how can this be distinguished from its meaning?
>
> We should give up the generalities & take particular cases, from which we started. Take conditional propositions: when are they nonsense, & when are they not?
>
> "If I were king" is nonsense, for either I, or the job, would have to be entirely different. I know exactly what the reasons are that makes this unthinkable & I see that the modifications required to make it thinkable would be so great, that I would not recognise myself so transformed, nobody would say that the job, as adopted to my present self, is that of a king.
>
> "If I were a lecturer" has sense. For I was last year, & I don't think I have changed much since, nor has the job. The difference is small, or rather I cannot see it. I don't' know exactly in what I have changed since last year. There is nothing repugnant to me in the idea.
>
> But does this simply depend on my knowledge? (For a difference is big or small, according to whether I see it clearly or not). If I knew enough they would all be nonsense.
>
> Then of course there are the propositions where "if" stands for "when": i.e., the name stands for a class, & the proposition is true (or thinkable, as supra) for one member at least of the class. (PSP n.d., D3/12/7, also quoted in Sinha 2006)

The question is 'how to distinguish 'sense' from 'non-sense'. Sraffa's first point is that understanding the rules of a language by mere observation of the occasions when participants of a language speak identifies the meaning of a word with the *cause* of its occurrence. But in this way one can never distinguish when something 'sensible' or 'nonsensical' has been spoken as there is always some cause for 'nonsensical' expressions as there is always some cause for sneeze. Sraffa seems to suggest that the criterion or the rule for distinguishing sense from non-sense cannot be determined in such generalized contexts. One needs to take up particular examples of certain expressions and see on what criterion we judge such expressions either as sensible or nonsensical. The example he takes up is the case of conditional counterfactual expressions, as it seems was the case under discussion: '[w]e should give up the generalities & take particular cases, from which we started'.

This point that rules of language cannot be constructed by simple observation was apparently not accepted in its entirety by Wittgenstein. As in *Philosophical Investigations* Wittgenstein writes:

> ...—Or a rule is employed neither in the teaching nor in the game itself; nor is it set down in a list of rules. One learns the game by watching how others play. But we say that it is played according to such-and-such rules because an observer can read these rules off from the practice of the game—like a natural law governing the play.—But how does the observer distinguish in this case between players' mistakes and correct play?—There are characteristic signs of it in the players' behaviour. Think of the behaviour characteristic of correcting a slip of the tongue. It would be possible to recognize that someone was doing so even without knowing his language. (p. 27[e])

It should, however, be noted that Wittgenstein's point seems to be not about the *causal* theory of meaning. As a matter of fact, his point seems to be that the grammar of language is completely implicated in our practice of language, including the mistakes that we make, and can be discovered by mere observation as the laws of physics can be discovered by observation. The argument that the 'mistakes' or 'non-sense' can also be identified by observation takes away Sraffa's main concern that it identifies 'meaning' of a word with the cause of its occurrence. In fact, in his copy of the *Philosophical Investigations*, Sraffa puts a double straight line in the margin of the line, '—like a natural law governing the play' and a single straight line running from the beginning of the quotation to that point; apparently showing his full agreement with the quotation, particularly the idea of grammar of a language being completely implicated in the practice of the language as a 'natural law' governing the play.

Sraffa then takes up a conditional proposition such as 'if I were king' and judges it to be 'non-sense'. But, of course, this is not a full sentence. The full sentence has the structure: 'if I were king then ...'. Now, let's say we complete the sentence with 'if I were king then I *would wear the crown*'. There appears to be nothing unthinkable or nonsensical about that. But then this is not a typically counterfactual conditional sentence. 'Wearing the crown' is simply a description of what it is to be king. In such cases one can easily change 'if' to 'when' without changing the meaning of the sentence, e.g., 'if I were king then I would wear the crown' can also be expressed as 'when I will be king then I will wear the crown'. Here 'I' stands for any man who, when he will be king then he will wear the crown. But Sraffa is mainly concerned with a counterfactual conditional such as 'if I were king then *I would sign the peace treaty*'. Here the consequent is an action of *a* king, which the king may or may not make. Sraffa's claim is that such a sentence is nonsensical. Not because anything is wrong with its 'logical form' but because it is nonsensical to speculate what Sraffa (in early 1932) would do if he were king. This is because the real Sraffa of 1932 cannot know what a king Sraffa would think, feel or do as the circumstances of the real Sraffa is so greatly different from the circumstances of the king. Therefore, speculation of this sort is meaningless. But what about the antecedent of a counterfactual: '*if I were a lecturer* then I would criticize Marshall's theory of value in my lectures'? Sraffa says that this has 'sense' because he was a lecturer only last year and he knows that neither the job nor he has changed much since last year or at least he cannot see any change. So, is it okay to make counterfactual reasoning at the margin of the real? Sraffa's answer turns out to be non-affirmative. He thinks that the idea of sensibleness of the antecedent 'if I were a lecturer' turns out to be based on his lack of knowledge of all the changes that have taken place in him as well as in the job of a lecturer since last year. If he knew them all then perhaps this would also be nonsensical.

The next item is a letter from Wittgenstein to Sraffa written on 31.1.1934:

> Dear Sraffa,
>
> The following are some remarks I've put down on the topic of our last conversation. I hope they won't be too disconnected and I hope that you'll read them to the end.
>
> You said: "The Austrians *can* do most of the things the Germans did". I say: How do you know? What circumstances are you taking into account if you say they can? "He *can* remove the wedding ring from his finger". True, it is not too heavy and it doesn't stick to his finger. *But*: he may be ashamed of doing it; his wife may not allow it, etc.
>
> You say: "Learn from what has happened in Italy". But what should I learn from them? I don't know exactly how things happened in Italy. So the

only lesson I could draw from it is, that things one doesn't expect sometimes happen.

I ask: How will this man whose face I cann't imagine in a rage look like when he gets into a rage? And *can* he get into a rage? What shall I say when I see him in a rage? Not only "ah, so he could get into a rage after all", but: "so this is the *way he* can be in a rage; so this is how it connects up with former appearance."

It is a fact that I can easily imagine some features turning angry (remember what I said about Germans in Bohemia) and not others. This of course does not mean that I deny that the latter can look angry too; but I am puzzled as to the *kind* of anger they are capable of. And perhaps I will make a conjecture about the sort of anger I imagine it to be.

You say to me: "If a man is in a rage the muscles a, b, c of his face contract. This man (Austria) has the muscles a, b, c, so why shouldn't they contract. In fact if you Wittgenstein wish to know what he will look like in a rage just imagine him with these muscles contracted. What will Austria look like when it turns Nazi? There will be no socialist party, there won't be jewish {sic} judges, etc, etc, etc. That's what it will look like." I reply: This gives me no picture of a face; appart {sic} from the fact that I don't know enough about workings of things to know whether all these changes which you point out will *happen* together. For I understand what it means to say that the muscles a, b, c, will contract, but what will become of the many muscles etc. between them. Cann't the contraction of the one in this particular face prevent the contraction of the others? Do you know how in this particular face things interact?

You may say: Surely the only way to tell the future physiognomy is to know more and more exactly the contractions etc. of all (not only the main) muscles.

I say: I don't think this is the only way; there is another one, although the two ways meet. I may ask a *physiologist* what the face will be like, but also a *painter*. The two will give different answers (the painter by drawing the angry face) although if they both are correct they will agree. Of course I know that painters have to study anatomy. *I* want to know the painters {sic} answer, and I also want to know what the physiologist can tell me to check the painters {sic} answer.

I am interested to know what phrases the Austrians will use when they'll have turned Nazi. Supposing their patriotism is only *talk* then I'm interested *in their future talk.*

...

The crucial dispute between what Sraffa was supposedly arguing and what Wittgenstein is resisting is that Sraffa claims that Nazism has certain features or characteristics that would be repeated no matter which country turns Nazi. Wittgenstein is resisting this by maintaining that there is something called 'national character' or mentality of a people, which may not allow a country to turn Nazi in the first place or even if it turns Nazi, its Nazism may look very different because of its peculiar 'national character'. There is something *internal* or *qualitative* about a people or a person that an artist, a painter, can catch but a physiologist may not.[6]

After this there has been at least one conversation between Sraffa and Wittgenstein where this letter must have been discussed, as in another letter to Sraffa dated 21.2.1934, Wittgenstein refers to their conversation and admits his mistake. He argues that people usually associate a change in fashion with change in taste, i.e., a mental state (taste) first changes that *causes* fashion in turn to change. In other words, Wittgenstein argues: 'one presupposes a mental reservoir in which real causes of our actions are kept'. This way of reasoning, he thinks, is flawed and argues that his mistake in the previous letter was to presuppose that change in government was not related to any changes in the mental reservoir of primary causes: 'Now this connects up with our first question because one is tempted to think of such a reservoir, I.E. "the mentality of a people" and when one speaks of changes which the government of a country might undergo one imagines this thing, the mentality, not to alter'. So Wittgenstein realized that his mistake was to assume that there is something called 'national character' or mentality of a people that would remain constant when a country changes from being a non-Nazi to being a Nazi. This was unreasonable. This connects with Sraffa's point in the earlier note of 1932 that it is illegitimate to assume that the human character remains the same in real and a counterfactual condition.

In a short note written on 4.3.1934, we find Sraffa responding to some note from Wittgenstein (the exact note is not known), where the issue of human action and its relation to justification is raised:

> The error is to regard intuition as a provisional substitute for science: "when you will produce a satisfactory science, I shall give up intuition".—now the two things cannot be set against one another they are on entirely different planes. Intuitions are a way of acting, science one of knowing (Physician)

[6] Incidentally, in his copy of the *Tractatus*, in the margin of Wittgenstein's observation in 2.01231 that 'In order to know an object, I must know not its external but all its internal qualities', Sraffa wrote: 'what does this mean?'.

> Actions do not require a rational justification—they are *objects* of explanation.
>
> You try to rationalise intuition{s}—and say they are pisaller {*sic*, *pis-aller*-lesser evil} for science.

Let us suppose an instructor gives her pupil a mathematical series such as 2, 4,... and ask him to fill the three following dots. Suppose the pupil fills the dots by 6, 8, 10. Now, to the question: why did the pupil fill them by 6, 8, 10 and not by 8, 16, 32? The answer could be that the pupil *intuitively* saw the series to be adding 2 to the previous digit rather than doubling it. This answer is a *rationalization* of the action of the pupil. Such rationalizations have a causal structure. The act of the pupil of completing the series is logically independent of the pupil's 'intuition' or his 'state of mind' and it is the pupil's state of mind that is supposed to have *caused* him to write 6, 8, 10. Sraffa rejects the problem of 'rationalization' of human action as non-scientific. A scientific explanation of human action must not be built on the basis of the actor's 'state of mind', which cannot be observed.

In all these exchanges we see one constant theme, Sraffa is constantly attacking all explanations of human action based on psychology or mental state. All such explanations are built on the notion of 'reason' as the 'cause'. The lesson that Wittgenstein learns from these exchanges, which is apparent in both his *Brown Book* as well as *Philosophical Investigations*, is to abandon 'reason' as an explanation taut court, which also explains why his 'picture theory of meaning' of the *Tractatus* had also to be abandoned. For example, apparently directly responding to Sraffa's note discussed above, Wittgenstein in his *Brown Book* writes:

> It is no act of insight, intuition, which makes us use the rule as we do at the particular point of the series. It would be less confusing to call it an act of decision, though this too is misleading, for nothing like an act of decision must take place, but possibly just an act of writing or speaking. And the mistake which we here and in thousand similar cases are inclined to make is labelled by the word "to make" as we have used in the sentence "it is no act of insight which makes us use the rule as we do", because there is an idea that "something must make us" do what we do. And this again joins on to the confusion between cause and reason. We need have no reason to follow the rule as we do. The chain of reasons has an end. (1958, p. 143)

And in *Philosophical Investigations* Wittgenstein writes:

> Try not to think of understanding as a 'mental process' at all.—For *that* is the expression which confuses you. But ask yourself: in what sort of case, in what kind of circumstances, do we say, "Now I know how to go on," when, that is, the formula *has* occurred to me?—
>
> In the sense in which there are processes (including mental processes) which are characteristic of understanding, understanding is not a mental process. (p. 61[e])

> Misleading parallel: psychology treats of processes in the psychical sphere, as does physics in the physical.
>
> Seeing, hearing, thinking, feeling, willing, are not the subject of psychology *in the same sense* as that in which the movements of bodies, the phenomena of electricity etc., are the subject of physics. You can see this from the fact that the physicist sees, hears, thinks about, and informs us of these phenomena, and the psychologist observes the *external reactions* (the behaviour) of the subject. (p. 151[e])

Further on, in Part II of the book, Wittgenstein, in his detailed discussion of the psychological problem presented by the 'duck-rabbit' image derived from Jastrow's *Facts and Fable in Psychology*, clearly lays out a position that, in my opinion, closely represents what Sraffa set out to do with the modern or the 'subjective' theory of value:

> Imagine a physiological explanation of the experience. Let it be this: When we look at the figure, our eyes scan it repeatedly, always following a particular path. The path corresponds to a particular pattern of oscillation of the eyeballs in the act of looking. It is possible to jump from one such pattern to another and for the two to alternate. (Aspects A) Certain patterns of movements are physiologically impossible; hence, for example, I cannot see the schematic cube as two interpenetrating prisms. And so on. Let this be the explanation.—"Yes, that shews it is a kind of *seeing*."—You have now introduced a new, a physiological, criterion for seeing. And this can screen the old problem from view, but not solve it.—The purpose of this paragraph however, was to bring before our view what happens when a physiological explanation is offered. The psychological concept hangs out of reach of this explanation. And this makes the nature of the problem clearer. (p. 212[e])

## References

Hanfling, Oswald. 1989. *Wittgenstein's Later Philosophy*, London: Macmillan.

McGuinness, Brian. 2002. *Approaches to Wittgenstein*, London: Routledge.

McGuinness, Brian. 2008. *Wittgenstein in Cambridge Letters and Documents 1911–1951*, Oxford: Blackwell Publishing.

Monk, R. 1990. *Ludwig Wittgenstein the Duty of Genius*. London: Vintage.

Pears, D. 1969 (1985). *Ludwig Wittgenstein* (First Edition), New York: The Viking Press.

Pears, D. 1985. *Wittgenstein* (Second Edition), London: Fontana Press.

Sinha, Ajit. 2006. 'A Comment on Sen's "Sraffa, Wittgenstein, and Gramsci"', *Journal of Economic Behavior and Organization*, vol. 61, 504–512.

Sraffa, Piero. n.d. 'Sraffa papers', Cambridge: Wren Library, Trinity College, University of Cambridge.

Von Wright, Georg Henrik. 1955. 'Ludwig Wittgenstein, A Biographical Sketch', *Philosophical Review*, 64 (4): 527–545.

Wittgenstein, L. 1958. *Blue and Brown Books*, ed. R. Rhees, Oxford: Blackwell.

Wittgenstein, L. 1953. *Philosophical Investigations*. Oxford: Basil Blackwell.

Wittgenstein, L. 1974 [1922]. *Tractatus Logico-Philosophicus* (Tr. D.F. Pears and B.F. McGuinness). London and New York: Routledge.

Comment

# A Comment on Sinha on the Relationship between Sraffa and Wittgenstein

*John B. Davis*

Ajit Sinha's discussion of the relationship between Sraffa and Wittgenstein (Sinha in this volume; also, 2016, pp. 92–109) notes the closeness of their friendship, the intensity of their intellectual exchanges, and Wittgenstein's acknowledgement of his considerable debt to Sraffa for the shift in his thinking from the *Tractatus* to the *Investigations*.[7] Sinha also notes that though there is relatively little evidence regarding the nature and extent of their exchanges, there nonetheless exists something of a consensus among scholars that Wittgenstein's shift to his later key idea of meaning as use was influenced by what he is said to have called Sraffa's 'anthropological' way of thinking. In part, Sinha shows, this influence appears to have involved Sraffa arguing to Wittgenstein that meaning as use could not be given a psychological interpretation—in effect, the notion that how people thought about how language is used determined how it is used—a view Wittgenstein had explored but whose later rejection became central to his important idea of 'language-games.'

This elucidation of what Sraffa's 'anthropological' way of thinking may have involved makes sense when we think of Sraffa's philosophical critique of subjectivism that was motivated by his rejection of the Marshallian marginalist theory of value and effort to rehabilitate a Classical cost of production theory of value based on 'objectivist' philosophical foundations.[8] An 'anthropological' way of thinking does not derive how people interact in social settings from how they individually understand their interaction but

[7] Wittgenstein returned to Cambridge in 1929 and soon after met Sraffa. Their friendship and interaction continued, with interruptions, to the end of Wittgenstein's life in 1951.

[8] I discuss Sraffa's 'objectivist' philosophical thinking at length in Davis (2012).

rather derives from how they understand their interaction directly from their common experience of it. If we think of Sraffa's friendship with Antonio Gramsci and his knowledge of the class and labour struggles in Turin (see Naldi in this volume), the term 'anthropological' could be seen as a mild way of expressing Sraffa's rejection of subjectivist views of the world. Wittgenstein could not but have been aware of Sraffa's connection to Gramsci, and 'anthropological' may have been a term they used as a general (and innocent) reference for a philosophical view they came to share.

What I wish to focus on, however, is whether Sraffa and Wittgenstein shared a particular, real world instantiation of this 'anthropological' way of thinking. We know what that instantiation was for Wittgenstein: his idea of 'language games.' A 'language-game' (in German, a 'Sprachspiel') is a rule-governed form of social interaction in which the rules of use for a term establish its meaning. People learn a language by learning these rules just like they learn how to play games by playing them. Thus one doesn't begin with a pre-determined set of meanings in one's mind—a 'private language' as he called it—and then apply them in one situation after another. Indeed Wittgenstein thought the idea of a 'private language' was incoherent because he thought language is an essentially public activity. The rules we learn in interacting with others in learning language and in learning how to do all sorts of things are intrinsically social.

Thus, a 'language-game' or any other rule-governed interaction between people identifies a relatively integrated, distinguishable realm of social activity defined by the set of rules that apply to it. Sets of rules in any setting, in order to give guidance, need to be broadly consistent with one another in regard to what they apply to, and this serves to mark out and distinguish different realms of activity. For example, the 'language-game' governing social greetings marks out that realm of language and social activity from others, say, a speech or a lecture where other rules apply. At the same time, however, any given 'language-game' is only partially closed in this way because language meanings are transferred across social contexts and across different 'language-games.' Wittgenstein's idea of a 'language-game' thus refers to a domain of social activity that is, we can say, both closed yet open at the same time.

In fact, Sraffa employed the same, real world instantiation of this 'anthropological' way of thinking in his 1931 unpublished "Surplus Product" manuscript (Sraffa 1931). Noting a problem he found he encountered in his determining the value of surplus product using his cost of production explanation of commodity values, Sraffa explicitly adopted an open-closed interpretation of the relationship between the system of commodity values and how distribution acted on that system (cf. Davis 2012, 2018, and Davis in this volume). Specifically, he maintained that commodity production constituted a 'closed system' that was acted upon by causes external to that it associated with the distribution of the surplus. Referring to that 'closed system' as the 'economic field' he concluded:

> There must be a leak at one end or the other: the 'closed system' is in communication with the world.

Here, 'the world' refers to distribution of surplus product, and reflected the historical and social conflict over the shares of wages and profits in national income. He then added:

> When we have defined our 'economic field', there are still outside causes which operate in it; and its effects go beyond the boundary. (D3/12/7: 161 (3–5), see Sinha 2016 for the complete note)

Much as for Wittgenstein 'language-games' are both relatively closed and yet still open, then, for Sraffa "outside causes" associated with distribution "operate *in*" the 'economic field' and produce "effects that go *beyond* the boundary" created by 'the economic field' (my emphases). His later *Production of Commodities by Means of Commodities*, does not refer to the open-closed distinction, but it can be argued it retains this same structure since there also the determination of commodity values is influenced by the division of the surplus.

The analogy I am suggesting, then, is between Wittgenstein's idea of a 'language-game' and Sraffa's 'economic field' conception (later his system of commodity values). Both refer to relatively integrated, distinguishable realms of social activity that also communicate with what lies outside them. Both set aside subjectivist accounts of the world for objectivist 'anthropological' ones. Both reason in terms of different types of forces operating in the real world—one set that operates in a relatively tight knit way within a set of social boundaries and another set that operates more broadly across social boundaries. Both used this conceptual framework to reject reductionist, subjectivist theories of the world—Wittgenstein in rejecting the idea that terms always refer to and name things we privately recognize, and Sraffa in rejecting the idea that all value is determined in an individualist subjectivist manner. Sinha has alerted us to this connection in emphasizing Wittgenstein's reference to Sraffa's 'anthropological' way of thinking. My argument is that Sraffa and Wittgenstein shared a particular social instantiation of this way of thinking that made use of the open-closed distinction.

It's worth commenting that this view gives us a new view of the intellectual relationship between Sraffa and Wittgenstein. Wittgenstein's acknowledgement of his debt to Sraffa in the 'Preface' to his *Investigations* suggests influence went primarily from Sraffa to Wittgenstein. That may still be true, but it might also be the case that they converged on mutually shared views as a result of their exchange and interaction. It also points us toward a potentially important contribution they may have jointly made, arguably deserving further investigation, to how an 'anthropological' way of thinking may combine careful analysis of a particular type of phenomena with its place in a wider social, historical process.

## References

Davis, John (2012) "The Change in Sraffa's Philosophical Thinking," *Cambridge Journal of Economics* 36 (6): 1342–1356.

Davis, John (2018) "Sraffa on the open vs. 'closed systems' distinction and causality," *Research in the History of Economic Thought and Methodology* 35B: 153–170.

Sinha, Ajit (2016) *A Revolution in Economic Theory: The Economics of Piero Sraffa*, Cham: Palgrave Macmillan.

Sraffa, Piero (1931) "Surplus Product," D3/12/7, August, Wren Library, Trinity College, Cambridge.

Reply

# A Reflection on John Davis's Comments

*Ajit Sinha*

I thank Professor Davis for his highly thought provoking comments on my paper (Sinha 2021). There are three conceptual points raised in his comment: (1) the anthropological way of thinking, (2) the question of open and closed systems and (3) the status of 'surplus' in economic theory. Davis relates the first and the third points to the second, where he finds the common meeting ground for Sraffa and the later Wittgenstein. He first relates the question of the 'anthropological way' with Sraffa's early critique of Marshall's subjectivism as well as the real world instantiation of this way of thinking. He argues that Wittgenstein's conceptual frame of the 'language game' is an example of his real world application of the 'anthropological way of thinking'. A 'language game' is a system governed by a set of rules and every new learner of a language learns those rules to make sense of the words used in such language games—in this framework there is no room for the idea that true meaning of a word resides in the psyche of the speaker or the mental picture in the mind of the speaker. Davis argues that though this framework of rules appears to be a closed system, the very fact that 'language meanings are transferred across social contexts and across different "language-games" makes it open as well—'Wittgenstein's idea of a "language-game" thus refers to a domain of social activity that is, we can say, both closed yet open at the same time' (Davis 2021). He then relates this idea with Sraffa's 1931 note on 'Surplus Product'. Davis argues that in this note, Sraffa, like Wittgenstein, argues that the value of the surplus product is determined in a closed system of production that defines the 'economic field'; however, the economic field of production is acted upon by outside forces such as the distribution of the surplus—'Much as for Wittgenstein "language-games" are both relatively closed and yet still open, then, for Sraffa "outside causes" associated with distribution "operate *in*" the "economic field" and produce "effects that go *beyond* the boundary" created by "the economic field" (author's emphases)' (Davis 2021).

Without contradicting Davis, I would like to take this opportunity to think further about the idea of open/closed systems and to what extent this could be made a central or common ground for Sraffa and the later Wittgenstein's way of thinking. Strictly speaking, there is no system that can be characterized as *completely* closed or isolated except for the universe itself, which has no 'outside'. But still, in many fields of pure sciences the distinction between open and closed systems have precise meaning. For example, 'in thermodynamics, a closed system can exchange energy (as heat or work) but not matter, with its surroundings', 'in chemistry, a closed system is where no reactants or products can escape, only heat can be exchanged freely' (Wikipedia). Unfortunately, in economics, or social sciences in general, we have not yet given it a precise meaning—therefore, when we say 'closed yet open system at the same time', it, at best, remains vague as to what we want to say. Let us take an example from economics. Many heterodox economists consider the orthodox general equilibrium theory to be a 'closed system' since it determines all prices as well as income distribution from the closed set of information it admits at the outset. But it cannot be denied that the information on 'given endowments' and 'preference orderings' are the data that connect it to the outside of the system. One can always argue that the initial endowments and preference orderings are determined in a larger socio-economic context and changes in those have effects on the variables determined inside the system and these variables in turn have effect on the larger socio-economic context. So how is this any different from Sraffa's 'economic field' of his 1931 note and the problem of the determination of the 'surplus' as explained by Davis? However, if we define 'economic field' as a 'system' that determines primarily prices of commodities and the variables of income distribution such as rates of wages, profits and rents then one can legitimately classify the general equilibrium system as a relatively closed system since it determines all economic variables within the system. On the other hand, if we compare this with Sraffa's 'system' in *Production of Commodities by Means of Commodities* then we find that the two nodes of the 'general equilibrium system' that connect it to the outside of 'economic field', namely, the initial endowments and the preference orderings are made irrelevant. In this case, we find the economic variables of income distribution such as rates of wages, profits and rents are placed in the open space—hence the 'economic field' as such is no longer 'closed' but remains open.

On the question of Wittgenstein learning from Sraffa the 'anthropological way' of looking at philosophical problems, my sense is that it is not much concerned with the question of open and closed systems. Wittgenstein's philosophical questions were mainly related to the problem of whence a word or a proposition gets its meaning or what criteria can distinguish sense from non-sense? Now, as I have argued in my paper (Sinha 2021), the young Wittgenstein of the *Tractatus* thought that every proposition can be broken down to or analysed into independent elementary propositions. These elementary propositions are made of string of words that represent 'simples' or things. If these strings of words or things that they represent are related in a 'logical form' then the proposition becomes meaningful or sensible otherwise it is non-sense. I think Sraffa's attack must have been on the metaphysical or the unscientific foundation of this thesis. This way of thinking implies that the true meaning of a proposition resides in the picturing of the things and the relations in which they stand in the mind of the speaker. Now we have enough evidence to be confident that Sraffa must have argued that since the analyst cannot have access to the speaker's 'mental images', Wittgenstein's criterion to distinguish 'sense' from 'non-sense' becomes metaphysical. The 'anthropological way' in this context must have meant a pointer to look for observable data. Thus, instead of analysing 'propositions' or 'words' in abstract or in its generality, look for particular cases or concrete examples in which a proposition is made. It was the genius of Wittgenstein to develop the idea of 'language games' to contextualize such particular cases or utterances and then show that most of the words in a language do not represent simples or things—the same word in different language games can convey different meanings; furthermore, these 'language games' themselves cannot be generalized as having some common property—they only resemble each other as faces in a family. Thus meaning of a word or a proposition cannot be separated from a way of life.

In this context, it does not appear to me that the conceptual distinction between open and closed systems is particularly helpful in understanding the distinction between the *Tractarian* thesis and the thesis proposed in the *Investigations* by Wittgenstein. As I have pointed out in my paper, 'the fundamental concerns of both the *Tractatus* and the *Investigations* was to draw a line between sense and non-sense. In the *Tractatus* Wittgenstein drew an outer limit of language, which pushed all other propositions except the factual propositions to the other side of the limit. In the *Investigations,* however, there is no single outer limit of language but several internal limits are drawn. The "language games" represent those internal limits. A word may have sense within a particular language game but an attempt to drag it from one language game to another produces non-sense. For example, words such as God or Soul may have well-understood meanings in a religious language game but produce non-sense in a scientific language game' (Sinha 2021). One can draw a parallel to Wittgenstein's 'language game' with Sraffa's Standard system and the Standard commodity. Every Sraffa's system of 'basic goods' has a unique Standard system and a Standard commodity. It is within this context only that changes in prices with respect to variations in the distribution of income can be meaningfully interpreted since the Standard commodity as the standard of measure of prices is the only commodity that is not itself affected by the variations in the distribution of income; but once the 'system of basic goods' changes even slightly, 'a comparison of the prices by the two methods become meaningless since its result appears to depend on which commodity is chosen as the standard of prices' (Sraffa 1960, p. 82).

Now coming to Sraffa's note of 1931 on 'Surplus Product', it seems to me that Professor Davis has mistaken Sraffa's characterization of the Physiocratic interpretation of the 'surplus product' as Sraffa's own, since in his interpretation of Sraffa's quotation the crucial reference to it by Sraffa, expressed as, 'Consider, e.g., the so-called "natural causes" of rent' is missing (see Sinha 2016, pp. 82–89, for the complete note of Sraffa and my detailed commentary on it). This interesting note by Sraffa is yet again not primarily concerned with the question of open and closed systems. Its problem is: how to explain the existence of a 'surplus product' in economics if one relies on the 'principle of sufficient reason' or takes a *deterministic* scientific position. The problem is that conceptually the 'surplus product' is something that has no cost associated with it but if we take the deterministic scientific point of view then its existence must be explained by the sufficient cause that gave rise to it and in economic terms that sufficient cause must turn into a 'cost', negating the very concept of surplus. It was in this context, that Sraffa suggested that one way out of it would be to limit the economic field in such a way that the cause of 'surplus product' has no economic value and hence zero cost—the Physiocratic explanation of the 'surplus product' as a 'gift of nature' fits this way out; that is, the surplus product being the free energy of nature converted into economic goods in agricultural production. But this was not Sraffa's own position as it is evident from the position adopted by him in the *Production of Commodities by*

*Means of Commodities*. In this case, the surplus product is defined as the output over and above all the commodity inputs used in the production process—its origin or cause is not sought in anything freely available. The problem with the Physiocratic perspective lies in strictly measuring the cost of labour input. Is the contribution of energy by the labourer in the production process exactly equal to the calorie or the total energy content of the wages? If the energy content of wage is more than the energy contributed by the labourer in production then how does one account for the 'surplus product' appropriated by the labourer? Even if we suppose that one can identify exactly all the commodities that determine the minimum subsistence of the worker, once it is accepted that the wages are higher than the specified minimum subsistence it is no longer possible to draw a clear line between the subsistence wage and the surplus; as Joan Robinson had nicely put it, 'we could hardly imagine that, when the workers had a surplus to spend on beef, their physical need for wheat was unchanged' (1961, p. 54). This creates an indeterminacy in the measurement of either the necessity or the surplus. Sraffa's way out of it was to adopt a technocratic subject's standpoint who observes a reproducing system going round and round where it no longer is necessary to treat wages in a linear fashion as an input that produces human energy that in turn goes into the production process—the entire wage could be treated as a payment made after the production or a deduction from the 'surplus' produced, which can be determined by deducting all the commodity inputs from the commodity outputs. This is a measure that is determined from the standpoint of an observer just as the position of an electron is determined only in relation to an observer. This perspective rejects the Newtonian determinism in favour of indeterministic perspective of the quantum physics.

## References


Davis, John. 2021. 'Comment on Sinha on the relationship between Sraffa and Wittgenstein', in A. Sinha (ed.), *A Reflection on Sraffa's revolution in Economic Theory*, Cham: Palgrave Macmillan.

Robinson, Joan. 1961. 'Prelude to a critique of economic theory', *Oxford Economic Papers* 13: 53–58.

Sinha, Ajit. 2016. *A Revolution in Economic Theory: The Economics of Piero Sraffa*, Cham: Palgrave Macmillan.

Sinha, Ajit. 2021. 'Sraffa and Wittgenstein', in A. Sinha (ed.), *A Reflection on Sraffa's revolution in Economic Theory*, Cham: Palgrave Macmillan.

Sraffa, Piero. 1960. *Production of Commodities by Means of Commodities*, Cambridge: Cambridge University Press.

CHAPTER 8

# Sraffa Versus Keynes on the Method of Economics: Measurement, Homogeneity and Independence

*Anna Carabelli*

I investigate the methodology Sraffa and Keynes apply in their critique of economic theories and the development of their own theory.[1] The main focus is on their views on the measurement of economic magnitudes and

This paper was first published in *Annals of the Fondazione Luigi Einaudi. An Interdisciplinary Journal of Economics, History and Political Science*, Fondazione Luigi Einaudi, Torino (Italy), vol. 52(2), pp. 137–168, December, 2018. Here it is reprinted with the permission of the publisher.

[1] Here I use 'method' and 'methodology' interchangeably, but the terms have different meanings. The first is used more to refer to technique and how concepts are used, applied, and developed. The second refers to epistemology and the philosophy of science basis on which explanations are made. The ideas of objectivism/physicalism, rationalism, and empiricism are epistemological concepts. Heterogeneity, independence, etc. are method concepts. Here I make use of both meanings, because method and methodology are strictly connected, but the purpose of the paper is ultimately to examine the latter.

A. Carabelli (✉)
Università del Piemonte Orientale, Vercelli, Italy
e-mail: anna.carabelli@uniupo.it



A. Sinha (ed.), *A Reflection on Sraffa's Revolution in Economic Theory*, Palgrave Studies in the History of Economic Thought,
https://doi.org/10.1007/978-3-030-47206-1_8

the assumptions required. There are striking points of similarity in the methodology of their critique, but also contrasting approaches and divergences when they set about positively constructing their own new theory. I will consider both their published and unpublished writings.

My examination of Sraffa's manuscripts (held at Trinity College, Cambridge), which are less known to the general public, concentrates on the 1920–1930 period, which was contemporary with Keynes.

Sraffa read most of Keynes's published writings, including—surprisingly—some writings which I find relevant to this comparison. I do not know exactly how much Keynes knew of Sraffa's manuscripts which have now become available and are examined here. It is an established fact that they exchanged ideas, and Keynes certainly read or at any rate discussed some of the points under investigation here. I have no evidence that Keynes actually read Sraffa's unpublished papers, apart from those I refer to later on. In 1928 Keynes and Pigou read a text written by Sraffa, which may be document D3/12/6 (in particular items 4, 6–7 and 9–15), covering more or less chapters 1 and 2 of *Production of Commodities by Means of Commodities*, where the model of a system with three commodities and surplus is investigated. A letter by Pigou to Sraffa dated January 1928 has survived (Sraffa C239).

Again, I have no evidence that Sraffa read or at any rate knew of some of Keynes's manuscripts (held at King's College, Cambridge) now available to current readers, which have cast further light on Keynes's method. Keynes might have shown some of them to Sraffa. Among Sraffa's books we find three interesting items written by Keynes in 1908–1913: reprints of Keynes's 'Board of Trade Index-Numbers of Real Wages', *Economic Journal* Dec. 1908 (Sraffa 4565)[2]; 'Tables showing for each of the Years 1900–1911 the estimated value of the Import and Exports of the United Kingdom at the Prices prevailing in 1900', *Economic Journal*, Dec. 1912 (Sraffa 4576) and Keynes's reviews of Barbour D. *The standard of Value*

[2] I thank both the publisher Olschki and the Einaudi Fondazione for permission to republish the article already published in the Annals of Fondazione Einaudi, 2018/2, a. 52, pp. 137–168. I also thank the Keynes Trustees and the Sraffa Trustees for permission to quote respectively from Keynes's manuscripts held in King's College Library, Cambridge, and Sraffa's manuscripts held in Wren Library, Trinity College, Cambridge. Books in Sraffa's library are indicated with brackets containing the number of the book under consideration as catalogued in Trinity College Library, Cambridge. I also thank the two anonymous referees for their useful comments and suggestions.

(1912) and Hobson J.A. *Gold, Prices and Wages*, (1913), *Economic Journal*, Sept. 1913 (Sraffa 4577). Furthermore, tucked into one of the two copies of Keynes's *The End of* Laissez-Faire (1926) in Sraffa's library (Sraffa 2661),—a copy which seems to have previously belonged to C. R. Fay—there is a loose sheet written by Keynes containing a list of Keynes's readings on 'Averages and Laws of Error' ('Pigou, J.N.K, Norton, Berry, Johnson, Whitehead, Russell, Kapteyn, Pearson, Edgeworth, Yule, Hardy, Richmond, Sanger').

I pay particular attention to Keynes's *A Treatise on Probability* (*TP*, hereafter) (1921), which I consider his major work on method (Carabelli 1988). In particular, I do not know if Sraffa read the *TP* before 1925. We know for a fact that by the end of 1928 Sraffa had carefully read, taken notes on and annotated his own copy of Keynes's *A Treatise on Probability* (Sraffa 2645), and in particular he had drafted separate notes on the 'Appendix on Cause'. In his own copy of *A Treatise on Probability* his annotations on the last page regard in particular the atomic character of natural law and legal atoms (*TP*: 249; CW VIII: 278); dependence for knowledge and for causation (*TP*: 164–165; CW VIII: 181–183); independent variety (*TP*: 251; CW VIII: 279); Notes on Part III On the use of the term Cause (*TP*: 275–277; CW VIII 306–308).

As an exercise in comparison of methodologies, this chapter does not investigate on whether Sraffa derived his negative methodological approach, i.e. his methodology of criticism, from Keynes's approach, or vice-versa. I simply argue that there are significant similarities on relevant points. Further, I argue that they moved in opposite directions when constructing their own theories; thus we find sharp contrast in their positive method.

Section 1 deals with Sraffa's and Keynes's views on the measurement of economic magnitudes. Section 2 deals with the two economists' criticism of the assumptions of homogeneity (Sect. 2.1) and independence (Sect. 2.2) in economic theory. Section 3 deals with Sraffa's and Keynes's constructive approach to economic theory, discussing Sraffa's search for a common physical unit of measure for aggregates and Keynes's adoption of money as measure. Section 4 concludes.

## 1 Sraffa and Keynes on Measurement

Both Sraffa's and Keynes's investigation into measurement starts with definitions of concepts. This procedure helps to avoid ambiguity and vagueness in expression, and the risk of adopting methods that beg the question, using words to which no definite concepts necessarily correspond.

In defining concepts to be used in science, one main problem is their quantitative measurement, be it numerical or ordinal. The concept of quantity itself is interrelated with that of measure. Measurement is to be understood at the theoretical level—i.e. from the point of the philosophy of measure—and not that of the actual measuring of magnitudes. Sraffa, like Keynes, was not interested in the difficulties involved in actually measuring quantities (the so-called statistical problems) but in the theoretical problems of measurement.

Their interest was in the existence of a unit of quantity which has characteristics relevant to the purposes of the theory chosen and of a unit (or standard) of measurement which is independent of the quantity to be measured. In particular, their attention was directed to the system as a whole and not to the individual elements—an attitude spelt out by Keynes ('the choice of the units of quantity appropriate to the problems of the economic system as a whole', CW VII: 37). Sraffa was further interested in enquiring as to whether these units are absolute.

Economic quantities can be expressed (measured) in value; not all economic quantities can be expressed (measured) in physical terms. Moreover, quantities can be expressed or measured in physical terms only under restricted conditions. In fact, when difficulties in their physical measurement arise, economists usually have recourse to value measures, either in price or in money terms. And yet, problems may also arise with value measures.

### 1.1 *Sraffa on Measurement*

In the 1950s, at the Corfu conference on capital theory, organised by the International Economic Association (4–11 September 1958) (Lutz and Hague 1961), Sraffa distinguished the requirements for theoretical measurement from those specific to statistical analysis. From the report of Sraffa's interventions in the discussion of Hicks' paper 'The Measurement of Capital in relation to the Measurement of Other Economic Aggregates' (ibid., pp. 18–31):

> *Mr Sraffa* thought one should emphasize the distinction between two types of measurement. First, there was the one in which the statisticians were mainly interested. Second there was measurement in theory. The statisticians' measures were only approximate and provided a suitable field for work in solving index number problems. The theoretical measures required absolute precision. Any imperfections in these theoretical measures were not merely upsetting, but knocked down the whole theoretical basis. One could measure capital in pounds or dollars and introduce this into a production function. The definition in this case must be absolutely water-tight, for with a given quantity of capital one had a certain rate of interest so that the quantity of capital was an essential part of the mechanism. One therefore had to keep the definition of capital separate from the needs of statistical measurement, which were quite different. The work of J.B. Clark, Böhm-Bawerk and others was intended to produce pure definitions of capital, as required by their theories, not as a guide to actual measurement. If we found contradictions, then these pointed to defects in the theory, and an inability to define measures of capital accurately. It was on this—the chief failing of capital theory—that we should concentrate, rather than on problems of measurement. (Report of Sraffa's interventions in Lutz and Hague 1961: 305–306)

According to the report, Sraffa went on to say that:

> the usefulness of any theory lay in its explanatory value [...]. [He] took the view that if one could not get the measures required by the theorists' definitions, this was a criticism of theory, which the theorists could not escape by saying that they hoped their theory would not often fail. If a theory failed to explain a situation, it was unsatisfactory. (ibid.: 306)

In *Production of Commodities* Sraffa used the Standard commodity as the measuring yardstick. He constructed his concept of a Standard commodity in order to make visible the inverse relationship under specified conditions between the share of wages and the rate of profits. The use Sraffa made of the Standard commodity was to solve Ricardo's problem that the value of the surplus to be distributed varies with changes in the distribution itself between wages and profits: assuming that wages are paid at the end of the year there is a linear relationship between the share of wages in the surplus and the rate of profits.

In his manuscripts, Sraffa approved Sidgwick, who stressed in his *Principles* that 'the importance of seeking the best definitions' is far greater than 'the importance of finding it'. This was done by analysing the

'Assumptions on which the theory of value is generally based' (Sraffa D1/20 3–4). Sraffa pointed out Sidgwick's statements on the 'necessity of unit of measurement, not for measurement, but for conception' (D1/20 5) and stressed that 'we must find a unit of measure for cost: the necessity for this unit arises, not from a desire of actually measuring,—it is prior to it, and is required even for thinking of cost' (D3/12/3 46). As we will see, this attitude is similar to Keynes's. The differences between Sraffa and Keynes rest on the question of which magnitudes ought to be taken into consideration in economic theory and which not. In a note written in the Summer-October of 1929, Sraffa distinguished between three groups of quantities. I quote the whole passage:

> 1) Those which cannot possibly be measured, because they are not defined in terms of the method of measuring them, e.g. marg.[inal] utility and sacrifice. (No definition at all is given for measuring them in the case of several individuals: in the case of one individual, they are defined as being proportional to certain quantities, i.e. prices, but this is, as Cairnes says, 'merely giving a name to the unknown causes of price'.) Such quantities must be excluded altogether: at the worst, they may be used as a fictitious device for solving problems, but must not appear either in the premises nor in the conclusions.

The second group consists of those quantities which Sraffa believed economic analysis should take into consideration:

> 2) At the opposite extreme there are quantities which can be, and in fact are, statistically measured. These quantities have an objective, independent existence at every or some instants of the natural (i.e. not interfered with by the experimenter) process of production and distribution; they can therefore be measured physically, with the ordinary instruments for measuring number, weight, time, etc. Such are quantities of various materials used or produced, of lands[,] quantities of labour, lengths of periods, etc. These are the only quantities which must enter as constants in economic theory, i.e. which can be assumed to be "known" or "given". [In parenthesis he added: The "extensive" theory of rent, and the labour theory of value only assume this kind of knowledge].
>
> Then there is a third group:

> 3) Finally, there is the class of quantities, which form the basis of Marshall's theory (or, rather, of Pareto's), such as demand & supply curves, marginal productivities, (i.e. rate of growth of total) indifference curves, etc. Here the constant quantities have no names—they are the parameters of curves. The several quantities represented by these curves do not exist at any one moment, nor during any period of the recurrent steady process of production or consumption. They are alternatives, only one of which can exist in any one position of equilibrium, all the others being thereby excluded (even the one does not really exist if there is no change, since it is the rate of growth of a quantity, i.e. marginal product: it can be inferred from price, but so can marginal utility, which under (1) we have agreed does not exist). Therefore, they cannot be found by merely observing the process or state of things, and measuring the quantities seen. They can only be found out by means of experiments—and these quantities in effect are always defined in terms of such experiments (successive doses applied to land; alternatives offered to the consumer; etc.).

He went on to point out:

> These experiments cannot be carried out (and never have been, as a matter of fact) for various reasons, 1) the practical difficulties, 2) the lack of definition of the conditions to be required, which are always summed up in the absurd "other things being equal". (D3/12/13: 2–3, 5)

For Sraffa economic magnitudes belong to group (2). They are concrete, tangible and visible; they can be physically measured, in number, weight (tons or gallons) and time. These quantities are the only ones that must enter as constants in economic theory, i.e. which can be assumed to be 'known' or 'given'. His concept of magnitudes and of their measurement was informed by the method of natural sciences, especially physics, mechanics and thermodynamics. This approach shaped his general conceptions about the method of economics and the nature of its subject matter. He was in favour of the 'objectivisation' (D3/12/7 46) of economics, accepting a materialist point of view and an empiricist epistemology. For him, the positivist method of hard sciences had to be applied to economics. For the development of science, there is one method, namely that of physics. The different disciplines, such as physics, biology, economics and psychology had to converge towards a scientific synthesis. He described his method as being concerned with establishing 'an entirely objective point of view', a 'natural science point of view' or 'an atomic analysis' (D3/12/7: 161 (3); D3/12/13: 16 (9), 18). Let us remember that the twenties and thirties of the last century were characterised by the dominance of the neo-positivism of the Vienna Circle and logical atomism.

As Kurz and Salvadori have shown (2005b: 425–426), in the second half of the twenties Sraffa read books on physics, the natural scientists and neo-positivist epistemology, in particular Hertz and Whyte.[3] Sraffa also read and made notes on A. S. Eddington's 1927 Gifford Lectures on *The Nature of the Physical World* (Eddington 1928). In April 1928 he studied and again made notes on A. N. Whitehead's *Science and the Modern World* (Whitehead 1926). Kurz and Salvadori judge from Sraffa's annotations that he agreed to a large extent with Whitehead's materialism and with what he had to say on science becoming quantitative; he also subscribed to the search for measurable elements among phenomena. In my view Sraffa also shared Whitehead's empiricism.[4] In November 1927 Sraffa read, among other texts, *The Trend of Economics* (Tugwell 1924), a collection edited by Rexford Tugwell of papers by leading American economists including F. H. Knight and J. M. Clark with the aim of reconsidering the method and content of economics in the light of developments in physics. Sraffa read the book, took notes on it and, according to Kurz and Salvadori (2005b), referred to it repeatedly in his papers (D3/12/11, dated November 1927). Knight denied that it was possible 'to construct in thought a world of real objects in purely objective terms' and that human phenomena were 'amenable to treatment in accordance with the strict canons of science' (Knight 1924: 242, 251). According to him 'it is impossible to discuss value in purely objective terms' (Knight 1924: 229). Kurz and Salvadori argue that Sraffa's notes show him disagreeing with Knight but concurring with some of the observations made by J. M. Clark, who advocated a fundamentally different point of view.[5]

[3] Remarkably, Sraffa took notes on Heinrich Hertz's (1899) *The Principles of Mechanics*, focusing attention on the physicists' concepts, and in particular on 'cause' and 'interdependence'.

[4] It is worthwhile to remember that Whitehead was one of the two referees of Keynes's 1907 Dissertation *The Principles of Probability*; his empiricism was one of the reasons for his rejection of Keynes's Dissertation, where an attack on the empiricist interpretation of probability (frequency) was launched.

[5] See also the recent debate over the meaning of Sraffa's objectivism/physicalism. Davis (2018) argues that Sraffa's 1931 'Surplus Product' paper (D3/12/7161: August 1) shows him questioning his pre-1928 view, primarily because it eliminated the concept of a measurable surplus and was inconsistent with including the role distribution played in determining commodity values.

### *1.2 Keynes on Measurement*

Keynes's approach to the measurement of economic magnitudes was the opposite of Sraffa's. For him, economics is a moral science and economic magnitudes are not analogous to those of the physical sciences (CW XIII: 299–300). Keynes explicitly criticised J.M. Clark's method and applied to economic magnitudes his philosophy of the measurement of probability (Keynes 1907, 1908, 1921; see Carabelli 1988, 1992, 1994).

In Chapter 3 of *A Treatise on Probability*, Keynes tackles the logical and theoretical difficulties in the measurement of probability, and not the statistical-mathematical problems due to our inability to measure it. He raises the question of the existence of a measure of probability, i.e. the existence of a general or '"natural" unit of measure and of quantity' (CW VII: 33). His discussion is more relevant to theoreticians (be they economists or otherwise) than to applied scientists or statisticians: 'That such comparison is theoretically possible, whether or not we are actually competent in every case to make the comparison, has been the generally accepted opinion' (CW VII: 21). The theoretical difficulty of numerical measurement is intrinsic and inherent in probability. We will see, how this difficulty is equally intrinsic and inherent in macroeconomic magnitudes and does not depend on our inability to measure them (CW XI: 52, 135). This intrinsic indeterminateness is stressed in *A Treatise on Probability*:

> It is not the case here that the method of calculation, prescribed by theory, is beyond our powers or too laborious for actual application. No method of calculation, however impracticable, has been suggested. Nor have we any *prima facie* indications of the existence of a common unit to which the magnitudes of all probabilities are naturally referrable [...] probabilities do not all belong to a single set of magnitudes measurable in terms of a common unit. (CW VII: 32–33, emphasis in original; on common unit see also 35–36)

For Keynes, the same difficulty arises in the case of nearly all the quantities dealt with by science:

> There is no unique unit of measurement [...]. The ground of the element of arbitrariness in the unit of measurement of most physical quantities is easily explained. The objective quality measured does not really possess numerical quantitativeness, although it may possess some of the properties necessary for numerical quantitativeness [...]. In general, the values which it can assume are capable of being ranged in order of magnitude and, in addition,

> it will sometimes happen that the series which is thus formed is continuous, that is to say, for any two selected values there can be found a value whose order in the series is between the two selected values. But it does not follow from this that there is any meaning in the assertion that one value is twice another value. The relations of continuous order can exist between the terms of a series of values, without the relations of numerical quantitativeness existing at the same time. But we can nevertheless measure the values after a manner which is somewhat arbitrary but which yields results sufficiently satisfactory for many purposes, those, for instance, of mathematical physics, though not for those of probability. This method is to select some other quantity or ratio [...]. For instance, the series of values of the quantity measured by a specific volume have this relation [...] to a conventional measurement [...]. The intervals which we regard as equal depend upon this arbitrary choice. (Keynes 1907: 97–98, emphasis in original)

In his discussion of the measurement of probability in the first version of the *Principles of Probability* (1907), Keynes considered the current state of the philosophy of measurement and, in particular, of the measurement of relations. His main interest was in the 'philosophy of magnitude', not in the mathematics of magnitude (Keynes 1907: 119; see also *TP*, CW VIII: 37). He devoted particular attention to Russell's *Principles of Mathematics* (1903). The probability relation is a quantity (Keynes 1907: 52, CW VIII: 21)—in general, a non-numerical quantity. Keynes argues that a quantity is not necessarily a number, 'the numerical view of probability is partly traceable to the frequent definition of probability as a number and not as a quantity' (Keynes 1907: 56). Only probability relations which are of the same kind and in the same unit of quantity are numerically measurable and therefore numerically comparable (ibid.: 62). Difficulties in finding a common unit of quantity arise when these quantities of probability are similar to quantities which may, at the same time, be characterised by a collection of attributes, such as a class of qualities rather than by one single quality: 'the comparison of the probability of such propositions as "this book is red", and "this book is blue". By "red" not one single quality is meant, but a class of qualities. The book is to have one out of a class of qualities which we may call "the reds"' (Keynes 1907: 120).

Keynes distinguishes between non-comparable, comparable and numerically comparable probabilities. There are

> three possible degrees of intimacy between the magnitudes of any pair of probabilities. Some cannot be compared at all in this respect; between oth-

> ers the relations of more or less can be established; and others again seem to be capable of numerical measurement in terms of some common unit. (ibid.: 60–61)

This heterogeneity in kind and in the units of quantity depends on a multidimensionality of the magnitude of probability. 'Yet the base units of the different strands (kinds of probability) are not altogether in different dimensions, as, for instance, are those of beauty and goodness' (ibid.: 65). Later, in his *Treatise on Money*, he became more radical as regards economic quantities, speaking of 'things being [...] in different dimensions [...] and capable of variations of degree in more than one mutually incommensurable direction' (CW V: 87–8).

The probability relation has an intrinsic quantitative indeterminateness. The impossibility of a numerical measurement of probability is not a product of mental incapacity or lack of knowledge, but arises from the nature of the case itself. This impossibility is 'absolute and inherent in the subject matter' (Keynes 1907: 65) because there is no determinacy of the units of quantity. These units belong to different kinds of magnitudes.

## 2 Sraffa's and Keynes's Methodology of Critique: Investigation into the Assumptions of Economic Theory

In connection with the problem of the existence of units of quantity and measure necessary "for conception", we find Sraffa's and Keynes's investigation of the various *logical* requirements for measurement. Crucial among these requirements were the assumptions of homogeneity and independence. Again, there are great similarities between Sraffa and Keynes in these *destruens* approaches.

Sraffa builds his critique of Marshall and Pigou—what I call the *destruens* part of his approach—on the role played by the assumptions of economic theory. He examined the assumptions that marginal economists had introduced into their theory with the aim of making their tacit assumptions explicit. His methodology shows great similarities to Keynes's criticism of the so-called 'classical economists' (Carabelli 1991). Sraffa and Keynes sought to understand whether these assumptions were necessary for the theory not to be logically fallacious.

Keynes's criticism of what he called the 'classical economists' was of a methodological nature, based on the search for a logical flaw. He set out to make the tacit assumptions of the "classical economists" explicit. The *General Theory* was addressed to 'his fellow economists' and had as principal object the study of 'difficult questions of theory', with the aim of 'persuading economists to re-examine critically certain of their basic assumptions' (CW VII, Preface xxi). Keynes's attitude towards the search for tacit assumptions in theories and his attempt to make them explicit, showing the limits to their validity, is the same that emerged in *A Treatise on Probability*; on this constant attitude, see also Carabelli 1988: 75, 267–268).

Keynes wanted to locate and identify the logical flaw in the classical theory (CW XIII: 489). He thought that the relationship between premises and conclusions in the classical theory was perfectly consistent: the superstructure of the classical theory was painstakingly built 'for logical consistency' (CW VII: xxi; see also 33, 192). So, if the fault of the classical theory lay neither in the empirical unacceptability of the conclusions nor in the logical inconsistency between premises and conclusions, where did it lie? Keynes even went as far as saying that if the basic system of the classical theory was unassailable, then, consistently, one had to accept its conclusions (CW XIII: 491). However, the classical system of thought was, he held, assailable, and should be attacked at its heart. The flaw was in the premises of the classical theory, so the attack should be moved against them, rather than against the conclusions or against the connection between premises and conclusions. The premises lacked 'clearness' (CW VII: xxi; also 33,192). The classical theory did not make explicit—as it should have done—some assumptions upon which its conclusions stood and upon which its arguments depended for generality and domain of validity.

Keynes's search for the existence of tacit assumptions in the classical theory permeated the whole of his writings. In 1933, in his essay 'A Monetary Theory of Production', Keynes wrote of tacit assumptions:

> One of the chief causes of confusion lies in the fact that the assumptions of the real-exchange economy have been tacit, and you will search treatises on real-exchange economics in vain for any express statement of the simplifications introduced or for the relationship of its hypothetical conclusions to the facts of the real world. (CW XIII: 410)

In 1936, in the *General Theory*, he wrote: 'this strange supposition [...] is what all members of the orthodox school are tacitly assuming [...] the classical school have slipt in an illicit assumption' (CW VII: 13; see also xxv). [6]

### 2.1 On Homogeneity

Sraffa's pre-1928 manuscripts include thorough discussion of the requirements for the measurement of economic quantities, with analysis of both homogeneity and relevance (D1/20 'General', mainly brief notes extracted from books (7 docs.) and 'On Economic Quantities: homogeneousness and measurement' (D1/5).

Homogeneity means that there is no difference (or variety) in the kind (or substance) of quantities: 'their action is so much different in degree as almost to be a difference in kind' (D3/12/3 Notes London, Summer 1927 Physical Real Costs, p. 6). In the case of the quantities of labour, of land, of capital and of products, for example, it must be determined whether the respective quantities are homogeneous. The existence of a *common* standard (a unit of measure) is involved here. In contrast with quantities of labour, land, products, capital and utility, quantities of money are homogenous by definition or convention.

In his search for the requirements of homogeneity, Sraffa not only investigated the assumptions of differences and variety in the substance, but also stressed the assumption of *relevant* differences or variety. Heterogeneity and variety represented his starting point, so homogeneity was not an intrinsic quality in general, but a logical assumption. Introduction of the assumption of homogeneity depended on judgements of relevance. For Sraffa relevance implied that although there exist differences and variety in the quantities they can be judged negligible: relevance was not empirical but logical, i.e. relevance to the problem under investigation, 'relevant (to the problem at hand)', as he wrote. He argued that the difficulty was 'to find a criterion for determining the limits' of the

[6] In 1939, in the Preface to the French Edition of the *General Theory*, Keynes again stressed the point: 'Say was implicitly assuming that [...]' (CW VII: xxxv). Keynes's criticism of Pigou, in the appendix to Chapter 19 of the *General Theory*, is structured around the search for the existence of tacit assumptions in Pigou's theory (CW VII: 272–275): 'Since the tacit assumptions, which govern the application of the analysis, slip in near the outset of his argument, I will summarise his treatment up to the crucial point' (CW VII: 272; see also 274–275 and 277).

assumption of homogeneity when non-homogeneous quantities are dealt with in general (D1/5 p.1).

For Keynes too, economic quantities are in general non-homogeneous except under restricted conditions which should be investigated. He applied to them his view on probability: probability is a non-homogeneous object. Economic magnitudes are 'non homogeneous complex which cannot be measured [...] except in certain special cases' (CW VII: 38) and 'incommensurable collections of miscellaneous objects' (ibid.: 39). Probability is intrinsically a non-precise quantity; the *General Theory* 'vague and non quantitative concepts' (p. 40) are similar to probability. Keynes maintained that probability and the material of probability are, in general—except in particular limited cases—non—homogenous and non-atomic: 'A degree of probability is not composed of some homogeneous material, and is not apparently divisible into parts of like character with one another' (*TP*, CW VIII: 32). Economic magnitudes and the material of economics are the same.

We will be looking at Sraffa's and Keynes's analysis of the homogeneity of quantities of labour, products, capital, utility and money in detail.

*Labour and Products*

In dealing with the measurement of economic quantities, great attention was paid by Sraffa to the criticism made by orthodox economists (Bailey, Sidgwick and Böhm-Bawerk in particular) of the non-homogeneous substance of the quantity of labour and of products. For these economists, the quantity of labour was an ambiguous expression (D1/5, p.1). Sraffa also discussed the analogy between measurement of labour and of land (D1/5, p. 7).

In the twenties Sraffa shared with the critics of the labour theory of value the view that labour was a non-homogeneous quantity, which therefore could not be used as the ultimate standard for the theory of value. As Kurz and Salvadori (2005b) argue, he later changed his mind: non-homogeneous quantities of labour can be homogenised.

With respect to the assumption of homogeneity of the quantities of products, Sraffa criticised Marshall's treatment of the varying composition of the dose and variety of products. Marshall introduced in different contexts either a tacit assumption of homogeneity or the money unit (D1/5, p. 6).

Like the later Sraffa, Keynes argued that labour is a non-homogeneous quantity but that it can easily be made homogenous through a process of reduction. In a well-known passage in the *General Theory*, Keynes argues his use of the unit of labour as 'the sole physical unit' (CW VII: 213–214). He proposed 'to make use of only two fundamental units of quantity, namely, quantities of money-value and quantities of employment' (CW VII: 41). Further, 'We shall call the unit in which the quantity of employment is measured the labour-unit; and the money-wage of a labour-unit we shall call the wage-unit'. The labour-unit can be reduced to a homogenous unit by 'taking an hour's employment of ordinary labour as our unit and weighting an hour's employment of special labour in proportion to its remuneration; i.e. an hour of special labour remunerated at double ordinary rates will count as two units' (ibid.). The labour-unit provides an unambiguous measure of output; 'we shall measure changes in current output by reference to the number of hours of labour paid for (whether to satisfy consumers or to produce fresh capital equipment) on the existing capital equipment, hours of skilled labour being weighted in proportion to their remuneration' (ibid.: 44). He argued that any other units of measurement lead to 'unnecessary perplexity' due to heterogeneity of inputs and outputs.

*Capital*

In his early notes made in the twenties Sraffa also paid great attention to the difficulty involved in the measurement of capital and the role of the assumption of homogeneity of the quantities of capital. Capital is regarded by economists—including Sraffa—as a collection, an aggregate of the most diverse objects. On the theory of capital, Sraffa read Fisher, Cannan and also analysed Marshall's discussion of capital in his *Principles* (D 1/2; D1/5, pp. 6–7).

Later, on 27 October 1936, in a letter to Joan Robinson, Sraffa explained why he thought the neoclassical theory of capital did not meet the requirements for theoretical measurement:

> Many thanks for your letter—it is a valuable addition to my museum and shall hang it next to an extract from Sidgwick where, after lecturing Ricardo on a quantity of labour, he goes on cheerfully himself to talk of quantities of utility.
>
> If one measures labour and land by heads or acres the result has a definite meaning, subject to a margin of error: the margin is wide, but it is a question of degree. On the other hand if you measure capital in tons the result is purely and simply nonsense. How many tons is, e.g., a railway tunnel?

> If you are not convinced, try it on someone who has not been entirely debauched by economics. Tell your gardener that a farmer has 200 acres and employs 10 men—will he not have a pretty accurate idea of the quantities of land & labour? Now tell him that he employs 500 tons of capital, & he will think you are dotty—(not more so, however, than Sidgwick or Marshall). (quoted from Bradford and Harcourt 1997: 131)

In his reply to Harrod in the *Economic Journal* in 1962, Sraffa asked: 'what is the good of a quantity of capital [...] which, since it depends on the rate of interest, cannot be used for its traditional purpose [...] to determine the rate of interest?' (Sraffa 1962: 479).

In Keynes's early writings, explicit references to the problems of the measurement of the real stock of capital and to the difficulties of finding a physical unit of capital are hard to come by. Owing to the difficulties in measuring capital, Keynes usually dealt with money capital. Only in 1936, in the *General Theory*, do we find two explicit passages on the measurement of capital, one in Chapter 4 ('the stock of real capital') (CW VII: 37) and one in Chapter 11 ('difficulties as to the definition of the physical unit of capital, which I believe to be both insoluble and unnecessary') (CW VII: 138).

*Utility*

On the assumption of homogeneity of the quantities of utility, Sraffa thought that if the quantity of labour was not homogeneous, neither was utility. He noted that if it was 'meaningless' to talk of 'physical real cost', it was 'equally meaningless' to talk of 'psychological real cost', of utility and also of diminishing returns (D1/20 5). Measuring real cost by utility entailed many problems; for example, the heterogeneity of individuals raises difficulties in aggregating their utility (D3/12/3 49).

Keynes's analysis of the measurement and comparison of utility was developed in considerable detail as from 1905. It runs through his 1909 *Essay on Index Numbers* on to *A Treatise on Money* and later writings, covering the problems of the non-homogeneity of utility, the non-interpersonal comparison of utility, the measure of the amount of utility which a given sum of money purchases and the non-exact measurement of utility due to organic unities. It is interesting to read Keynes's early comments on the measurement of utility in his 1909 *Essay on Index Numbers*:

> The measure of the amount of utility, which a given sum will purchase, is intrinsically and from the nature of the case beyond our reach. Since the total utility of a commodity is not proportional to its quantity, we have no means of comparison between two different total utilities which are part of differing wholes, and the total utility of a given amount of wealth depends upon its distribution. Even if we know the distribution of wealth, there is no *measure* of the aggregate of individual utilities. The aggregate exists and is perfectly determinate, but we must not infer from this that two such aggregates can be measured in terms of a common unit. It may be true that there is always a determinate answer to the question as to which of two aggregates of distributed utilities corresponds to the greater amount of total well-being. We are denying, not the existence of such an answer, but the possibility of obtaining it by measuring and applying the formulae of addition to the individuals of the aggregate. There is an aggregate of utilities, we may say, but not a sum. (CW XI: 59–60)

Later on, in 1930, Keynes re-stressed his point of view, in particular examining a number of possible methods of arriving at approximations of the relative purchasing power of incomes, distinguishing between the direct method of comparing incomes of similar persons and various indirect methods of comparing prices of equivalent composite commodities. But there were limits to all of these methods:

> We are not in a position to weigh the satisfactions for similar persons of Pharaoh's slaves against Fifth Avenue's motor cars, or dear fuel and cheap ice to Laplanders against cheap fuel and dear ice to Hottentots [...]. We cannot hope to find a ratio of equivalent substitution for gladiators against cinemas, or for the conveniences of being able to buy motor cars against the conveniences of being able to buy slaves. (CW V: 104–109)

*Money, Money Aggregates and Index Numbers*

Money is homogeneous by definition or convention; therefore units of money are homogeneous by themselves. Money is the common standard of measure in economics; problems may arise either in the stability over time of the various price levels, or of *the* price level, if it exists, or, in the comparability in space and time of the purchasing power. Problems may also arise on the money side or on the side of the commodities under measurement.

Two main difficulties arise in the money measurement of magnitudes: *theoretical* and *practical* difficulties. They are commonly addressed with index number theory, which Sraffa took an interest in, as did Keynes, as we will see.

In his pre-1928 manuscripts, Sraffa's search for the requirements of measurement included money, in particular the measure of money value and index numbers. He took notes of Walker's book on money and paid close attention to the Chapter on "Stabilità dell'unità di misura". Sraffa noted that 'Il denaro sorge per costume: varie merci-denaro: oro, arg.'. Money is 'Regolato dalla legge: poi, carta moneta', 'Il denaro è una unità di misura, definizione arbitraria dell'unità [stock esistente di oro]' (Sraffa D2/1 3).

Sraffa argued that Marshall distinguished strictly between 'expenses of production' which are measured in money—giving the money measure of real cost—and real cost: 'Expenses of production', reckoned in money, are 'the prices that have to be paid in order to call forth the supply' of those efforts and waitings, and are the money measure of real cost' (D2/4 3 (54)). Commenting on Pigou's 1927 article 'Laws of increasing and diminishing costs', Sraffa noted:

> If by real costs P.[Pigou] means expenses of production [...]. Since the reduction to unity is made through their exchange value (and not according to any other, psychological or physical standard) the obvious unit of measure is money (i.e. purchasing power of commodities in general). (D3/12/3 49)

Sraffa's interest in money also entailed an interest in the value of money and index numbers. Sraffa noted Marx's comment in his History 'Sono "indici"' (D1/9 10). In his pre-1928 'Notes on probability' Sraffa commented on Fisher's unit of measure, pointing out that, like labour, it was variable (D1/49 16). His interest was mainly in the fact that index number theory deals with non-homogeneous complex magnitudes, i.e. composite heterogeneous commodities. Complex magnitudes are incommensurable, apart from some special cases—special cases which Sraffa was particularly interested in. To explain these special cases of complex magnitudes such as the community output, let me quote Keynes's words in the *General Theory*, 'the community's output of goods and services is a non homogeneous complex which cannot be measured, strictly speaking, except in certain special cases, as for example, when all the items of one output are included in the same proportions in another output' (CW VII 38).

In addition to reading books by Walter and various other authors on money, Sraffa also read Keynes's writings on the subject. In his manuscripts we find a note on an earlier table of contents of Keynes's *A Treatise on Money*, which seems to be close to the table of contents dated 22 September 1927 (CW XIII: 48–50), where Keynes deals with index numbers, weighting and the problem of measurement. Sraffa also read and took notes on Keynes's *Tract on Monetary Reform*, as well as translating it into Italian. On the other hand, I have no evidence as to whether Sraffa read Keynes's 1909 *Essay on Index Numbers*, where he first raised the problems of measurement, weighting and proportions. The 1909 essay reappears with changes both in the 1923 *Tract on Monetary Reform* and in the 1930 *Treatise on Money*.

Sraffa's analysis of weights and proportions merits detailed examination since it is at the basis of his Standard commodity. In his pre-1928 'Notes on interdependence and causality' he pointed out:

> I may be measuring a group of quantities without knowing their identity, i.e. when I measure another group of quantities, that are really distinct from them, but within a certain range are appreciably proportional to them: within these limits the theory holds good, but it is bound to be falsified when the two groups become differentiated, disagree and move independently of one another. This is very common in economics. (D1/9 10)

In his 'Notes on language', in dealing with the rate of wage, Sraffa argued that 'the definitions of the rate of wages and rate of interest are similar in many respects, and different in one'. Sraffa's stress on their difference more than on their similarities is relevant here: 'The difference is that rate of interest is 'money per unit of money'; rate of wages 'commodity per unit of man'. Hence the dimensions of money and rate of wage are different. He concluded:

> It is necessary to reduce the two rates to a common basis. This can be done by regarding all the quantities concerned in both rates as quantities of labour. In order to do this however we must first be able to measure the quantities of capital and of product concerned in terms of labour. This may be done provided we assume that the total price of the aggregate of commodities that compose wages is made up in the same proportion of wages and profits as the total price of the aggregate of commodities that compose profits. (D 3/12/7 141)

Sraffa, then, held it necessary to introduce the assumption that the two aggregates of commodities were made up in the same proportion of wages and profits to reduce them to a common basis. On 13 and 19 February 1931, dealing with capital and the standard chosen for measuring the value of capital, Sraffa came back to the weights and proportions argument, as regards 'the value of a mass of articles' and the value of non proportional quantities. He explicitly referred to the problem of weights in index numbers:

> The problem turns entirely upon the standard chosen for measuring the value of capital [...]. Since the inquiry is on the rate of interest (five per cent), the measure must be such as to equate the "five" (consumption goods) with the "cent" (capital); therefore capital must be evaluated in terms of consumption goods. The unit of measure is a parcel of cons. goods, combined in the same proportions as they are produced: these proportions are assumed not to vary; and each capitalist and each worker consumes them in these proportions: no saving takes place.
>
> The problem of measuring the value of a mass of articles is identical with that of an index No of prices: only, the weights are automatically determined by the quantity of each article [...]. Consequently, the two total values (or index numbers), being differently weighted, will not vary proportionally when wages are raised. (D 3/12/7/157 3, 4)

On 22 July 1931, commenting on Keynes's distinction between fixed and circulating capital in his *Treatise on Money* and on his 'way of regarding machines as a factor of production which has nothing to do with circulating capital', Sraffa then concluded that 'It also amounts to acknowledging that capital is really only measurable in terms of value. Marshall (however muddled) as opposed to Böhm Bawerk was not so wrong after all. Nor were Ricardo, Marx, etc. who always measured capital in money' (D3/12/7/159). In 1932, the problem of weighting also underlay the discussion with Hayek:

> For each possible index number of price (i.e. the price of composite commodity) there is a rate of interest on money which, if adopted by the banks (who lend freely at it) will keep it stable: each rate is an average of rates on the various commodities included in the index number, at any one moment, weighted as they are in the index number. (D3/9 7)

Sraffa continued to work on the hypothesis of balanced proportions in the forties. In his notes dated from 23.1.44 to 20.3.44 (D3/12/36) he worked on the Standard system equations. His hypothesis was to assume that the proportions between the sum of the inputs of each commodity and of output of the same commodity be equal in all the commodities. He called this hypothesis 'statistical'. In his Note 70 he stressed that the elements of the reduction to dated quantities of labour: 'are in constant proportions, such that the ratio of any one of them to the sum of all preceding ones (i.e. to the left of it) is R'. Sraffa's 1960 standard commodity is a composite commodity, with 'balanced' proportions; that is, with the same proportions in all its 'layers': 'The 'balanced' commodity which we have just considered (§21) would present no peculiarities of this type, since the same proportion would be found in all its "layers"'(Sraffa 1960: 18).

Keynes's views on the measurement of the value of money, and of other complex economic magnitudes in general, were formulated in considerable complexity over a long period of time. There is nothing new in Keynes's approach to complex magnitudes and their units of quantity and measure as found in Chapter 4 of the *General Theory*. He had already considered them in at least five of his previous writings: in 1905, in his papers on index numbers written for Marshall, in 1907, in the first version of his Dissertation *The Principles of Probability* (later *A Treatise on Probability*) whilst considering the measurement of the relation of probability; in his 1909 *Essay on Index Numbers* while considering the measurement of price level, in his analysis of the price level in the *Tract on Monetary* Reform (1923) and in *A Treatise on Money* (1930) (see Carabelli 1992). Central to his thought is the problem of weighting and proportions (CW XI: 72–94; CW V: Chap. 4 and Chap. 8).

### 2.2 *On Independence*

Sraffa paid great attention to the notions of independence, interdependence and causality (D1/9). Dependence means that some type of causality is implied, but there is no agreement in philosophy and in the history of thought on what causality means. On this topic, Sraffa's readings were De Quincey on 'Malthus on the measure of value', Pigou's *Economics of Welfare* and his 1913 article on 'The interdependence of different sources of demand and supply in a market', Wicksteed's *Common Sense*, Sidgwick's *Principles* and Keynes's *A Treatise on Probability* (1921).

Sraffa's criticism of partial equilibrium theory was based on making explicit the assumption of mutual independence of the demand and supply curves of products. The argument that in economic theory nothing was to be clarified as regards the theory of value had, he observed, already been stated by John Stuart Mill in 1849, Pantaleoni in his obituary of Pareto and John Maynard Keynes in his 1923 *Introduction to Cambridge Economic Handbooks.* Sraffa was of the contrary opinion. He stressed that for the economic theory of partial equilibria the hypothesis of independence was a necessary condition for it to stand in terms of logic without falling into fallacies. In particular, he went on, the theory of partial equilibria required that the demand and the supply curves be mutually independent (D1/20 7).

In his Notes on Marshall's condition 'Other things being equal' in *Principles of Economics* (8th edition) dated April 1923, Sraffa stressed that this is not one condition, but many conditions, or rather *infinite conditions* (D 1/2). In his 1928–1931 advanced lectures, while discussing partial and general equilibrium, Sraffa noted that the tacit ('concealed') assumption of independence had to be introduced: prices and quantities of all the other commodities do not remain unchanged (D2/4 3 (84)). In particular, when dealing with the production as a whole, the assumption of independence may give rise to logical fallacies; for example, the demand for corn is not independent of variation in national income (D1/22 17).

Sraffa also analysed 'the conditions in which the "existence of substitutes" blows up every economic theory', as it means that there is interdependence between commodities (D1/20 5). Marshall's 'exceptional cases' were the target of his investigation. Exceptions 'are the rule' (D1/22/1). On the interdependence of demand curves, Sraffa noted that Wicksteed and Cairnes had already raised the question of the interdependence of supply and demand curves (D2/4 21(3); D2/4 24; D1/22 11).

Sraffa's discussion of the interdependence of commodities, prices and demand and supply is to be associated with his interest in general economic equilibrium. In his 'Lecture notes on economic theory' Sraffa asked what economic theory actually was, and argued that reality was extremely complex and the price of each commodity was influenced by the prices of all other commodities, concluding that to know the causes of the price of a commodity would require so much knowledge as to fill 10.000 books—beyond the realm of possibility (D2/1 1). On this topic, Sraffa referred to Walras', Pareto's and Cassel's theories of general equilibrium, with Cassel's version considered inferior (D2/4 4(3)), 5(6)).

Keynes took great care in detecting the different nature of the tacit assumptions of independence in classical theory. The various assumptions he considered can be classified in three types:

- the assumption of independence from changes in the value of money;
- the assumption of independence from changes in the value of output and unemployment;
- the assumption of independence from changes in the level of income.

The assumption of independence from changes in the value of money carried with it the idea of neutral money and allowed the classical theoretician to pass, without any change in reasoning, from a real exchange economy to a money economy, and brought with it the false analogy between the two. So, truly and consistently, under this assumption a money economy was equal to a real economy. According to Keynes, the assumption took on different forms within the classical theory: that of neutral money, that of a uniform purchasing power of money (as if money were a 'mean sum') and that of neglecting possible changes in the general purchasing power of money.

Keynes took Marshall's and Pigou's writings into consideration with regard to this type of assumption. Marshall, he observed, supported Cournot's assumption of a standard of uniform purchasing power (CW XIII: 409). As regards Pigou, Keynes referred to his introduction of the tacit assumption that the supply of labour was independent of changes in the value of money (CW XIII: 409–410).

The assumption of independence from changes in the value of output and unemployment implied that the economic system was operating to its full capacity, which meant an independence from the level of output or employment. According to the classical theory, introduction of the assumption allowed for straightforward transition from argumentation based upon a full-capacity economy to argumentation based upon a situation characterised by unemployment. In the Preface to the French Edition of the *General Theory*, Keynes suggested that the use of Say's basic assumptions, such as that demand is created by supply, meant independence from the level of output (CW VII: xxxii-xxxiii; see also p. 21 and CW XIII: 278).

Upon this was grounded Keynes's criticism that the classical saving and investment schedules could not 'shift independently of one another' and that there was interdependence between the two (CW VII: 179). Similarly, supply and demand curves for loanable funds were not independent. One 'could not obtain a determinate conclusion without introducing some additional equation or datum' (CW XXIX: 228). Keynes linked the interdependence of supply and demand with his money theory of value (CW VII: xxii–xxiii).

In comparison with Sraffa's methodological attitude, it is also interesting to note Keynes's methodology of criticism in his *A Tract on Monetary Reform*, a book well known to Sraffa. In fact, a very similar criticism based on the introduction of tacit assumptions of independence had already been raised by Keynes against the quantity theory of money in his *Tract on Monetary Reform* (1923):

> the [quantity] theory has often been expounded on the further assumption that a mere change in the quantity of the currency cannot affect k, r, and k'—that is to say, in mathematical parlance, that n is an *independent variable* in relation to these quantities. It would follow from this that an arbitrary doubling of n, since this in itself is assumed not to affect k, r, and k', must have the effect of raising p to double what it would have been otherwise … Now 'in the long run' this is probably true […] But this *long run* is a misleading guide to current affairs. In *the long run* we are all dead […] In actual experience, a change of n is liable to have a reaction both on k and k' and on r. (CW IV: 65)

In a similar vein were Keynes's criticisms of the theory of purchasing power parity ('the theory requires a further assumption for its validity') and the presumed intrinsic stability of the value of gold ('The independent variety of the influences determining the value of gold has been in itself a steadying influence […] The value of gold is no longer the resultant of the chance gifts of nature and the judgement of numerous authorities and individuals acting independently') (CW IV: 75, 133–134). In *A Treatise on Money* (1930), the point was re-emphasised by Keynes: 'the purpose of isolating "changes on the side of money"'; 'subject to "independent" influences'; 'the non-independence of relative price changes' (CW V: 73, 75, 77).

## 3 Sraffa's Versus Keynes's Constructive Approach

Having discussed the similarities between Sraffa and Keynes's critical approaches to economic theory based on the assumption of homogeneity in the quantities of labour, products, capital, utility and money, and on the role played by the assumption of logical independence, let us now consider their constructive approaches. Here they move in opposite directions, along with their contrasting choices of magnitudes to be considered in economics. Sraffa, in his constructive approach, searched for an absolute physical standard of value and measure, while Keynes stuck to money measures and to the unit of labour as the sole physical unit.

### 3.1 *Sraffa on the Theory of Value. Real Costs: Disutility or Physical Real Cost?*

In theory of value Sraffa searched for the ultimate standard of value and defended cost as physical real cost against disutility. He noted that, according to economists such as Marshall, Clark, Scharling and Böhm Bawerk, real cost is 'an aggregate of the unpleasant feelings of various sorts felt by the individuals connected with production' (D2/4 3 (21)) and that 'All the ultimate standards we have considered up to the point are psychological [...] Real costs interpreted as disutility' (D3/12/3 42; see also D1/22 5–6).

In his analysis of Marshall's theory of value—called by him the 'single theory of value'—Sraffa held that the fundamental step made by Marshall was the reduction of cost to a quantity of disutility (D2/4 3 (18, 20–21). Two heterogeneous quantities were reduced to a common homogeneous one (D1/28 2–3; D1/30). In this way Marshall was able to bring together two quantities, utility and disutility, which are heterogeneous—not of 'the same kind', as Sraffa wrote (D2/4 3 (66))—to a homogeneous quantity: disutility was made of 'the same nature' as utility. By reducing cost to a common measure based on utility Marshall was able to take it as a negative magnitude which could be 'added or subtracted and balanced against one another' (D2/4 3 (17)); thus the laws of addition and subtraction and the idea of a common balance on which to weight cost and utility could be applied. Sraffa pointed out the central role that the measurement of motives played in Marshall's method and theory of value (D2/4 3 (22)). Marshall's use of symmetry in economic theory and his idea of the scissor theory of value rested, then, on the tacit hypothesis of the homogeneous nature of quantities.

### 3.2 *Sraffa's Search for a Physical Unit of Measure: Food Rather than Labour?*

Sraffa aimed at a theory of cost which was grounded on physical terms, a real physical costs theory. As we have seen, he held that utility could not be a common standard: the psychological standard was to be rejected. He was interested in searching for a common physical unit of quantity and a common physical standard of measure: 'we must find some ultimate standard, independent from the variables we are considering, such [as] utility, disutility or labour' (D3/12/3 Notes London, Summer 1927 Physical Real Costs, p. 13).

The quantity of labour was inadequate because it was not homogenous; money as measure carried for Sraffa a flavour of utility. Even less was he interested in the money wage of labour. In the Ricardo-Malthus controversy, Sraffa sided with Ricardo, defending Ricardo's labour theory of value against Malthus's theory based on the money wages received by labour (D3/12/3 47; D/3/12/4).

Petty's theory of physical quantities was Sraffa's benchmark. Petty's common measure of value was 'the days food of an adult Man'. A theory of cost which was 'concrete, tangible and visible', that could 'be measured in tons or gallons'. It was the opposite of Marshall's cost, which was individually based, and could only be measured in money terms. Sraffa quoted, with approbation, Petty's well known passage in his preface to Political Arithmetick (Petty 1670, I: 244):

> The method [...] to express my self in terms of Number, Weight or Measure; to use only Argument of sense, and to consider only such Causes, as have visible foundation in Nature; leaving those that depend upon the mutable Minds, Opinions, Appetites and Passions of particular Men to the Consideration of others. (Sraffa D2/4 3 (21))

Sraffa's aim was to find an absolutely necessary commodity; this necessary commodity represents one of the bases of his 1960 *Production of Commodities by Means of Commodities*. Was corn this ideal and absolutely necessary commodity? In the history of economic thought economists went 'from food to labour'. Sraffa thought 'The right notion of cost as "loaf of bread". Then somebody started measuring it in labour, as every day's labour requires the same amount of food' (D/3/12/4 Nov 1927). His answer to the question whether corn is that necessary commodity, was negative. Corn was not:

> an absolutely necessary commodity found, the difficulty of reducing to a common measure the various things entering into real cost would solve by itself [...] by going back enough in the genealogy of production [...] we might find exactly the total amount of corn (if this were the ideal necessary commodity which is not). (D3/12/3 44)

Rather than corn, 'the objective, physical, necessaries of existence' consisted in the amount of various commodities:

> we must find a unit of measure for cost: the necessity for this unit arises, not from a desire of actually measuring,—it is prior to it, and is required even for thinking of cost. The best measure available is the amount of various commodities that is required to support during an hour a average labourer [...] the amount of necessaries. (D3/12/3 46)

### 3.3 *Sraffa's Question: Is There Any Real Need of a Common Natural/Physical Unit of Measure?*

In his pre-1928 notes Sraffa raised the question as to whether physical aggregate quantities—always and in general—require a common unit of measure and standard. This seemed to be indispensable for defining real costs. But, as we will see, this was not what Sraffa thought. He quoted and commented on Sidgwick on the necessity of a common standard for the conception itself of physical real costs. In contrast with Sidgwick, Sraffa thought that, under limited conditions, there is no need to have a common natural/physical unit of measure:

> Some of the most eminent (economists) have not always seen that it is impossible to think definitely of the quantity of an aggregate of diverse elements, except so far as these elements admit of being reduced to a common quantitative standard, and that unless this is done, when we speak of such an aggregate having increased or decreased in amount, or of something else varying "in proportion to" it, we are using word to which there are necessarily no definite thoughts corresponding.

Sraffa commented on Sidgwick:

> The first sentence seems right, but the second wrong. The conception of physical real costs, the replacement of exactly those things that have been used in production, is a clear and definite conception and does not require

> a "common standard". This necessity of a "common standard" is a prejudice: if we don't vary the quantity, or if we vary the physical quantity of all elements in the same proportion, no common unit is required. (D1/22 10)

He advanced the idea—which we now know underlies his interpretation of Ricardo's theory of profits (Sraffa 1951) and the main assumptions of his *Production of Commodities by Means of Commodities* (1960), that if the quantities do not vary or if the quantities of all elements vary in the same proportion, no common unit or standard is required. In Sraffa's interpretation of Ricardo's corn model, there is no problem of aggregation of heterogeneous commodities, corn being the only commodity; therefore homogeneity is assumed by definition. When there are many commodities (two commodities are enough), if the quantities do not vary (or if they are kept constant by assumption) or if the quantities vary in the same proportion in the two aggregates under consideration or comparison; or if the quantities of the numerator and of the denominator are in the same proportion, it is as if there is homogeneity. In his *Production of Commodities by Means of Commodities* (1960) the hypothesis that quantities do not vary is explicitly present; and the "standard commodity" shows proportional variations in the quantities of commodities:

> The 'balanced' commodity which we have just considered (§ 21) would present no peculiarities of this type, since the same proportion would be found in all its 'layers'. It is true that, as wages fell, such a commodity would be no less susceptible than any other to rise or fall in price relative to other individual commodities; but we should know for certain that any such fluctuation would originate exclusively in the peculiarities of production of the commodity which was being compared with it, and not in its own. If we could discover such a commodity we should therefore be in possession of a standard capable of isolating the price-movements of any other product so that they could be observed as *in a vacuum*. (Sraffa 1960: 18)

Particular proportions, such as the Standard ones, may give transparency to a system and render visible what was hidden, but they cannot alter its mathematical properties (Sraffa 1960: 23).

Sraffa's idea that if the quantities do not vary or if the quantities of all elements vary in the same proportion, no common unit or standard is required, seems to me to have been borrowed from index numbers analysis and applied to the theory of prices and distribution. As seen in the section on money, Sraffa was interested in index number theory. In this latter theory, 'the community's output of goods and services is a non homogeneous complex which cannot be measured, strictly speaking, except in certain special cases, as for example when all the items of one output are included in the same proportions in another output' (CW VII: 38). The special case is Sraffa's case, a case which Keynes did not like.

### 3.4 *Keynes's Monetary Standard and His Money Theory of Value*

In the *General Theory*, Keynes argues that the choice of units was one of 'the three perplexities which most impeded my progress in writing this book, so that I could not express myself conveniently until I had found some solution for them (CW VII: 37). The central concepts and magnitudes involved in his analysis are complex. These concepts are the volume of real output/income, the volume of real net output, the stock of real capital, aggregate goods and services and the general price level (ibid.). Keynes defines them as 'incommensurable collections of miscellaneous objects' (ibid.: 39). In particular, 'the community's output of goods and services' is 'a non-homogeneous complex which cannot be measured' (ibid.: 38). In the *General Theory* there is also another complex magnitude, that of probability. Keynes considers probability separately in chapters 5 and 12. For Keynes, these concepts are theoretically vague. He refers to 'the well-known [...] element of vagueness which [...] attends the concept of the general price level' (ibid.: 39). Theoretical vagueness does not mean that these logically indefinable concepts are also vague and ambiguous from an ordinary language point of view or in business practice. In ordinary language their meaning is easily grasped and they can be used for practical purposes.

Complex magnitudes are indefinable in logical terms since it is impossible to reduce them to more simple terms. They cannot be reduced to simple terms without falling into logical fallacies and paradoxes. The fallacy of composition is well known to economists. Complex magnitudes also possess attributes which belong to different dimensional scales. They are multidimensional magnitudes, inasmuch as they can move simultaneously in more than one direction. This calls for a theory of the units of quantities appropriate to them and it also means having an appropriate theory of measurement. For Keynes, complex magnitudes are theoretically troublesome as far as measurement is concerned. They raise logical "conundrums" rather than mathematical or statistical difficulties (CW VII: 39). The classical economists had committed logical errors by ignoring this problem: 'the units, in terms of which economists commonly work, are unsatisfactory' (ibid.: 37). Being non-numerical quantitative concepts (ibid.: 40), the quantities of real output, real net output, real capital, aggregate goods and services and the general price level do not exist, any more than do absolute and homogeneous units of measure appropriate to these quantities. For this reason Keynes stated that complex concepts 'cannot in themselves provide the material for a quantitative analysis' (ibid.: 39). Thus, the endeavour by classical economists 'to erect a quantitative science' (ibid.: 38) upon these concepts was a logical failure. The outcome was 'a mock precision', as it tried 'to use such partly vague and non-quantitative concepts as the basis of a quantitative analysis' (ibid.: 40). As a result, he wrote:

> It is my belief that much unnecessary perplexity can be avoided if we limit ourselves strictly to the units, money and labour, when we are dealing with the behaviour of the economic system as a whole; reserving the use of units of particular outputs and equipments to the occasions when we are analysing the output of individual firms or industries in isolation; and the use of vague concepts, such as the quantity of capital equipment as a whole and the general level of prices, to the occasions when we are attempting some historical comparison which is within certain (perhaps fairly wide) limits avowedly unprecise and approximate. (CW VII: 43)

For Keynes, money is the only standard and unit of measure. Economic magnitudes are not 'concrete, tangible and visible'; they cannot be 'physically measured, in number, weight (tons or gallons)', as Sraffa required. Furthermore, as 'psychological uncertainties', they are not directly observable (CW XIV: 300). The hypothesis of 'equal proportions' cannot be applied as quantities change in a non-proportional way.

## 4 Conclusion

The main focus of this chapter is on measurement in economics. I have investigated two authors, Sraffa and Keynes, who were in close contact in Cambridge in the twenties and thirties, but for whom at least up to now we have no methodological comparisons.

We have seen that there are significant similarities in Sraffa's and Keynes's methodology of critique, the so-called *pars destruens* of their thought, based on making explicit the logical tacit assumptions in the economic theory under attack. Two tacit assumptions are fundamental here with respect to measurement: the assumption of homogeneity in kind and logical independence. Both Sraffa and Keynes accepted that economics was characterised by heterogeneity and variety; both believed that the material of economics was complex and interdependent and require a theory and concepts with similar characteristics. Both believed that investigation into measurement was necessary for the purpose of conceptual reasoning rather than for actual measurement.

An original conclusion we draw in this chapter is that, in Sraffa's development of the standard commodity, a central role is played by the assumption of 'equal proportions' between components or layers—a hypothesis he seems to have borrowed from index number theory, and which is connected with the problem of weighting. When quantities do not change or when they change in the same proportions between two complex aggregates under measurement or comparison, we do not need a common unit of measure. Sraffa began by searching for a physical measure, a commodity. Initially it was corn rather than labour, labour not being homogeneous. Subsequently he moved towards a hypothetical composite commodity which showed equal proportions—an abstract commodity necessary at the conceptual level and to understand what happens in economic reality, without actually picturing it.

Notwithstanding these striking similarities, the differences between Sraffa and Keynes in their constructive approaches were considerable. They showed contrasting attitudes in the *pars construens* of their economic theory: Sraffa moved towards an objectivisation of economic theory based on physicalism and empiricism, while Keynes always remained faithful to a money vision of the economy which opposes physicalism and empiricism and defends a weak form of rationalism, stressing that some economic magnitudes are intrinsically non-measurable, but that this precluded neither work on economic theory nor justification of economic intervention. Money measures were inevitable. For Sraffa, on the contrary, physical measurement was still necessary and possible.

## References


Bellofiore, R. and Carter, S. (eds) (2014) *Towards a New Understanding of Sraffa: Insight from Archival Research*, Basingstoke: Palgrave Macmillan

Blankenburg, S., Arena R. and Wilkinson F. (2012) 'Piero Sraffa and "the True Object of Economics": the Role of the Unpublished Manuscripts', *Cambridge Journal of Economics, Special Issue: New Perspectives on the Work of Piero Sraffa*, 36, 6, 1267–1290

Bradford, W. and Harcourt, G.C. (1997) 'Units and Definitions', in G.C. Harcourt and D.A. Riach (eds) *A Second Edition of the General Theory*, Vol. 1, London: Routledge, 107–131.

Bridgman, P. W. (1927) *The Logic of Modern Physics*, New York: Macmillan. [Sraffa 3617].

Carabelli, A. (1988) *On Keynes's Method*, London: Macmillan.

Carabelli, A. (1991) 'The Methodology of the Critique of the Classical Theory: Keynes on Organic Interdependence', in B. Bateman and J. Davis (eds) *Keynes and Philosophy*, Aldershot: Elgar, 104–125.

Carabelli, A. (1992) 'Organic Interdependence and Keynes's Choice of Units in the *General Theory*', in B. Gerrard and J. Hillard (eds) *The Philosophy and Economics of J.M. Keynes*, Aldershot: Elgar, 3–31.

Carabelli, A. (1994) 'Keynes on Mensuration and Comparison', in K. Vaugh (ed.) *Perspectives on the History of Economic Thought*, Aldershot: Elgar, 204–238

Carabelli, A. (1995) 'Uncertainty and Measurement in Keynes: Probability and Organicness', in S. Dow and J. Hillard (eds) *Keynes, Knowledge and Uncertainty*, Aldershot: Elgar, 137–160.

Carabelli, A. (2003) 'Keynes: economics as a branch of probable logic', in J. Runde and S. Mizuhara (eds) *The Philosophy of Keynes's Economics*, London: Routledge, 216–226.

Chiodi, G. and Ditta, L. (2010) *Sraffa or an Alternative Economics*, London: Macmillan

Chiodi, G. and Ditta, L. (2013) 'Sraffa and Keynes: Two Ways of Making a "revolution" in Economic Theory', in Levrero, E.S., Palumbo A. and Stirati, A. (eds) Sraffa and the Reconstruction of Economic Theory: volume one, London: Macmillan, 218–240

Ciccone, R, Gerhrke, C. and Mongiovi, G. (eds) (2011) *Sraffa and Modern Economics*, New York: Routledge

Davis, J. (2018) "Sraffa on the Open vs. 'Closed Systems' Distinction and Causality," *Research in the History of Economic Thought and Methodology* 35B, 153–170.

Eddington, A. S. (1928) *The Nature of the Physical World*, Cambridge: Cambridge University Press. [Sraffa 3080].

Hertz, H. (1899) *Principles of Mechanics*, London: Macmillan.

Keynes, J. M. (1972–1989) *The Collected Writings*, D. Moggridge and E. Johnson (eds), Macmillan: London, Vol. VII *The General Theory of Employment, Interest and Money* (1936); Vol. VIII *A Treatise on Probability* (1921); vol. XIII. *The General Theory and After: Part I. Preparation*, 1973; vol. XIV. *The General Theory and After: Part II, Defence and Development*, 1973.

Keynes, J. M. (1908) *The Principles of Probability*, submitted as Fellowship Dissertation to King's College, Cambridge in Dec. 1908; Keynes MSS, King's College Library, Cambridge

Keynes, J. M. (1907) *The Principles of Probability*, submitted as Fellowship Dissertation to King's College, Cambridge in Dec. 1907; Keynes MSS, King's College Library, Cambridge.

Keynes, J. M. MSS, King's College Library, Cambridge.

Knight, F. H. (1924) 'The Limitations of Scientific Method in Economics', in R. G. Tugwell (ed.) (1924).

Kurz, H. D. (1998) 'Against the current: Sraffa's unpublished manuscripts and the history of economic thought', *European Journal of the History of Economic Thought*, 5, 437–451.

Kurz, H. D. and Salvadori, N. (1998) 'Given quantities', in H. D. Kurz and N. Salvadori (eds) *The Elgar Companion to Classical Economics*, Vol. I., Cheltenham: Edward Elgar.

Kurz, H. D. and Salvadori, N. (2004). 'Man from the moon'. On Sraffa's objectivism, *Economies et Societes*, 35, 1545–1557.

Kurz, H. D. and Salvadori, N. (2005a) 'Removing an "insuperable obstacle" in the way of an objectivist analysis: Sraffa's attempts at fixed capital', *European Journal of the History of Economic Thought*, 12, 493–523.

Kurz, H. D. and Salvadori N. (2005b) 'Representing the Production and Circulation of Commodities in Material Terms: On Sraffa's Objectivism', Review of Political Economy, 17, 413–441. Reprinted in H. Kurz, L. Pasinetti, & Salvadori (eds.), *Piero Sraffa: The man and the scholar*). London: Routledge, 249–277.

Lutz, F. and Hague, D. C. (eds) (1961) *The Theory of Capital*, London: Macmillan.

Marshall, A. (1920) *Principles of Economics*, 8th edn, London: Macmillan.

Petty, W. (1899) [1691] Political Arithmetick, London in Petty, W. *The Economic Writings of Sir William Petty*, Vols I–II (C. H. Hull, Ed.), Cambridge: Cambridge University Press. [Sraffa 1587].

Ricardo, D. (1951–1973) *The Works and Correspondence of David Ricardo*, Vols I–XI (ed. by P. Sraffa with the collaboration of M.H. Dobb), Cambridge: Cambridge University Press.

Signorino, R. (2000a) 'Method and analysis in Piero Sraffa's 1925 critique of Marshallian economics', *European Journal of the History of Economic Thought*, 7, 569–594.

Signorino, R. (2000b) 'The Italian debate on Marshallian (and Paretian) economics and the intellectual roots of Piero Sraffa's "Sulle relazioni fra costo e quantità prodotta". A note', *History of Economic Ideas*, 8, 143–157.

Signorino, R. (2001a) 'An appraisal of Piero Sraffa's "The laws of returns under competitive conditions"', *European Journal of the History of Economic Thought*, 8, 230–250.

Signorino, R. (2001b) 'Piero Sraffa on Utility and the Subjective Method in the 1920s: a Tentative Appraisal of Sraffa's Unpublished Manuscripts', *Cambridge Journal of Economics*, 25, 749–763.

Sraffa, P. (1925) 'Sulle relazioni fra costo e quantità prodotta', *Annali di Economia*, 2: 277–328; English translation in: L.L. Pasinetti (ed.) (1998) *Italian Economic Papers*, Vol. III, Bologna: Il Mulino and Oxford: Oxford University Press.

Sraffa, P. (1926) 'The laws of returns under competitive conditions', *Economic Journal*, 36, 535–550.

Sraffa, P. (1930) 'A criticism' and 'Rejoinder' (contributions to the Symposium on 'Increasing Returns and the Representative Firm'), *Economic Journal*, 40, 89–93.

Sraffa, P. (1931) Surplus Product, D3/12/7 161: August 1–5. Sraffa MSS, Wren Library, Trinity College, Cambridge.

Sraffa, P. (1932) 'Dr. Hayek on Money and Capital', *Economic Journal*, 42, 42–53.

Sraffa P. (1951) *Introduction to Ricardo's Works*, Cambridge: Cambridge University Press.

Sraffa, P. (1960) *Production of Commodities by Means of Commodities. Prelude to a Critique of Economic Theory*, Cambridge: Cambridge University Press. Italian edn: *Produzione di Merci a Mezzo di Merci. Premesse a una Critica della Teoria Economica*. Torino: Einaudi.

Sraffa, P. (1962) 'Production of Commodities: A comment', *Economic Journal*, LXXII, 477–479.

Sraffa, P. MSS, Wren Library, Trinity College, Cambridge.

Tugwell, R. G. (ed.) (1924) *The Trend of Economics*, New York: Alfred A. Knopf.

Whitehead, A. N. (1926) *Science and the Modern World*, Cambridge: Cambridge University Press. [Sraffa 662].

Whyte, L. L. (1928) *Archimedes or The Future of Physics*, London: Kegan, Paul, Trench, Trubner & Co.

XIII The General Theory and After: Part I. Preparation.

XIV The General Theory and After: Part II, Defence and Development.

Comment

# A Comment on Carabelli on Sraffa and Keynes on Measurement

*John B. Davis*

Anna Carabelli's (2018) excellent paper comparing Sraffa and Keynes on the measurement of economic magnitudes addresses both how the issue of measurement figured into their critiques of other economic theories and how it provided grounds on which they developed their own theories. She sees strong similarities in their critiques of existing theories, but she also sees significant differences between them in regard to how they each went on to construct their own theories. What they shared was the belief that "economics is characterised by heterogeneity and variety … [and] that the material of economics is complex and interdependent and require[s] a theory and concepts with similar characteristics" (166). Where they differed concerned the philosophical foundations of the theories they themselves developed: "Sraffa moved towards an objectivisation of economic theory based on physicalism and empiricism, while Keynes always remained faithful to a money vision of the economy which opposes physicalism and empiricism" (ibid.). To make this all clearer, let us begin by distinguishing Sraffa and Keynes in terms of both methodology and method.

Consider first their common ground regarding the critique of existing theories. Here the issue is what kinds of concepts are used in economics. A problem they agreed existing theories often encountered was their recourse to vague concepts in the sense of concepts not susceptible to quantitative measurement, whether in value terms or physical terms. Vagueness in concepts, then, not only precludes theories' application to the world but also their empirical evaluation. We can characterise this as a specifically methodological problem because, as Carabelli notes, methodology concerns "epistemology and the philosophy of science basis on which explanations are made" (ftn. 1). Since this issue concerns the adequacy of existing theories, the problem of vagueness provides a basis for their critique of existing theories.

Consider second where they differed and in particular the issue of the unit or standard of measure, or that by which measurement is carried out. Clearly a unit or standard of measure needs to be independent of the quantities it is used to measure since otherwise it fails as a standard of measure. On this they agreed. In this case, we can characterise the issue here as a once concerning method because, as Carabelli notes, method concerns "technique and how concepts are used, applied, and developed" (ibid.). Then, since which particular standard of measure one adopts affects the foundations on which new theories may be reliably constructed, the issue bears on the question of where Sraffa and Keynes went in different directions theoretically.

Thus, in regard to the measurement of economic magnitudes, Sraffa and Keynes largely agreed with one another on methodological grounds and in their critiques of much existing theory, but when it came to the theories they each developed they disagreed with one another on grounds of method regarding how economic magnitudes should be identified. An implication of this is that Sraffa's "objectivisation of economic theory based on physicalism and empiricism" and Keynes's "faithful[ness] to a money vision of the economy" derived not from methodological differences between them regarding what theories needed to do in order to be adequate but from differences over the technique and practice in how theories could be developed. That is, once the problem of vagueness is addressed, what techniques one employed depended on what types of theory one intended to develop.

Carabelli, then, nicely captures the dilemmas Sraffa and Keynes each encountered—indeed the economist encounters—regarding determining a method for measurement of economic magnitudes:

> not all economic quantities can be expressed (measured) in physical terms ... [and often] quantities can be expressed or measured in physical terms only under restricted conditions. In fact, when difficulties in their physical measurement arise, economists usually make recourse to value measures, either in price or in money terms. And yet, problems may also arise with value measures. (142)

One might prefer physical measures, then, but they may not be feasible given the goals of one's type of analysis. At the same time, recourse to value measures encounters its own difficulties. Thus, there is no simple path forward when it comes to constructing theory, and one has no alternative but to exercise judgment regarding how one proceeds.

In Sraffa's case, this came out in his remarks at the Corfu conference on capital theory, where he distinguished between measurement as statisticians understand it and measurement in theory. The latter is the basis for the former and in itself requires absolute precision. This depended upon having an adequate definition of capital, which Sraffa believed Böhm-Bawerk and J.B. Clark had failed to achieve. For him, this was thus a methodological failing that helped him ultimately develop his understanding of the Standard Commodity, which allowed him to advance his view that there exists a linear relationship between the share of wages and the rate of profits in the surplus. Carabelli reviews, then, Sraffa's account and evaluation of three types of quantities possible in economics, and his rejection of two of the three and recommendation that economic theory be based on the third, specifically, quantities that have an objective, independent existence at points in time or throughout the processes of production and distribution.

In Keynes's case, how he proceeded stemmed from his view of economics as a moral science. This in turn derived from his early thinking about the logical and theoretical difficulties involved in thinking about the nature of probability, which carried over to his economic thinking about the macroeconomy where expectations play an important role in a probabilistic world. For Keynes, economic magnitudes are not like those employed in the physical sciences, because economic magnitudes exhibit an intrinsic indeterminacy. Thus for him there is no unique unit of measurement. Indeed, as Carabelli puts it: "The impossibility of a numerical measurement of probability is not a product of mental incapacity or lack of knowledge, but arises from the nature of the case itself" (148). Needless to say, this implies a quite different approach to economic theory than the one Sraffa adopted.

However, despite their differences regarding method and theory construction, Sraffa and Keynes agreed methodologically about the centrality of the assumptions of homogeneity and independence—or rather their heterogeneity and variety—when it came to how we think about the phenomena that economics investigates. At root in their thinking was their shared conviction "that the material of economics is complex and interdependent (166). Here I only briefly summarise their thinking about homogeneity and heterogeneity, as reviewed by Carabelli, since the degree to which we can ascribe independence to economic phenomena depends first on whether and when we can judge them as homogeneous.

As she points out, then: "Homogeneity means that there is no difference (or variety) in the kind (or substance) of quantities" (150). Thus for Sraffa, the heterogeneity of economic phenomena was the starting point, though importantly under particular conditions homogeneity might be assumed. That is, the issue for him are the "*relevant* differences or variety" (ibid.). As Carabelli explains, the

> [I]ntroduction of the assumption of homogeneity depended on judgements of relevance. For Sraffa relevance implied that although there exist differences and variety in the quantities they can be judged negligible: relevance was not empirical but logical, i.e. relevance to the problem under investigation. (ibid.)

Similarly, for Keynes, economics quantities are generally not homogeneous except under quite specific conditions. For him, economic magnitudes are a

> non homogeneous complex which cannot be measured [...] except in certain special cases ... [they are an] incommensurable collections of miscellaneous objects. (CW 7: 38, 39)

For both, consequently, the goals of theory construction consequently determined when homogeneity in economic magnitudes might be introduced or assumed. This was in contrast to the theories they criticised, in which homogeneity of economic magnitudes was mistakenly assumed to be the case by default, thus mis-representing the nature of economic phenomena from the start.

Carabelli's paper thus goes on to review in careful detail how on these grounds Sraffa and Keynes reasoned about the concepts of labour, capital, utility, and money. This discussion is much recommended for careful study. Indeed, it bears directly on not only how we understand the relationship between Sraffa and Keynes, but also why they moved in "opposite directions" to construct quite different types of theories—Sraffa with his search for "an absolute physical standard of value and measure" and "Keynes stuck to money measures and to the unit of labour as the sole physical unit" (160).

Of course there is much debate about the relationship between Sraffa and Keynes's ideas and about the differences between the theories they constructed. Yet generally these debates have ignored how their methodological views and understanding of method were foundational to their critiques of existing theories and the theories they developed. However, as Carabelli nicely shows, their common ground effectively created a division of labour between them regarding their respective positive contributions. In this way, she opens the door to deeper study of the commonalities and distinctions within the Cambridge School.

## References

Carabelli, Anna (2018) "Sraffa versus Keynes on the Method of Economics: Measurement, Homogeneity and Independence," *Annals of the Fondazione Luigi Einaudi. An Interdisciplinary Journal of Economics, History and Political Science*, Fondazione Luigi Einaudi, Torino (Italy), 52(2): 137–168.

Keynes, J.M. (1936; 1973) *The General Theory of Employment, Interest and Money*, vol. 7, *The Collected Writings of John Maynard Keynes*, ed. by E. Johnson and D. Moggridge, London: Macmillan.

CHAPTER 9

# The Role of Sraffa Prices in Post-Keynesian Pricing Theory

*Geoffrey Colin Harcourt*

## 1 Introduction

Alessandro Roncaglia has been analyzing Piero Sraffa's contributions and how they fit into the context of the developments of economic theory since the time of the classical political economists on. In particular, he has concerned himself with how the structure of Sraffa (1960) prices in *Production of Commodities by Means of Commodities* are to be interpreted in relation to both Sraffa's (and our) predecessors and as a contribution to modern economic theory. In his comprehensive volume on Sraffa in Tony Thirlwall's important series, *Great Thinkers in Economics*, Roncaglia (2009) identifies three major interpretations of what Sraffa has done and where he may have wished developments of his contributions to go: a

This chapter was first published in *Classical Economics Today: Essays in Honor of Alessandro Roncaglia*, 2018, Anthem Press. It is here reprinted with the permission of the publisher.

G. C. Harcourt (✉)
School of Economics, NSW Business School, University of New South Wales, Sydney, Australia
e-mail: gch@unsw.edu.au



A. Sinha (ed.), *A Reflection on Sraffa's Revolution in Economic Theory*, Palgrave Studies in the History of Economic Thought,
https://doi.org/10.1007/978-3-030-47206-1_9

Smithian interpretation, a Ricardian interpretation and a Marxian interpretation. Along with his mentor, Paolo Sylos-Labini, Roncaglia places himself in the Smithian stream. He identifies Heinz Kurz and Neri Salvadori especially in the Ricardian stream and Pierangelo Garegnani especially in the Marxian stream.

While I greatly respect his arguments and provision of evidence in the public domain, I want to argue here that the Marxian stream is the most appropriate one, both for interpreting Sraffa's own views and inclinations and for providing relevant theoretical developments to aid our understanding of the structure of, and processes at work in, the modern capitalist world. In arguing this I am comforted by the fact that, for example, the late Krishna Bharadwaj, whose understanding of Sraffa and his works was second to none, Giancarlo de Vivo, the late Pier Luigi Porta and Luigi Pasinetti are in overall agreement with such a take on these important matters.

Many years ago I published an exploratory and speculative paper, "Marshall, Sraffa and Keynes: Incompatible Bedfellows?" (Harcourt 1981). Its principal objective was to examine the role of the concept of a center of gravitation in the contributions of these three great political economists. I pointed out that Alfred Marshall was trapped in a dilemma partly of his own making, partly because his principal method of analysis was akin to that of classical physics, while his "vision" of the society he was observing and analyzing was of an evolving organic system. The Mecca of economists was therefore biology, not physics.

Marshall wished to have a theory of long-period normal prices that included as ingredients normal profits and normal wages. His time periods—market period, short period, long period—were used to establish his theory. He argued that they could merge imperceptibly one into another and could be short or long according to the immediate purpose at hand, and actual or potential according to the realism of the factors locked up in the ceteris paribus pound in any particular case. He vacillated between whether they were actual or potential, sometimes naming periods of calendar time as illustrative of what he had in mind, wishing to have it both ways. That is to say, he wished his normal prices to be real centers of gravitation, making sense of observations in historical time, yet he knew that if he simultaneously had in mind an economy moving forward through time with technical progress and accumulation occurring, there were puzzles associated with arguing that they could be revealed from observations on actual prices. He in fact says that it is only in the stationary state that the actual and normal coincide, can be "convertible terms" (Marshall 1890, p. 372).

He never solved his conundrum—hence his use of time-period analysis became less and less satisfactory and illuminating, the longer the period (or run) he had in his sights. Moreover, he was not able to draw on a consensus of views from the evolutionary biologists of his time because there was none. After his death, as Neil Hart has shown (2012, 2013), evolutionary theory itself moved more toward a consensus among evolutionary biologists and allowed great strides forward to be made by modern evolutionary economists.

While John Maynard Keynes was very much a Marshallian in method, even in *The General Theory*, he was not as handicapped as was his mentor because for most of the time he was analyzing short-period systemic problems. There, I argued, the application of the concept of a center of gravitation provided a useful shortcut for establishing illuminating theory and providing sensible and realistic policy proposals. The concept had a part to play in the Keynesian context of short-period rest states (and, possibly, even longer sustained rest states associated with deep and sustained depressions), a view I believed could be supported from Keynes's own writings and by Jan Kregel's (1976) classic interpretation of them.

When I wrote in 1981, there was an ongoing debate concerning whether the concept should or should not be maintained in a Marxian and/or post-Keynesian analysis of a modern capitalist economy moving forward through time. I argued that their natural place was in the theory of pricing that characterized important sections of such economies, in the normal cost-pricing hypothesis and the connection of price making with the investment decision. In particular, I thought then, and I do so now, that there may be a role for the concept of the natural rates of profits and of wages—that they are macroeconomic concepts associated with the working of the system as a whole, imposing themselves as norms on group behavior within the system.

Similarly, if we accept the snapshot view of what Sraffa was doing—as Roncaglia has long argued and which the emerging evidence from the unlocking of the Sraffa archives more and more establishes—Sraffa too is not bedeviled by Marshall's conundrums.[1] Recently, Ajit Sinha (2013) has

[1] I pointed out that Sraffa did not object to the use of Marshall's method in appropriate settings, but he did object to Marshall's use of it in conjunction with supply and demand functions and curves.

argued that the uniform rate of profits in Sraffa's system is not necessarily based on Smithian (natural prices), Marxian (prices of production) and Marshallian (normal long-period equilibrium prices) centers of gravitation, but is instead a mathematical property of Sraffa's system. I think there is merit in this argument; however, it is certainly not one that has been accepted by other Sraffa watchers. Indeed, there is much hostility to it.

Be that as it may, I want to argue that there is a similarity between the role that the dominance of the sphere of production over the sphere of distribution and exchange in Karl Marx's analysis, and the role that Sraffa prices play in modern price theory, especially if allied with the promising methods and advances associated with the development of the cyclical growth models of Richard Goodwin (1967) and (late) Michał Kalecki (1968). Put together, I shall argue that we potentially have a structure that overcomes the inability of the mainstream structure to link the short period with the long period because the theory of the medium period in between is incoherent, to say the least.[2]

Moreover, it is unclear whether the long period is an actual state but a theoretical concept, as Marshall sometimes had it, or a stretch of historical time in which changes in the capital stock, methods of production and the supplies of different types of skilled labor are occurring. Suppose, though, that the role of the long period is interpreted as the gathering together of expected long-term forces and factors that have an impact on decisions made in the short period. Then, if this is accompanied by the impacts of expected short-term factors, we may analyze how activity and so on is established short period by short period, run by run, over actual historical time.[3]

The proponents of the long-period method—Pierangelo Garegnani, Heinz Kurz, Neri Salvadori and others—while they have always stressed that they want an apparatus with which to handle the interrelationships of persistent forces, have never been able to incorporate the two most characteristic sustained and persistent forces—an inescapable environment of fundamental uncertainty and continuous technical changes—into their formal analytical structures. In contrast, the approach outlined here does, I argue, allow us to start to tackle these problems, and Sraffa's contributions are a vital, central part of what may evolve.

[2] On this, see Nevile et al. (2015, especially pages 109–1 0), and Solow's (1997, 2000) wise comments on the issues.

[3] For the distinction between the concepts of period and run, so often now regarded as interchangeable, see Harcourt (2012).

## 2 Marx's Role

First, let me sketch the relevant role of Marx's analysis for our discussion. The person who has set this out most explicitly is Donald Harris (1975, 1978). Drawing on Marx's distinction between the sphere of production and the sphere of distribution and exchange, and on the synthesis associated with Joan Robinson and her circle's writings on generalizing *The General Theory* to the long period, Harris provided a diagrammatic analysis that clearly brought out the logical dominance of the sphere of production over the sphere of distribution and exchange. His analysis also revealed how the realization problem identified by Marx, then by Keynes and, independently and in a more appropriate setting of Marx's schemes of reproduction, by Michał Kalecki, is solved in the sphere of distribution and exchange.

Common to Adam Smith, David Ricardo, Marx and Sraffa is the central core concept of the surplus—its creation, extraction, distribution and use—and the implications that follow from these processes. Harris shows how the *potential* surplus available for accumulation and other uses is the outcome, period by period, of the relationship between the current state of the class war between the capitalists and the wage earners and the existing methods of production in the current capital stock, itself the outcome of past accumulation associated with the use of previous realized surpluses. The potential surplus is the ultimate source of the potential size of profits and the rate of profits realized in the sphere of distribution and exchange—that is one of the core meanings of Marx's labor theory of value (Harcourt and Kerr 1996). On the one hand, realization of the surplus depends on the factors determining aggregate investment expenditure (a relationship between expected profitability and planned accumulation), and, on the other hand, the relationship between actual accumulation and actual profitability, itself associated with the impact of the distribution of income between profits and wages on aggregate saving.

For the present purposes, the moral to be drawn is that the relationships in the sphere of production dominate and determine what may happen in the sphere of distribution and exchange without there being any need for a complete one-to-one mapping from the actual size of the potential surplus onto the corresponding sizes of profits and the rate of profits. I argue that exactly the same features are present in Sraffa's system of prices in relation to the determination of the sizes of markups on costs and their corresponding prices in the multinational oligopolistic market structures of the modern world.

In Harcourt (1981, p. 261), I noted the puzzles that hound the concept of prices of production when trying to incorporate them as operational concepts in the analysis referred to above. The dynamic nature of capitalist development, with the embodiment of technical changes through investment expenditure, may be so rapid in most periods (runs) as to not allow sufficient historical time for centers of gravitation of a lasting nature to be formed. There is not the time, as Joan Robinson put it, for traders to become familiar with what is the norm through actual experience, so that when their bearings are cut loose, they are all at sea, rudderless, as are also the overlooking economists. The factors needed to be held constant theoretically so as to allow the centers of gravitation they imply to be ultimately struck are in fact changing so fast that there is not the time needed for their ultimate values to be established. I still do not see how John Eatwell's (1979, p. 2) statement that the forces that determine the centers of gravitation are the most dominant, systematic and persistent, and that '[whether] this centre of [gravitation] is a temporal constant, or takes different values through time, does not affect the essence of the method,' overcomes this puzzle.

## 3 Effects of Systemic Constraints

Many political economists have analyzed the overall systemic constraints imposed by macroeconomic characteristics of the economy, constraints that create the environment in which individual decision makers have to make decisions about prices and especially about the sizes of markups. These systemic constraints relate to both overall levels of aggregate demand prevailing at any moment of time (arising from the interplay of real, monetary and financial factors) and from the relationships in the sphere of distribution and exchange outlined above. They result in the formation of norms, of ranges of behavior within which it is possible to make choices.

Within this aspect of the analysis, I have in mind the links between discretion over price setting and the need to raise internal finance for planned accumulation. These ideas are especially associated with Jim Ball (1964), Al Eichner (1973, 1976), Adrian Wood (1975) and Peter Kenyon and myself (1976). My conjecture is that the Sraffa prices at each moment of time are both the systemic and individual industry and firm constraints within which these decisions have to be made.

In the long-period interpretation of Sraffa prices, the long-period structure of prices is seen to be in the outcome of persistent processes. Is this not inconsistent with, or at least unable to cope with, the impact of continuous technical change and fluctuations in aggregate demand on the economy and on the structure of Sraffa prices? The snapshot view overcomes this but, of course, carries with it the danger that if the characteristics of the snapshots vary widely from period to period because of the impact of changing levels of activity and technical variations in investment to labor and investment to output ratios, the establishment of general norms that guide behavior will not occur.

This last consideration brings to the surface a problem that plagued Keynes, his ultimate despair of ever being able to find a definite unit of time with which he could handle the analysis of all the interrelated processes he had emphasized in the development of his revolutionary theory of the workings of a monetary production economy in *The General Theory* and afterwards. He decided therefore never to push any particular piece of analysis very far past its starting point, so as to get the central message across. In his 1937 lectures, as Kregel (1976, p. 213) has documented, Keynes said that if he were to write the book again, he would start with the factors responsible for existence, as we would say now, in order "to distinguish the forces determining the position of equilibrium from the technique of trial and error by means of which the [entrepreneurs discover] where the position is" (1937, p. 182). The upshot is that instability and indeterminateness may persist in certain historical time periods, in a sense the analogue of modern general equilibrium theorists finding multiple positions of equilibrium in many parts of their analysis.

That certain time periods are characterized by instability or even crisis is, of course, not a surprising inference to be drawn from any relevant theoretical structure and approach to the analysis of economic systems. (That it was *not* an inference of the dominant theoretical structure is a major criticism of its failure to predict the recent global financial crisis, Harcourt 2010). Certainly, I do not think Sraffa would have been either surprised or dismayed by such an inference. I also conjecture that these suggestions fit well into, indeed supplement sensibly, the structures of Goodwin's and (late) Kalecki's cyclical growth models.

The contributions of other great thinkers from the last century are also relevant here. I start with Dennis Robertson. One of his criticisms of what he saw as the Keynesians attempting to sustain full employment and attain a steady rate of growth arose from his understanding of the real business cycle he argued formed the foundations for the processes at work in modern economies. He pointed out that when we consider a interrelated production economy—which has different lengths of gestation periods for the construction of machines in different sectors and their accompanying methods of use, and different economic lifetimes of their operation—it is just not possible to conceive that the aggregate outcome of their operation would be either full employment of labor over time or steady growth of output over time (Anyadike-Danes 1985). In his 1975 book on Keynes, Hyman Minsky discerned in Keynes's approach an endogenous cyclical mechanism that not only took in Robertson's insights but also added to the real story a sophisticated analysis of the interrelationship of real monetary and financial factors associated with Keynes's insistence that the latter be included right from the start of the analysis. These contributions are complementary to the Goodwin/Kalecki growth cycles, which contain both theories of accumulation and the distribution of income that are as much Marxian as Keynesian and that spell out the implications in the sphere of distribution and exchange of the dominance of the sphere of production mentioned above.

Moreover, Goodwin ultimately brought together the aggregate analysis of Keynes with the production interdependence insights of Sraffa and Wassily Leontief (see Goodwin and Punzo 1987; Goodwin had made major separate contributions to both these two approaches over his lifetime). I understand that Sraffa, unlike some of his followers, had great respect and liking for Kalecki, as did Maurice Dobb. In the late 1930s, Sraffa worked with Kalecki on pricing in UK manufacturing industries. Goodwin (1983, chaps. 7 and 8) also was influenced by and wrote on Sraffa's approaches, so it is pleasing that these three great original political economists may and should be agreeably brought together in a deeper understanding of the processes at work in modern capitalism.

## References

Anyadike-Danes, M. K. 1985. "Dennis Robertson and Keynes's *General Theory.*" In *Keynes and His Contemporaries*, edited by G. C. Harcourt, 105–23. London: Macmillan.

Ball, R. J. 1964. *Inflation and the Theory of Money.* London: Allen and Unwin.

Eatwell, J. L. 1979. "Theories of Value, Output and Employment." Thames Papers in Political Economy. London: Thames Polytechnic.

Eichner, A. S. 1973. "A Theory of the Determination of the Mark-up under Oligopoly," *Economic Journal* 83, no. 332: 1184–200.

———. 1976. *The Megacorp and Oligopoly.* Cambridge: Cambridge University Press.

Goodwin, R. M. 1967. "A Growth Cycle." In *Socialism, Capitalism and Economic Growth: Essays Presented to Maurice Dobb*, edited by C. H. Feinstein, 54–58. Cambridge: Cambridge University Press.

———. 1983. *Essays in Linear Economic Structures.* London: Macmillan.

Goodwin, R. M., and L. F. Punzo. 1987. *The Dynamics of a Capitalist Economy: A Multi-Sectoral Approach.* Cambridge: Polity Press; Oxford: Basil Blackwell.

Harcourt, G. C. 1981. "Marshall, Sraffa and Keynes: incompatible bedfellows?", *Eastern Economic Journal,* 7 (1), January, pp. 39–50; Reprinted in On Political Economists and Modern Political Economy: Selected Essays of G. C. Harcourt, edited by C. Sardoni, 250–64. London and New York: Routledge, 1992.

———. 2010. "The Crisis in Mainstream Economics." *Real-World Economics Review* 53, June, 47–51. Rev. version in *Macroeconomics and the History of Economic Thought: Festschrift in Honour of Harald Hagemann*, edited by H. M. Kramer, H. D. Kurz and H.-M. Trautman, 183–90. London and New York: Routledge, 2012.

———. 2012. "On the Concepts of Period and Run in Economic Theory." In *Classical Political Economy and Modern Theory: Essays in Honour of Heinz Kurz*, edited by C. Gehrke, N. Salvadori, I. Steedman and R. Sturn, 257–65. London and New York: Routledge, 2012.

Harcourt, G. C., and P. Kenyon. 1976. "Pricing and the Investment Decision" *Kyklos* 29, no. 3: 449–77. Reprinted in *On Political Economists and Modern Political Economy: Selected Essays of G. C. Harcourt*, edited by C. Sardoni, 48–66. London and New York: Routledge, 1992.

Harcourt, G. C., and P. Kerr. 1996. "Marx, Karl Heinrich (1818–1883)." In *International Encyclopedia of Business and Management*, edited by M. Warner, 3388–95. London: Routledge. Reprinted as "Karl Marx, 1818–83," in G. C. Harcourt, *50 Years a Keynesian and Other Essays*, 157–68. Houndmills, Basingstoke, Hampshire: Palgrave Macmillan, 2001.

Harris, D. J. 1975. "The Theory of Economic Growth: A Critique and a Reformulation," *American Economic Review, Papers and Proceedings* 65, no. 2: 329–37.

———. 1978. *Capital Accumulation and Income Distribution.* Stanford, CA: Stanford University Press.

Hart, N. 2012. *Equilibrium and Evolution: Alfred Marshall and the Marshallians.* Houndmills, Basingstoke, Hampshire: Palgrave Macmillan.

———. 2013. *Alfred Marshall and Modern Economics: Equilibrium Theory and Evolutionary Economics.* Houndmills, Basingstoke, Hampshire: Palgrave Macmillan.

Kalecki, M. 1968. "Trend and Business Cycle Reconsidered," *Economic Journal* 78, no. 2: 263–76.
Keynes, J. M. 1936. *The General Theory of Employment, Interest and Money*. London: Macmillan. Reprinted as *The General Theory of Employment, Interest, and Money*, vol. 7 of *The Collected Writings of John Maynard Keynes*. London: Macmillan, 1973.
———. 1937. "'Ex Post and Ex Ante.' Notes from Keynes' 1937 Lectures." Reprinted in *The General Theory and After. Part II: Defence and Development*, vol. 14 of *The Collected Writings of John Maynard Keynes*, 179–83. London: Macmillan, 1973.
Kregel, J. A. 1976. "Economic Methodology in the Face of Uncertainty: The Modelling Methods of Keynes and the Post-Keynesians," *Economic Journal* 86, no. 342: 209–25.
Marshall, A. 1890. *Principles of Economics*. London: Macmillan.
Minsky, H. P. 1975. *John Maynard Keynes*. New York: Columbia University Press.
Nevile, J. W., G. C. Harcourt, and P. Kriesler. 2015. "Macroeconomic Policy for the Real World: A Post-Keynesian Approach," *Economic Papers* 34, no. 3: 108–17.
Roncaglia, A. 2009. *Piero Sraffa*. Basingstoke: Palgrave Macmillan.
Sinha, A. 2013. "On the Notion of Equilibrium or the Center of Gravitation in Economic Theory." In *The Oxford Handbook of Post-Keynesian Economics*, Vol. 1: *Theory and Origins*, edited by G. C. Harcourt and P. Kriesler, 101–21. New York: Oxford University Press.
Solow, R. M. 1997. "Is There a Core of Usable Macroeconomics We Should All Believe In?" *American Economic Review* 87, no. 2: 230–32.
———. 2000. "The Neoclassical Theory of Growth and Distribution," *Banca Nazionale del Lavoro Quarterly Review* 53, no. 215: 349–81.
Sraffa, P. 1960. *Production of Commodities by Means of Commodities: Prelude to a Critique of Economic Theory*. Cambridge: Cambridge University Press.
Wood, A. 1975. *A Theory of Profits*. Cambridge: Cambridge University Press.

Comment

# Sraffa and Kalecki: One More Time

*John King*

It is good to see Geoff Harcourt still reflecting on the relationship between Sraffian and Post Keynesian theory, almost four decades after his paper on Marshall, Sraffa and Keynes (Harcourt 1981). In fact, if Fred Lee is to be believed, Geoff's interest in these matters goes back even further, to a 1965 article in the *Economic Record* in which 'he combined a corn model with a non-basic good and mark-up pricing to examine the issues of income distribution and employment in the short run' (Lee 1998, p. 171, citing Harcourt 1965). In this note I want to complement Geoff's chapter in three ways. First, I comment briefly on the interest taken in these questions in the final decades of the twentieth century by theorists outside Cambridge (and Italy). Then I reflect on the brief but ferocious controversy that erupted in the early 1990s on the precise nature of the relationship between Sraffian and Kaleckian economics. Finally, I document the almost complete separation of the two schools that was evident by the early 2010s and speculate on the reasons why the earlier efforts to integrate them had failed and why such efforts may yet prove to be successful.

Lee was convinced that 'by the early 1980s the integration of the Sraffian circular production model with mark-up (or normal cost) pricing equations had been completed, laying the foundation for a model of general microeconomic analysis' (Lee 1998, p. 171). His argument is worth quoting at some length. After Harcourt, Lee wrote, Paolo Sylos-Labini had shown that non-uniform rates of profits due to differential degrees of monopoly could be easily incorporated in Sraffa's price model (Sylos-Labini 1971); his arguments were subsequently supported by Alessandro Roncaglia (1978) and Ian Steedman (1977). Later Lynn Mainwaring

(1977) applied mark-up pricing with differential profit rates to a corn model with a non-basic good and ran that model alongside a Sraffian price model with differential rates of profit. Finally, beginning in the mid-1970s various economists began inserting mark-up and normal cost pricing equations directly into Sraffian price models in order to produce an alternative Post Keynesian model of general equilibrium, to examine economic wage-price spirals, and to examine the conditions under which actual prices in the market converged to long-period prices. (ibid., p. 171). In support of these claims he cited one of his own early publications (Lee 1985), and also complementary work by Luciano Boggio, Alfred Eichner, David Kotz, S.J. Moss, Hukukane Nikaido and Susumu Kobayashi, Joan Robinson, Malcolm Sawyer, Willi Semmler and Fausto Vicarelli.

In this context the paper by Lynn Mainwaring deserves special attention. Mainwaring begins by quoting Maurice Dobb as doubting the relevance of Sraffa's analysis to the monopoly stage of capitalism, where it cannot be assumed that a uniform rate of profit applies across the entire economy, and endorsing Kalecki's theory of distribution as 'having the situation of modern capitalism evidently in mind' (Mainwaring 1977, p. 674, quoting Dobb 1973, p. 268). 'It is surprising therefore', Mainwaring continues, 'that there has been little attempt to adapt a Kaleckian approach to the Sraffian system at the level of price determination' (ibid., pp. 674–5). He sets out a series of short-run models in which each firm is assumed to have a specific individual rate of profit and the notion of an 'industry' is redefined as 'a collection of firms producing slightly differentiated products and, therefore, using slightly differentiated techniques (and with different efficiencies)' (ibid., p. 675). There are two broad sectors, one producing fixed capital goods and the other producing consumer goods. In both sectors prices are determined by Kaleckian markups, so that in the capital goods sector 'monopoly power extends to both sides of the market' and a firm's mark-up is also 'limited by the extent of *monopsony* in its market' (ibid., p. 679; original stress).

Mainwaring illustrates his simplest two-sector model by means of a three-quadrant diagram showing the relationships between the real wage, the rates of profit in the two sectors and the relative prices of capital goods and consumer goods (ibid., Figure 1, p. 680). He then extends the analysis to allow for the internal financing of a proportion of each firm's investment expenditure, leading him to conclude that the 'firms' mark-ups must be such that at normal rates of capacity use the resulting "target" rates of profit are sufficient to finance the required investment' (ibid., p. 682).

Next Mainwaring considers the effects of an overall increase in the degree of monopoly (ibid., pp. 683–5) and the employment of different types of labor with different degrees of unionization and therefore also of union monopoly power (ibid., pp. 685–6). He notes that uncertainty is dealt with in his model 'by the assumption that firms operate at less than full capacity; this is *their* response to an uncertain world' (ibid., p. 687; original stress). Mainwaring concludes by insisting that the short-run nature of his analysis requires no assumption to be made concerning constant (or non-constant) returns to scale; he was to deal with the theoretical consequences of non-constant returns in a later article (Mainwaring 1979).

The brief bibliography appended to Mainwaring's very impressive paper includes work in a similar vein by Tom Asimakopoulos, Alfred Eichner, Wynne Godley, Richard Goodwin, Jan Kregel, Edward Nell, Peter Riach and Adrian Wood, several of whom are also cited by Geoff Harcourt in his chapter above. Together with the authorities referred to by Fred Lee, it might be thought that Lee's judgement on the successful integration of Sraffian and Kaleckian theory was indeed well-founded. However, there were by the early to mid-1980s already signs of growing tension between the Sraffians and the Post Keynesians, reflected in disputes at the Trieste summer school that involved both 'divisions in the organizing committee and in acrimonious conflicts in the classroom', though the principal protagonists on the Post Keynesian side seem to have been fundamentalist Keynesians rather than Kaleckians (King 1992, p. 158, citing Arena 1987 and Parinello 1988).

Open warfare with the Kaleckians broke out in 1992, when Ian Steedman launched a lengthy attack on them in the recently established heterodox journal *Review of Political Economy*. Steedman complained that Kaleckian pricing models ignored input-output relations and made no reference to the use of fixed capital or to the problems posed by joint production. 'The usual presentation of Kaleckian pricing theory', he argued, 'is not only short period but also short-sighted, focusing myopically on the single industry torn from its wider setting and the industrial system' (Steedman 1992, p. 130). He noted that in an economic system with non-basic commodities the inverse relationship between mark-ups and the real wage, which was fundamental to the Kaleckian system, was no longer necessarily valid (ibid., pp. 135–6). Moreover, 'if various mark-ups change but not all in the same direction as one another, the wage share could *fall*, while all relevant measures of real wage rates *increase*' (ibid., p. 136; original stress). Steedman was also highly critical of the way in which Kaleckians had used the assumption of vertical integration to evade these problems (ibid., p. 139). As for fixed capital, this was conspicuous by its absence from Kaleckian mark-up pricing models, and the serious problems posed by joint production had also been ignored. 'How are direct costs of providing a passenger service journey from Stop 3 to Stop 37 to be separated', he asked, 'from those of providing that from Stop 13 to Stop 73?' How, then, could mark-up pricing be applied by bus companies? (ibid., p. 141).

Steedman's conclusion was ambiguous on the crucial issue of whether Kaleckian pricing theory could be saved, if its advocates responded seriously to the issues that he had raised, or whether it was beyond redemption (and, if so, exactly what he would replace it with). The two final sentences of his paper read as follows:

> Once Kaleckians have firmly set their theory in the context of input-output relations and the importance of joint-products (having jettisoned their attachment to vertically integrated theorizing), they may find—who knows—that their underlying concept of mark-up pricing has new, positive contributions to offer. If this article provides overtly a set of questions, it may also be found to suggest, *sotto voce,* that there are many interesting implications of mark-up pricing in a modern economy which have yet to be fully explored. (ibid., p. 150)

At the very least, there needed to be some answers to his list of fifteen 'questions for Kaleckians' (ibid., pp. 147–50).

Answers did soon come from Peter Kriesler (1992, 1993), Lynn Mainwaring (1992), Malcolm Sawyer (1992) and Josef Steindl (1993). Easily the most hostile response was that of Kriesler, who saw 'the essence of Steedman's paper' as 'an attempt to contort one type of analysis into the framework of another'. But the Sraffian and Kaleckian approaches 'use very different frameworks to consider very different questions' (Kriesler 1992, p. 163). While Sraffa's concern was with price theory, Kalecki's interest lay in the laws of motion of the capitalist economy. The Polish theorist 'was not so much interested in the formal properties of models as he was in understanding the world' (ibid., p. 166). In this context, Steedman's emphasis on the problems posed by joint production was nothing more than 'a red herring' (ibid., p. 168).

The same point was made, but without the same antagonistic tone, by some of the other participants in the debate. Thus Sawyer noted that the purpose of Kalecki's microeconomics was quite different from that of

Sraffa, since he was primarily concerned with the influence of the firm's pricing procedures on the profit share and on the real wage. Given the difference in purpose, 'different assumptions and frameworks' were inevitable (Sawyer 1992, p. 161). Steindl also found that 'there exist strong differences in aims and idiosyncrasies' between the two schools, 'which often make it difficult for the contestants to understand each other' (Steindl 1993, p. 121). Kalecki's work, he concluded, had much more empirical relevance than Sraffa's (ibid., p. 123). In his reply Steedman argued that Kriesler had engaged in obfuscation, while Kriesler complained that Steedman had largely ignored the important issues that he had raised (Steedman 1993, p. 115; Kriesler 1993, pp. 117–18).

Just four years earlier, Geoff had co-authored with Omar Hamouda a widely read paper in which Sraffian economics had been identified as one of the three principal streams of Post Keynesian thought, along with the Fundamentalist Keynesians and the Kaleckians (Hamouda and Harcourt 1988). By the end of the century this was looking as dubious as Lee's optimistic 1998 verdict on the relations between Sraffians and Kaleckians. As I noted in my book-length history of Post Keynesian economics, Joan Robinson had turned against the Sraffians just before her death in 1983. She now believed that they 'played down the importance of uncertainty... and ignored the precarious basis on which expectations were formed'. In addition, she maintained, the Sraffians 'paid too much attention to static and timeless conceptions of equilibrium, and in particular made a fetish out of long-period positions' (King 2002, p. 209). Indeed, the Garegnani version of Sraffian economics was metaphysical, not scientific (ibid., p. 209, citing Robinson 1979, pp. 179–80).

Hyman Minsky was even more critical. 'At the arid level of Sraffa', he wrote, the Keynesian view that effective demand reflects financial and monetary variables has no meaning, for there is no monetary and financial system in Sraffa'. 'At the concrete level of Keynes', he continued, 'the technical conditions of production, which are the essential constructs of Sraffa, are dominated by profit expectations and financing conditions' (Minsky 1990, p. 363). In the real world, capitalist economies had no tendency to oscillate around long-period centers of gravitation, as the Sraffians claimed. Uncertainty was not an imperfection, as they seemed to believe, but rather 'a fundamental aspect of nature in an accumulating capitalist economy' (ibid., p. 366). The Australian Post Keynesians Joseph Halevi and Peter Kriesler made similar criticisms, objecting both to the Sraffian insistence on a uniform profit rate and their emphasis on long-period rather than short-period analysis (Halevi and Kriesler 1991, pp. 86–7). As we have seen, Kreisler dismissed the entire Sraffian framework as nothing more than a 'thought experiment', with little relevance to the real-world analysis of output, employment and growth (Kriesler 1992, p. 169). In similar vein, when Philip Arestis, Stephen Dunn and Malcolm Sawyer identified three distinct sub-schools of Post Keynesian theory they were seen as 'emanating from Keynes, Kalecki, and institutionalists' (Arestis et al. 1999, p. 527). As I noted at the time, in this account the Sraffians 'had disappeared from the Post Keynesian pantheon, to be replaced by institutionalists', on the grounds that they were 'closed-system thinkers' (King 2002, p. 219).

There were good methodological reasons for this. Kaleckians in particular criticized Sraffian economic models for having no agents, no entrepreneurs, no profit expectations, no financial constraints, and no financial instability. Moreover, since in Sraffian models there is always a monotonically downward-sloping relationship between the real wage and the rate of profit, it is difficult to see how such models might incorporate the wage-led growth regime that has been identified in much recent Kaleckian macroeconomic modelling (see, e.g., Stockhammer 2017). As Kriesler emphasized, in a recent paper co-authored with another Australian, Neil Hart, 'recent optimism concerning the establishment of coherence between the Keynesian-Kaleckian and Sraffian perspectives is misplaced, largely because of a failure to understand the substantial methodological issues which divide the two traditions' (Hart and Kriesler 2016, p. 188). The Sraffians place their analysis 'in what Joan Robinson called logical time, and not in the historical time that Post Keynesian economists believe is the appropriate method to use' (ibid., pp. 197–8). This led them to neglect the role of uncertainty and expectations, and to 'ignore the importance of institutions and of social and political forces in shaping economic events' (ibid., p. 198). Hart and Kriesler end by quoting the Marxian political economist Maurice Dobb, for whom Sraffian theory was 'to be regarded as primarily constructive *anti-kritik*, and not as a new theoretical system, replacing or mediating between its predecessors' (ibid., p. 199, citing Dobb 1975–6, p. 468).

And yet, and yet. Any Kaleckian reading Geoff's chapter in the present volume must surely pause for thought. The fact that the two theoretical systems are different does not mean that they are entirely inconsistent, or that there is nothing to be gained from attempting to reconcile them. As Geoff notes, the systemic constraints identified by Sraffa create the environment in which pricing and investment decisions have to be made, and hence have an influence on effective demand and the distribution of income. Lynn Mainwaring had made the same point in responding to Steedman: there is no necessary inconsistency between Kalecki's short-run price theory and Sraffa's long-run analysis, and the former can be seen as an important component of any systematic analysis of the gravitation of prices to their long-period values. Mainwaring had concluded by affirming 'the possibility of a synthesis between Kalecki and Sraffa' (Mainwaring 1992, p. 176).

A similar approach was taken, a decade later, by Gary Mongiovi. In his entry on 'Sraffian economics' in the first edition of the *Elgar Companion to Post Keynesian Economics*, Mongiovi began by arguing that 'Sraffa's analysis and Keynes's theory of effective demand intersect at several junctures' (Mongiovi 2003, p. 319). 'Throughout most of the formative period of Post Keynesian economics', he suggested, 'from 1960 until the mid-1980s, the affinities between the Keynes-Kalecki effective demand mechanism and Sraffa's work were not a matter of dispute'. Subsequently, he admitted, 'the Post Keynesian project of integrating Keynes, Michał Kalecki and Sraffa has lost momentum, and something of a rift appears to have developed between the Sraffian camp and many of those who identify themselves as Post Keynesians' (ibid., p. 320). But this was a mistake: 'There is no necessary incompatibility on methodological grounds between the Post Keynesian and Sraffian frameworks; the two complement each other, each providing a different set of tools for different sorts of analytical problems' (ibid., p. 321). Ten years later, when he revised his entry for the second edition, Mongiovi left the entire text unchanged but added a substantial paragraph on the recently developed 'Sraffian supermultiplier', which 'can accommodate the possibility that aggregate demand drives macroeconomic processes even in a long-period context' (Mongiovi 2012, p. 504). This reinforced his belief that the two traditions were indeed compatible, as Geoff Harcourt continues to maintain.

## References


Arena, R. 1987. 'L'éçole internationale d'été de Trieste (1981–5): vers une synthèse classic-keynesienne?' ('The international Trieste summer school (1981–5): towards a classical-Keynesian synthesis?'), *Oeconomica*, 7, March, pp. 205–38.

Arestis, P. Dunn, S.P. and Sawyer, M. 1999. 'Post Keynesian economics and its critics', *Journal of Post Keynesian Economics*, 21 (4), Summer, pp. 527–49.

Dobb, M.H. 1973. *Theories of Value and Distribution since Adam Smith*. Cambridge: Cambridge University Press.

———. 1975–6. 'A note on the Ricardo-Marx-Sraffa discussion', *Science and Society*, 39 (4), Winter, pp. 468–70.

Halevi, J. and Kriesler, P. 1991. 'Kalecki, classical economics and the surplus approach', *Review of Political Economy*, 3 (1), January, pp. 71–92.

Hamouda, O. and Harcourt, G.C. 1988. 'Post Keynesianism: from criticism to coherence', *Bulletin of Economic Research*, 40 (1), January, pp. 1–33.

Harcourt, G.C. 1965. 'A two-sector model of the distribution of income and the level of employment in the short-run', *Economic Record*, 41 (93), March, pp. 103–17.

———. 1981. 'Marshall, Sraffa and Keynes: incompatible bedfellows?', *Eastern Economic Journal*, 7 (1), January, pp. 39–50; Reprinted in On Political Economists and Modern Political Economy: Selected Essays of G. C. Harcourt, edited by C. Sardoni, 250–64. London and New York: Routledge, 1992.

Hart, N. and Kriesler, P. 2016. 'Keynes, Kalecki, Sraffa: coherence within pluralism?', in J. Courvisanos, J. Doughney and A. Millmow (eds), *Reclaiming Pluralism in Eco nomics*, London and New York: Routledge, pp. 186–202.

King, J.E. 2002. *A History of Post Keynesian Economics Since 1936*. Cheltenham, UK and Northampton, MA, USA: Edward Elgar.

Kriesler, P. 1992. 'Answers for Steedman', *Review of Political Economy*, 4 (2), April, pp. 163–70.

———. 1993. 'Reply to Steedman', *Review of Political Economy*, 5 (1), January, pp. 117–18.

Lee, F.S. 1985. 'Full cost prices, classical price theory, and long period method analysis: a critical evaluation', *Metroeconomica*, 37 (2), June, pp. 199–219.

———. 1998. *Post Keynesian Price Theory*. Cambridge: Cambridge University Press.

Mainwaring, L. 1977. 'Monopoly power, income distribution and price determination', *Kyklos*, 30 (4), November, pp. 674–90.

———. 1979. 'The wage-profit relation without constant returns', *Metroeconomica*, 31 (3), October, pp. 337–62.

———. 1992. 'Steedman's critique: a tentative response from a Kaleckian', *Review of Political Economy*, 4 (2), April, pp. 171–7.

Minsky, H.P. 1990. 'Sraffa and Keynes: effective demand in the long run', in K. Bharadwaj and B. Schefold (eds), *Essays on Piero Sraffa: Critical Perspectives on the Revival of Classical Theory*, London: Unwin Hyman, pp. 362–9.

Mongiovi, G. 2003. 'Sraffian Economics', in J.E. King (ed.), *The Elgar Companion to Post Keynesian Economics*, Cheltenham, UK and Northampton, MA, USA, pp. 318–22.

———. 2012. 'Sraffian Economics', in J.E. King (ed.), *The Elgar Companion to Post Keynesian Economics*, Cheltenham, UK and Northampton, MA, USA, second edition, pp. 499–505.

Parinello, S. 1988. 'Il ruolo di una scuola estiva di economia' ('The role of a summer school in economics'), *Economia Politica*, 5 (3), December, pp. 335–41.

Robinson, J. 1979. 'Garegnani on effective demand', *Cambridge Journal of Economics*, 3 (2), June, pp. 178–9.

Roncaglia, A. 1978. *Sraffa and the Theory of Prices.* New York: Wiley.

Sawyer, M.C. 1992. 'Questions for Kaleckians: a response', *Review of Political Economy*, 4 (2), April, pp. 152–62.

Steedman, I. 1977. *Marx after Sraffa.* London: New Left Books.

———. 1992. 'Questions for Kaleckians', *Review of Political Economy*, 4 (2), April, pp. 125–51.

———. 1993. 'Points for Kaleckians', *Review of Political Economy*, 5 (1), January, pp. 113–16.

Steindl, J. 1993. 'Steedman versus Kalecki', *Review of Political Economy*, 5 (1), January, pp. 119–24.

Stockhammer, E. 2017. 'Wage-led versus profit-led demand: what have we learned? A Kaleckian-Minskyan view', *Review of Keynesian Economics*, 5 (1), Spring, pp. 25–42.

Sylos-Labini, P. 1971. 'The theory of prices in oligopoly and the theory of growth', in P. Sylos-Labini, *The Forces of Economic Growth and Decline.* Cambridge, MA: MIT Press, 1984, pp. 123–45.

CHAPTER 10

# Garegnani's Surplus Equation and Marx's Falling Rate of Profits

*Nuno Ornelas Martins*

## 1 Introduction

Adam Smith (1776) had already argued that there is a tendency for a fall in the rate of profits with the accumulation of capital, when writing in chapter 9 of the *Wealth of Nations* that the mutual competition that arises with accumulation reduces profits. Smith was criticised by David Ricardo in chapter 21 of his *Principles of Political Economy and Taxation*. Ricardo (1821: 194) argues that the existence of lower profits caused by increased production and its demand 'depends solely on the rise of wages; and the rise of wages, excepting for a limited period, on the facility of producing the food and necessaries of the labourer.'

Ricardo assumed that if the prices of all commodities are reduced through competition, a rate of profits will still exist as long as there is a sector of the economy, which for Ricardo was the agricultural sector, in which outputs are also used directly or indirectly as inputs. In that sector, the reduction of the prices of outputs will also entail a reduction of the

N. O. Martins (✉)
Universidade Católica Portuguesa, Porto, Portugal
e-mail: nmartins@porto.ucp.pt



A. Sinha (ed.), *A Reflection on Sraffa's Revolution in Economic Theory*, Palgrave Studies in the History of Economic Thought,
https://doi.org/10.1007/978-3-030-47206-1_10

prices of inputs, and a rate of profits will still exist. Competition between sectors will ensure that this rate of profits will exist in other sectors too. Ricardo saw agriculture as the sector which shows these characteristics, and thus the profits of the farmer would regulate the profits of other sectors.

Ricardo saw the tendency of the rate of profits to fall as a result of diminishing returns on land, which reduce profits in the agricultural sector. The diminishing returns of land lead to higher rents, defined as the difference between the productivity of a given land and the productivity of the worst land, a difference which increases as worse lands start to be used. The introduction of worse lands also leads to an increase in wages given the higher cost of production of wage-goods for the workers. It is this increase in wages that Ricardo sees as the cause for the reduction of profits, as noted above.

The introduction of machinery into this scheme of analysis raises some interesting questions. As Ajit Sinha (2014: 185) argues, Ricardo (1821) saw the introduction of machinery as a cause for the reduction of the demand for labour, which is replaced by machines. This, in turn, leads to a reduction in wages, thus attenuating the tendency of the rate of profit to fall. Karl Marx, in contrast, believed that the introduction of more machinery which replaces labour is a long term cause of the rate of profits to fall (Sinha 2014: 185). But an understanding of Marx's argument requires further scrutiny of his notion of the composition of capital (Fine and Saad-Filho 2016; Sinha 2014).

I will explain Marx's tendency for the rate of profit to fall as a part of his overall view of the process of economic reproduction, and argue that important details in Marx's analysis can be fruitfully addressed through Piero Sraffa's equations, and the use that Pierangelo Garegnani makes of such equations in Garegnani's (1984) own surplus equation. This approach is in line with Garegnani's (2018) own view, according to which the perspectives of Marx and Sraffa are best seen as complementary approaches, rather than opposed ones, as they are usually interpreted.

It must be noted that I shall not engage in a mechanical analysis of how changes in the composition of capital bring about a fall in the rate of profits. Rather, my more modest aim is to show certain properties of Garegnani's surplus equation can be seen as an expression of the underlying tendency of the rate of profits to fall. Sraffa's analysis took the economic system to be an interconnected whole, where a certain line of causation between event X and event Y cannot be analysed in isolation assuming everything else to remain constant, contrarily to what Alfred Marshall (1923 [1919]: 677–678) held—see Martins (2013: 41) for a discussion based on Sraffa (D3/12/11/11).

Many misunderstandings have arisen due to the belief that Sraffa's equations are a direct characterisation of what goes on in the economy in terms of mechanical causation, believing, for example, that because they are linear equations, the underlying reality responds to isolated changes linearly, thus presupposing constant returns to scale. But as Amartya Sen (2003) notes, Sraffa's equations can be seen as Ludwig Wittgenstein's (1963 [1953]) language games, in the sense that they merely express an underlying reality which need not be linear, even if causality must be involved at some point, for example when determining wages or the rate of profits, as Garegnani (2012: 1422–1423) notes when commenting on Sen's (2003) position. But in Sraffa's view what we have is causality within an interconnected system, rather than mechanical causation between event X and event Y while presupposing everything else to remain constant within a linear system (Martins 2013: 38–45)—see also Sinha (2016: 194: fn. 9) on Sen's (2003) position.

It is the ordinary and conventional practice of accounting values in terms of summation that leads to the use of linear equations, rather than the underlying nature of the economy system being studied. It may be noted in passing that Sraffa (1960: 48) also suggests seeing his own equations in similar terms (as accounting units) when addressing the difficulties associated with the emergence of negative multipliers and negative quantities in a standard system with joint production, imagining units to be assets and liabilities in a company, or obligations to deliver a certain quantity of commodities, and the like.

I shall also be concerned with how Sraffa's system, and Garegnani's surplus equation, can be used to express an underlying reality, the tendency of the rate of profits to fall due to the accumulation of capital and the changes in the composition of capital it brings. But I shall not be concerned with any mechanical relation of causation between the composition of capital and the tendency of the rate of profits to fall. Rather, all the equations presented below are mere expressions of conventional accounting practices, and the way in which they lead to various forms of expressing capitalisation, the composition of capital, and the rate of profits.

## 2 Labour Values and Prices of Production in Marx's Analysis and Sraffa's System

Marx (1867, 1885, 1894) starts his analysis presupposing different individual values for the commodities produced by different producers who operate under different conditions (Martins 2013: 21–22). Market value arises in a given sector or industry through the competition of individual producers, and is influenced by the dominant conditions of production in the sector or industry. More surplus value will be obtained in sectors or industries where the proportion of labour employed is higher. This is so because workers are paid for their labour-power (as they compete with each other in the labour market), but perform more labour than what is necessary for reproducing their labour-power. The difference between labour performed and the labour that is necessary for reproducing labour-power is a surplus that originates profits, which is higher in sectors or industries where there is more labour-power to be exploited.

When competition between sectors or industries arises the market values that were formed through intra-sectoral competition give way to prices of production that emerge through inter-sectoral competition, as the different rates of profits of various sectors and industries get equalised across sectors or industries, leading to the formation of a general (inter-sectoral) rate of profits. Sectors or industries where a greater proportion of labour is employed see their prices reduced from the original market value, and sectors where a lower proportion of labour is employed see their prices increase from the original market value.

These various stages of Marx's analysis have been subject to criticisms, concerning the formation of market value through a process of gravitation of individual values, the transformation of market values into prices of production, whether there is ever perfect competition between capitals, as Marx seems to presuppose, and the idea that the exploitation of labour is the origin of the surplus. I shall argue that none of these potential problems undermines Marx's analysis. On the contrary, they can easily be dealt with drawing on Sraffa's (1960) equations, without undermining Marx's overall description of capitalism.

The problem of gravitation of market prices around market values within one sector, or around prices of production, must be dealt with by noting the impact of variations of supply and demand given heterogeneous conditions of production. As Marx explains in chapter 10 of Volume III of *Capital*, there are different producers with different costs of production, and as demand increases (or decreases) or supply decreases (or increases) producers with worse (or best) conditions of production and thus higher (or lower) costs are brought to the market, increasing (or decreasing) the market values in a given sector, or the prices of production across sectors or industries. It is difficult to be more specific than this (Martins 2013: 21–22, 2019), and Sraffa (1960) actually leaves the topic aside—see Sinha (2016) for a critique of more ambitious views of gravitation.

Marx's transformation of market values into prices of production has also received much criticism, as the market values which Marx considers as inputs are not necessarily equal to the prices of production at which outputs are valued, and the value or price at which commodities are bought must be equal to the prices at which they are sold in exchange. And even if we want to focus on the production process only, the prices of inputs and outputs must be equal if we want to study the condition of reproduction at a given moment in time (Martins 2013: 43–45). This problem can be dealt with drawing on Sraffa's (1960) system of equations, where the prices of inputs and outputs are equal, as in the equation below:

$$A\mathbf{p}(1+r)+w\mathbf{l}=X\mathbf{p} \tag{10.1}$$

"$A$" is the inputs matrix with "$k$" lines and "$k$" columns, with each element $A_{ij}$ in line "$i$" and column "$j$" being the quantity of commodity "$j$" necessary to produce commodity "$i$". Vector "**p**" is constituted by the prices of the "$k$" commodities, "$r$" is a scalar that represents the rate of profit, "$w$" is a scalar that represents wages, **l** is a vector constituted by the quantity of labour necessary to produce each of the "$k$" commodities, and "$X$" in a diagonal matrix with "$k$" lines and "$k$" columns where each element of the diagonal is one of the "$k$" outputs. Sraffa (1960) further defines that net output (the difference between the total value of outputs and the total value of inputs) is equal to one, and the sum of labour quantities is also equal to one, since labour quantities are defined as fractions of total labour.

If we set "$r$" equal to zero and "$w$" equal to one, the value of the commodities forming the net product will be proportional to the quantity of labour that enters directly or indirectly into its production. In this case the prices obtained correspond to what Marx calls market values, that is, the value of a product within a given sector is expressed in terms of the quantity of labour required (directly or indirectly) to produce it. As soon as "*r*" acquires a positive value, a general rate of profits emerges across sectors, and the values of the commodities can be best interpreted as Marx's prices of production, the prices which enable the continuous reproduction of the economic system while allowing for a common rate of profits across sectors.

Exploitation can be seen as the source of the surplus to the extent that when "*r*" acquires a positive value, workers as a whole are not receiving the full surplus. This is a technical, rather than an ethical or political, description of exploitation. In this regard, it is important to note that Marx himself did not take the quantity of labour performed by individual workers as an ethical or political criterion of distribution, contrarily to Smith. Smith argued that workers should receive the full surplus of their labour so that an efficient organisation of economic activity can arise (Smith 1776: 336–337), while believing that this was the case in primitive societies, and that if things had remained this way, there would have been greater improvements in the division of labour (Smith 1776: 63).

In this regard, Smith's ethical approach is very much in line with the natural rights tradition stemming from John Locke. Marx, in contrast, thought that production should take place according to the possibilities of each individual, and distribution should be made taking into account the needs of each individual, since Marx saw no ethical or political reason why individuals should benefit from being more productive than others. The labour theory of value was, for Marx, used for a scientific description of the economy, not as an ethical or political theory for assessing distribution, as it is often interpreted (Garegnani 2018).

Sraffa's economic system can be used in different ways depending on the ethical or political perspective we want to adopt. Smith's analysis suggests that we should view the case when wages receive the full surplus as what we may term a ***natural system*** (Martins 2013: 421–427), which existed in earlier stages of society, and constitutes an ethical goal to be achieved again. Marx's analysis, in contrast, suggests that we consider the case when wages receive the full surplus in merely technical terms, as what we may term the ***value system*** (in which values are proportional to the labour employed), and the case where a general (inter-sectoral) rate of profits exists as what we may term the ***price system***. In ethical or political terms, for Marx, the value system reflects the underlying relations of production expressed in terms of power relations between the class of capitalists and the class of workers, which are in this case the power to command the labour of others (Smith 1776). Even when a general rate of profits emerges in the price system, those power relations still exist, and there is an implicit value system, which can be obtained by making wages receive the full surplus at the existing conditions of production.

Marx's assumption of perfect competition between capitals across sectors is also a possible target for criticism. A possible reply to this criticism is that the competition for money capital between firms at more advanced stages of capitalism means that competition is an increasingly important force, as David Harvey (2006 [1982]: 137–155) notes (Martins 2013: 8). But it is not a necessary assumption for, as Sraffa (D3/12/9: 10&11) notes, the common rate of profits can emerge simply because when an industry has higher profits, it accumulates capital at a higher pace and the rate of profits decreases. Conversely, when an industry has lower profits, it accumulates capital at a lower pace and thus its relative price rises compared to the industry accumulating capital at a higher pace.

There are also other effects acting towards an equalisation of the rate of profits, besides the effects in prices, which can act through changes in quantities, which we may consider if we want to engage in what Sraffa calls a "mechanical" analysis of changes in quantities, rather in a merely "geometrical" analysis with given quantities—see Martins (2013, 2019) and Sinha (2016) on this distinction. As Sraffa's (1926: fn. 3) notes, Alfred Marshall (1923 [1919]: 188) writes that "the economies of production on a large scale can seldom be allocated exactly to any one industry: they are in great measure attached to groups, often large groups, of correlated industries." This means that economies of scale tend to be internal from the point of the individual firm, but not from the point of view of an industry (one could say they are external from the point of view of the industry since they draw on various industries). Thus, whenever a single industry accumulates capital at a higher (lower) pace, it tends to do find decreasing (increasing) returns comparatively to other industries, thus setting a force for the equality of rates of profits that acts through quantities ("mechanically").

So Sraffa's system can be used to address various criticisms of Marx's theory. I will now argue that doing so does not change substantially the core tenets of Marx's analysis, including his analysis of the tendency of the rate of profits to fall, and its countervailing tendencies, which is often seen as a key component of Marx's analysis.

## 3 Marx's Law of Tendency of the Rate of Profits to Fall and Sraffa's System

Drawing on the analysis above, we can now address Marx's law of the tendency of the rate of profit to fall. To understand this tendency, we must first take into account Marx's distinction between the technical composition of capital, the organic composition of capital and the value composition of capital, which are different forms of expressing the relationship between constant capital (i.e., material inputs such as fixed capital and circulating capital in the forms of commodities) and variable capital (i.e., human labour).

Fine and Saad-Filho (2016) and Sinha (2014: 187–188) provide different elaborations of this topic. Fine and Saad-Filho (2016) note that the technical composition of capital reflects technology, and thus the quantities of inputs and outputs used. When there is a change in technology, the technical composition of capital will change. However, since capital includes heterogeneous units such as heterogeneous commodities, there is the need of converting them into homogeneous units so as to assess the aggregate effect of this change. A possibility is to use the value of the commodities as the unit to do so. But the technical change will also lead to changes in the values of commodities. If commodities are assessed at the new values, thus capturing both the effect of the technical change on quantities and on values, we reach the value composition of capital. If we want to focus only on the effect of technical change on quantities disregarding its effects on value, the commodities must be valued at their old values, and we reach the organic composition of capital.

We can interpret this distinction taking into account that the technical composition of capital corresponds to whatever is given by matrices "*A*" and "*X*" in Sraffa's system above. The organic composition of capital, for Fine and Saad-Filho (2016), corresponds to a case where there is a change in the technical composition of capital, that is, in matrices "*A*" and "*X*", but prices remains the same. Since the old prices are not necessarily those that ensure a common rate of profits under the new technical composition of capital (under the new matrices "*A*" and "*X*"), profits across industries will not be necessarily equal, and there will typically be what Sraffa (1960: 13) calls (when addressing a different problem,) "deficit" industries and "surplus" industries." Only when prices are adjusted to the new technical composition of capital (the new matrices "*A*" and "*X*"), so as to ensure again a common rate of profits across the system, we reach the value composition of capital.

The tendency of the rate of profit to fall is caused by increases in the organic composition of capital, that is, a greater proportion of constant capital relative to variable capital caused by a technical change. The reason why the increase in the organic composition of capital causes the tendency of the rate of profits to fall is because profits arise due to the exploitation of human labour, which can be defined in technical way in terms of the existence of a positive rate of profits that prevents workers as a whole from receiving the full surplus, as noted above. But because technological change also leads to cheaper commodities, the value of inputs also decreases, acting as a countervailing tendency through changes in the value composition of capital.

The analysis of the tendency of the rate of profits to fall and the countervailing tendencies was translated above into Sraffian terms without undermining the overall interpretation of Marx's theory. Sraffa's system constitutes an instantaneous photograph of a dynamic economy, but he leaves the topic of economic change unelaborated, albeit a study of the phases of capitalism was sketched by Sraffa himself (D3/12/11:50), who addresses briefly how capitalism evolves through various stages drawing on his economic system (Martins 2014: 123–125). But it is still necessary to explain whether Marx's analysis of the tendency of the rate of profits to fall is compatible with Sraffa's own analysis. There are other ways to address this problem, and some were developed drawing directly on Sraffa's system.

Sinha (2014), for example, defines a change in the organic composition of capital as a change to an already known technique due to a change in wages that causes substitution of capital for labour or vice-versa. In this case a change in technique does not represent an increase in industrial productivity, since it is merely a switch between already known techniques. A change in value composition reflects a change in technology which represents an increase in industrial productivity, and thus a change to a previously unknown technique.

I shall now provide another way to conceptualise the composition of capital that can take these various issues into account, drawing on Sraffa's analysis of the reduction of the value of commodities to dated quantities of labour, which refer to the quantity of labour that is presupposed in the current production of each commodity. To do so, I shall use Garegnani's surplus equation, which is developed based on Sraffa's reduction of the value of commodities to dated quantities of labour.

Drawing on Garegnani's (1984) surplus equation, I shall provide a way to express the composition of capital while separating it from changes in industrial productivity within the context of Sraffa's system, without the need of presupposing "deficit" industries and "surplus" industries". I do not intend to provide a view which is superior to Fine and Saad-Filho's (2016), or Sinha's (2014), from any specific vantage point. Rather, my more modest aim is to provide yet another angle for accounting for the composition of capital, while connecting it to the rate of profits and the rate of surplus value in a way that allows one to identify the tendency for the rate of profits to fall. The advantages of each different angle depends on the specific problem we want to address.

## 4 Garegnani's Surplus Equation

Sraffa studies the formation of a general rate of profits showing how it depends only on "basic commodities" that enter directly or indirectly into the production of all commodities in the economic system. Garegnani's (1984: 313) notion of the 'vertically integrated sector of the wage-goods' captures the same idea while drawing our attention to the importance of human labour, and the fact that any "basic commodities" sector must include the wage-goods, integrated with the other goods that contribute to the reproduction of human labour, a topic which was elaborated in more detail by Marx.

It must be noted that Garegnani's (1984) use of a vertically integrated sector that is reduced to pure labour constitutes an approximation only. For Sraffa (1960: 34–40), the process of reducing commodities to dated quantities of labour presupposes that there must remain always a residue of commodities. The latter can be made negligible with sufficient iterations of the reduction algorithm Sraffa provides so that, as Sraffa writes:

> we shall obtain the 'reduction equation' for the product in the form of an infinite series
>
> $$L_A w + L_{A1} w(1+r) + \ldots + L_{An} w(1+r)^n + \ldots = Ap_a.$$
>
> How far the reduction need be pushed in order to obtain a given degree of approximation depends on the level of the rate of profits: the nearer the latter is to its maximum, the further must the reduction be carried. Besides

> the labour terms there will always be a 'commodity residue' consisting of minute fractions of every basic product, but it is always possible, by carrying the reduction sufficiently far, to render the residue so small as to have, at any prefixed rate of profits short of $R$, a negligible effect on price. It is only at $r = R$ that the residue becomes all-important as the sole determinant of the price of the product.
>
> (Sraffa 1960: 35)

This means that this procedure of reducing to dated quantities of labour is acceptable only when the rate of profits "$r$" is below the Maximum rate of profits "$R$", which is a reasonable assumption since workers must receive a wage. But it constitutes only an approximation. So vertically integrated systems such as the one presented by Garegnani (1984) here, or by Luigi Pasinetti (1993), where commodities are reduced to labour, must be seen with this qualification in mind, as otherwise they will make us lose sight of the interconnected nature of Sraffa's (1960) system (Martins 2013; Sinha 2016).

Garegnani focuses on the wage-goods sector, which must produce a quantity of commodities that is bought through the wages paid to the labourers of the whole economy:

> $L$ is the known number of labourers employed in the economy. The wages paid in the sector, on the other hand, will be those of the $Lv$ labourers required for the direct and indirect production of the "necessary consumption", a known magnitude, since the real, wage and the technical conditions of the direct and indirect production of its constituents are known. These wages will therefore "command" $Lv$ labour-years.
>
> It follows that in the integrated wage-goods sector, the amount of profits in terms of "commanded labour" constitutes a "surplus value" ($L - Lv$) which is known before the rate of profit and the relative prices are determined. This surplus value coincides numerically with Marx's own social surplus value $s$ (since $L = s + v$ and $Lv = v$), though it differs from the latter in conception because: (i) it is the surplus value of the wage-goods sector, and not that of the entire economy; (ii) it is expressed in terms of labour commanded and not in terms of labour embodied.
>
> (Garegnani 1984: 314)

It must be noted that "$Lv$" consists of the labour years that the wages paid to the workers in the wage-goods sector can command, which also provides a surplus (measured as "$L - Lv$") in this sector which sustains the workers of other sectors, so that "*L*" are the labour years that the wages paid to the workers of the whole economy can command, even if Garegnani does not clarify this in the quote above.

Garegnani seems to be thinking in terms of Marx's reproduction schemes, and assuming that the profits of the wage-goods sector are identical to the wages of other sectors. This value is $L - Lv = s$. Michal Kalecki (1971: 80) makes a similar assumption drawing on Marx's reproduction schemes. It may be noted that Sraffa certainly saw an analogy between his equations and Marx's reproduction schemes, and there is even a dispute on whether his equations are inspired on the latter, as Giorgio Gilibert (2003) and Giancarlo de Vivo (2003) argue, or not, as Heinz Kurz and Neri Salvadori argue (2015). Kurz and Salvadori (2015: 2–3) argue that Garegnani changed his mind on whether Sraffa's equations are inspired in Marx's reproduction schemes. Regardless of whether this is the case or not, the idea that Sraffa's equations play a similar role to Marx's reproduction schemes (regardless of whether the equations were originally inspired by them or not) seems consistent with Garegnani's views of the matter.

Garegnani provides an example where we divide the workers in the wage-goods sectors (commanding *Lv* labour years) into two equal groups (commanding *Lv*/2 labour years), the first group producing ploughs, the second using the ploughs to produce corn in the next period. Wages consist only of corn, and the produce obtained command "*L*" labour years, leading to the equation:

$$L = \left[\frac{Lv}{2}(1+r) + \frac{Lv}{2}\right](1+r) \quad (10.2a)$$

In this equation, we see that the value "*Lv*/2", that refers to the value of the wages of workers producing the "ploughs", must be multiplied by $(1 + r)$, that is, must be capitalised one period (at rate of profits, or interest, "*r*") before summing it to the value "*Lv*/2" that corresponds to the wages of the workers employed in the next period's production of corn. This sum can then be multiplied by $(1 + r)$ to obtain the value of the final produce, which will be ready only after another year of production. Multiplying the last factor $(1 + r)$ by the parcels within the square brackets, we have:

$$L = \frac{Lv}{2}(1+r)^2 + \frac{Lv}{2}(1+r) \qquad (10.2b)$$

This shows that labour spent producing the ploughs (which happened 2 periods ago) is capitalised two times, that is, multiplied by $(1 + r)^2$, while the labour spent producing corn (while using the ploughs) is capitalised only one time, that is, that is, multiplied by $(1 + r)$. A simple algebraic manipulation (namely, expanding the polynomial, subtracting $Lv$ from both sides of the equation, and dividing both sides by $Lv$) will lead to the equation Garegnani (1984: 316, Eq. 10.13) provides, which he calls the *surplus equation*:

$$\frac{L - Lv}{Lv} = \frac{3}{2}r + \frac{1}{2}r^2 \qquad (10.2c)$$

Garegnani calls the right-hand side of the surplus equation the *profit function*, a function of '$r$' that rises monotonically with '$r$', and thus has only one intersection with the horizontal line $(L - Lv)/Lv$. Once we solve the surplus equation for the point of this intersection, we obtain the rate of profits. Garegnani (1984: 317) notes that this reasoning can applied even in cases of fixed capital of constant efficiency, since a 'profit function' that rises monotonically with '$r$' can still be constructed (using what Sraffa calls the 'annuity approach').

Note also that the analysis above assumed a reduction to (dated) labour of all commodities. But in a circular process of production of commodities by means of commodities, there will be a Maximum rate of profits, as Garegnani (1984) explains, due to a residue of commodities, as noted above. This fact does not change the analysis above, and the main consequence of it is that as '$r$' approaches the Maximum rate of profits '$R$', the surplus value of the wage-goods sector must rise to infinity—see Garegnani (1984: 318).

In this analysis, Garegnani shows the connections between Sraffa's (1960) analysis and the surplus theories of Ricardo and Marx. But Garegnani (1984: 319) also argues that 'the proportion of labour to means of production on which the rate of profits depends is that of the wage-goods sector, and not that of the economy as a whole as Marx thought.' However, the idea of an integrated wages-goods sector must also include the sectors that provide the means of production for the wage-goods sector. So the only sectors left aside in the determination of the rate of profits are those that produce what Sraffa (1960) calls non-basic commodities (those that do not influence the prices of the whole economy), typically luxury goods or capitalists' consumption (Kalecki, 1971). But even such sectors influence the integrated wage-goods sector to the extent that capital can flow between sectors changing quantities. Garegnani's point is that these changes in quantities must be seen separately from the determination of prices, which is why Garegnani (1984: 319) leaves them aside while taking quantities as given, as Sraffa (1960) also does.

Garegnani (1984: 319) also notes that 'it is impossible to measure capital by a single magnitude independent of distribution', as the Cambridge controversies in the theory of capital have shown. Thus the composition of capital 'must instead be expressed by the proportionate distribution over time of the labour necessary to produce the wage commodity, or by the quantities of the several means of production, in the price equations' (Garegnani 1984: 319). Using this different form of expressing Marx's ideas, Garegnani (1984: 319) notes how this analysis confirms Marx's conclusions 'that the rate of profit depends on two factors only, the rate of surplus value $s/v$ and the proportions between means of production and labour.'

I shall now argue that Garegnani's surplus equation can be used to express Marx's analysis of the tendency for the rate of profits to fall. Garegnani does not mention this relationship when presenting his surplus equation. Rather, the relationship between Garegnani's surplus equation and Marx's tendency for the rate of profits to fall is referred to in Martins (2013: 76–83). But the relationship can be seen quite clearly through Garegnani's surplus equation.

The idea that capital from older periods reduces the rate of profits can be further scrutinised by imagining that $Lv$ is divided only in two groups (each containing $Lv/2$) as in Garegnani's example, but that the first period of production happens 'n' periods ago, to be resumed only in the beginning of this year. At the end of the year, we reach the following amount of commanded labour:

$$L = \frac{Lv}{2}(1+r)^n + \frac{Lv}{2}(1+r) \qquad (10.3)$$

Equation (10.3) will lead to a lower rate of profits than Eq. (10.2b) for the same level surplus value of the wage-goods sector, provided "*n*" is greater than 2. The higher "*n*" is, the lower must the rate of profits "*r*" be. This happens because the composition of capital is different now, since the first quantity of labour is now capitalised '*n*' years in eq. (10.3), rather than '2' years as in Eq. (10.2b). Thus, it constitutes a larger proportion of capital, due to the passage of time. For capital in this context is just past labour capitalised through time. So "*r*" must be lower when capitalised "*n*" times if the rate of surplus value, or exploitation ratio, $(L - Lv)/Lv$, is to remain the same. But this fact is merely an expression of the tendency of the rate of profit to fall with the accumulation of capital, provided there is no increase in industrial productivity that would lead to an increase in the rate of surplus value, or exploitation ratio, $(L - Lv)/Lv$.

## 5 Generalising Garegnani's Example

The example provided by Garegnani (1984: 313–320) includes only two periods (the first for the production of the 'ploughs', the second for the production of 'corn'). But it can be generalised to '*n*' periods of production, leading to the transformation of Eq. (10.2b) into:

$$L = \frac{Lv}{n}(1+r)^n + \frac{Lv}{n}(1+r)^{n-1} + \cdots + \frac{Lv}{n}(1+r)^2 + \frac{Lv}{n}(1+r) \quad (10.4a)$$

It may happen that there are periods where no production took place. In that case, we simply omit the corresponding terms of the summation in the right-hand side of the above equation. It may also happen that the proportions of labour "*Lv*" undertaken in each period are not equal, and cannot be assumed to be $Lv/n$ for all periods as assumed above. In that case, different proportions can be used. The omission of some terms or a different set of weights does not affect the key results to be obtained below, which will be obtained assuming that production took place in every period, and that labour is equally distributed over the various periods.

Since labour is distributed equally over the '*n*' periods, the wages paid in the wage-goods sector, '*Lv*', are divided equally by '*n*' periods (rather than 2 periods as in Garegnani's example). We can expand the various polynomials $(1 + r)^k$ of various orders '*k*' from 1 to '*n*', and we will find that the various rows of Pascal's triangle will appear in the expansion of each polynomial:

$$L=\frac{Lv}{n}(1+r)^{n}+\frac{Lv}{n}(1+r)^{n-1}+\cdots+\frac{Lv}{n}(1+r)^{2}+\frac{Lv}{n}(1+r)=$$

$$=\frac{Lv}{n}\left[\binom{1}{0}+\binom{1}{1}r\right]+\frac{Lv}{n}\left[\binom{2}{0}+\binom{2}{1}r+\binom{2}{2}r^{2}\right]$$

$$+\frac{Lv}{n}\left[\binom{3}{0}+\binom{3}{1}r+\binom{3}{2}r^{2}+\binom{3}{3}r^{3}\right]+\cdots$$

$$+\frac{Lv}{n}\left[\binom{n-1}{0}+\binom{n-1}{1}r+\cdots+\binom{n-1}{n-2}r^{n-2}+\binom{n-1}{n-1}r^{n-1}\right]$$

$$+\frac{Lv}{n}\left[\binom{n}{0}+\binom{n}{1}r+\cdots+\binom{n}{n-2}r^{n-2}+\binom{n}{n-1}r^{n-1}+\binom{n}{n}r^{n}\right]=$$

$$=\frac{Lv}{n}\left\{\left[\binom{1}{1}+\binom{2}{1}+\binom{3}{1}+\cdots+\binom{n-1}{1}+\binom{n}{1}\right]r\right.$$

$$+\left[\binom{2}{2}+\binom{3}{2}+\cdots+\binom{n-1}{2}+\binom{n}{2}\right]r^{2}+\cdots+\left[\binom{n-1}{n-1}\right.$$

$$\left.+\binom{n}{n-1}\right]r^{n-1}+\left.\left[\binom{n}{n}\right]r^{n}\right\}$$

The expression above can be simplified using Pascal's triangle. In Pascal's triangle, each element at row '*n*' has two elements below it in row '*n* + 1', and two elements above it in row '*n* - 1' (in fact the triangle is generated by summing the two elements in row '*n* - 1' in order to obtain the element in row '*n*'). Thus the first row (row 0) is constituted by 1, the second row (row 1) by 1 and 1 (implicitly, we have summed the previous 1 to the zeros that surround it), the third row (row 2) is 1, 2 and 1, the fourth row is 1, 3, 3 and 1, and so on. So we get the following triangle:

1
1 1
1 2 1
1 3 3 1
1 4 6 4 1
1 5 10 10 5 1

1 6 15 20 15 6 1
1 7 21 35 35 21 7 1
..........................

Since the first row of Pascal's triangle has a single element (the number 1) and is usually designated as the row zero, we find that the row of order '*k*' appears in the expansion of the polynomial of the same order '*k*'. Each element of row '*k*' of Pascal's triangle is multiplied by '*r*' to an increasing power, that is, the first element of row '*k*' of Pascal's triangle, which is always '1' for any row, is multiplied by $r^0$ (which is also '1'), the second element of row '*k*' of Pascal's triangle will be multiplied by $r^1$, the third element of row '*k*' is multiplied by $r^2$, and so on.

If we collect the terms that are multiplied by each $r^k$ for each different '*k*', we find that each $r^k$ is multiplied by the sum of the terms of the *k*th diagonal of Pascal's triangle up to row '*n*'. It follows from the construction of Pascal's triangle that the sum of the elements of a diagonal up to row 'n' is obtained by looking at one of the two elements in row '$n + 1$', *not* the element that would follow the line drawn by the diagonal, but the *other* element below the last element of the diagonal (forming thus the pattern of an hockey stick rather than a straight line, hence the name 'hockey rule').

If we subtract '*Lv*' from both sides of the equation, and divide both sides by '*Lv*', in order to obtain $(L - Lv)/Lv$ as in Garegnani's example, the terms corresponding to the first diagonal of Pascal's triangle are eliminated. The hockey rule makes us look another diagonal further down, and thus to the elimination of another diagonal. Once this is done (that is, after the two first diagonals are eliminated), we reach a subtriangle within Pascal's triangle, where the first row (row zero) is constituted by 1, the second row (row 1) by 3 and 1, the third (row 2) by 6, 4 and 1, and so on. This is the following subtriangle:

1
3 1
6 4 1
10 10 5 1
15 20 15 6 1
21 35 35 21 7 1
....................

Subtracting and dividing by *Lv* we get:

$$\frac{L-Lv}{Lv}=\frac{1}{n}\left[\binom{n+1}{2}r+\binom{n+1}{3}r^2+\binom{n+1}{4}r^3+\cdots+\binom{n+1}{n}r^{n-1}+\binom{n+1}{n+1}r^n\right]$$

(10.4b)

The elements of row '$n$ - 1' of this new subtriangle (which are in row '$n$ + 1' of the original Pascal triangle from which two diagonals were eliminated, as indicated in the expression above) appear multiplied by each $r^k$ (where '$k$' goes from '1' to '$n$'), and the result above can be re-written in the following generalised form of Garegnani's surplus equation:

$$\frac{L-Lv}{Lv}=\frac{1}{n}\sum_{k=1}^{n} r^k \frac{(n+1)!}{(n-k)!(k+1)!} \qquad (10.4c)$$

Equation (10.4c) is a generalisation of Eq. (10.2c) above, that is, of Garegnani's equation, from '2' periods to 'n' periods—this can be seen by setting $n = 2$, for this makes the first two terms of the summation above equal to those in Eq. (10.2c).

The profit function in Equation (10.4c) increases monotonically as '$n$' increases. To see this, note that for each value of '$n$', we have that each $r^k$ (from '1' to '$n$') is multiplied by each element of row '$n$ - 1' of the subtriangle which emerges after we eliminate the first two diagonals of Pascal's triangle. But due to the properties of Pascal's triangle, the first diagonal of the subtriangle defined above (which is the third diagonal left in Pascal's triangle after we eliminate the first two) is the sequence of the triangular numbers, while the second diagonal of the subtriangle (the fourth in Pascal's triangle) is the sequence of the tetrahedral numbers, the third diagonal of the subtriangle (the fifth in Pascal's triangle) is the sequence of pentatope numbers, and so forth.

This means that as '$n$' increases, $r^1$ will be multiplied by a number that increases following the sequence of triangular numbers, $r^2$ will be multiplied by a number that increases following the sequence of tetrahedral numbers, $r^3$ will be multiplied by a number that increases following the sequence of pentatope numbers, and so forth. Note that the expression is divided by '$n$', but since '$n$' increases following the sequence of natural numbers (as does the second diagonal of Pascal's triangle), which grows at a lower rate than any of the other sequences by which each $r^k$ is multiplied, then Eq. (10.2c) can be transformed into a summation of '$k$' terms from

'1' to '*n*' which consist of each $r^k$ multiplied by a fraction that grows with '*n*'. That is, the profit function of the surplus equation increases monotonically with '*n*', and with '*r*'.

But then, as '*n*' increases, '*r*' must decrease for the rate of surplus value given by $(L - Lv)/Lv$ (which we may also call the exploitation ratio, following Marx) to remain the same. But this simply means that when the proportion of capital to labour increases (where we may call the proportion of capital to labour the composition of capital, following Marx), the rate of profits decreases for a given ratio of surplus value $(L - Lv)$ to labour $Lv$. For the increase in '*n*' simply means the increase in labour performed at previous periods, which constitutes capital in the present period. The results obtained by Marx can then be maintained, although they must be reformulated in order to take into account the results of the Cambridge controversies in the theory of capital.

Different distributions of '$Lv$' over the '*n*' periods (here an equal distribution was assumed) complicate the analysis above, but do not change the tendency of the rate of profits to fall. The omission of terms (for example, years when no labour was performed) may also complicate the analysis, but the result is maintained: we simply need to look at Pascal's triangle while removing some elements corresponding to those years. Omission of some terms or a different set of weights will influence the tendency of the rate of profits to fall depending on how the composition of capital is affected, but the general result is maintained: a relative increase in the number or weights of terms with older dated quantities of labour (that is, an increase in constant capital, i.e., older labour terms, relative to variable capital, i.e., current labour terms) will be associated with a greater tendency for the rate of profits to fall.

## 6 Discussion

Using this analysis, we can confirm Marx's analysis of how the proportions between the means of production and labour (which is expressed in the distribution of past labour over time) lead to a tendency for the rate of profits to fall with the process of accumulation. The introduction of new technologies acts as a countervailing tendency that is expressed in the introduction of more recent labour terms at the expense of older labour terms in the profit function, and in the surplus equation both through this change in the profit function and through an increase in the surplus ratio $(L - Lv)/Lv$, as new technologies can reduce the prices of wage-goods necessary for the reproduction of labour power in the integrated wage-goods sector, or increase productivity in this sector, in both cases raising the horizontal line $(L - Lv)/Lv$.

Here it is important to distinguish between the case when there is an introduction of a new technique, and the case when there is merely a change to a different technique. A new technique introduced by a given capitalist can become widely known, but it is still a new technique for the capitalists that still have not introduced it (Sinha 2014: 188), so its effect as a countervailing mechanism will still exist as long as the new technique is still in the process of being adopted by various capitalists. It is only after all capitalists have introduced the new technique, so that it achieves the status of being the dominant technique, that we reach the final outcome of the introduction of a new technology, which can be expressed as a technical composition of capital that uses less labour, and there is an increase in the composition of capital expressed in the profit function. As Sinha (2014: 188) argues, both Ricardo and Marx had the tendency to believe that the effect of this widespread diffusion of the new technique would be relatively fast, but this need not be so. Since this change is permanently taking place, the impact on the composition of capital and its effect on the falling rate of profit is a long term tendency permanently disrupted by countervailing tendencies caused by the introduction of new techniques.

We can also confirm Ricardo's conclusion that the accumulation of capital reduces the rate of profits. Remember that for Ricardo, this happens due to the inclusion of less fertile lands in production. Since the less fertile lands determine the rate of profits, the rate of profits would decrease, and the rents given by the more fertile lands would increase. For rent depends on the difference between the productivity of the more fertile lands, and the productivity of the less fertile lands, and this difference increases as accumulation brings less fertile lands into cultivation.

This effect manifests itself in the reduction of the surplus value produced in the integrated wages-good sector, or through an increase in wages (since agricultural goods used for subsistence will be more expensive) which Sraffa saw as a way to conceptualise the accumulation of capital (Martins 2013: 67–68). It will then lower the horizontal line given by $(L - Lv)/Lv$, which will then intersect the profit function of '*r*' at a lower point, leading to a lower '*r*' (since the surplus equation increases monotonically with '*r*').

The effects of accumulation on rent are also an important part of Marx's analysis of the dynamics of capitalism, especially of how the distribution of the surplus impacts on accumulation itself. To see this, it important to note that Marx's analysis of rent went beyond Ricardo's analysis, by distinguishing different types of rent other than those considered by Ricardo. As Marx notes, the producers that possess access to advantages not available to every producer (such as the natural resources or locational advantages given by a given land) obtain a differential rent (Fine 1979) due to their lower cost of production. The advantages that a given land provide may be caused by improvements achieved through the investment of capital on land, so differential rent can also arise because of unequal capital investment on equal lands, which constitutes a second type of differential rent, that stands in contrast with the first case of differential rent Marx considers where lands are unequal from the very beginning (Fine 1979). These cases are also considered by Sraffa (1960).

If some producers in sectors or industries where market value is above the price of production are able to evade the competition between sectors by possessing advantages due to landed property that allows the price to be maintained close to or at a market value above the price of production, an absolute rent emerges (Fine 1979). Absolute rent can also be conceptualised through Sraffa's (1960) system comparing what was called above the value system (where prices are proportional to quantities of labour) and the price system. Marx presupposes that the product is sold at a monopoly price which arises as a consequence of the absolute rent. But a monopoly price can also arise because the final product is a rare one in short supply, in which case a monopoly rent arises, which is a consequence of the monopoly price, rather than a cause of it as in absolute rent.

These increases in rent must be seen together with the overall process of accumulation (Fine 1979). For as the process of accumulation proceeds the increase in rent and the raw materials associated with it leads to a redirection of the surplus away from industrial activities (Martins, 2016), which may lead to a lack of investment and periodic crises (Martins 2018: 25), as Marx (1910) notes in his *Theories of Surplus Value*.

Crises are also exacerbated by the lack of effective demand caused by the fact that workers do not receive the full product of their labour, and so the consumption of all production depends on the decisions of capitalists for using the surplus value they appropriated in consumption or investment (Kalecki 1971; Martins 2011). Indeed, just as one can fruitfully identify the notion of an organic composition of capital causing the tendency for the rate of profits to fall, it is also possible to identify an organic composition of income (Martins 2011), where income is divided between consumption and saving, and where an increase in saving comparatively to consumption causes a tendency for effective demand to fall when savings are not fully reinvested, which is likely to be the case during a crisis, and especially so when increases in the prices of raw materials divert income away from productive investment (Martins 2016). These problems can get articulated with the interplay between the tendency of the rate of profits to fall and the countervailing tendencies addressed above, leading to a complex situation.

## 7 Concluding Remarks

The interaction between the tendency of the rate of profits to fall and the countervailing tendencies must be accounted as a key factor in the generation of crises that lead to the destruction of capital, thus restoring the rate of profits by reducing the organic composition of capital whenever it increases too much without being sufficiently compensated by variations in the value composition of capital.

The tendency for a rate of profits to fall identified above must be seen in conjunction with Marx's remarks on the countertendencies that act against it, which are connected to the role of technical change. In various unpublished manuscripts (e.g., Sraffa D3/12/35/12), Sraffa notes that the economy is a dynamic system which in general is not in a self-replacing state. Rather, accumulation requires new techniques, and the discontinuation of others, leading to technical change. This means that the result above is a mere tendency, and must be seen in the broader context set by the continuous introduction of new techniques which can lower the cost of capital and reestablish an value composition of capital that helps restoring again a higher rate of profits.

As Sinha (2014: 188) explains, the widespread adoption of new techniques by capitalists can take quite long, leading to a slowing down of the tendency of the rate of profits to fall. Thorstein Veblen (1904: Ch. VII), in contrast, argued that the introduction of new techniques brings a permanent tendency for depression, since it forces capital invested in older technologies to compete with new capital invested in new technologies. This would mean that rates of profits would be lower for older capital, further aggravating the results obtained above concerning the impact of older capital present in the production process in lowering the rate of profits. Nevertheless, Veblen also notes how the combinations between businessman aimed at reducing competition counteract this tendency.

Wesley Mitchell (1913: 569, fn.42) adds further countertendencies, such as the fact that older technologies are often more able to compete with newer technologies that usually thought, and noting also that the 'increase of demand for construction work' also exercises 'a powerful influence in restoring conditions which promise a fair margin of profit.'

The tendency for the rate of profits to fall must then be seen as one tendency amongst others, in a context where the economy is an open system where tendencies, even when exercised, may not be actualised due to the countertendencies that are exercised at the same time (Lawson, 1997). Luigi Pasinetti (2005: 847) notes the need of 'combining ′pure theory' with 'institutional analysis", following the example set by John Maynard Keynes. A more technical analysis, such as the one undertaken above, can be useful for seeing some tendencies with more transparency, but does not replace a fuller analysis of the overall institutional context, such as the one undertaken by Marx, Veblen, Mitchell, and many other economists centred on institutional analysis.

## References


De Vivo, G. 2003. Sraffa's path to *Production of Commodities by Means of Commodities*: an interpretation, *Contributions to Political Economy*, 22, 1–25.

Fine, B. 1979. On Marx's Theory of Agricultural Rent, *Economy and Society*, 9(3): 241–278.

Fine, B. and Saad-Filho, A. (2016) *Marx's Capital*, 6th edition, London: Pluto Press.

Garegnani, P. 1984. Value and Distribution in the Classical Economists and Marx. *Oxford Economic Papers*, 36, 291–325.

Garegnani, P. 2012. On the present state of the capital controversy, *Cambridge Journal of Economics*, 36(6), 1417–1432.

Garegnani, P. 2018. On the Labour Theory of Value in Marx and in the Marxist Tradition. *Review of Political Economy*, 30(4): 618–642.

Gilibert, G. 2003. The equations unveiled: Sraffa's price equations in the making, *Contributions to Political Economy*, 22, 27–40.

Harvey, D. 2006 [1982]. *Limits to Capital*. London: Verso.

Kalecki, M. 1971. *Selected Essays on the Dynamics of the Capitalist Economy*, Cambridge: Cambridge University Press.

Kurz, H. and Salvadori, N. 2015. *Revisiting Classical Economics: Studies in long-period analysis*, London, Routledge.

Marshall, A. 1923 [1919]. *Industry and Trade*, London: Macmillan.

Martins, N. 2011. Globalisation, Inequality and the Economic Crisis, *New Political Economy*, 16(1), 1–18.
Martins, N. O. 2013. *The Cambridge Revival of Political Economy*, London and New York, Routledge.
Martins, N. O. 2014. Sraffa on Fixed Capital, Money and the Phases of Capitalism, *Review of Political Economy*, 26(1), 111–127.
Martins, N. O. 2016. Political Aspects of the Capital Controversies and Capitalist Crises, *Journal of Post Keynesian Economics*, 39(4), pp. 473–494
Martins, N. O. 2018. Marx, Lefebvre and Rhythmanalysis, *International Journal of Philosophy and Social Values*, 1(2), pp. 13–30.
Martins, N. O. 2019. The Sraffian Methodenstreit and the Revolution in Economic Theory, *Cambridge Journal of Economics*, 43(2), 507–525.
Marx, K. 1976 [1867]. *Capital*, Vol. I., Pelican Books, New York, NY, (translated by B. Fowkes), London: Pelican Books.
Marx, K. 1978 [1885]. *Capital*, Vol. II., Pelican Books, New York, NY, (translated by D. Fernbach), London: Pelican Books.
Marx, K. 1981 [1894]. *Capital*, Vol. III, Pelican Books, New York, NY, (translated by D. Fernbach), London: Pelican Books.
Marx, K. 1910. *Theorien über den Mehrwert, Edited by Karl Kautsky, 3 vols*, J.H.W. Dietz, Stuttgart.
Mitchell, W. C. 1913. *Business Cycles*. Berkeley: University of California Press.
Pasinetti, L. L. 1993. *Structural Economic Dynamics*. Cambridge: Cambridge University Press.
Pasinetti, L. L. 2005. The Cambridge School of Keynesian Economics, *Cambridge Journal of Economics*, 29: 837–848.
Ricardo, D. 1821 [1817]. *On the Principles of Political Economy and Taxation*, London: John Murray: Albemarle-Street.
Sen, A.K. 2003. Sraffa, Wittgenstein, and Gramsci, *Journal of Economic Literature*, 41, 1240–1255.
Sinha, A. 2014. On Marx's Law of the Falling Rate of Profit: Disentangling Some Entangled Variables, *Review of Radical Political Economics*, 46(2): 184–189.
Sinha, A. 2016. *A Revolution in Economic Theory: The Economics of Piero Sraffa*, Cham: Palgrave Macmillan.
Smith, A. 1993 [1776]. *An Inquiry into the Nature and Causes of the Wealth of Nations*, Oxford: Oxford University Press.
Sraffa, P. 1926. The laws of returns under competitive conditions, *Economic Journal*, 36, 535–550.
Veblen, T. 1904. *The Theory of Business Enterprise*, New York: Charles Scribner's Sons.
Wittgenstein, L. 1963 [1953]. *Philosophical Investigations*, trans. G. E. M. Anscombe. Oxford: Blackwell.

Comment

# The Falling Rate of Profit and General Equilibrium—Some Methodological Considerations

*Naoki Yoshihara and Roberto Veneziani*

## *Introduction*

The long-run trends in capital accumulation and in the distribution of wealth and income in capitalist economies have been central questions both for classical economists such as Smith, Ricardo, Mill, and Marx, and for early neoclassical economists such as Walras, Schumpeter, and Hicks. As Piketty (2014: 15–16) has aptly noted, "The economists of the nineteenth century deserve immense credit for placing the distributional question at the heart of economic analysis and for seeking to study long-term trends. Their answers were not always satisfactory, but at least they were asking the right questions." Yet, long-run trends in distribution and capital accumulation have long since lost their prominence in academic economics: "For far too long, economists have neglected the distribution of wealth" (Piketty (2014: 16).

A central topic in classical and early neoclassical analysis of the long-run trends of capitalist economies concerned the dynamics of the rate of profit, or interest, which was widely considered to display a tendency to fall. According to Ricardo, Mill, and Walras, the long-run tendency of the equilibrium profit rate to fall was caused by decreasing returns to scale due to the scarcity of a primary factor of production, such as land. According to Marx (1954 [1867], 1959 [1894]), instead, capital-using and labour-saving technical progress—arguably the type of technical progress that has characterised much of the history of capitalist economies—is the cause of the tendency of the rate of profit to fall, if the real wage is constant.

Marx's theory of technical change, and the resulting fall in profitability, has sparked significant controversy ever since. A refutation of Marx's theory has been famously provided by Nobuo Okishio in a seminal paper (Okishio 1961; see also the extensions by Roemer 1977, 1979). The so-called "Okishio Theorem" (OT) demonstrates that if the real wage rate is fixed at the (historically and culturally determined) subsistence level, then any cost-reducing technical innovation—the technical changes that would indeed be adopted by maximising capitalists—would *increase* the equilibrium profit rate.

A number of authors in the Marxian camp have sought to refute OT and to establish the validity of Marx's theory of the falling rate of profit. From a broad theoretical perspective, the paper by Nuno Martins (2021) belongs to this class of contributions. One important innovation in Martins' (2021) analysis is that it builds on the formal approach pioneered by Garegnani (1984), and in particular on his so-called 'surplus equation', linking the profit rate to surplus value, constant capital, and variable capital, all measured in terms of labour commanded. This is quite interesting, and thought provoking, as the Sraffian approach is often considered to be inadequate to capture Marxian insights.

In this note, we reconsider Garegnani's (1984) analysis and generalise his 'surplus equation'. Then we discuss its implications in the analysis of the distributive effects of technical change.

### *The Model and Notation*

Consider an economy in which a set $N = \{1, \dots, n\}$ of commodities can be produced by means of themselves and labour. Production can be activated using a standard Leontief production technique, $(A, L)$, where $A$ is a $n \times n$ nonnegative and productive matrix of material input coefficients, and $L$ is a $1 \times n$ positive vector of labour coefficients. We rule out joint production so that each sector produces only one good. The coefficient $a_{ij}$ represents the amount of good $i$ necessary to produce one unit of good $j$ and is measured in the relevant physical unit. The coefficient $L_i$ measures the amount of labour necessary to produce one unit of good $i$. Let $x$ be a $n \times 1$ vector of activity levels describing the level at which production is activated.

The price vector is $(p, w, r)$ where $p$ denotes the $1 \times n$ vector of commodity prices, $w$ the wage rate, and $r$ the profit rate. Assuming wages to be paid ex ante, as in Garegnani (1984), the standard production price equations are:

$$p = (1+r)(pA + wL) \tag{10.5}$$

Following Garegnani (1984) and Martins (2021), suppose that the set of goods $N$ can be partitioned into two subsets $B = \{1, \ldots, b\}$ and $D = \{b+1, \ldots, n\}$ indicating, respectively, the vertically integrated sectors producing wage goods (the sectors producing wage goods plus the sectors producing inputs used in the production of the former, and so on), and all other sectors, such that $N = B \cup D$ and $B \cap D = \emptyset$. Let $p_B$ and $p_D$ denote, respectively, the $1 \times b$ vector of the prices of the goods in $B$ and the $1 \times (n - b)$ vector of the prices of the goods in $D$, where $p = [p_B, p_D]$.

The sector of vertically integrated wage goods constitutes a self-contained subsystem. Therefore we can define the nonnegative and indecomposable $b \times b$ matrix $A_B$ which describes the material input coefficients used (directly or indirectly) in the production of wage goods, and the $1 \times b$ vector of labour coefficients $L_B$. We can write $L = [L_B, L_D]$ and a similar notation holds for the vector of activity levels $x = [x_B, x_D]$.

### *The Surplus Equation*

Garegnani (1984) starts from Marx's definition of the general rate of profit as the ratio of aggregate surplus value, $s$, to the sum of aggregate constant and variable capital, $c$ and $v$, respectively. Formally

$$r = \frac{s}{c+v}, \tag{10.6}$$

where all quantities are measured in terms of labour time. In the standard interpretation, variable capital corresponds to the amount of labour embodied in the subsistence bundle that workers can afford, while constant capital is equal to the amount of labour embodied in the capital goods used in production.[1] Surplus value is the difference between the total labour embodied in the net product and variable capital.

[1] See Mohun and Veneziani (2017) for a critical discussion of the standard interpretation, and a survey of alternative views.

This definition of the profit rate is notoriously problematic, especially in the context of the determination of production prices from labour values. One of the contributions of the Sraffian approach has been precisely to point out the inconsistency of the standard approach to the so-called 'transformation problem', and to suggest that there is nothing really to transform: prices and distribution can be determined simultaneously and independently from quantities of embodied labour.

In order to avoid such inconsistencies, and at the same time, partially rectify the negative verdict on Marx's theory of prices and distribution, Garegnani (1984) argues that the surplus Eq. (10.6) still provides a useful tool to understand the determinants of the profit rate and relate the functional income distribution to labour amounts. To that purpose, he implements two fundamental interpretational shifts: all quantities in (2) are reinterpreted as amounts of labour *commanded*, rather than labour *embodied*, and the variables are not economy-wide aggregates but refer to magnitudes *in the (vertically integrated) wage goods sector*.

Formally, variable capital in the wage goods sector is equal to the amount of labour that the wages paid in that sector can command: this is just the amount of labour performed in the sector, $v = L_B x_B$.[2] Similarly, surplus value in the wage goods sector is equal to the amount of labour that the net output of that sector can command minus variable capital. Because the net output of the integrated wage goods sector is equal to the total wages paid in the economy, $p_B(I - A_B)x_B = wLx$, it follows that $s = Lx - L_B x_B$.

Therefore, two of the three variables on the right hand side of the surplus Eq. (10.6) can be determined independent of information on prices: they are known amounts of labour (commanded). The problem is the third variable, constant capital $c$, which cannot be similarly computed.

Garegnani (1984) proposes a way out of this problem by looking at the production of the means of production and considering "it as the result of previous stages in the production of the final commodity: the stages between moments (-2) and (-1), in which the means of production have been produced; the stage between (-3) and (-2) in which the means of production of those means of production have been produced, *and so forth*" (Garegnani 1984: 315, emphasis added). In other words, capital

[2] Here and everywhere else in the paper, including in quotes from Garegnani (1984) and Martins (2019), we shall use the notation laid out in section 2 for consistency.

can be reduced to amounts of dated labour and this approach can be "applied to the means of production of the integrated wage-goods sector to obtain a 'Surplus equation' determining the rate of profit" (Garegnani 1984: 315).

Garegnani (1984) illustrates the formal relation between surplus value and the rate of profit in a special case. Consider an economy where wages consist only of one consumption good, corn, which is produced using one unit of labour and one unit of an intermediate good, a plough, which is in turn produced only using one unit of labour. "The [$L_B x_B$] labourers employed in the integrated wage-goods sector will therefore be distributed half in (directly) producing the 'corn' and, and half in reproducing the 'ploughs' " (Garegnani 1984: 315). Going back only one period, Garegnani then derives the following surplus equation:

$$Lx - L_B x_B = rL_B x_B + r\frac{L_B x_B}{2} + r^2 \frac{L_B x_B}{2}, \tag{10.7}$$

or, equivalently

$$\frac{Lx - L_B x_B}{L_B x_B} = r + \frac{r}{2} + \frac{r^2}{2}. \tag{10.8}$$

Equation (10.8) provides the desired link between amounts of labour (commanded), on the left hand side, and the profit rate, on the right hand side, and a positive relation can be derived between the rate of surplus value in the integrated wage-goods sector, $\frac{Lx - L_B x_B}{L_B x_B}$, and the profit rate (Garegnani 1984: 317).

### *The Falling Rate of Profit*

Building on previous work (Martins 2013: 76–83), Martins (2021) argues that "Garegnani's surplus equation can be used to express Marx's analysis of the tendency for the rate of profits to fall." The argument is as follows. First, note that Eq. (10.7) can be equivalently be written as follows:

$$Lx = \frac{L_B x_B}{2}(1+r) + \frac{L_B x_B}{2}(1+r)^2. \tag{10.9}$$

The latter expression derives from the fact that "The example provided by Garegnani … includes only two periods (the first for the production of the 'ploughs', the second for the production of 'corn')" (Martins 2021). According to Martins, however, it can be generalised to $t$ periods of production, leading to the transformation of Eq. (10.9) into the following expression (Martins 2021):

$$Lx = \frac{L_B x_B}{t}(1+r) + \frac{L_B x_B}{t}(1+r)^2 + \ldots + \frac{L_B x_B}{t}(1+r)^t . \tag{10.10}$$

From the latter expression, and after some algebraic manipulation, Martins (2021) derives his final version of the surplus equation:

$$\frac{Lx - L_B x_B}{L_B x_B} = \frac{1}{t}\Sigma_{i=1}^{t} r^i \frac{(t+1)!}{(t-k)!(i+1)!}. \tag{10.11}$$

Equation (10.11) provides the basis of the main analytical conclusion. For Martins argues that the right hand side is monotonically increasing in $t$,

> [b]ut then, as '[$t$]' increases, '$r$' must decrease for the rate of surplus value given by $\left[\frac{Lx - L_B x_B}{L_B x_B}\right]$ … But this simply means that when the proportion of capital to labour increases …, the rate of profits decreases for a given ratio of surplus value [$Lx$ - $L_B x_B$] to labour [$L_B x_B$]. For the increase in '[$t$]' simply means the increase in labour performed at previous periods, which constitutes capital in the present period. The results obtained by Marx can then be maintained, although they must be reformulated in order to take into account the results of the Cambridge controversies in the theory of capital. (Martins 2019: 24)

This conclusion is intriguing and it follows from a very thorough analysis of Garegnani's (1984) celebrated article, and from a scholarly discussion of classical political economy. Nonetheless, in our view, it is unclear that Marx's theory of the falling rate of profit, or more generally his theory of technical change, can indeed be vindicated based on Garegnani's (1984) surplus equation.

### *The Surplus Equation Reconsidered*

The first point to note is that although Eq. (10.11) can in principle be interpreted as a generalisation of Garegnani's (1984) surplus equation, it is not fully general. Quite aside from the rather special assumptions on the distribution of labour across sectors,[3] Eq. (10.11) does not provide a complete reduction of all magnitudes into dated labour.

To see this, as a first step, substitute the expressions for surplus value and variable capital into Eq. (10.6):

$$r = \frac{Lx - L_B x_B}{c + L_B x_B}. \tag{10.12}$$

Multiplying both sides by $c + L_B x_B$ one obtains

$$rc + rL_B x_B = Lx - L_B x_B. \tag{10.13}$$

Next, note that in terms of (production) prices, constant capital is equal to $p_B A_B x_B$. Given that the sector of wage goods is vertically integrated, we can consider its $b$ production price equations independently:

$$p_B = (1+r)(p_B A_B + wL_B). \tag{10.14}$$

Solving for prices

$$p_B = (1+r)wL_B\left[I - (1+r)A_B\right]^{-1}. \tag{10.15}$$

[3] And over time. It is not entirely clear to us, in fact, why the term $\frac{L_B x_B}{2}$ is replaced by $\frac{L_B x_B}{t}$. According to Martins (2019: 20), "Since labour is distributed equally over the '[$t$]' periods, the wages paid in the wage-goods sector, '[$L_B x_B$]', are divided equally by '[$t$]' periods (rather than 2 periods as in Garegnani's example)." However, in Garegnani's example, the term $\frac{L_B x_B}{2}$ has nothing to do with the idea of labour being equally divided across production periods: it derives from the structure of production, and in particular from the assumption that, during every production period, half of direct labour employed in the integrated wage goods sector is used to produce corn, whereas the other half is used to produce ploughs, an assumption that Martins does not explicitly drop. Yet, as none of our main arguments depend on this, we will not press the issue here.

The usual method can then be used to express prices as amounts of dated labour (commanded)

$$\frac{p_B}{w} = L_B \Sigma_{i=0}^{\infty} (1+r)^{i+1} (A_B)^i. \tag{10.16}$$

Finally, multiplying both sides of Eq. (10.16) by $A_B x_B$ we obtain an expression for constant capital in terms of commanded labour:

$$c = \frac{p_B A_B x_B}{w} = L_B \Sigma_{i=1}^{\infty} (1+r)^i (A_B)^i x_B. \tag{10.17}$$

Substituting the latter expression into Eq. (10.13) and dividing both sides by $L_B x_B$ gives us the general formula of the surplus equation:

$$\frac{Lx - L_B x_B}{L_B x_B} = r \left[ 1 + \Sigma_{i=1}^{\infty} (1+r)^i \frac{L_B (A_B)^i x_B}{L_B x_B} \right]. \tag{10.18}$$

Several points can be made based on Eq. (10.18). First of all, if constant capital is properly reduced into an infinite sum of dated amounts of labour, then it is not possible to represent technical change as the addition of further terms in Eq. (10.11). Such additions allow one to have a better approximation of the general formula by adding further terms (of vanishingly small magnitude) in the summation but are *not* equivalent to a lengthening of the production process.

Second, it is very unclear that one can derive any insights on the distributive effects of technical change by directly looking at Eq. (10.18). On the one hand, there is no obvious way of modelling technical change *directly in* Eq. (10.18). For example, one cannot assume that capital using, labour saving technical change corresponds to an increase in $\frac{L_B (A_B)^{i+1} x_B}{L_B x_B}$.

On the other hand, one cannot hold the left hand side of the equation constant. One of the key insights of OT is precisely that even if one were to assume capital using labour saving technical change to correspond to an increase in $\frac{L_B (A_B)^{i+1} x_B}{L_B x_B}$, such technical change would also yield an increase in the rate of surplus value.

Third, OT is based on a standard production price framework and compares production prices, and the corresponding rate of profit, for a given (subsistence) wage, before and after technical change. The framework adopted by Garegnani (1984) to derive the surplus Eq. (10.6), and then extended by Martins (2021) is the same as Okishio's (1961). In this framework, and under the assumptions of the Okishio Theorem, it is not possible to escape the conclusion that (cost-reducing) capital using labour saving technical change yields an increase in the profit rate, if the wage rate remains constant.

## *Conclusions*

Within the theoretical framework normally used to prove the Okishio Theorem (OT), there is no way of escaping its conclusions. If one focuses on production prices, and analyses technical change in terms of a comparative statics exercise with a given real wage, then one has to conclude that cost-reducing capital-using labour-saving technical change yields an increase in the profit rate.

The real question, in our view, is whether a mere focus on production prices, and the associated notion of equilibrium as profit rate equalisation, is adequate to capture the key features of Marx's theory of technical change. Interestingly, in later contributions, Okishio (2000) himself has cast some doubts on the relevance of OT as a characterisation of the long-run tendencies of capitalist economies. On the one hand, Okishio (2000) has argued that it is unrealistic to assume a constant real wage rate in the dynamic process of capital accumulation. On the other hand, he has pointed out that OT is essentially a comparative statics result that compares two equilibrium rates of profits (before and after the introduction of a new technique).

More generally, we would argue that in order to properly capture Marx's theory of technical change, and the distributive effects of innovations in capitalist economies, it is necessary to consider a more complex model of the economy.[4] First, a key element of Marx's analysis lies in the dialectic between individually rational choices and unintended, suboptimal aggregate outcomes. It is difficult to capture this dynamics without explicitly modelling individual behaviour, and capitalists' profit maximising choices of techniques. Second, a capitalist economy is a complex network of markets, as well as production and consumption activities. This is partly captured in the Sraffian model, and its notion of capital as a vector of produced goods (rather than a single aggregate factor, as in neoclassical macro models). Nonetheless, it is important to consider equilibrium notions that go beyond the mere equalisation of profit rates and to consider explicitly the general equilibrium effects of technical change.

## References

Garegnani, P., 1984. Value and Distribution in the Classical Economists and Marx. *Oxford Economic Papers* 36, 291–325.

Martins, N. O., 2013. *The Cambridge Revival of Political Economy*, London and New York, Routledge.

Martins, N. O., 2021. Garegnani's Surplus Equation and Marx's Falling Rate of Profits, in this volume.

Marx, K., 1954 [1867]. *Capital*, Vol.I. Lawrence & Wishart, London.

Marx, K., 1959 [1894]. *Capital*, Vol.III. Lawrence & Wishart, London.

Mohun, S., Veneziani, R., 2017. Value, price and exploitation: the logic of the transformation problem. *Journal of Economic Surveys* 31, 1387–1420.

Okishio, N., 1961. Technical Change and the Rate of Profit. *Kobe University Economic Review* 7, 86–99.

Okishio, N., 2000. Competition and Production Prices. *Cambridge Journal of Economics* 25, 493–501.

Piketty, T., 2014. *Capital in the Twenty-First Century*, Harvard University Press, Cambridge, MA.

Roemer, J.E., 1977. Technical change and the 'tendency of the rate of profit to fall'. *Journal of Economic Theory* 16, 403–424.

Roemer, J.E., 1979. Continuing controversy on the falling rate of profit: fixed capital and other issues. *Cambridge Journal of Economics* 3, 379–398.

Yoshihara, N., Veneziani, R., 2019. Technical progress, capital accumulation, and distribution. Working Paper 899, School of Economics and Finance, Queen Mary University of London.

[4] A thorough analysis of the distributive effect of innovations along the lines described here is developed in Yoshihara and Veneziani (2019).

Reply
# Causality and Causation: Some Methodological Considerations

*Nuno Ornelas Martins*

I must start by thanking Naoki Yoshihara and Roberto Veneziani (2021) for their thoughtful and important comments on my chapter "Garegnani's surplus equation and Marx's falling rate of profits", made in their chapter titled "The falling rate of profit and general equilibrium: some methodological considerations". As the title of their contribution highlights, and the last sentence of their contribution also stresses, the approach they outline is one where the aim is to go beyond the analysis of the equilibrium of profit rates, so as to consider the general equilibrium effects of technical change. Building within this framework, they provide a generalisation of Pierangelo Garegnani's (1984) surplus equation.

I take this opportunity to also address some methodological considerations. In particular, I would like to note that the careful analysis undertaken by Yoshihara and Veneziani (2021) is built on methodological foundations that are entirely different from the ones I presuppose in my own take on this problem (Martins 2021). To put it in a single sentence, I do not undertake in any kind of equilibrium analysis. Equilibrium analysis typically presupposes constant conjunctions of the form "if event X then event Y" (Lawson 1997). And I try to provide a conception which does not presuppose this type of constant conjunctions, which are the defining characteristic of what Tony Lawson (1997) calls closed systems.

Let me elaborate on this topic drawing on an author who seems to have influenced Piero Sraffa on these issues, as Ajit Sinha (2016: 75–76) argues, namely Arthur Stanley Eddington. Indeed, Sraffa's own references to the geometrical approach he employs (Martins 2013, 2019; Sinha 2016) are consistent with the geometrical approach adopted by Eddington in chapter VI of his book *The Nature of the Physical World* (Eddington 1948: 111–137). Eddington, in chapter XIV of this book (Eddington 1948: 293–315) uses the term ***causation*** to refer to an asymmetrical causal relation where event X must precede event Y. Causation can be contrasted with ***causality***, which is a symmetrical causal relation which runs in both directions, from event X to event Y, and also from event Y to event X.

Sraffa's (1960) analysis, in the way I interpret it, presupposes causality, rather than causation, in Eddington's (1948) sense. The various variables in Sraffa's system are potentially all interconnected (Martins 2013), in a context where causality can run in various directions. An econometrician would readily point out that this constitutes a possible endogeneity problem between the variables, and exogeneity tests are necessary in order to identify a relation of causation with a time arrow running from event X to event Y. The reply I would give is that the social world is characterised by open systems, rather than by closed systems with constant conjunctions of the form "if even X then event Y" (Lawson 1997). Causality and endogeneity are an everyday occurrence, not an econometric problem to be solved in an attempt to develop a closed system.

Sraffa rejects Alfred Marshall's (1920) partial equilibrium analysis for assuming that we can focus on the influence of event X on event Y while taking everything else to be constant (Martins 2013). Rather, Sraffa focuses on the conditions for the reproduction of the whole economic system at a given moment in time, rather than on relations of causation between event X and event Y. In Sraffa's (1960) system, there is no functional unidirectional relationship between inputs and outputs. It is for this reason that both the inputs and the outputs in Sraffa's equations are exogeneous variables. The same goes for the surplus equation, hence the possibility of keeping constant the rate of surplus value while changing the distribution of labour through time.

If we want to take into account causality, we need to provide a theory that expresses the interconnections of the system at a given moment in time, as Sraffa (1960) does. Once we reach such a theory, as Sraffa (1960) does, we have the tools for expressing an underlying reality such as a tendency of the rate of profits to fall. But those tools cannot provide any proof for or against the existence of the tendency of the rate of profits to fall. Rather, they simply allow for an expression of an interconnected totality, rather than a representation of relations of causation of the form "if event X then event Y".

To express a reality mathematically, as Sraffa (1960) does, is not to model it while imposing functional relations of causation. Rather, it is something more akin to an accounting exercise, which merely measures something objectively without trying to impose a deterministic view of causation to the phenomena under analysis. This also means that certain mathematical characteristics of Sraffa's (1960) system, such as the use of linear equations, are best interpreted as a result of the conventional practice of accounting in terms of summation (Martins 2021), rather than an attempt to impose linearity on the conditions of production, in which case we would be presupposing constant returns to scale, as it is usually the case when linearity is imposed in the context of functional relations.

This provides a broad framework that can be fruitfully combined with Karl Marx's analysis of the mode of production. The explanation of why there is a tendency for a falling rate of profit (or the possible rejection of the existence of this tendency) does not take the form of a mathematical proof. Rather, it must be found in the reality under analysis, not in the tools that are used to express it, or measure it objectively. In Marx's analysis of the capitalist mode of production, capitalists have an equal bargaining power vis-à-vis other capitalists, and capitalists as a class have a higher bargaining power than labourers. Adam Smith had already pointed out how workers could hold on for less time without receiving wages than their employers when not receiving profits. This means that capitalists can make use of their greater bargaining power to pay wages that enable the social reproduction of labour power, but do not cover the surplus produced by labourers, which is appropriated by the capitalists. Because the surplus is produced by labour, a reduction of the quantity of labour in production leads then to a reduction in the surplus.

This reduction of the surplus would, one expects, reduce the rate of profits. But to take this step, we need a theory that successfully transforms labour values into prices, so that profits can be determined. Sraffa's (1960) system shows the connection between labour and prices, noting how prices correspond to labour values when the total surplus is paid as wages, and profits do not exist. When wages are reduced, we are led to a situation where prices do not correspond to the quantity of labour, and a rate of profit emerges. But the magnitude of the profit that can be achieved depends not only on the level of wages (which allows for the social reproduction of labour power), but also on how much labour exists to be exploited.

Garegnani's surplus equation can be used, in this context, to express the reality that the reduction of the quantity of labour presupposed in the methods of production must correspond to a lower rate of profit, for a given rate of surplus value. What leads to the reduction of the rate of profit is not, of course, the degree to which the reduction to dated labour is undertaken, as Yoshihara and Veneziani (2021) note. The reduction to dated labour is merely a form of expressing an underlying reality. The reduction of the rate of profit depends on the time structure of the labour terms that is presupposed in the production process, and is expressed after this reduction in the surplus equation. The structure of production takes place in time nonetheless, and thus it is not possible to distinguish the structure of production from its time characteristics. This is also connected to a geometrical conception of time that is presupposed in Eddington's (1948) approach.

Yoshihara and Veneziani (2021) note that the reduction I undertake does not enable a full reduction to labour. It is important to note that my goal is not to reach a full reduction to labour, but only an approximation. Sraffa's (1960) system provides a circular conception of the production of commodities by means of commodities, so the reduction can be carried on indefinitely, and still there is always a residue of commodities left. It is this residue of commodities that ensures that there is a Maximum rate of profits, as Sraffa (1960) notes. So the reduction to dated labour is always an approximation. If we could engage in a full reduction to dated labour, we would not have a circular and interconnected process of production governed by causality, and there would no Maximum rate of profits, as Sraffa (1960) notes.

Furthermore, my goal is to reach a surplus equation that expresses the specific technology employed in a specific mode of production. And to do so, we need a surplus equation that depends on the specific mode of production. The generalisation I suggest is a generalisation in the time dimension, with implications for the structure of production (and time structure of production), but made in order to highlight certain properties of the accumulation of capital through time with greater transparency. But it also means that the surplus equation needs to be constructed anew for each specific technology. Doing so is certainly in line with a constructivist approach to mathematics which seems to be the one adopted by Sraffa (Martins 2013). And it also fits with the overall purpose of engaging in what I was attempting to do: an accounting exercise that measures an objective reality, rather than an attempt to establish functional relations of causation.

## References

Eddington, A.S. 1948. *The Nature of the Physical World*. Cambridge: Cambridge University Press.

Garegnani, P. 1984. Value and Distribution in the Classical Economists and Marx. *Oxford Economic Papers* 36, 291–325.

Lawson, T. 1997. *Economics and Reality*. London: Routledge.

Marshall, A. 1920 (1890). *Principles of Economics* ($8^{th}$ edition), London: Macmillan.

Martins, N. O., 2013. *The Cambridge Revival of Political Economy*, London and New York, Routledge.

———. 2019. The Sraffian Methodenstreit and the Revolution in Economic Theory, *Cambridge Journal of Economics*, 43(2), pp. 507–525.

Martins, N., 2021. Garegnani's surplus equation and Marx's falling rate of profits. In (A. Sinha, ed.): *A Reflection on Sraffa's revolution in Economic Theory*, London, Palgrave Macmillan.

Sinha, A. 2016. *A Revolution in Economic Theory: The Economics of Piero Sraffa*, AG Switzerland: Palgrave/Macmillan.

Sraffa, P. 1960. *Production of Commodities by Means of Commodities: Prelude to a Critique of Economic Theory*, Cambridge: Cambridge University Press.

Yoshihara, N. and Veneziani, R., 2021. The falling rate of profit and general equilibrium: some methodological considerations. In (A. Sinha, ed.): *A Reflection on Sraffa's Revolution in Economic Theory*, London, Palgrave Macmillan.

CHAPTER 11

# Sraffa's Silenced Revival of the Classical Economists and of Marx

*Guglielmo Chiodi*

## 1 Introductory Remarks

Over the years, I have repeatedly asked myself the question: Why has Sraffa's 1960 seminal contribution *Production of Commodities by Means of Commodities* (*PCMC*) been so little influential, if at all, in the *way of thinking* in contemporary economic theory? After all, it paved the way for an *alternative* paradigm to the neoclassical one, which triumphantly dominated over the twentieth century and is still dominating even

I want to express my deep gratitude to Kumaraswamy Velupillai for his extremely useful comments and suggestions on a preliminary draft of this chapter. I am also most grateful to Leonardo Ditta, Geoffrey Harcourt and Ajit Sinha for their critical and beneficial remarks. Needless to say, none of them is responsible for misleading interpretations and infelicities still contained therein. I thank also Mr Jonathan Smith for the kind assistance given at the Trinity College Library (Cambridge) whilst working through Sraffa's Papers.

G. Chiodi (✉)
Sapienza Università di Roma, Rome, Italy
e-mail: guglielmo.chiodi@uniroma1.it



A. Sinha (ed.), *A Reflection on Sraffa's Revolution in Economic Theory*, Palgrave Studies in the History of Economic Thought,
https://doi.org/10.1007/978-3-030-47206-1_11

now—certainly in the western economies and (perhaps and quite sadly in my opinion) all over the world.

The concise and only apparently easy-readable little book by Sraffa does contain a neat representation of an economy within a logically well founded theoretical framework, reminiscent of the point of view taken by the classical economists and thus by Marx as well. Sraffa, in fact, made a very great breakthrough: on the one hand, he rehabilitated the 'vision' of the classical economists and of Marx; on the other, he proved their analysis *amendable*, viz. not 'irremediably wrong', up to the point of structuring a new theory of value *from that same standpoint* in an analytically robust framework.

In the sequel, an attempt will be made by showing that that part of Sraffa's contribution, which rehabilitated the 'vision' of the classical economists and of Marx, has been severely *silenced* by most Sraffians who used Sraffa's 'prelude to a critique of economic theory' in the early 1960s and in the following years for mainly criticising specific *tools* and *concepts* typical of neoclassical theory (as, for example, the production function and the notion of 'capital').

By contrast, since the publication of *PCMC*, a few scholars have indeed explored and extended Sraffian suggestions to *different directions* from those taken by the 'capital theory' people. Notable examples, in this regard, are Bharadwaj (1976), Parrinello (1983, 2001), Velupillai (1989, 2008, 2010), Velupillai and Zambelli (1993), Pasinetti (1993), Schefold (1997), Zambelli (2004), Edwards (2008), Cesaratto (2008), Fredholm (2009), Sinha (2012, 2016), Harcourt (2018).

However, the kind of critique made by most Sraffians had incontestably the merit of putting to the fore some *irremediable flaws* which the neoclassical theory was supposed to contain, and thus it was essentially addressed to the *logical inconsistencies* characterizing that theory. That approach, to be fair, turned out to be extremely useful in the very first instance, for it went quite deeply into the roots of the neoclassical theory, thus shaking quite strongly the theoretical foundations on which that theory was based.[1]

Supposedly, at the bottom of the approach adopted by most Sraffians there exists their firm confidence that by destroying the theoretical foundations of the neoclassical theory *from a logical point of view* would have been sufficient to destroy the entire paradigm outright, thus making

[1]Cfr. the 'classic' review by Harcourt (1972), and Harcourt (1976), Pasinetti and Scazzieri (1990).

possible, in that way, the triumphant coming into being of a new way of thinking in economic theory. In effect, the logical structure of a theory is extremely important in *any* discipline, and the economic discipline should not be an exception in this respect.

But *that* kind of critique, however, was evidently not enough, because in the economic discipline (which is basically a *social* discipline) *any* logical structure would be altogether *meaningless* irrespective of the 'vision' it wants to convey. Having a 'vision' of a society, in fact, essentially means possessing a specific key-reading of it, characterized by a variety of cultural, social, institutional and political elements, all entering, and contributing to forging, that 'vision'. This same thing might equivalently be expressed by saying that the economic discipline cannot be but *dependent upon* value-judgements, and that therefore ethical and moral values must ultimately come in, explicitly or implicitly, in any kind of discussion.[2]

The critique advanced by most Sraffians had—in the opinion of the present writer—the serious drawback of being altogether *detached* from the far more general new 'vision' of the society brought about by *PCMC*, as contrasted to the market-centred 'vision' of the neoclassical theory, or, at most, of leaving it at the very background of the issues discussed.

As is widely well known, the actual development of economic theory, and the corresponding policy prescriptions based on it over the last fifty years, went in a completely different direction with respect to that envisaged by the critics of the neoclassical theory. Supporters of the latter, in fact, although partially acknowledging most of the critiques, put them quite rapidly under the carpet by holding the strong conviction that they were ultimately not-influential on the tenet of the neoclassical paradigm. In other words, the neoclassical economists maintained the strong belief that the core of their theory was still robust enough in explaining the working of a market economy in the most appropriate way than any other theory, and thus it continued undaunted its dominance in the economic *culture* all over the world.

## 2 Community vs Automatons

Before discussing some general features of Sraffa's 'vision', as it seems to emerge from his *PCMC*, whose approach, as already been said, 'is reminiscent of certain points of view taken by the old classical economists from

[2] Cfr. Chiodi and Ditta (2008), Putnam (2003).

Adam Smith to Ricardo',[3] it should be duly emphasized that the sub-title of the book is: *Prelude to a Critique of Economic Theory*. This immediately recalls the analogous sub-title of Marx's Capital: *Critique of Political Economy*.[4]

The opening page of *PCMC* starts by considering 'an extremely simple *society*', p. 3, italics added. Adam Smith's book *Wealth of Nations* (1970) does have the 'nation' from the very beginning as the main subject of his enquiry and then this *collective* subject is at the centre of the whole book. Ricardo's opening sentence of the 'Preface' to his *Principles* refers to the division of the produce of the earth as 'divided among three classes of the *community*', (1951a, p. 5, italics added). Marx's opening sentence of his *Capital* Vol. I is: 'The wealth of those *societies*...' (1974, p. 43).

In each of the above references a 'collective' subject is taken into account, just because each of the above analysis refers to a *community* of human beings—in opposition to the micro perspective of neoclassical theory, which constantly takes into account an 'automaton' (being a single individual or a single agent) and contemplates at most 'the interaction of the agents', Debreu (1959, p. ix)—whose subjects are basically considered as a set of 'automatons' on the basis of a purely 'additive' conception.

## 3 Some Characterizations of Adam Smith's *Wealth of Nations*

The above 'word-characterization' is not at all trivial or unimportant.[5] It acquires a profound meaning in the context of the analysis performed by each author. In the case of Adam Smith, he wanted to investigate the *nature* and the *causes* of the wealth of a nation, which substantially means—as Hicks aptly put it—'what the social product is, what is meant by its being large or small; what is meant by its growing [...] why the social product is large or small, and why it grows', Hicks (1976, p. 7). Adam Smith (1970), in the opening sentence of his book, writes:

> The annual labour of every nation is the fund which originally supplies it with all the necessaries and convenience of life which is annually consumed, [...]. (p. 1)

[3] See the flyleaf of Sraffa's book (1960).
[4] This is the English literal translation from the original German sub-title of *Das Kapital*.
[5] As Lakoff (2014, p. xii) put it: "All words are defined relative to conceptual frames".

It contains key-words such as 'labour' and 'necessaries and conveniences of life' which reflect the essential characteristic of all his inquiry, ultimately devoted to investigating the wellbeing and the welfare of a community.

A huge number of books and papers have been written on Adam Smith's *Wealth of Nations*, most of them centred on his 'labour theory of value', on the much quoted 'invisible hand' and on the superb merit of the 'division of labour'.

Viewed from a different perspective, one can equally well say that Adam Smith's *Wealth of Nations* brought systematically to the fore the very notion of 'social product'—a key-concept characterizing the subsequent analyses distinctly made by Ricardo, Marx and Sraffa. The 'social product' was intended as a common denominator *linking* the human beings as a *community* and one of the basic elements of their life.

The notion of 'social product' immediately evokes the analogous notion of *produit net*, first introduced by Quesnay and the other Physiocrats in the middle of the eighteenth century. But—as is well known—not only did they refer to that notion exclusively with respect to the agricultural sector, but they did not possess a proper theory of value for quantitatively defining it in a rigorous way.[6]

Adam Smith felt the necessity of extending the notion of 'social product' to *all* the sectors of the economy and, at the same time, to provide a general theory of value—imperfect and wrongly circular in so many respects—to the effect of giving analytical form and substance to the notion of 'social product'.

An important feature of Adam Smith's *Wealth of Nation* has been generally undervalued by economists, if not practically ignored, so much attracted they have been by the most immediate 'analytical' parts of Adam Smith's work.[7] This consists in the equally important ethical and moral

[6] In an unpublished manuscript, however, Sraffa writes that "It was only Petty and the Physiocrats who had the right notion of cost as «the loaf of bread», Sraffa Papers D3/12/4. And, in another unpublished manuscript titled "Degeneration of cost and value", he writes: "A. Smith and Ricardo and Marx indeed began to corrupt the old idea of cost—from food to labour. But their notion was still near enough to be in many cases equivalent", ibidem. It should be noted that in *PCMC*, in the initial tables representing the economy, Sraffa includes "the necessaries for the workers" as *physical quantities*, and therefore he is there adopting the Petty-Physiocrats notion of cost. On the full significance of this, I will return later in the paper.

[7] For an exception see Sen (1987).

content, inextricably disseminated throughout the whole *Wealth of Nation*, up to the point of pervading all the arguments and discussions therein contained. It is hardly possible, in fact, to make sense of many Smithian concepts and sentences without systematically taking it into account.

The notion of 'wage' in Adam Smith, for example, as in any other classical economist, like Ricardo, as well as in Marx and in Sraffa, is taken to be as 'given'. This obviously means, as far as the effect on the analytical framework is considered, each physical component of the wage is known beforehand. Adam Smith, however, is far more specific in this respect. After having defined in Book I of his *Wealth of Nations* the 'real recompense of labour' as 'the real quantity of the necessaries and convenience of life which it can procure to the labourer', p. 69, in Book V he comes back again on the concept of 'necessaries' by specifying:

> By necessaries I understand not only the commodities which are indispensably necessary for the support of life, but whatever the custom of the country renders it indecent for creditable people, even of the lowest order, to be without. (Smith 1970, pp. 351–352)

This is to be considered apart from the *effective* determination of the wage. Two aspects are worth putting forward in this respect.

The first of these aspects regards the determination of the *money* price of labour which, according to Adam Smith, is regulated by 'the demand for labour, and the price of the necessaries and conveniences of life', (p. 76) whose eventual changes, according to the circumstances considered, must not have any effect whatsoever on the purchasing of the commodities composing workers' subsistence. This aspect is duly emphasized by Adam Smith towards the end of his *Wealth of Nation*:

> As the wages of labour are everywhere regulated, partly by the demand for it, and partly by the average price of the necessary articles of subsistence, whatever raises this average price must necessarily raise those wages so that the labourer *may still be able to purchase that quantity of those necessary articles* which the state of the demand for labour, whether increasing, stationary or declining, requires that he should have. (Smith 1970, p. 352, italics added)

The second aspect of the effective determination of the wage regards the actual process of its negotiation, from which Adam Smith is explicitly careful in bringing about the ***conflicting*** as well as the ***asymmetric*** positions of the two parties, the workers and the employees. He writes:

> What are the common wages of labour, depends everywhere upon the contract usually made between the two parties, whose interests are by no means the same. [...] It is not, however, difficult to foresee which of the parties must, upon ordinary occasions, have the advantage in the dispute, and force the other into a compliance with their terms. (Ibidem, pp. 58–59)

However, note, in the first of the two last sentences quoted above, the *use* Adam Smith is making of the concept 'the demand for labour'. As in any other analogous circumstance, when dealing with 'demand and supply of a commodity' as elements causing ***deviations*** from the natural price, he is always referring—like Ricardo and the classical economists and Marx—to ***proportions of known and given quantities*** and ***never to functions*** (in the sense of 'standard mathematics'), as it was subsequently made instead by the neoclassical tradition.

It is also important to emphasize the careful attention devoted by Adam Smith to the social and institutional characterization of the wage—a circumstance which, once more, underlines the ***external*** determination of the wage, as being independent of any problem of value and independent also of any problem concerning the production process. An echo of this view can later be traced in a letter by Ricardo to McCulloch dated 13 June 1820 in which he maintains that 'the proportions in which the whole produce is divided between landlords, capitalists and labourers [...] are not essentially connected with the doctrine of value'. Quoted in Sraffa (1951, p. xxxiii).

## 4 Ricardo's Continuity and Discontinuity with Adam Smith's *Wealth of Nations*

Ricardo, in the 'Preface' of his book, although 'found it necessary to advert more particularly to those passages in the writing of Adam Smith from which he sees reason to differ', specifies that 'he hopes it will not, on that account, be suspected that he does not, in common with all those who acknowledge the importance of the science of Political Economy, participate in the admiration which the profound work of this celebrated author so justly excites' Ricardo (1951a, p. 6).

It is well known that among the 'passages' from which Ricardo differs from Adam Smith one encounters the theory of value. On this issue, in fact, he was tenaciously adherent to an *embodied* labour theory of value. He summarizes his own view by stating 'I know no other criterion of a thing being dear or cheap but by the sacrifices of labour made to obtain it'. Ricardo (1951b, p. 397). But, more importantly, Ricardo's main preoccupation—as clearly evidenced by Sraffa's 'Introduction' to his *Principles*—was to state a *rigorous* theory of value because in his view the division of the 'social product' between classes was the 'principal problem in Political Economy' Ricardo (1951a, p. 5). To successfully perform that task, he however needed to find a *measure of value* having the property of being invariant to changes in distribution. These changes, in fact, by 'distorting' the very object of distribution, would have rendered impossible an *exact* measure of the shares between classes.

Measuring distribution *exactly*, however, was not an analytical problem *as such*: the most profound problem hidden in that task was to bring about the *conflicting relation* existing between employers and employees, i.e., from an analytical perspective, the existence of a *negative* relation between wages and profits—once the role of rents has been made passive. As Ricardo (1951a) concisely put:

> I have endeavoured to shew, first, that a rise of wages would not raise the price of commodities, but would invariably lower profits. (p. 127)

From the point of view here considered, it might be said that between Adam Smith and Ricardo there coexist a continuity and a discontinuity. The former, in fact, had already made evident, in a *qualitative* mode, the inherent conflicting views between employers and employees on the settlement of the wage contracts—as has been made evident in Adam Smith's quotations reported above. The latter, instead, wanted to *quantitatively* establish the *exact measure* of that conflict. Thus, Ricardo's approach on the issue of value turns out to be not only far more complex and articulated than Adam Smith's but also far more 'refined'.

On many other issues, Ricardo does *not* significantly differ from Adam Smith's *Wealth of Nation*. In the case of wages, for example, he seems to follow Adam Smith almost to the letter. Although the opening passages of chapter V, On Wages, 'appear to be derived from Torrens's *Essay on the External Corn Trade* (1815, p. 62), as Sraffa (1951, p. 93, fn.) noted. Nevertheless the substance of the whole chapter recalls quite evidently what Adam Smith had already said on wages. The 'natural price of labour', in fact, continues to be 'that price which is necessary to enable the labourers, one with another, to subsist and perpetuate their race', its physical composition being given by 'the quantity of food, necessaries, and conveniences [...] essential to him from habit', Ricardo (1951a, p. 93). And the conceptual distinction and the numerical difference between the 'market' and the 'natural' price of labour continue to come about 'from the natural operation of the *proportions* of the supply to the demand' (ibidem, p. 94, italics added).

What Ricardo emphasizes more, compared with Adam Smith, is the important 'dynamical' component concerning the labourer that is his *family*. That component, on reflection, reflects the function attributed to the wage, viz. that of making possible the *social reproduction* of the working class, besides the most obvious, but not exclusive, function of 'natural recompense' of labour, using the words of Adam Smith (1970, p. 57).

## 5 The Relevance of Marx's *Capital* Nowadays

Too much weight has often been given to Marx's labour theory of value, as if it were the only pivotal element of *Capital* Vol. I and, more generally, of the whole Marx's thought. Witness the much debated 'transformation problem', still alive occasionally in the literature.

One of the consequences of that approach has been that of limiting or, even worse, that of reducing drastically the very importance of Marx's contribution to the social and political disciplines generally, and to the economic one, in particular.

In other words, the acceptance or rebuttal of Marx's thought, especially within the profession of the economic discipline, has crucially been taken dependent upon the 'rightness' or the 'wrongness' of his labour theory of value.

The final judgment still prevailing today on Marx's thought—looking at the latter from the above perspective—might be a combined result of an *analytical* and *ideological* judgment mixed together, sometimes with the prevailing weight of the one, sometimes of the other.

One of the theses of the present chapter is that Marx's contribution to the understanding of the capitalist society should instead be looked at independently of the validity of the labour theory of value *as a theory of prices*. Marx's perspective and insights, in fact, go far beyond the restricted conceptual boundaries imposed by any social discipline and, particularly, far beyond the highly disputable conceptual framework designed by traditional economic thought. His own *language* and the new *terminology* employed in *Capital* and elsewhere are all designed to give a new conceptual form to ideas, to bring about aspects of social life otherwise hidden or hardly noticed.

There are at least two aspects of Marxian thought traceable through *Capital* Vol. I (with obvious extensions to his many other writings as well) which are worth considering today because they are still able to cast light on many structural phenomena of the present-day world economies:

1. his notion of 'vulgar economy',
2. his perspective on 'labour'.

The above two aspects are strictly linked together for they are highly suggestive of a new perspective in representing the economy. They are also of an utmost importance and relevance for interpreting and evaluating some critical changes which occurred in most economies over the last century, in the light of the ever-increasing pervasiveness and dominance of present day neoclassical economic theory, which can properly be dubbed *neo-vulgar economy*.

In *Capital* Vol. I, Marx (1974) neatly introduces the distinction between 'classical Political Economy' and 'vulgar economy':

> [B]y classical Political Economy, I understand that economy which, since the time of W. Petty, has investigated the *real relations of production* in bourgeois society, in contradiction to vulgar economy, *which deals with appearances only*, ruminates without ceasing on the materials long since provided by scientific economy, and there seeks plausible explanations of the most obtrusive phenomena, for bourgeois daily use, but for the rest, confines itself to systematizing in a pedantic way, and proclaiming for everlasting truths, the trite ideas held by self-complacent bourgeoisie with regard to their own world, to them the best of all possible world. (Marx 1974, p. 85, italics added)

It may be of some interest, in this connection, to consider some Sraffa's notes. In an unpublished manuscript, he writes:

> 'Classical Political Economy (The age of Ricardo) or A. Smith? from Petty to Ricardo. Right conception fundamental assumptions primitive, rudimentary technique'. Then 'Vulgar Political Economy (the age of Mill) from

> Malthus to Stuart Mill. All wrong here: they have the wrong conception of modern economics and the rudimentary technique of the classical'. Finally, 'Economics, (the age of Marshall) since Jevons and Co., and Marshall. Highly refined technique, rotten conceptions and fundamental assumptions. But technique so highly perfected that sometimes compels them unconsciously to modify their conscious assumptions (justly contradicting themselves) and thus reaching partially true conclusions'. Note that at the end of the classics developed socialism (Owen, Hodgskin) and caused Vulgar P.E. At the end of vulgar period came Marx and caused economics. (Sraffa Papers D3/12/4, underlined in the original)

In the previous quotation by Marx, although he credits classical Political Economy with having investigated 'the real relations of production', in opposition to vulgar economy, he remains unsatisfied with the former. And the reason to be unsatisfied lies basically in the fact that, for him, the capitalist mode of production is not 'eternally fixed by Nature for every state of society' (Marx 1974, p. 85). The most dangerous consequence of this point of view, so explicitly criticized by Marx, would be that of perilously considered Political Economy as a ***neutral*** science, that is as completely detached by any ***value-judgement*** (be it social, political, religious, ethical or whatsoever) as well as by any historical context, much like disciplines such as mathematics or physics (and even in this case, the latter disciplines should not be considered 100% neutral, as commonly believed).

Unfortunately, however, what happened last century at the beginning of 1930s was even worse than that potential danger, for Robbins (1932) published a sort of 'new constitution' for the economic discipline, by proclaiming it a 'Science' and by showing its foundational characteristics as essentially based on the individual rationality of optimal choice, maximization principle, efficiency and general equilibrium—in explicit opposition to the classical economists' approach and by openly criticizing the use they have been made of notions such as 'social product' and income distribution, which sounded completely nonsense to him.

More recently, the widespread use of econometric techniques in economic analysis and the uncritical reference to macroeconomic parameters having no logical foundations at all may perhaps induce one to think the economic discipline as being characterized by a technical and neutral structure. But the issue of the identification problem in econometrics and the generally catastrophic results produced by the austerity measures adopted in most European countries in recent years are both examples of facts which bring to the fore the very impossibility of disposing of value-judgements and that the social relations among people are not indeed governed by 'laws of Nature', as are the planets in the universe.[8]

## 6 Marx's Steps Ahead of Adam Smith and Ricardo

As has been seen in Sects. 3 and 4 above, Adam Smith's and Ricardo's works were characterized (1) by the *centrality of labour* within their respective theories of value, and (2) by the *antagonism* between classes. The 'quantity of labour' (commanded for Adam Smith, embodied for Ricardo), was in fact the most reliable instrument for valuing the 'social product', and thus for measuring, respectively, the wealth of a nation and the distribution of income between classes. And Ricardo's endless search for an invariable measuring rod was precisely directed *to exactly measuring* the antagonism between workers and capitalists.

Marx considered the classical economists' analysis *incomplete*. The invariably negative relation between the wage rate and the rate of profit, which Ricardo thought of having put forward rigorously, was judged by Marx insufficient for *completely* revealing the 'real relation' existing between workers and capitalists in a capitalist economy.

He wanted to go far deeper than what the classical economists had achieved with their analysis of capitalism and, above all, he wanted to go far beyond the analysis of 'vulgar economy', whose *mystification* in representing the economy through the glasses of *market* relationships he wanted to unravel.

In this connection Marx should be credited for having perceived long before the initial signs of an extremely dangerous transformation in the economic thought through the coming into being of a 'vulgar' representation of the economy, which later in the years would have produced not only the oblivion of the classical economists' thought but also, on practical grounds, the adoption of inappropriate and even dangerous policies for entire communities of human beings—as has been hinted at in the immediately previous section.

It is again worth pointing out some of Sraffa's considerations on features characterizing the transition from Classical Political Economy to what Marx called 'Vulgar Political Economy'. Sraffa points out that after the 'corruption' of the old idea of cost made by the Classical economists,

[8] Cfr. Pasinetti (1998), Nuti (2013).

and by Marx (as reported in the quotation from Sraffa's manuscripts in Sect. 5 above):

> a much bigger step was taken in the process of shifting the basis of value from physical to psychical processes: Jevons, Menger, Walras. This was an enormous breach in the tradition of Pol. E.; in fact, this has meant the destruction of the classical P. E. and the substitution for it, under the old name, of the calculus of Pleasure and Pain (Hedonistic). [...] When the Jevonians turned back to write their own history, they found with pride (it ought to have been with dismay) that they had no forerunners amongst P. E.; their forerunners were mainly two or three cranks, an engineer, Dupuit, a mathem., Cournot, a Prussian Civil servant Gossen, who had only cultivated P. E. as a hobby. [...] They had not the slightest knowledge of the works of the Classical economists. They drew it out of their fancy. [...] It is unfortunate that so much time has been taken to change the name of P. E. into Economics: but it is appropriate: it marks the cleavage, or rather the abyss, between the two. (Sraffa Papers, D 3/12/4)

Marx repeatedly pointed out that the workers, *as human beings*, although have constantly been *exploited* by other human beings in different historical contexts and periods, have nonetheless been exploited in a *visible* way, in contrast to the capitalist economy whose *exploitation*, instead, has been rendered *invisible* by the complex set of market exchanges. It was precisely *this* which Marx found *peculiar* to the capitalist economy, and the use he made of the labour theory of value was *uniquely directed to bring about capitalist exploitation and not to explain exchange relationships.*

At the end of *Capital* Vol. I, Marx illustrates the historical origin of that *peculiarity*, which came into being since the advent of 'the so-called primitive accumulation', defined by him as 'the historical process of divorcing the producer from the means of production' (Marx 1974, p. 668).

Notions like 'labour force' as distinguished from 'labour', 'constant capital' as distinguished from 'variable capital', 'necessary labour' as distinguished from 'surplus labour' were all devised to make evident the fact that the workers, in a capitalist economy, are *forced* to sell the only commodity they possess (labour power) and *forced* to perform labour inside the economic process, because they are all deprived of any decision concerning *what* society should produce, *which methods of production* it should employ, if *any surplus* should be produced at all and *for whom*. Decisions regarding all these circumstances are taken by people who have instead full control of the economic processes.

The feature of capitalist exploitation of being *obscured* and *hidden* by market relationships can perhaps be put to the fore by considering the mechanism of demand and supply through which the wage is determined—a mechanism still characterizing any model pertaining to what I called before *neo-vulgar economy*. Within this context *it appears* that *quantities* of labour demanded and *quantities* of labour supplied (and therefore *objective* entities) jointly determine in the market 'the most appropriate value' of the labour power, completely free of any value-judgement. In fact, the most immediate consequence of the market mechanism determining the wage is that the latter loses any kind of connection with the necessaries of life of the workers and of their families, and therefore also any ethical principle behind wage setting is definitely lost as well.

Interpreted in this way, Marxian labour theory of value, in the way he used it in *Capital* Vol. I, should only be an *analytical instrument* to bring about *capitalist exploitation*, a phenomenon which no theory of prices, not even that characterizing the classical economists' analysis, could be able to put forward.

The analytical effort he made in trying to build a coherent transition from the 'system of values' to the 'system of prices' is amply documented in his *unpublished* notes during his life, which only *after* his death were published as to be materialized as *Capital* Vols. II and III.

The main and ultimate purpose of Marx, however, was *not* that of providing a correct theory of prices, to the effect of furnishing a proof of the 'validity' of his theory of exploitation made in *Capital* Vol. I. His analysis of the capitalist economy and of capitalist exploitation stands on its own, and any further proof to validate it turns out to be altogether *unnecessary*.

Marx's 'system of values' of *Capital* Vol. I should properly be considered the required *complement* to the 'system of prices', for each of them casts a different light on the workings of an economy from different perspectives. It would then be incorrect to look at his effort to possibly build a transition from one system to the other as a 'step forward', or as 'an almost complete approach', to the 'right solution' to the price system—as Garegnani (1981), for example, maintains. This way of looking, in fact, would drastically scale down the positive contribution of Marx's *Capital* Vol. I.

## 7 Sraffa's Breakthrough

The connection of Sraffa's *PCMC* to the classical economists and Marx has been amply recognized and analysed in the literature, thanks also to the explicit statements by Sraffa himself (1960, p. v) and to the highly informative Appendix D of his book.

It is also well known that long before Sraffa, at the end of the nineteenth century, Dmitriev (1974) had successfully reshaped the Ricardian theory of labour value and von Bortkiewicz (1906, 1907) has also reshaped Marx's equations, whereas von Neumann (1937) had already published a model whose structure captures with clear evidence the basic features of the classical economists' standpoint, and that therefore it seems formally to anticipate some features of the analogous structure of *PCMC*.

The works by Dmitriev, von Bortkiewicz and von Neumann, however, have all been designed to analytically take up and solve *specific* problems, until then not yet taken up or solved, whereas Sraffa's *PCMC* was *intentionally* designed to lay the foundations for a *critique* of economic theory from the classical economists' standpoint by means of *a genuinely new way of thinking* in the economic discipline and *a completely new vision* of the society.

Sraffa's *PCMC* should thus be looked at as an extremely complex work for more than one reason.

In the first instance, it shows that the classical economists' *standpoint* can be brought to the light again as the basis of a meaningful representation of the economy, without incurring in any sort of analytical and conceptual infelicity. This is because the inescapable and fundamental theory of value, which characterizes that vision (and which was neither successfully accomplished by Adam Smith, nor by Ricardo), can instead be shaped and solved meaningfully.

The solution to the system of production prices provided by Sraffa seems to imply *no necessity* to make explicit reference to *labour values*. This circumstance, at first sight, might have two conflicting interpretations. The first—which I personally reject—would state that Sraffa's approach seems to have definitely cut any connection whatsoever with Marx, because labour values, having no explicit analytical role to play within the framework considered, are altogether redundant. Quite reasonably, therefore, they are bound to disappear completely from the scene.

A different and opposite interpretation of Sraffa's approach in providing the correct solution to the system of production prices seems however possible.

In shaping his own framework, Sraffa takes as *given* the quantities of commodities used and produced in the production processes of the system, not more, nor less, than the classical economists and Marx did.

This initial feature should be seen not only in patent contrast to the 'marginal approach', which instead 'requires attention to be focused on change' (Sraffa 1960, p. v). It should also be viewed as a radical change of perspective. In *PCMC*, in fact, the 'given quantities' standpoint reflects the very important circumstance that some crucial *decisions* must be taken *before* the production processes gets started, which implies that they are completely divorced from *any outcome of production* as from *any price-based valuation*. Therefore, being situated *outside* the 'core', they are essentially characterized, by their own nature, by social, political and institutional elements. This is in striking contrast to the *simultaneous determination* of quantities *and* prices by the same anonymous and undifferentiated 'market laws', typical of neo-vulgar economy.

Among those decisions, the sustenance, in dimension and composition, for those who work must already be fixed; decisions regarding the methods of production to be used and which and how much commodities shall be produced must be taken; if any surplus shall be produced at all must also be known; and the basic institutional features characterizing the economy must obviously be known as well. The quantitative relations between those 'given quantities' thus summarizes the most relevant flow of events which occurs in the system till the moment considered.[9]

Thus, production prices—which necessarily do need 'given quantities' for their determination—are *unable* to convey or to reveal anything concerning the above decisions. The essential function of production prices rests on their making possible the *reproduction of the system*. Only in this case the system can be said to be 'viable', which means the possibility of the system to at least reproduce those commodities necessary *for the livelihood of the community* and in making possible the distribution of the surplus, if it exists, according to criteria reflecting the antagonism between classes and essentially based on social and political elements *outside* the production system.[10] This is actually the *meaning* and the *function* Sraffa

[9] Cfr. Chiodi (2010), Chiodi and Ditta (2013).
[10] Cfr. Chiodi (1992, 1998, 2010).

attributes to production prices, which thus merely appear as numerical reflections of far deeper social and political relationships among the people.

In an unpublished manuscript Sraffa writes:

> The old conception of cost and value was designed to solve this fundamental problem: what is the relation, if any, between *the importance to the community* [...] *of a commodity* and *its price* (exchange value). They assumed that there was a relation of some sort between the two, and they expressed this mysterious relation by such words as 'cause' or 'measure'—but obviously they did not mean what is their appropriate sense (analyse). [...] The modern conception is derived from the erroneous belief that the classics were literally looking for the 'causes' of value, and not its relation with importance, i.e. its 'meaning'. (Sraffa Papers D3/12/4: 3, italics added)

Thus, Sraffa finely and rigorously traces the boundaries of what should be the 'core' of the economy to be analysed, by showing, at the same time, the *impossibility* of its own autonomy, *without making necessary reference to other elements outside the 'core'*. The *invisible* elements outside the 'core' are thus as important as the *visible* ones inside it.

## 8 Sraffa's 'Core'

Sraffa's reconstruction of a 'core', with its inherent 'openness', thus seems to show quite evidently that the economic discipline does need flows of information and sets of decisions based on principles coming necessarily from other different disciplines, and that therefore no universal law or objective rule can ever be envisaged for the economic discipline in particular, with the exception for the logical relations and statements which can only be drawn *from within* the 'core'. This is in sharp contrast to neoclassical theory, which has repeatedly stated the existence of universal economic laws, independently of the historical context considered. By contrast, for this very reason, Sraffa's viewpoint here under discussion exactly reflects the classical and Marxian 'vision' of the society. At the same time, it implicitly emphasized the very importance and necessity of the Marxian 'system of values', as contrasted to the 'system of prices', each of them being able to cast light on *different* crucial aspects of production.

The possible existence of a 'surplus wage'—a novel characteristic of *PCMC* which Sraffa with great care takes *conceptually* and *practically* distinct from the 'subsistence wage'—reflects a remarkable aspect of the antagonism between classes. The very existence of a 'surplus' and its origin cannot be revealed by the system of production prices. In so far that it does exist, it potentially belongs to the community as such, ***whatever be its class division and group segmentation***. In fact, even in the case in which the workers were the owners of the means of production ***and*** labour were of different qualities, there would necessarily be 'antagonism' between the workers, because to solve the system of production prices, ***either*** differences in quality must previously be reduced to differences in quantity, ***or*** proportions between different level of wages must be fixed in advance. In the former as well as in the latter case, decisions of ***genuinely institutional character*** must be taken ***before*** any determination of prices.

Sraffa's *PCMC*, however, does also possess a 'pedagogical' feature which in my opinion has passed almost unnoticed. Concepts, definitions notions which are found in *PCMC* are all in perfect synchrony with the classical economists' and Marx's languages. Any single word, which might even evoke some feeble connection with neoclassical theory, is carefully avoided in their use, as in the case with the term 'cost of production' and especially with the much-debated tricky term 'capital' (Sraffa 1960, p. 9). It is worth noticing, in this connection, Sraffa's view on this expressed in the late 1920s:

> It is terrific to contemplate the abysmal gulf of incomprehension that has opened itself between us and the classical economists. Only one century separates us from them: [...] how can we imagine to understand the Greeks and the Romans? [...]
>
> The classical economists said things which were perfectly true, even according to our standards of truth: they expressed them very clearly in terse and unambiguous language, as is proved by the fact that they perfectly understood each other. (Sraffa Papers D3/12/4:14)

The hidden pedagogical feature of *PCMC* is perceived in the way in which Sraffa, chapter by chapter, introduces and explains his propositions. Simple numerical examples are generally given before any generalization. This style of writing is also in perfect line with the mathematical language used by Sraffa and with his own way of reasoning in proving the statements contained in *PCMC*. The mathematics employed by Sraffa is 'constructive mathematics', as Velupillai (1980, 1989, 2008), who was the first scholar to rightly draw attention to this very important aspect, has persuasively shown. Space does not allow us to go further on this interesting and important issue. Suffices it to say that through his 'constructive proofs' Sraffa wanted to take due distance even from the way in which until then mathematics was used in neoclassical theory. On reading, for example, the long sequence of lemmas and definitions in the seminal paper by Arrow and Debreu (1954), it seems that economic theory should *conform to* mathematics, whereas Sraffa wanted to do exactly the other way round. In this connection, in a passage in his unpublished manuscript Sraffa writes:

> Our metaphysics is in fact embodied in our technique; the danger lies in this, that when we have succeeded in thoroughly mastering a technique, we are very liable to be mastered by her. (Sraffa Papers D3/12/4:15)

The 'pedagogical' approach of Sraffa here put forward gives the strong perception that he wanted *to completely reset* from the very beginning the entire setting of the economic discipline and of its teaching as well. In the flyleaf of his book it is written that '[a]lthough the reasoning is abstract it requires no knowledge of mathematics beyond elementary algebra' and, in the *Preface*, he states that the 'unusual' mathematical notation adopted goes in the direction of helping 'the non-mathematical reader' (p. vii).

The use of that kind of mathematics by Sraffa sounded obviously quite strange to Solow (2014), as soon as he read Sraffa's book in 1960, as he expresses 'some astonishment that Sraffa has managed to learn so much about Leontief Models *with such primitive tools*' (p. 63).

It is worth pointing out that Sraffa is also very careful in making a *direct* criticism of neoclassical theory. The long paragraph devoted to criticising economic theory for the wrong attempt it made by using the 'period of production' as a measure of 'the quantity of capital' is entirely written in parenthesis, to testify that it should be taken apart from the main part of the book. It should also worth pointing out that chapter XII 'Switch in the Methods of Production', which most inspired the initial capital theory controversy in the 1960s and 1970s, is the last chapter of *PCMC*.

## 9 Some Criticisms of Economic Theory Independently of *PCMC*

It might perhaps be useful to recall that criticisms of *specific* and crucial aspects of neoclassical economic theory have been made by several authors over the last century and some of them even before Sraffa's *PCMC*.

Long before Robinson's (1953) critique of the neoclassical production function, Wicksell (1901)—who should be considered a heterodox and quite a peculiar neoclassical economist—made very clearly his critique of the notion of 'capital' employed in neoclassical theory.[11]

He writes:

> 'Whereas labour and land are measured each in terms of its own technical unit [...] capital, on the other hand, [...] is reckoned, in a common parlance, as a sum of exchange value'—whether in money or as an average of products. [...] it is a theoretical anomaly which disturbs the correspondence which would otherwise exist between all the factors of production. (Ibidem, p. 149)

If capital also were to be measured in technical units, the defects would be remedied and the correspondence would be complete. But, in that case, productive capital would have to be distributed into as many categories as there are kinds of tools, machinery, and materials, etc., and a unified treatment of the role of capital in production would be impossible (Ibidem).

He directed his criticism against von Thünen, for having put on the same footing wages and interest as being determined on the same symmetrical principle of 'marginal productivity' (ibidem, p. 147) but also against Walras, *for arguing in a circle*, in so far as capital goods were not treated with extreme precaution:

> Again, it is futile to attempt—with Walras and his followers—to derive the value of capital-goods from their own cost of production or reproduction; for in fact these costs of production include capital and interest, whereas our analysis of the laws of the cost of production has hitherto proceeded on the assumption that production is non-capitalistic. *We should, therefore, be arguing in a circle.* (Ibidem, 149, italics added)

Wicksell then constructed a model with 'dated' quantities of *non*-produced inputs as 'representatives' of capital goods.

Suffices it to take a simple stationary economy with only one single final product and dated quantities of labour and land over $n$ periods composed of $2n + 3$ independent equations: (1) production function (whose arguments are the dated quantities of labour and land over the $n$ periods), $2n$

[11] Cfr. Chiodi (2013). It is also worth reading Goodwin (1970) and Chiodi (2015).

'marginal productivity' conditions, and (2) demand and supply equality conditions for labour and land respectively.

The system of equations formulated in that way does have $2n + 4$ unknowns: the quantity of the final product, the $2n$ dated quantities of labour and land, the wage rate, the rent rate and the rate of interest (rate of profit). If the system is solvable, one of the unknowns can thus be fixed from *outside* the system of production, and—obviously enough—the most eligible unknown which might be fixed in this way is just the rate of interest. The latter possibility straightforwardly implies the unavoidable circumstance that distribution cannot find a consistent solution from within the model of production and that 'capital'—*however defined*—cannot be measured independently of distribution. It should also be added that the notion of 'marginal productivity' of *any* 'factor', although formally definable, turns out to be irrelevant in relation to the distribution of income.

A less known criticism—albeit far more profound than Joan Robinson's and with much Ricardian and Sraffian 'flavour'—was de Graaff's firm rebuttal of the conceptual separation between 'size' and 'distribution' of income.

In a one commodity world some definite meaning could be attached to a phrase like 'the size of the national income' [...] But as soon as we leave a one commodity world this ceases to be true. There is no unambiguous meaning which we can attach to 'the size of the national income' when we have a heterogeneous collection of goods and services. [...] [A Paretian welfare function] will only tell us what weights to use when the distribution of the goods among the members of the community is given. [...] [T]he size-distribution dichotomy is inconsistent with the basic Paretian value judgments that individual preferences are to count and that a cet. Par. increase in any one man's well-being increases social well-being. [...] Moreover, 'size' in this sense will generally change whenever the distribution change, even if the collection of goods and services distributed remains the same (de Graaff 1957, pp. 91–92).

Lastly, it is worth briefly mentioning a paper by Nuti (1970), in which a 'putty-clay' model of production, widely used in the economic literature in the 1960s, is used to investigate some properties of the choice of production techniques.

Nuti eliminates in his 'putty-clay' model two of the customary assumptions i.e. (1) that the transformation of putty into clay is costless and (2) that there are no gestation lags of investment. He is then able to show that the problem of 'reswitching' of techniques (the same technique might be eligible at more than one level or range of the rate of interest) arises straightforwardly. This implies that for specific critiques of economic theory it is utterly unnecessary to refer to the Sraffian framework.

## 10 Concluding Remarks

The main thesis of the present chapter can be summarized by saying that the most characterizing feature of *PCMC* lies in having brought about a *new way of thinking* in the economic discipline, by reshaping the classical economists' and Marxian standpoint in an analytically robust framework and by using a *lexicon* completely different from that introduced and used since the advent of neoclassical economics. *This* very feature of Sraffa's *PCMC* should then be considered the most important and *constructive* critique of the dominant economic theory.

Over the years, however, most Sraffians have pursued a different path in the critique of economic theory, essentially based on bringing about, here and there within the equilibrium-market-centred neoclassical standard model, mistakes and infelicities *on logical grounds*. This very approach—which nonetheless deserves appreciation and intellectual respect—has ultimately revealed itself too feeble to upset the consolidated and widespread 'vision' of the society conveyed by neoclassical economics. A litmus test of this can be seen by looking at the economic policies pursued around the world and at the consequences produced by them on the people.

## References

Arrow, K.J. and Debreu, G. (1954), "Existence of an equilibrium for a competitive economy", *Econometrica*, July, pp. 265–90.

Bharadwaj, K. (1976), *Classical Political Economy and Rise to Dominance of Supply and Demand Theories*, Bombay, Orient Longman.

Cesaratto, S. (2008), "The Classical 'Surplus' Approach and the Theory of the Welfare State and Public Pensions", in Chiodi, G. and Ditta, L. (eds.), pp. 93–113.

Chiodi, G. (1992), "On Sraffa's notion of viability", *Studi Economici*, n. 46, pp. 5–23.

Chiodi, G. (1998), "On non-self-replacing states", *Metroeconomica*, February, pp. 97–107.

Chiodi, G. (2010), "The Means of Subsistence and the Notion of 'Viability' in Sraffa's Surplus Approach" in S. Zambelli (ed.), *Computable, Constructive &*

*Behavioural Economic Dynamics, Essays in Honour of Kumaraswamy (Vela) Velupillai*, Routledge, Abingdon, pp. 318–330.

Chiodi, G. (2015), "On Richard Goodwin's *Elementary Economics from the Higher Standpoint*", *Cambridge Journal of Economics*, Vol. 39, No. 6, November, pp. 1533–1550.

Chiodi, G. and Ditta, L. (eds.) (2008), *Sraffa or An Alternative Economics*, Palgrave Macmillan, Houndmills, Basingstoke.

Chiodi, G. and Ditta, L. (2013), "Sraffa and Keynes: Two Ways of Making a Revolution in Economic Theory", in Levrero, E. S., Palumbo, A., Stirati, A. (eds.), *Sraffa and the Reconstruction of Economic Theory*, Vol. I, Palgrave Macmillan, Basingstoke, pp. 218–240.

Debreu, G. (1959), *Theory of value. An axiomatic analysis of economic equilibrium*, Yale University Press, New Haven and London.

de Graaff, J. V. (1957), *Theoretical Welfare Economics*, Reprint (1967), Cambridge at the University Press, London.

Dmitriev, V. K. (1974), *Economic Essays on Value, Competition and Utility*, translated by D. Fry and edited with an Introduction by D. M. Nuti, London: Cambridge University Press. (Originally published between 1898 and 1902.)

Edwards, P. (2008), "Sraffa: Notes on Moralizing, Money, and Economic Prudence", in Chiodi, G. and Ditta, L. (eds), pp. 211–26.

Fredholm, T. (2009), *Essays on the Theory of Production*, PhD dissertation, University of Aalborg, DK.

Garegnani, P. (1981), *Marx e gli economisti classici*, Einaudi, Torino.

Goodwin, R. M. (1970), *Elementary Economics from The Higher Standpoint*, Cambridge University Press, Cambridge.

Harcourt, G. C. (1972), *Some Cambridge controversies in the theory of capital*, Cambridge University Press, London.

Harcourt, G. C. (1976), "The Cambridge Controversies: Old Ways and Old Horizons—Or Dead End?", *Oxford Economic Papers*, 28, pp. 25–65.

Harcourt, G. C. (2018), "The Role of Sraffa Prices in Post-Keynesian Pricing Theory", in *Classical Economics Today. Essays in Honour of Alessandro Roncaglia*, Corsi, M., Kregel, J. and D'Ippoliti, C. (eds.), London and New York, Anthem.

Hicks, J. R. (1976), "'Revolutions' in Economics", in Latsis, S. (ed.), *Method and Appraisal in Economics*, Cambridge University Press, Cambridge, reprinted in Hicks, J. R. (1983), pp. 3–16.

Lakoff, G. (2014), *Don't Think of an Elephant*, Chelsea Green Publishing, VT, USA.

Marx, K. (1974), *Capital*, Volume I, Lawrence & Wishart, London. (Originally published in 1867).

Nuti, D. M. (1970), "Capitalism, socialism and steady growth", *The Economic Journal*, March, pp. 32–57.

Nuti, D. M. (2013), "Perverse Fiscal Consolidation", *Transition*, September.

Parrinello, S. (1983), "Exhaustible natural resources and the classical method of long-period equilibrium", in Kregel, J.A. (ed.), *Distribution, effective demand and international economic relations*, The Macmillan Press Ltd, London and Basingstoke. pp. 186–206.

Parrinello, S. (2001), 'The price of exhaustible resources', *Metroeconomica*, 52(3), 301–315.

Pasinetti, L. (1977), *Lectures on The Theory of Production*, London and Basingstoke, The Macmillan Press Ltd.

Pasinetti, L. L. (1993), *Structural Economic Dynamics*, Cambridge University Press, Cambridge.

Pasinetti, L. L. (1998), "The myth (or folly) of the 3% deficit/GDP Maastricht 'parameter'", *Cambridge Journal of Economics*, 22, pp. 103–116.

Pasinetti, L. L. and Scazzieri, R. (1990), "Capital Theory: Paradoxes", in Eatwell, J., Milgate, M., Newman, P. (eds.), The New Palgrave - Capital Theory, W. W. Norton & Company, New York, London.

Putnam, H. (2003), "For Ethics and Economics without the Dichotomies", *Review of Political Economy*, July, pp. 395–412.

Ricardo, D. (1951a), *On the principles of political economy and taxation*, Vol. I, *The works and correspondence of David Ricardo*, edited by Piero Sraffa with the collaboration of M.H. Dobb, Cambridge University Press, Cambridge. (Originally published in 1817.)

Ricardo, D. (1951b), *Pamphlets and Papers*, vol. IV, *The works and correspondence of David Ricardo*, edited by Piero Sraffa with the collaboration of M.H. Dobb, Cambridge University Press, Cambridge.

Robbins, L. (1932), *An Essay on the Nature and Significance of Economic Science*, Macmillan and Co., Ltd, London, 2nd ed. (1935).

Schefold, B. (1997), *Normal Prices, Technical Change and Accumulation*, Macmillan, Basingstoke and London.

Sen, A. (1987), *On Ethics and Economics*, Blackwell, Oxford.

Sinha, A. (2012), "Listen to Sraffa's Silences: A New Interpretation of Sraffa's *Production of Commodities*, *Cambridge Journal of Economics*, 36 (6), pp. 1323–1339.

Sinha, A. (2016), *A Revolution in Economic Theory. The Economics of Piero Sraffa*, Palgrave Macmillan, Springer International Publishing AG Switzerland.

Smith, A. (1970), *The Wealth of Nations*, London: J. M. Dent & Sons Ltd. (Originally published in 1776).

Sraffa, P. (1951), "Introduction", *The works and correspondence of David Ricardo*, edited by Piero Sraffa with the collaboration of M. H. Dobb, Cambridge University Press, Cambridge.

Velupillai, K. (1980), "Review" of Pasinetti, L. L. (1977), *Journal of Economic Studies*, vol. 7, pp. 64–65.

Velupillai, K. (1989), "The Existence of the Standard System: Sraffa's Constructive Proof", *Political Economy—Studies in the Surplus Approach*, Vol. 5, No 1, pp. 3–12.

Velupillai, K. (2008), "Sraffa's Economics in Non-Classical Mathematical Modes", in Chiodi, G. and Ditta, L. (eds.), pp. 275–94.

Velupillai, K. (2010), *Computable Foundations for Economics*, Routledge, London and New York.

Velupillai, K. with Zambelli, S. (1993), *The Economics of Production-Based Indicators and the Purchasing Power of Currencies for International Economic Comparisons*, for the World Bank, Socio-Economic Data Division, International Economics Department.

von Bortkiewicz, L. (1906), "Wertrechnung und Preisrechnung im Marxschen System", *Archiv für Sozialwissenschaft und Sozialpolitik*, XXIII.

von Bortkiewicz, L. (1907), "Wertrechnung und Preisrechnung im Marxschen System", *Archiv für Sozialwissenschaft und Sozialpolitik*, XXV.

von Neumann, J. (1937), "Über ein ökonomisches Gleichungssystem und eine Verallgemeinerung des Brouwerschen Fixpunktsatzes", *Ergebnisse eines mathematischen Kolloquiums*, (1935–1936), Heft 8, ed. by K. Menger, Franz Deuticke, Leipzig und Wien. pp. 73–83, English translation in *The Review of Economic Studies*, vol. 13, (1945), pp. 1–9.

Wicksell, K. (1901), *Föreläsningar i Nationalekonomi*, Första delen: Teoritisk Nationalekonomi, Lund. English edition: *Lectures on Political Economy*, translated from the Swedish by E. Classen and edited with an Introduction by Lionel Robbins, vol. I, The Macmillan Company, New York.

Zambelli, S. (2004), "The 40% neoclassical aggregate theory of production" *Cambridge Journal of Economics*, 28, pp. 99–120.

Comment

# A Comment on Sraffa's 'Silenced Revival of the Classical Economists and of Marx'

*Nuno Ornelas Martins*

Professor Guglielmo Chiodi provides in his chapter a very concise and interesting account of the key themes surrounding Piero Sraffa's revival of the standpoint of the classical political economists and Karl Marx. In this carefully argued chapter, Chiodi explains that Sraffa's *Production of Commodities by Means of Commodities* (PCMC) brought a new way of thinking, which reshaped the classical and Marxian standpoint in a robust analytical framework. Furthermore, PCMC is also radically different from neoclassical economics regarding the very language used, and the way mathematics is used. This new way of thinking is underpinned by a different vision to that advanced in neoclassical economics.

Chiodi notes, however, that Sraffa's PCMC has been interpreted and used essentially as a tool for destroying the foundations of neoclassical economics, rather than in a constructive way, as a basis for the development of the alternative vision it provides. He suggests a more constructive use of Sraffa's PCMC, while outlining some of the key topics addressed by the classical political economists whose standpoint is recovered by Sraffa. The starting point for such a project consists in the idea that the society or the community is the basis for the overall vision to be advanced. The starting point is not the single individual or "automaton" of neoclassical economics. Once this vision is adopted, we can then address notions such as the social product, a notion developed more systematically by Adam Smith, by generalizing the physiocrats' *produit net* to the whole economy (rather than circumscribing it to agriculture), and developing also a theory of value for measuring it.

The wage is taken as exogenous data from the point of view of economic theory, which means that it can be seen as ultimately determined by social and political factors. The wage plays a central role in the process of social reproduction, a process that, according to Chiodi, has been more emphasised by David Ricardo than by Adam Smith. Karl Marx, in turn, highlights the historicity of the capitalist mode of production, noting how the social and political factors that determine the social product, the wage, and the process of social reproduction, depend upon an historical context. As Chiodi notes, Marx identifies various forms of exploitation in various historical contexts, even if Marx's analysis is centred on the capitalist mode of production. In this context, Chiodi suggests that the labour theory of value should be seen only as an analytical instrument for understanding capitalist exploitation, rather than as a correct theory of prices.

Furthermore, Chiodi notes that Sraffa's PCMC shows how decisions regarding quantities can be taken before the production process starts, since quantities do not depend on prices. Thus, it is not necessary to address the problem of prices when discussing the social, political and institutional factors that determine quantities in an economic system, contrarily to what happens in neoclassical economics or, as Chiodi calls it, neo-vulgar economics. The reason for calling the approach "neo-vulgar" economics is because it is actually a continuation of what Marx called "vulgar economy", focused on superficial phenomena such as supply and demand, rather than on the underlying conditions for the reproduction of the socio-economic system, as the classical authors did. I must add that I agree with this terminology. In fact, I have suggested it too (Martins 2015, p. 163).

The significance of production prices, however, lies in the fact that they are the prices that enable the reproduction of the existing social and economic system. In this context, Chiodi argues that production prices have a deeper meaning in Sraffa's system, since they are connected to the livelihood of the community, while making possible the distribution of the surplus. This deeper meaning reflects the underlying vision of the society, as noted above, which Sraffa expresses using a language that is in synchrony with the one used by the classical political economists and Marx. The use of such a language is, for Chiodi, an important pedagogical feature of Sraffa's PCMC that has been relatively neglected. Another important aspect is the use of constructivist mathematics.

Chiodi is certainly right in pointing out that Sraffa's contribution provides a very significant breakthrough in various dimensions, including the theory it advances, the methods it employs, and the language it uses. All these elements are essential for recovering the standpoint of the classical political economists and Marx, and Sraffa's capacity for recovering all these elements into his unique contribution is an astonishing endeavour. Furthermore, it also provides the ground for the advancement of a constructive approach, as an alternative to neoclassical economics.

According to Chiodi, the reason why this constructive approach has not been as prominent as it should is because Sraffa's PCMC was used essentially as a tool for destroying the foundations of neoclassical economics on logical grounds. This seems to suggest that there is an excessive focus on a critique of neoclassical economics, rather than on a more constructive alternative. While I think there is much merit in this hypothesis advanced by Chiodi, I would like to add some details to this overall thesis. In particular, I think that it is important to understand why the critique of neoclassical economics was not followed by a revival of the classical political economists and Marx.

The role of language and methodology highlighted by Chiodi plays a central role here. In a nutshell, what happened is that the critique of neoclassical economics successfully identified the internal inconsistencies of neoclassical economics. But the neoclassical economists interpreted this critique in terms of their own methods and language. In so doing, they failed to note the problems associated with the underlying methodology and language used. This led, naturally, to the belief that once the internal inconsistencies of neoclassical economics would be acknowledged, as Paul Samuelson (1966) did, there would be no need of revising the underlying methodology and language.

Chiodi notes that Robert Solow (2014, p. 63) found the mathematics used by Sraffa a set of "primitive tools", while being surprised that Sraffa made so much progress in the study of "Simple Leontief Models" using those tools. Solow, like Samuelson, always interpreted Sraffa's (1960) contribution in terms of linear programming models. One must remember that Samuelson and Solow were students of Wassily Leontief, who were developing linear programming models with Robert Dorfman (Backhouse 2014, p. 256), leading to their book *Linear Programming and Economic Analysis* (Dorfman et al. 1958).

This helps understand why in the article mentioned by Chiodi, Solow (2014, p. 63) contrasts Sraffa's (1960) "primitive" use of Simple Leontief

Models with Generalised Leontief Models that allow for choice of technique within each industry. Solow's reference to Generalised Leontief Models that allow for choice of technique within each industry implies that Sraffa's (1960) contribution is a particular case where the same technique is applied to each industry. A similar misunderstanding had been made by Frank Hahn (1975, p. 362), who saw Sraffa's (1960) contribution as a particular case of general equilibrium theory instead, for example when he writes that "there is not a single formal proposition in Sraffa's book which is not also true in a General Equilibrium model constructed on his assumptions."

Hahn's comment is made in a subsequent stage of the Cambridge controversies in the theory of capital, when the debate had already moved from a discussion on the production function to general equilibrium theory (Cohen and Harcourt 2003). But the basic misunderstanding is the same as the one made by Solow. Both Solow and Hahn interpret Sraffa's equations in terms of a functional relationship between the quantities of inputs and the quantities of outputs, as if the quantities of inputs determined the quantities of outputs. But in Sraffa's (1960) analysis, there is no functional relationship between the quantities of inputs and the quantities of outputs. Rather, both the quantities of inputs and the quantities of outputs are taken as exogenous data which, together with the distribution of income between profits and wages, leads to a set of prices that are consistent with the reproduction of the economic system.

Sraffa's (1960) system is thus compatible with any technology, and there is no need of moving into generalised Leontief models for characterizing the technologies of each industry mathematically, or general equilibrium models. In fact, Sraffa's (1960) system is much more general than generalised Leontief models or general equilibrium models precisely because it does not assume that the relationship between the quantities of inputs and the quantities of outputs has to be determined mathematically. Neoclassical economists find this very difficult to understand because they presuppose that scientific analysis must assume the form of mathematico-deductivist models (Lawson 1997).

Samuelson's (2010) last stand on the controversies in capital theory is particularly symptomatic of this tendency. Samuelson (2010) simply assumes that the activity of self-interest optimizers would lead to a convexification of technologies that would make the neoclassical presuppositions adequate (Martins 2016), which again shows that no effort is made for understanding the methodological implications of the standpoint adopted by Sraffa. Joan Robinson (1970, 1975) still tried to emphasise the methodological aspects involved, but her methodological critique did not generate the same debate as her earlier critique of the neoclassical production function (Robinson 1953–1954).

So I would argue that the reason why the critique of neoclassical economics did not lead to a revival of the classical political economists and Marx is because the methodological presuppositions of Sraffa's conception have not been fully grasped by the neoclassical economists. This means that Chiodi's emphasis on the pedagogical nature of Sraffa's contribution is certainly a very important aspect, which has been neglected, as Chiodi rightly notes. But the reason why the more constructive critique advocated by Chiodi has not been pursued stems essentially from the fact that the more destructive critique was not as successful as it is usually believed. The more destructive critique of neoclassical economics succeeded in identifying inconsistencies in neoclassical economics, but failed to change the terms of the debate, while allowing the neoclassical side to interpret Sraffa's contribution in terms of their own methods and language. The revival of the classical political economists and Marx requires, as Chiodi rightly notes, taking into account the theory, methods and language carefully employed by Sraffa (1960).

## References

Backhouse, R. (2014), "MIT and the other Cambridge", *History of Political Economy*, 46 (suppl_1): 252–271.

Cohen, A. and Harcourt, G.C. (2003), 'Whatever Happened to the Cambridge Capital Theory Controversies?', *Journal of Economic Perspectives*, 17, 199–214.

Dorfman, R., Samuelson, P. A., and Solow, R. M. (1958), *Linear Programming and Economic Analysis*, New York, McGraw Hill.

Hahn, F. H., 1975. "Revival of Political Economy: The Wrong Issues and the Wrong Argument," *The Economic Record*, 51(135), pp. 360–36.

Lawson, T. (1997), *Economics and Reality*, London: Routledge.

Martins, N. O. (2015), "Why is this school called neoclassical economics? Classicism and neoclassicism in historical context", in Morgan, J. (ed.), *What is neoclassical economics: Debating the Origins, Meaning and Significance*, London: Routledge, pp. 149–167.

Martins, N. O. (2016), "Political Aspects of the Capital Controversies and Capitalist Crises", *Journal of Post Keynesian Economics*, 39(4), pp. 473–494.

Robinson, J. (1953–4), "The production function and the theory of capital", *The Review of Economic Studies*, vol. XXI, pp. 81–106.

Robinson, J. (1975), "The Unimportance of Reswitching." *Quarterly Journal of Economics*, 89, 32–39

Robinson, J. (1970), "Thinking about thinking", reprinted on pp. 5:110–119 in *Collected Economic Papers*, 1980, Cambridge MA, MIT Press.

Samuelson, P. A. (1966), "A Summing Up", *Quarterly Journal of Economics*, 80, pp. 568–83.

Samuelson, P. A. (2010), "Testing whether the 'capital reversal' syndrome mandates deadweight loss in competitive intertemporal equilibrium," in Vint, J., Metcalfe, J. S., Kurz, H., Salvadori, N., and Samuelson, A. (eds.), *Economic Theory and Economic Thought: Essays in honour of Ian Steedman*, London: Routledge, 167–173.

Solow, R. (2014), "Comment on Scott Carter", *Research in Political Economy*, 29, pp. 63–67.

Sraffa, P. (1960), *Production of Commodities by Means of Commodities: Prelude to a Critique of Economic Theory*, Cambridge: Cambridge University Press.

CHAPTER 12

# From Multipliers to the Distribution of Income: Connecting Leontief and Sraffa

*Albert E. Steenge*

## 1 Introduction

Wassily Leontief and Piero Sraffa belong to the great innovators of economic thought in the twentieth century. Both have brought entirely new concepts to the field and were at the forefront of a new literature. While both these authors have much in common, there also are substantial differences, sometimes puzzling ones.

In this contribution we shall focus on the role of Leontief and Sraffa in laying the foundations of what is generally referred to as multi-sectoral or input-output (I-O) analysis. This is a field of economic science in which production is described in terms of interconnected sectors that produce a characteristic product thereby employing fixed coefficients production functions. The multi-sector models presented by Leontief and Sraffa have both their similarities and their differences. For example, in Leontief's analysis the famous multipliers play a central role.

A. E. Steenge (✉)
Faculty of Economics and Business, University of Groningen,
Groningen, The Netherlands
e-mail: a.e.steenge@rug.nl



A. Sinha (ed.), *A Reflection on Sraffa's Revolution in Economic Theory*, Palgrave Studies in the History of Economic Thought,
https://doi.org/10.1007/978-3-030-47206-1_12

Multipliers are special constructs that enable us to calculate the total cost in terms of sacrificed resources in generating the net product of a region or country. Incomes in this analysis are distributed proportionally to the supplied inputs, with only limited attention being given to alternative income distribution mechanisms. On the other hand, Sraffa spends a major part of his analysis on the distribution of income between wages and profits including the corresponding prices. An example of an entirely new concept introduced by him is the Standard commodity, a commodity basket of specific proportions that is completely lacking in Leontief's models.

In this article we will also try to obtain a clearer picture of the similarities as well as the differences between Leontief and Sraffa. In particular, we will try to understand the contributions of both authors in terms of the analytical structure they have adopted, where we shall confine ourselves to a selection from their work. In the case of Leontief, this concerns his familiar open static model, probably the best known of his many models and sub-models. For Sraffa, our selection deals with the models proposed in Part I of his *Production of Commodities by Means of Commodities.*

We aim to show how the work of both authors can be understood by looking for concepts that they both have in common. More specifically, our goal is to see if a direct path can be found between the Leontief and Sraffa type of approaches at income distribution issues. If we would succeed, the possible gains would be quite clear: Leontief offers a wide range of models and access to a wealth of empirical data and experience, while Sraffa offers in many areas a wider range of insights, particularly a deeper insight into specific distribution mechanisms. We start from Leontief's early model and explore an extension based on a regularity in workers' consumer behaviour, expressed by means of so-called extended input coefficients. Regarding the Leontief based approach we conclude with a dual system of a price and quantity equation (Eqs. 12.17 and 12.18′ below) where especially the price equation has a remarkable resemblance to Sraffa's price equation. We conclude with some numerical exercises illustrating the role of the numeraire—including the Standard commodity—in both the Sraffian and the Leontief approach.

## 2 Leontief's Formulae

In this section we shall briefly discuss Leontief's static open model with a single primary vector, homogeneous labour.[1] This model is probably the best known of all of Leontief's models. The structure is familiar: There are $n$ industries or sectors, each sector producing a single, characteristic product. Input proportions for each sector are fixed, which means that there is no possibility of input substitution following a shift in prices or technology change. All workers are identical irrespective of the sector they are working in or how that sector is performing. The goal is to determine the resources (commodities and labour) to be sacrificed for producing a certain net output plus the corresponding prices.

The model consists of two parts, a real or 'physical' part explaining the production of commodities and a 'dual' part explaining prices. For the real model we have

$$\mathbf{x} = \mathbf{A}\mathbf{x} + \mathbf{f}, \tag{12.1}$$

where $\mathbf{A}$ is the $(n \times n)$ nonnegative and indecomposable matrix of direct input coefficients with dominant eigenvalue less than 1, $\mathbf{f}$ the final demand vector, also known as the final output or net output vector, and $\mathbf{x}$ the gross output vector.[2] $\mathbf{f}$ is assumed to be exogenously determined with positive demand for all goods, while $\mathbf{x}$ is endogenous. Solving for $\mathbf{x}$ gives

$$\mathbf{x} = (\mathbf{I} - \mathbf{A})^{-1}\mathbf{f}, \tag{12.2}$$

where matrix $(\mathbf{I} - \mathbf{A})^{-1}$ is the so-called multiplier matrix, also known as the Leontief inverse. We see that the columns of this matrix stand for the amounts of each of the $n$ goods required to produce one unit of final output of the good in question. For example, the first column of the inverse gives the quantities of each good required to produce a net output of one unit of good 1. We also have $\mathbf{x} > \mathbf{f}$, where we see the multipliers at work.[3]

[1] In this context the term 'primary' means that the production of this factor is not explained by the model.

[2] Matrices and vectors are in bold type, scalars in standard. Vectors are column vectors, row vectors are indicated by an apostrophe.

[3] See also the next section on additional properties of the system.

Labour requirements are given by

$$L = \mathbf{l}'\mathbf{x}, \tag{12.3}$$

where $L$ stands for total employment and $\mathbf{l}'$ for the (row) vector of direct labour input coefficients, that is the quantity of labour required per unit of output of each sector.[4] Together matrix $\mathbf{A}$ and vector $\mathbf{l}'$ define the technology of the system, both constant by assumption during the period under consideration. Because labour is the only primary factor, it receives the entire income or 'surplus', that is the commodity basket (or bundle) $\mathbf{f}$. Occasionally we shall refer to this vector as the real wage or the real wage basket; income per head thus equals $\mathbf{f}/L$.[5]

Corresponding equilibrium prices are given by

$$\mathbf{p}' = \mathbf{p}'\mathbf{A} + w\mathbf{l}', \tag{12.4}$$

where $\mathbf{p}'$ is the vector of commodity prices and $w$ the exogenously given wage rate in, say, Euro or \$ per unit of labour. From Eq. (12.4) we have

$$\mathbf{p}' = w\mathbf{l}'(\mathbf{I} - \mathbf{A})^{-1}, \tag{12.5}$$

where we meet the multiplier matrix again. The amounts of labour 'embodied' in each produced good are obtained by pre-multiplying the Leontief inverse by the row vector $\mathbf{l}'$. Equation (12.5) thus informs us that relative prices are given by these amounts of embodied labour while absolute prices are obtained by attributing a particular value to $w$. We directly have

$$\mathbf{p}'\mathbf{f} = w\mathbf{l}'\mathbf{x} = wL \tag{12.6}$$

That is, the consumption bundle $\mathbf{f}$ can be bought in its entirety out of the wage received by the work force, $L$. There are no profits; any investments are accounted for as part of final demand. For further details, see Miller and Blair (2009).

[4] We assume that each industry employs labour, thus $\mathbf{l}' > 0$.

[5] Each element of the vector $\mathbf{f}/L$ is equal to the corresponding element of the vector $\mathbf{f}$ divided by $L$.

## 3 Another View of Leontief's Open Static Model

Let us stay for a moment with the above model. Firstly, we observe that we have a fully functioning model. The model is complete in the sense that row sums equal corresponding column sums, prices and quantities are well-defined and the entire system is 'in equilibrium'. At the same time, the model is very limited in scope. The focus is on the multiplier matrix and its role in determining gross output and commodity prices. Clearly, many questions cannot be fruitfully addressed by employing the above framework, such as questions concerning the connection between prices and the distribution of income.

Over the years, many extensions of the basic model have been presented.[6] However, several fields or sub-fields have been neglected in this respect. An example of such a field is constituted by income distribution issues. Since this is a very important field, we will take another look at these issues in this chapter. In fact, we shall try to find out if there may be an extension of the basic model that will help us understand the differences between Leontief and Sraffa in terms of the fundamentals of multisector analysis. More concretely, our goal will be to see if the I-O model of Sect. 2 can be extended in such a way as to enable us to address specific distribution issues. As we shall see further below, this is indeed possible; there is a way to extend the model such that distribution issues can be addressed. The procedure involves a certain rearranging of the basic formulas and a close look at a familiar assumption.

The assumption we are referring to here is the so-called homogeneity assumption: the assumption that workers are identical and receive the same wage irrespective of the sector in which they are employed. In this chapter we will interpret this in the sense that workers also have identical tastes. That is, they consume the same commodity bundle, to be called the real wage or real wage basket. As we shall see immediately below, this interpretation of the term 'identical' allows us to work with what may be viewed as 'extended' input coefficients.[7] In fact, this type of coefficients allows the introduction of a distribution mechanism that embodies elements of both the Sraffa and Leontief tradition and can as such be seen as a bridge between both traditions.

[6] See Kurz and Salvadori (1995) or Miller and Blair (2009) for comprehensive presentations of historical background and current research lines.

[7] There are a few examples of work in which these commodity bundles are used. Although used in a different context, Quadrio Curzio (1967, pp. 43–44) provides an example. Seton (1977), in his analysis of Marxian price systems, provides another.

### 3.1 Extended Input Coefficients

Let us see how this works out. Section 2 answered the question: If **f** must be produced, how much gross output is required and what does that mean in terms of overall employment? However, we can look at the equations of Sect. 2 in a quite different way. To see this, we start by taking a second look at Eq. (12.6). We directly have

$$w = (\mathbf{p}'\mathbf{f})/L = \mathbf{p}'(\mathbf{f}/L), \tag{12.7}$$

which is the value of the net income per worker.[8] Returning to Eq. (12.4) and substituting our new expression for the wage, we obtain successively

$$\mathbf{p}' = \mathbf{p}'\mathbf{A} + [w]\mathbf{l}' \tag{12.8}$$

$$= \mathbf{p}'\mathbf{A} + \left[\mathbf{p}'(\mathbf{f}/L)\right]\mathbf{l}' \tag{12.8'}$$

$$= \mathbf{p}'\left[\mathbf{A} + (\mathbf{f}/L)\mathbf{l}'\right] \tag{12.8''}$$

We see here (following Eq. 12.4) that the price of the $i$-th good consists of two parts. That is, firstly a part consisting of the intermediate inputs used up in its production, represented by the $i$-th column of matrix **A**. The second part, following Eq. (12.8), consists of the wage part of the $i$-th good, here represented by the term $w\mathbf{l}'_i$, where $\mathbf{l}'_i$ stands for the $i$-th element of the labour input coefficients vector $\mathbf{l}'$.

However, we can say more, as indicated by Eqs. (12.8′) and (12.8″). Since workers do not save or borrow, the wage per worker must be equal to the value of the goods ***bought and consumed*** by that worker. So, the unit price of good $i$ must also cover the value of this basket of consumption goods which (again per unit) is equal to $[\mathbf{p}'(\mathbf{f}/L)]\mathbf{l}'_i$. In physical terms, this means that the price of good $i$ must consist of the price of *two* commodity baskets, i.e. one consisting of commodities as given by the $i$-th column of matrix **A** and one given by the consumption basket $(\mathbf{f}/L)\mathbf{l}'_i$. Equation (12.8′) thus also informs us immediately how these two types of inputs can be represented in terms of input coefficients.

A similar interpretation is valid for the real output model of Eq. (12.1). We have

[8] Note: if $\mathbf{p}'(\mathbf{f}/L) = 1$ *and* $L = 1$, $w$ stands for the value of the economy's net income.

$$\mathbf{x} = \mathbf{A}\mathbf{x} + (\mathbf{f}/L)L \tag{12.9}$$

$$= \mathbf{A}\mathbf{x} + (\mathbf{f}/L)\mathbf{l}'\mathbf{x} \tag{12.9'}$$

$$= \left[\mathbf{A} + (\mathbf{f}/L)\mathbf{l}'\right]\mathbf{x} \tag{12.9''}$$

We see that matrix $\mathbf{A} + (\mathbf{f}/L)\mathbf{l}'$ can be interpreted as the coefficients matrix of an economic system where commodities are produced (only) by commodities: we see very clearly that the wage in each sector has been replaced by the vector of consumption goods bought by the workers. Since the system is in equilibrium, these entities (i.e. the wage and the consumption expenditure) as said, are equal in value. Because the elements of the $i$-th column of matrix $(\mathbf{f}/L)\mathbf{l}'$ (as given by Eqs. 12.8″ and 12.9″) stand for a special type of input coefficients, we shall designate the elements of matrix $\mathbf{A} + (\mathbf{f}/L)\mathbf{l}'$ as 'extended input coefficients'. Matrix $\mathbf{A} + (\mathbf{f}/L)\mathbf{l}'$ then can be called an 'extended input coefficients matrix', representing in an alternative way the *technology* of the entire system. We should note here that Leontief's fixed coefficients assumption applies also here as long as workers do not diversify their consumption. (If we wish to do so, we can fix the value of the vectors **f** or $\mathbf{f}/L$ as we like. This can be 1, but also any value that fits the equations in question.)

The above equations lead to a most interesting interpretation. Looking at Eq. (12.8″), we see that the price vector $\mathbf{p}'$ as introduced in Sect. 2 can be interpreted in an entirely different way, namely as the *left-hand eigenvector* corresponding to the dominant or Perron-Frobenius eigenvalue of the matrix in brackets. Analogously, the gross output vector can be interpreted as the *right-hand eigenvector* of the dominant eigenvalue of the same matrix. Thus, Eqs. (12.8″) and (12.9″) provide an alternative way to view Leontief's open static model, the model we started with in Sect. 2.

### 3.2 Properties of the Extended Input Coefficients Matrix

A substantial part of the I-O literature consists of finding and interpreting properties of matrix **A** that guarantee that the multiplier matrix is positive (because otherwise problems of interpretation will arise and $\mathbf{x} > \mathbf{f}$ cannot be guaranteed). We already mentioned that matrix **A** has a dominant eigenvalue less than 1.[9] However, the extension presented in Eqs. (12.8″) and (12.9″) points to further properties of the matrices involved here.

To simplify notation, let us write

$$\mathbf{M} \equiv \mathbf{A} + (\mathbf{f} / L)\mathbf{l}' \tag{12.10}$$

We observe that this matrix **M** has a number of interesting properties. For example, from Eqs. (12.8″) or (12.9″) we have that it has a dominant eigenvalue $\lambda$ equal to 1. Under the conditions of Sect. 2, we also have that each element of **M** is positive, that is $\mathbf{M} > \mathbf{0}$, for short. This means that the eigenvectors corresponding to $\lambda$ also are positive. So, $\mathbf{p}' > 0$ and $\mathbf{x} > 0$.[10]

A question we should ask here is: Why should we be interested in the above extension in the first place? One answer is that the new formulations of Sect. 3 are important because they point to additional relations between the variables concerned. In fact, the 'extended input coefficients' allow us to study entirely different relations between the core variables **x**, **p′** and *L*. Rewriting Eq. (12.9″) we have

$$\left(\mathbf{I} - \left[\mathbf{A} + (\mathbf{f} / L)\mathbf{l}'\right]\right)\mathbf{x} = \mathbf{0} \tag{12.11}$$

From this we have the important result that the determinant of the matrix on the left-hand side of Eq. (12.11) must be equal to zero, that is

$$\left|\mathbf{I} - \left[\mathbf{A} + (\mathbf{f} / L)\mathbf{l}'\right]\right| = 0 \tag{12.12}$$

For us, Eq. (12.12) is particularly important because, given **f**, it can be interpreted as a single equation in one variable, *L*. And, once we have obtained *L*, we can find **x** (as the right-hand eigenvector of the matrix in question) and standardize it using $\mathbf{l}'\mathbf{x} = L$.

[9] These properties are basically self-evident. Indecomposability e.g. prevents that there are sub-systems capable of (re)production without inputs, directly or indirectly, from sectors not belonging to that sub-system. The full rank property means that there are no industries using the same technology. A dominant eigenvalue smaller than 1 guarantees that a surplus is possible. For further details, see e.g. Takayama (1974) or Seneta (1981).

[10] See also footnote 4. With $\mathbf{f} > 0$ and $\mathbf{l}' > 0$ also matrix $(\mathbf{f}/L)\mathbf{l}' > 0$ and, consequently, matrix **M**.

There is a second reason why the above extension is important. We have seen that 'equilibrium' in the sense of the first paragraph of Sect. 3 can be modelled employing an input coefficients matrix (i.e. matrix $\mathbf{M}$) that has a dominant eigenvalue equal to 1. However, matrix $\mathbf{M}$ consists of two parts. Suppose now that for some reason certain elements of matrix $\mathbf{A}$ *increase* in size. Adopting the symbol $\underline{\mathbf{A}}$ for this matrix, that would mean that, based on the continuity rules of standard matrix theory, matrix $\underline{\mathbf{A}} + (\mathbf{f}/L)\mathbf{l}'$ would have a dominant eigenvalue larger than 1 and equilibrium would be destroyed. However, equilibrium might be restored—in the sense of the relevant matrix $\mathbf{M}$ having a dominant eigenvalue equal to 1—if simultaneously some elements of matrix $\mathbf{f}/L)\mathbf{l}'$ would *decrease*. That is, if final demand for certain goods would decrease. Below, in Sect. 4, we shall further explore this connection.

## 4 The Distribution of Gross Output

In this section we continue with Eqs. (12.8″) and (12.9″). Suppose indeed that at the beginning of a new period, due to some unexpected event, more intermediate inputs (per unit of output) are required than before. Say, the new intermediate input requirements are given by a new matrix $\alpha\mathbf{A}$ with $\alpha > 1$.[11] With output $\mathbf{x}$ unchanged and for all intermediate inputs taken together, we then need the quantity $\alpha\mathbf{Ax}$ to be available. However, the right-hand side of Eq. (12.9″) (the 'destinations') is now larger than the left-hand side ('production'). So the economy's equilibrium is disturbed.

To see what may happen we will definitely need more context. There are various possibilities; so let us select one, and suppose that workers plan to buy consumption goods in the same quantities as before. The standard model of Sect. 2 informs us that this would imply that both $\mathbf{x}$ and $L$ would increase. However, what if circumstances do not allow that? Suppose that because of a shortage of labour, employment must remain the same. Because more intermediate inputs than before are required and only the same amount of primary inputs (i.e. labour) is available, the only way to proceed is to accept that workers' consumption, in one form or another, must *decrease*.

How might that happen? To proceed, we need a further assumption regarding the consumption behaviour of the workers. If we assume that

[11] That is, each element of $\mathbf{A}$ is multiplied by $\alpha$ in the new configuration.

workers consume goods in the *same proportions* as before but in lower quantities, each worker can buy, in the new circumstances, a real wage basket equal to $\beta(\mathbf{f}/L)$, where $0 \leq \beta < 1$ is a new parameter that captures the decrease in real income.

In terms of the 'extended' model of the previous section, this would mean that it changes correspondingly to

$$\mathbf{x} = \left[\alpha\mathbf{A} + \beta(\mathbf{f}/L)\mathbf{l}'\right]\mathbf{x}, \tag{12.13}$$

where $\beta$ now has to be determined such that the matrix on the right-hand side of Eq. (12.13) has dominant eigenvalue equal to 1 (which means that it models a circular flow). Writing

$$\mathbf{M}(\alpha,\beta) \equiv \alpha\mathbf{A} + \beta(\mathbf{f}/L)\mathbf{l}', \tag{12.14}$$

the next step is to determine the pairs of $\alpha$ and $\beta$ that satisfy Eq. (12.14) and then to select the value of $\beta$ that corresponds to the exogenously given value of $\alpha$. Following this, gross output $\mathbf{x}$ can be calculated as the right-hand eigenvector of $\mathbf{M}(\alpha, \beta)$ which gives the distribution of output after standardization via $\mathbf{l}'\mathbf{x}$.

Prices follow the same pattern, and will change according to

$$\mathbf{p}' = \mathbf{p}'\left[\alpha\mathbf{A} + \beta(\mathbf{f}/L)\mathbf{l}'\right], \tag{12.15}$$

$\mathbf{p}'$ here is the left-hand eigenvector corresponding to the dominant eigenvalue. This also means that the system determinant changes to

$$\left|\mathbf{I} - \left[(\alpha\mathbf{A} + \beta(\mathbf{f}/L)\mathbf{l}'\right]\right| = 0, \tag{12.16}$$

which is an equation in two variables, $\alpha$ and $\beta$. A question to be asked is if there is a combination of economically interpretable values of $\alpha$ and $\beta$ that satisfy Eq. (12.16). In fact, this is guaranteed by the model of Sect. 2. The basic question is about the behaviour of the multiplier matrix $(\mathbf{I} - \alpha\mathbf{A})^{-1}$. If $\alpha$ is such that the Perron-Frobenius eigenvalue of matrix $\alpha\mathbf{A}$ is smaller than 1, there generally will be no problem.[12]

So, summarizing, on the basis of the structure proposed above, we can straightforwardly extend the basic model of Sect. 2 to investigate the

[12] For further background, see Steenge and Serrano (2012), or Steenge (2015).

consequences of different distributions of gross output. The extension was based on the assumption that the consumption per head vector $\mathbf{f}/L$ stays the same and that the decrease in workers' consumption manifests itself in a lower value for $\beta$. A vice versa trajectory would apply for $\alpha < 1$. In the next section we shall focus on the distribution of ***net output***.[13]

## 5 The Distribution of Net Output

### 5.1 *Distributing Net Output in an Extended Leontief World*

The above model does not yet give us a clear income distribution picture. However, in line with Sects. 3 and 4, a way forward can be formulated. To this end, we go back to the assumption of homogeneous labour and the assumption that workers have identical preferences.

If we put $\alpha = 1 + r$, we have the beginning of a model with a rate of profit. However, as in the model of the previous section, one more step is required to interpret the model in terms of an income distribution model. That is, we need to determine if workers' consumption patterns will change if their wage increases or decreases. In order to provide a sufficiently comprehensive context, we again need to adopt one or more policy rules. If we assume that consumption proportions ***and*** employment remain unchanged, we effectively fix the vector $\mathbf{f}/L$ as exogenously given. Following the procedure of Sect. 4, that leaves the parameter $\beta$ to reflect changes in workers' consumption, and we can interpret shifts in the income distribution as shifts in $r$ and $\beta$. With $\beta \varepsilon [0, 1]$ we impose that the value of $\beta$ can fluctuate from 0 to 1. If $\beta = 0$, we have a situation of capital obtaining all profit. If, contrarily, $\beta = 1$, profits are zero.

Substituting $1 + r$ for $\alpha$ we have accordingly, for the real side

$$\mathbf{x} = \left[(1+r)\mathbf{A} + \beta(\mathbf{f}/L)\mathbf{l}'\right]\mathbf{x}, \qquad (12.17)$$

where the term $r\mathbf{Ax}$ stands for the profit part and $\beta(\mathbf{f}/L)\mathbf{l}'\mathbf{x}$ for the wage part.

We observe that each combination of $r$ and $\beta$ that satisfies Eq. (12.17), in general, generates a new right-hand side matrix, and, thus, a new gross

[13] However, we should point out again that the choices we have made in this section only are one example of possible extensions. Future work may be called for to explore further possibilities.

output vector. This vector subsequently can be standardized via the assumption of unchanged total employment.

So, we model our distribution of income as a simultaneous shift in $r$ and $\beta$ where the dominant eigenvalue of the matrix on the right-hand side of Eq. (12.17) should remain 1, the value we started with. This means, again ceteris paribus, that if $r$ increases, $\beta$ must decrease, and vice versa.

For corresponding prices we obtain

$$\mathbf{p}' = (1+r)\mathbf{p}'\mathbf{A} + \beta\left[\mathbf{p}'(\mathbf{f}/L)\right]\mathbf{l}' \tag{12.18}$$

$$= \mathbf{p}'\left[(1+r)\mathbf{A} + \beta(\mathbf{f}/L)\mathbf{l}'\right] \tag{12.18'}$$

With

$$\mathbf{M}(r,\beta) \equiv (1+r)\mathbf{A} + \beta(\mathbf{f}/L)\mathbf{l}', \tag{12.19}$$

we have straightforwardly

$$\left|\mathbf{I} - \mathbf{M}(r,\beta)\right| = \left|\mathbf{I} - \left[(1+r)\mathbf{A} + \beta(\mathbf{f}/L)\mathbf{l}'\right]\right| = 0, \tag{12.20}$$

which is a single equation in two unknowns, $r$ and $\beta$, and can directly be solved for $r$ and $\beta$. Next, given these values, we substitute them in (12.19), and calculate the left-hand Perron-Frobenius eigenvector. This vector gives the relative prices which subsequently can be standardized, see also the next sub-section. Thus, the model is quite clear. We first have to calculate the income distribution determining parameters $r$ and $\beta$. After that, the corresponding prices can be determined from the left-hand Perron-Frobenius eigenvector of matrix $\mathbf{M}(r, \beta)$.

### 5.2 *The Rate of Profits and the Wage Rate in the Sraffian Price Model*

In this section we shall juxtapose parts of the Leontief and the Sraffa tradition. We thereby shall focus on the price equation because the Sraffian real model is very much like Leontief's (chs. 1 and 2 of *Production of Commodities by Means of Commodities*).

So far we have followed the Leontief extension of Sects. 3 and 4. In that model we assumed that workers will have the same preferences over the entire range of $\beta$, the relevant parameter for workers. A lower wage (accompanied by a higher rate of profit) means a lower $\beta$ and thus less consumption, though proportions within the basket of consumed goods remain the same.

Let us now consider the Sraffian price equation (Sraffa, 1960, ch. 5), Kurz and Salvadori (1995, ch. 4) or Abraham-Frois and Berrebi (1997, ch. 2). We have

$$\mathbf{p}' = (1+r)\mathbf{p}'\mathbf{A} + w\mathbf{l}' \tag{12.21}$$

Matrix $\mathbf{A}$ and row vector $\mathbf{l}'$ have the same interpretation as above, i.e. as input coefficients matrix and vector of direct labour input coefficients, respectively.[14] $\mathbf{p}'$ is the price vector, with $r$ the rate of profits and $w$ the wage rate in terms of the national income, determined exogenously. Solving for $\mathbf{p}'$ we find

$$\mathbf{p}' = w\mathbf{l}'\left[\mathbf{I} - (1+r)\mathbf{A}\right]^{-1} \tag{12.22}$$

So, if $r$ changes, $\mathbf{p}'$ will change accordingly. However, there is no relation between a change in $r$ and the wage rate $w$ of the type discussed in Sect. 4 and sub-section 5.1. In Eq. (12.21), the scalar $w$ just establishes absolute prices, relative prices being given by the row vector

$$\mathbf{l}^*(r) = \mathbf{l}'\left[\mathbf{I} - (1+r)\mathbf{A}\right]^{-1}. \tag{12.22'}$$

Now let us return for a moment to the Leontief-based price Eq. (12.18′). If we put

$$\underline{w} = \beta\left[\mathbf{p}'(\mathbf{f}/L)\right], \tag{12.23}$$

with $\beta$ the workers' share of the national income and substitute in Eq. (12.18′) we obtain

$$\mathbf{p}' = (1+r)\mathbf{p}'\mathbf{A} + \underline{w}\mathbf{l}', \tag{12.24}$$

[14] Matrix $\mathbf{A}$ in this model is assumed to have the same properties as matrix $\mathbf{A}$ in the Leontief model of Sect. 2.

which is a price equation that is structured around the inverse relation between $r$ and $\beta$. As we observe, written in this way the Leontief-based Eq. (12.18′), combined with the wage variable $\underline{w}$ as in Eq. (12.23), is remarkably similar to the Sraffian price Eq. (12.21). Also, $\underline{w}$ can be interpreted as a flexible numeraire which can be given any acceptable value. Future work here will have to decide if additional points of similarity can be found, which also concerns the role of standardizations such as income or employment being fixed at unity.

### 5.3 *The Standard Commodity*

We have seen above that the relation between $r$ and $\beta$ can be found by solving Eq. (12.21) for these two variables. In general, this function will be non-linear. However, looking at Eqs. (12.17) and (12.19) we see that if $\mathbf{f}$ has the proportions of the Perron-Frobenius eigenvector of matrix $\mathbf{A}$, the relation between $r$ and $\beta$ is a linear one. In fact, we have, with $R$ the maximum rate of profit and using $\lambda = 1/(1 + R)$, after some manipulation,

$$\beta = 1 - r/R \tag{12.25}$$

This is the case of the well-known Standard system and Standard commodity. So, we meet here, not un-expected, the famous linear relation between rate of profits and the wage rate, here $\beta$.[15] In this case, $r$ and $\beta$ can be found *without* having first to calculate the functional relation as given by Eq. (12.21).

## 6 Numerical Illustration

### 6.1 *The Non-linear Case*

We shall illustrate the above-proposed procedure by means of a numerical example with a two-industries case and one primary factor, homogeneous labour, as interpreted before. The data are artificial, but satisfy the standard requirements of Sect. 2.

[15] For a derivation of this relation in the Sraffian system, see e.g. Pasinetti (1977, Sect. 12.2). For a recent discussion of Standard commodity and Standard system, see Sinha (2016, esp. ch. 7).

Let us say that the technology is given by the following input coefficients matrix[16]

$$\mathbf{A} = \begin{bmatrix} 0.25 & 0.40 \\ 0.54 & 0.42 \end{bmatrix}$$

and vector of direct labour input coefficients

$$\mathbf{l}' = \begin{bmatrix} 0.70 & 3.20 \end{bmatrix}$$

Suppose further that final demand **f** is

$$\mathbf{f} = \begin{bmatrix} 1600 \\ 800 \end{bmatrix}$$

Direct calculation gives[17]

$$\mathbf{x} = \begin{bmatrix} 5698.630 \\ 6684.932 \end{bmatrix}$$

and

$$L = 25{,}380.822$$

Furthermore, for the dominant eigenvalue we have

$$\lambda = 0.807,$$

so,

$$R = (1 - \lambda)/\lambda = 0.238.$$

[16] Symbols have the same interpretation as before.

[17] We observe that vectors **f** and **x** are not proportional to each other, so there is no role for the Standard commodity, see the next section.

Calculation gives

$$\varphi(r,\beta) = 0.2190 - 0.8920r - 0.2190\beta - 0.1110r^2 - 0.0740r\beta = 0.$$

Graphically, we have Figs. 12.1 and 12.2. Figure 12.1 gives the relation between $r$ and $\beta$ for the area relevant for our study. First of all, we observe the *non-linearity* of the curves. We also observe that $\beta = 1$ if $r = 0$. Furthermore, when $\beta = 0$, we find, as expected, $R = 0.238$. The right-hand figure below gives the wider picture of the curves where only a small part is relevant for us.

It may be useful to look at some particular numerical values. Selecting $r = 0$, we have $\beta = 1$. After having obtained matrix $\mathbf{M}(r, \beta)$ we obtain the relative commodity prices from the left-hand Perron-Frobenius eigenvector as $p_2/p_1 = 1.256$. For $r = 0.1$, we find $\beta = 0.568$. Substituting these values in $\mathbf{M}(r, \beta)$ and calculating its left-hand Perron-Frobenius eigenvector we find for the relative prices $p_2/p_1 = 1.154$. For $r = 0.2$, we have $\beta = 0.155$ and $p_2/p_1 = 1.064$. Absolute prices can be obtained by standardizing at any appropriate value.

It is straightforward to obtain the surplus and the part each party receives. For example, for $r = 0.2$ we have for the profits part

$$\alpha\mathbf{Ax} = 0.2\mathbf{Ax} = \begin{bmatrix} 798.215 \\ 1117.665 \end{bmatrix},$$

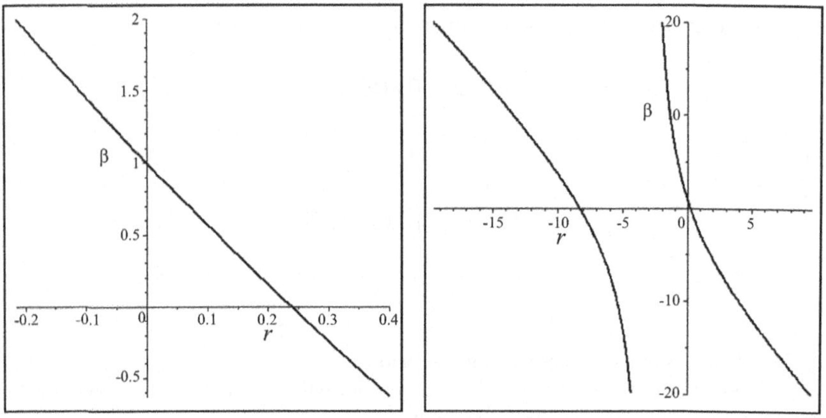


**Fig. 12.1** Relation between r and β; non-linear case. Left-hand: relevant area; right-hand: the wider picture.

and for the wage part,

$$\beta(\mathbf{f}/L)(\mathbf{l}'\mathbf{x}) = 0.155(\mathbf{f}/L)(\mathbf{l}'\mathbf{x}) = \begin{bmatrix} 247.456 \\ 123.728 \end{bmatrix},$$

adding up to

$$\begin{bmatrix} 1045.671 \\ 1241.393 \end{bmatrix}.$$

### 6.2 *The Standard System*

In this sub-section we take a look at the Standard system, where we keep employment $L$ the same, that is 25,380.822. We first calculate the Perron-Frobenius eigenvector of matrix **A**. If the final demand vector has the same proportions as this eigenvector, also the gross output vector has. With $L$ as above, and the symbol **f*** standing for the Standard commodity, calculation gives

$$\mathbf{f}^* = \begin{bmatrix} 947.0732 \\ 1319.9050 \end{bmatrix}$$

Direct calculation (via Eq. [12.1], substituting **f*** for **f**) gives

$$\mathbf{x}^* = \begin{bmatrix} 4919.0161 \\ 6855.4720 \end{bmatrix}$$

We now are looking for $r$ and $\beta$ such that

$$\left|\mathbf{I} - \mathbf{M}(r,\beta)\right| = \left|\mathbf{I} - \left[(1+r)\mathbf{A} + \beta(\mathbf{f}^*/L)\mathbf{l}'\right]\right| = 0$$

Direct calculation gives

$$\varphi(r,\beta) = 0.2190 - 0.8920r - 0.2190\beta - 0.1110r^2 - 0.0265r\beta = 0$$

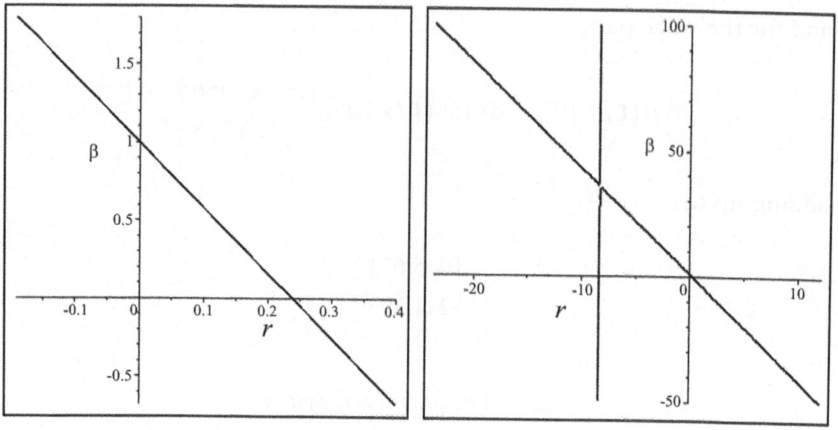


**Fig. 12.2** Relation between r and β; linear case. Left-hand: relevant area; right-hand: the wider picture.

We observe that the curve is *linear*, also in the area $0 \le r \le R$, see Fig. 12.2. (Minor differences are due to rounding.)

Also here, once combinations of $r$ and $\beta$ have been selected (from Eq. 12.24), commodity prices are obtained straightforwardly. For example, for $r = 0.1$, we have $\beta = 0.581$. Substitution in Eq. (12.19) gives $p_2/p_1 = 1.154$. For $r = 0.2$ we find $\beta = 0.161$ and $p_2/p_1 = 1.0641$. We note that prices are the same for the $r$ values as in the non-linear case. This is because relative prices obey Eqs. (12.21) and (12.18′) (in an appropriate standardization).

## 7 Final Remarks and Conclusion

In this chapter we have explored a new approach to understanding the contributions made by Leontief and Sraffa in the field of multi-sector analysis. We focused on Leontief's well-known open static model—probably the best known of his models—and on the model proposed by Sraffa in Part I of *Production of Commodities by Means of Commodities*. We observed that both models have a number of properties in common, but also show significant differences. Our objective in this contribution was to see if both models could be viewed from an alternative perspective from which their similarities and differences could be understood, see Sect. 3.

We showed that such a perspective does in fact exist; in fact, several ones can be found. We decided to concentrate on a particular perspective which is based on Leontief's earlier work. In doing so we particularly focused on the so-called 'homogeneity property' of employed labour. If we interpret this property in terms of identical consumer preferences, we obtain, in a situation of a changing income distribution, one additional parameter, that is the real wage, or, equivalently, the real wage per capita. By including this new parameter in the basic system equations employing the so-called extended input coefficients, we are able to present the distribution mechanism in terms of two equations: a real output equation and a price equation which are dual to each other, i.e. Eqs. (12.17) and (12.18) in Sect. 5. Clearly, this chapter should be seen as a first step. As mentioned, several types of extensions are possible, and future work will be required to elaborate on these.

Concluding, we have presented here an extension of Leontief's early work meant to include basic elements of an income distribution mechanism. We have shown that this extension can be an interesting building block in finding a theoretical background that bridges a number of differences between a Leontief and a Sraffian approach in multi-sector analysis, and may help us to understand the similarities and differences between the two. The main characteristics of this approach can be summarized as follows:

1. Inputs are registered in a new way which leads to a definition of technology in terms of extended input coefficients, i.e. coefficients that include both technical and consumption based elements.
2. Simultaneously, a new coefficients matrix is created with dominant eigenvalue equal to 1 which signals a situation of equilibrium; the Perron-Frobenius eigenvectors of this matrix are the equilibrium price and quantity vectors.
3. A disturbance of this equilibrium is characterized by this dominant eigenvalue becoming greater than 1 or less than 1.
4. Equilibrium can be restored by introducing mechanisms that increase profits and reduce the wage, or vice versa.
5. In this extension, the rate of profits and the wage rate are determined independently of prices, although prices are dependent on them.

As we stated, future work will be needed to explore the further possibilities offered by this approach.

## References

Abraham-Frois, G. and E. Berrebi (1997), *Prices, Profits and Rhythms of Accumulation*, Cambridge, Cambridge University Press.

Kurz, H.D. and N. Salvadori (1995), *Theory of Production: A Long Period Analysis*, Cambridge, Cambridge University Press.

Miller, R.E. and P.D. Blair (2009), *Input-Output Analysis; Foundations and Extensions*, Cambridge, Cambridge University Press.

Pasinetti, L.L. (1977), *Lectures on the Theory of Production*, London and Basingstoke, Macmillan.

Quadrio Curzio, A. (1967) *Rendita e distribuzione in un modello economico pluri-settoriale*, Milano, Giuffrè.

Seneta, E. (1981), *Non-negative Matrices and Markov Chains*, 2nd ed., New York, Springer-Verlag.

Seton, F. (1977), 'The Question of Ideological Obstacles to Rational Price Setting in Communist Countries', in A. Abouchar (ed.), *The Socialist Price Mechanism*, Durham, NC, Duke University Press.

Steenge, A.E. (2015), 'Limits, Resources and Distributional Trade-Offs: Structural Constraints and Opportunities', in M. Baranzini, C. Rotondi and R. Scazzieri (eds.), *Resources, Production and Structural Dynamics*, Cambridge, Cambridge University Press, pp. 107–25.

Steenge, A.E. and M. Serrano (2012), Income Distributions in Input-Output Models, Economic Systems Research, 24: 4, pp. 391–412.

Takayama, A. (1974), *Mathematical Economics*, The Dryden Press, Hinsdale, IL.

Comment

# A Comment on Steenge's 'From Multipliers to the Distribution of Income: Connecting Leontief and Sraffa'

*Roberto Scazzieri*

In his contribution to this volume (Steenge 2020) Professor Albert Steenge outlines an ambitious framework aimed at bridging the approaches to the economy as a circular flow due, respectively, to Wassily Leontief (1941, 1991 [1928]) and Piero Sraffa (1960). In so doing, Steenge provides the coordinates of a general scheme for the investigation of a circular economy, of which the contributions of Leontief and Sraffa appear to be special cases. Steenge's aim is 'to understand the contributions of both these authors in terms of the analytical structure they have adopted' by concentrating on Leontief's open static model and on the model that Sraffa outlines in Part One of *Production of Commodities by Means of Commodities.* At the root of Steenge's treatment is the consideration of the relationship $\mathbf{x} = \mathbf{Ax} + \mathbf{f}$, 'where $\mathbf{A}$ is the $(n \times n)$ matrix of direct [intermediate] input coefficients, $\mathbf{f}$ the final demand vector, also known as the final or net output vector, and $\mathbf{x}$ the gross output vector' (Steenge 2020, this volume). In this model, labour requirements per unit of output are given by vector $\mathbf{l}'$, which is 'the (row) vector of direct labour input coefficients' (Steenge 2020, this volume). An important consequence of Steenge's approach is that, since labour is the only primary (non-produced) factor of production, it 'receives the entire income or "surplus", that is the commodity bundle $\mathbf{f}$' (Steenge 2020, this volume). If we solve equation $\mathbf{x} = \mathbf{Ax} + \mathbf{f}$ for $\mathbf{x}$, we obtain $\mathbf{x} = (\mathbf{I} - \mathbf{A})^{-1}\mathbf{f}$, in which matrix $(\mathbf{I} - \mathbf{A})^{-1}$ is the mathematical operator known as the 'Leontief inverse'. It is at this point that Steenge's argument introduces an important twist concerning the role of labour in a circular production economy. For Steenge highlights that, if we assume that 'workers are identical and receive the same wage irrespective of the sector in which they are employed', then each unit of the labour input (say, each worker) may be assumed to consume 'the same commodity bundle, to be called the real wage or real wage basket' (Steenge 2020, this volume). This procedure entails introducing what Steenge calls 'extended input coefficients', that is, input coefficients including both the quantity of each intermediate input that is needed to deliver one unit of any given product as well as the quantity of that input entering the 'real wage basket' needed to deliver each unit of that product. This approach leads Steenge, with $L$ total employment, to the following expression for the wage basket: $w = (\mathbf{p'f}) / L = \mathbf{p'}(\mathbf{f} / L)$, and to the following expression for the price equation: $\mathbf{p'} = \mathbf{p'A} + w\mathbf{l'} = \mathbf{p'A} + \mathbf{p'}(\mathbf{f} / L)\mathbf{l'} = \mathbf{p'}[\mathbf{A} + (\mathbf{f} / L)\mathbf{l'}]$. In short, 'the price of the $i$-th good consists of two parts. That is, first, a part consisting of the intermediate inputs used up in its production, represented by the $i$-th column of matrix $\mathbf{A}$. The second part consists of the bundle of commodities consumed per unit of the labour input in each sector, here represented by the term $(\mathbf{f} / L)\mathbf{l'}_i$, where $\mathbf{l'}_i$ stands for the $i$-th element of the labour input coefficients vector $\mathbf{l'}$' (Steenge 2020, this volume). This approach to labour inputs entails that 'the unit price of good $i$ must cover the value of this basket of consumption goods which (again per unit) is equal to $[\mathbf{p'}(\mathbf{f} / L)]\mathbf{l'}_i$' (Steenge 2020, this volume). In turn, this means that the unit price of each produced commodity *must* cover not only the cost of intermediate inputs needed to deliver any unit of that commodity, but also the costs of what may in fact be considered as the 'necessary consumption' needed to produce each unit of that commodity via the unit labour input requirements for that commodity.

As Steenge points out, this treatment of labour inputs has already been followed by a few previous authors, such as Alberto Quadrio Curzio and Francis Seton. Quadrio Curzio introduced extended coefficients in a discussion of a 2-commodity Leontief closed system in which 'consumption only takes place by way of the labour employed in production' (Quadrio Curzio 1967, p. 43), so that the following system is obtained:

$$\begin{bmatrix} a_{11} + l_1 c_1 & a_{12} + l_2 c_1 \\ a_{21} + l_1 c_2 & a_{22} + l_2 c_2 \end{bmatrix}' \begin{bmatrix} p_1 \\ p_2 \end{bmatrix} = \begin{bmatrix} p_1 \\ p_2 \end{bmatrix}$$

Quadrio Curzio argued that solving the above system would lead to determining a 'subsistence wage' $c_1 p_1 + c_2 p_2$, which would be the value of a wage basket consisting of the necessary consumption of commodities 1 and 2 entering each unit of labour input (Quadrio Curzio 1967, p. 44). In a similar vein, Francis Seton called attention to the possibility of representing the production system by means of an 'augmented technology' including both intermediate input coefficients as well as the coefficients of necessary consumption (Seton 1977). As Steenge noted in a previous contribution, this approach 'opens up the possibility to formulate also this model [the Leontief open static model] in terms of 'commodities being produced by commodities', since 'labour inputs are still present [...] but expressed in terms of the remuneration labour receives in exchange for its productive inputs' (Steenge 2015, p. 113).

The consequences of switching from a 'mixed' representation of production technology separately considering intermediate inputs and labour inputs to an 'augmented' representation of technology embedding labour inputs (via real wage baskets) into the physical representation of commodity flows, are far reaching. First of all, Steenge's extended coefficients allow the introduction of an augmented technology matrix with the remarkable property that a value of 1 of its dominant eigenvalue expresses a condition of uninterrupted circular flow. This means that the viability (self-replicability) condition of the circular economy is now expressed by explicitly considering 'final consumption' (defined as the consumption of final goods entering the workers' real wage basket) as a constitutive component of the circular flow, and therefore as a condition for the economy to be in a self-replacing state (i.e. as a condition for the whole production process to be repeated from one period to another). In addition, by attributing 'weights' $\alpha$ and $\beta$ to matrices $\mathbf{A}$ and $(\mathbf{f} / L)\mathbf{l}'$, under the condition that the dominant eigenvalue of matrix $\alpha\mathbf{A} + \beta(\mathbf{f} / L)\mathbf{l}'$ must be equal to 1, consequences of varying the distribution of income can be followed straightforwardly in a situation of uninterrupted circular flow.

Steenge's analysis suggests several remarks and several questions. The consideration of extended technical coefficients makes it possible to *generalize the point of view of the circular flow* beyond the domain of conventional intermediate goods (means of production needed for the formation of means of production, including themselves) and to include within the circular flow itself the final consumption goods needed for the economy to be in a self-replacing state. This is clearly a return to the classical concept of 'necessary consumption', which is not in the foreground when considering Leontief's open static input-output model or Sraffa's treatment of 'the whole of the wage as a variable' (Sraffa 1960, p. 10).[18] Steenge's emphasis on extended technical coefficients leads to a more comprehensive theory of the circular flow that might be of considerable heuristic value. For example, an augmented technology matrix with dominant eigenvalue greater or less than 1 would point to an 'interrupted' circular flow, which could in turn signal that some necessary product flows have been disregarded, or that the economy is following a technical change trajectory associated with bottlenecks and emergent scarcities. This point of view suggests the formulation of a 'flexible' circular model in which multiple different closures of the circular flow may be possible depending on different representations of the circular flow and on different policy questions to be addressed. For example, a stationary or semi-stationary economy may require a type of closure through necessary consumption ('necessary' final demand) that might not be required when considering a dynamic economy growing at a sustained rate triggered by intermediate flows internal to the production system.[19] Or, the closure of a circular flow governed by internal demand may be different from the closure of a circular flow driven by external demand (export-led closure). Steenge's approach to system closure leads to a remarkable twist in the analysis of distribution between wages and profits, which Steenge investigates by focusing on the physical circular system. Here too the starting point of Steenge's contribution is the consideration of extended technical coefficients including workers' necessary consumption. This representation of technology highlights the (aggregate) physical commodity basket going to workers and the (aggregate) physical commodity basket entering the 'take out' (what is left after replacing the means of production and remunerating workers).[20] In this case, wages and the aggregate 'take out' (profits in Sraffa's terminology) are

[18] Sraffa is of course aware of the possibility that wages may at least in part consist 'of the necessary subsistence of the workers' and would thus enter 'the system on the same footing as the fuel for the engines or the feed for the cattle' (Sraffa 1960, p. 9). However, he decides to 'follow the usual [post-classical] practice of treating the whole of the wage as variable' (Sraffa 1960, p. 10).

[19] Different types of closure in the two cases of stationary (or semi-stationary) conditions and of sustained growth are discussed in Quadrio Curzio and Scazzieri (1986), who introduce a distinction between the representations of production technology corresponding to the two above states of the economy (*transformation apparatus* and *structural apparatus*, respectively).

[20] The concept of 'Take Out' has been introduced in John Hicks's *Capital and Time* as 'the difference between value of output and value of input' (Hicks 1973, p. 30). Here it is used as a general expression to denote what is left after subtracting from the gross product the physical quantities needed to cover the self-replacement of means of production and workers' consumption.

determined independently of prices, and an inverse relationship may be envisaged between the aggregate wage basket and the aggregate 'take out' (aggregate profits), expressed in terms of the physical commodity shares assigned to the workers' wage basket and to the 'take out' respectively. This result should be assessed against Sraffa's own formulation of the relation $r = R(1 - w)$ between wages (as a proportion of the Standard net product) and the rate of profits (Sraffa 1960; Sinha 2016). Steenge does not provide a detailed discussion of this issue but highlights that his analytical formulation leads to an inverse (and generally non-linear) relationship between the workers' share of the national income, here denoted by $\beta$, and the rate of profits $r$.[21] He also argues that if the final output or net output vector $\mathbf{f}$ 'has the proportions of the Perron-Frobenius eigenvector of matrix $\mathbf{A}$, the relation between $r$ and $\beta$ is a linear one' (Steenge 2020, this volume). This latter result replicates, in a different setting, Sraffa's own formulation within the Standard system. Steenge's argument suggests a possible generalization, which however raises new questions concerning the interpretation of net output under an 'augmented' production technology (extended technical coefficients) and, correspondingly, the interpretation of the rate of profits under that technology.

## References


Hicks, J. (1973) *Capital and Time. A Neo-Austrian Theory*, Oxford, Clarendon Press.

Leontief, W. (1941) *The Structure of American Economy, 1919–1929*, Cambridge, Harvard University Press.

Leontief, W. 1991 [1928] 'The Economy as a Circular Flow', *Structural Change and Economic Dynamics*, 2 (1, June), pp. 181–212.

Morishima, M. (1966) 'Refutation of the Nonswitching Theorem', *The Quarterly Journal of Economics*, 80 (4, November), pp. 520–25, https://doi-org.ezp.lib.cam.ac.uk/10.2307/1882913 (accessed 25.2.2020).

Quadrio Curzio, A. (1967) *Rendita e distribuzione in un modello economico plurisettoriale*, Milano, Giuffrè.

Quadrio Curzio, A. and Scazzieri, R. (1986) 'The Exchange-Production Duality and the Dynamics of Economic Knowledge, in M. Baranzini and R. Scazzieri (eds.), *Foundations of Economics. Structures of Inquiry and Economic Theory*, Oxford and New York, Basil Blackwell, pp. 377–407.


[21] See Morishima (1966, p. 521) for an alternative proof of this proposition.

Seton, F. (1977) 'The Question of Ideological Obstacles to Rational Price Setting in Communist Countries', in A. Abouchar (ed.), *The Socialist Price Mechanism*, Durham, NC, Duke University Press.
Sinha, A. (2016) *A Revolution in Economic Theory: The Economics of Piero Sraffa*, Basel, Springer.
Sraffa, P. (1960) *Production of Commodities by Means of Commodities*, Cambridge University Press.
Steenge, A. (2015) 'Limits, Resources and Distributional Trade-Offs: Structural Constraints and Opportunities', in M. Baranzini, C. Rotondi and R. Scazzieri (eds.), *Resources, Production and Structural Dynamics*, Cambridge, Cambridge University Press, pp. 107–25.
Steenge, A. (2020) 'From Multipliers to the Distribution of Income: Connecting Leontief and Sraffa', this volume.

Reply

# A Response to Scazzieri's Comment

*Albert E. Steenge*

I have very little to add to Roberto Scazzieri's remarks on my contribution to this volume. The chapter, as pointed out excellently by Scazzieri, tries to find a bridge between the works of the two great contributors to multi-sectoral analysis, Wassily Leontief and Piero Sraffa. Such efforts have been made before, naturally, but a gap has remained, in the sense that none of these efforts has been fully successful in providing an *overall* conceptual framework. The present contribution should be seen as belonging to that same category. However, the premise from which we start is less familiar, since we proceed by endowing the standard model of one of the two, Leontief, with a less well-known attribute, that is the concept of extended input-output (I-O) coefficients. We then proceed to explore the potential of this concept.

First of all, obviously these two directions or 'schools' (if you like) have quite a bit in common. Here we may think of the strict distinction between produced and non-produced goods, and the distinction between economies with a surplus and without a surplus. Nonetheless, the focus of both these schools of thought has been quite different. In Leontief-based research and applied work, the focus has largely been on policy and policy analysis based on multipliers. However, in terms of theoretical underpinnings, quite a few structural aspects still require further study; for instance in cases of multiple primary resources or multiple final destinations for the produced products.

In comparison, the Sraffa-based research direction has a different focus. In addition to structure, much attention is given to the distribution of income between the owners of capital and labour, and the corresponding price movements. Furthermore, Sraffian analysis includes a significant role for standards of value, a concept that is absent in the Leontief world.

A promising research direction here is offered, in my view, by the concept of *extended input coefficients*. Standard input coefficients stand for the amount of produced goods used as intermediate inputs in unit production of produced goods. To these intermediate inputs, extended input coefficients (partly or wholly) add the standardized elements of the consumption bundle that are consumed by the primary non-produced factor—labour—and paid for out of the wage (there are no savings). A shift in this consumption bundle, given some additional conditions, is then interpreted as a shift in income.

As we have shown, these extended input coefficients can straightforwardly be introduced in a Leontief framework. If there is equilibrium in the sense of each industry being in equilibrium, that is being in a state where it sells its goods for the same total amount as it paid for its inputs (including the labour input), all coefficients can be assembled into a coefficients matrix with a number of special properties. In our contribution, this matrix consists of two non-negative matrices to which 'weights' can be attributed. In terms of this structure, shifts in these weights can then be interpreted as shifts in the distribution of income, per industry.

Here we are building on the work of authors such as Quadrio Curzio (1967) or Seton (1977), who utilized the concept of the so-called extended I-O coefficients. How to proceed precisely in special cases is basically a matter of choice, and of selecting a required context. In fact, quite a few questions then can be formulated while still remaining in a Leontief-based methodology. The present contribution just provides the outlines of an early exploration in this area. Formal 'bridges' can be constructed along the way.

A number of challenges definitely remain. Roberto Scazzieri correctly suggests a closer look at the possible role of 'flexible' circular models, thereby partly returning to classical thought. In fact, as he points out, economic theory may suggest several ways of 'closing' the model, each of which is motivated by economic or political necessity. Proceeding along these lines may also generate intriguing further links between economic theory and linear algebra, in particular Perron-Frobenius theory, which are now found mainly in multiplier and growth theory.

## References

Quadrio Curzio, A. (1967) *Rendita e distribuzione in un modello economico pluri-settoriale*, Milano, Giuffrè.

Seton, F. (1977) 'The Question of Ideological Obstacles to Rational Price Setting in Communist Countries', in A. Abouchar (ed.), *The Socialist Price Mechanism*, Durham, NC, Duke University Press.

CHAPTER 13

# Definitions, Assumptions, Propositions and Proofs in Sraffa's *PCMC*

*K. Vela Velupillai*

## 1 Introductory Thoughts

"Sraffa wrote English prose of remarkable purity and elegance, but the subject matter of his published work did not give much scope for the *wit* and *the penetrating observations* on public and private affairs that were the constant delight of his friends." ***Times Obituary***, 6/9/83; italics added.

Sinha (2016, p. xi; italics added)—and emphasized by Harcourt, in his *Foreword*—thinks, and makes an excellent case for, taking into account the

**PCMC**: **Production of Commodities by Means of Commodities—Prelude to a Critique of Economic Theory** by Piero Sraffa (Cambridge University Press, Cambridge, 1960). My 'interpretations' of **PCMC** are based 'on what [is] available in the public domain', to use Geoff Harcourt's felicitous words on p. vii, in the *Foreword* to Sinha (2016). I am greatly indebted to my friend, *Dr. Ragupathy Venkatachalam*, for constructive (sic!) comments on the first draft; he is definitely not responsible for any of the remaining errors.

K. V. Velupillai (✉)
Solna, Sweden



A. Sinha (ed.), *A Reflection on Sraffa's Revolution in Economic Theory*, Palgrave Studies in the History of Economic Thought,
https://doi.org/10.1007/978-3-030-47206-1_13

*philosophical* nature of Sraffa's *PCMC* (in addition, of course, to the central economic theoretical aspect to it):

> The common mistake, that the 'Sraffians' as well as the leading orthodox theorists have made with regard to the nature of Sraffa's contribution, is that he is read within the well-established frame of discourse in economic science *without any concern for the philosophical underpinnings of his contribution to economic theory.* This has led to a profound misunderstanding of the nature of Sraffa's contribution.[1]

I wish to make a case, in this brief chapter, for the *mathematical*[2] nature of *PCMC*. I have been greatly helped by many insightful interpretations by Sinha (ibid) of Sraffa's *oeuvre*, including—in particular—his (i.e., Sraffa's) 'notes' deposited in the Wren Library of Trinity College, Cambridge. Among Sinha's many appropriate characterisations of *PCMC*—always well-founded—one that struck me vividly (at least from the mathematical point of view) was that on p. ix, (italics added):

> [*PCMC*] is less than 100 pages long and reads like an example of *minimalist art in economic prose.*

Nothing superfluous should be in a *precise*[3] mathematical tract and, thus, *PCMC* viewed as a particular mathematical tract, as I shall do so, is *minimalist*, although almost all of its rich contents are *economic.*[4]

For at least the past forty years, even as I have read and re-read *PCMC*, and increased my understanding of various aspects of *constructive mathematics* and its *intutionistic* underpinnings, one constant remains intact: I have always maintained that *PCMC* is a mathematically impeccable text and, in particular all the (explicit) *proofs* of *propositions* in it are *constructive.*

[1] Sinha (2016, *op. cit.*, italics added), substantiates his case by quoting Sen 2003, p. 1240:

> Sraffa's economic contributions ... cannot, in general, be divorced from his *philosophical understanding*

[2] *Not*, perhaps, as good a *philosophical* case for Sraffa, in general, and *PCMC* in particular, as made by Sinha (and Sen; see also McGuiness 2008, pp. 15–16 & 229).

[3] I use 'precise' instead of 'rigorous', mainly because I have *never* read a definition of *rigour* or *rigorous*, in the standard mathematical literature.

[4] *Minimalist* does not imply *minimum*.

On the other hand, I cannot—and do not—claim that I neither (completely) understand *PCMC*, nor do I have a complete grasp of ***constructive mathematics*** and, above all, its ***intuitionistic*** underpinnings.

I have no doubts whatsoever that Sraffa had an admirably complete understanding of the ***assumption*** and ***definitions*** that resulted in the ***propositions*** and their ***proofs***, in *PCMC*. As for ***constructive mathematics*** and its ***intuitionist*** underpinnings, I am not sure there is a 'uniform' view or interpretation, but my own visions—at least as regards enlightened economics are concerned—have been moulded by the writings of Brouwer (1975, 1976, 1981), Bishop (1967, 1970, 1985), Martin-Löf (1984, 1998), Gandy (1959, 1980, 1995), Bridges (1982) and Watson (1938)—among ***many*** others.

This chapter is organized as follows: In the next section I outline my forty-year odyssey with the mathematics, and the mathematicians, of *PCMC*. Assumptions, definitions and propositions, of a more-or-less classical economic vein, are given numerical content and meaning in the strictly mathematical framework of *PCMC*. The propositions are, then, proved, in *PCMC*, I contend, in a strictly constructive way. Some, like Joan Robinson, may call the logic of the proofs in *PCMC* to be in ***economic logic***; I am sure, however that the proofs are ***not*** in terms of ***classical logic***. For various reasons that are outlined, all of them 'circumstantial', the new element in my contention is that the constructive mathematics of *PCMC* is underpinned by intuitionistic type theory.

The 'circumstantial' evidence, as almost always (particularly for a person relying only on publicly available material), is in terms of the mathematics and mathematicians that I document in the section.

*PCMC* is challengingly subtitled, ***Prelude to a Critique of Economic Theory***. It is possible that Sraffa refers by ***Economic Theory*** to some aspect of neoclassical or, perhaps, Austrian ***theory***. Sraffa, in my opinion, was a master of economic theory—in its measurement (theoretical) and returns to scale aspects, in particular (but much else, both micro- and macroeconomics).

Indeed, in the *Preface* to *PCMC* (p. vi, italics added), Sraffa explicitly states:

> It is, however, a peculiar feature of the set of propositions now published that, though they do not enter into any discussion of the marginal theory of value and distributions, they have nevertheless been designed to serve as a basis of a critique of that theory.

Especially because. p. v (italics added):

> In a system in which, day after day, ***production*** continued ***unchanged*** in this respect [i.e. ***change***], the marginal product of a factor (or alternatively the marginal cost of a product) would not merely be hard to find—it just would ***not be there to be found***.

Alas, I do not have Sraffa's mastery of the methodology of marginal economics. Thus, in the concluding section, §3, my reading of ***prelude*** is based on the ***narrow*** non-numerical aspect of every kind of economic theory; *PCMC* was unique in considering the numerical content and meaning of the formal propositions, the assumptions and definitions underlying them, and—above all—the ***effectivity***[5] of the ***proofs*** in the book.

It remains for me to say—even 'confess'—that *PCMC* is an ***algorithmic theory of production***. ***Algorithms*** are intrinsically ***numerical***, in content and meaning. Petty's protestations notwithstanding, or even simple counting of equations and variables in any kind of mathematical economic model, ***economic theory*** is devoid of numerical content or meaning.[6]

Therefore, *PCMC* can be written in terms of computer implementable algorithms, especially—of course—***all*** the definitions, assumptions, propositions and proofs. I subscribe wholeheartedly to Knuth's view that (1996, p. ix: italics added):

> ***Science*** is what we understand well enough to explain to a ***computer***.[7] ***Art*** is everything else we do.

I am convinced that *PCMC* is scientific, but I also feel that science is not a given, immutable, activity, evolving according to Darwinian precepts. It is in this sense that my own ***critique*** is nothing other than a

[5] In the ***intuitive*** sense customary, for example, in constructive and computable mathematics.

[6] Economic theory, of whatever ideology, is—at most—syntactical, hardly semantical. I shall, however, not develop this line of thought further (in this paper). It will take me away from the main task of studying, from an algorithmic point of view, *PCMC*, to consider Jevons, Walras, Fisher, Pareto, Barone, von Mises, von Hayek, Lange, and even Scarf (see, in particular, Sreenivasan 2005, p. 92).

[7] I suspect that Knuth confines the notion of computers to the ***digital*** kind; I am more eclectic, allowing the notion to refer to ***analog*** and hybrid computers, as well (especially with *PCMC* in mind).

***prelude***, but in the full meaning of *PCMC* as an uncompromising scientific treatise.

Any attempt to 'shackle' *PCMC*, within any theoretical framework, such as the Neo-Ricardian or the kind of mathematics of the theory of non-negative square—or even non-square—matrices, goes against the tenets of the subtitle of *PCMC*. I do not think *PCMC* tries to resurrect, or make a case for, a revival of any kind of economics—classical, Ricardian, Marxian, Keynesian, Neoclassical or other hybrid versions.

For example, neither the classical theory of production, whatever the period it refers to—but particularly the so-called long-period theory of production—nor the activity analysis model of production are ***algorithmic*** theories. Knuth (1985) is particularly illuminating on constructive mathematical algorithmic thinking as distinct from classical, non-algorithmic, mathematical thinking.

Moreover, any kind of ***algorithmic thinking***, particularly mathematically constructive or computable, is ***non-unique*** (Gurevich 2011); even a simple consideration of ***non-deterministic Turing Machines*** is enough to dispel this misconception and associated with this is the concept of *Oracle*—or *relative*—*computation*.

## 2 Some Notes on *Mathematics* & *Mathematicians* in *PCMC*

> As the nineteenth century began, virtually all mathematical research was of a concrete, ***constructive***, ***algorithmic*** character. By the end of the nineteenth century much abstract, ***non-constructive***, ***non-algorithmic*** mathematics was under development. What happened, how did it happen, and why? *Metakides & Nerode* (1982, p, 319; italics added)

I was convinced, for over forty years, that *PCMC* was an intuitionistic constructive mathematical text—in addition to being intensional (rather than extensional[8]), dawned upon me somewhat later—was simple. Sraffa, in *PCMC*, did not base his mathematics on axioms (or the axiomatic system); he, before the time was apposite, used just computation and infer-

[8] I mention this aspect only because Alister Watson refers to Frank Ramsey as adhering to *extensions* (Watson 1938, p. 444), who are—according to Sraffa, in *PCMC*, pp. vi-vii—two of three mathematicians, the third being Besicovitch, to whom he is indebted for 'invaluable mathematical help.' I shall have more to say about these three mathematicians, in the context of my *PCMC* thesis on intuitionistic constructive mathematics.

ence rules to prove the propositions in *PCMC*, based on definitions and assumptions of an economic nature.

In particular, Sraffa did *not* use, in deriving any of the *propositions*, from the definitions and assumptions, and *proving* them in *PCMC*:

- *reductio ad absurdum* method;
- *tertium non datur* (law of the excluded middle);
- *law of double negation*;
- any notion of the *axiom of choice* (or Zorn's lemma)—any axioms or axiomatic systems, at all (infinity, reducibility, etc.);
- the concept or notion of *theorem*;
- the *well-ordering principle;*
- any version of the *Bolzano-Weierstrass Theorem*[9];
- any notion of the *Heine-Borel theorem* or (its mathematical generaisation as) *compactness*[10];
- the *Perron-Frobenius theorem* on non-negative square matrices;
- any notion of the *Brouwer* or *Kakutani fixed point theorems*;
- any notion of the *Hahn-Banach*, or *separation*, theorems;
- any notion of the formal definition of *relations* (total, partial or equivalent);

Sraffa, especially in *PCMC*, never proves any proposition using the method of *reductio ad absurdum*. By contrast, it is widely used in classical mathematical and mathematical economics proofs. For example, the Frobenius theorem is proved, in classical (non-constructive) mathematics, using this method in the widely quoted volume II of Gantmacher (1960)—see, p. 56 and for, mathematically competent economists, in Nikaido (1970, p. 114), or Arrow and Hahn (1971, Mathematical Appendix A).

[9] I suspect that Sraffa knew of *Specker sequences*, recursive (un)decidability, undecidable disjunctions, the distinction between Cauchy's and Dedekind's definitions of real numbers—and, in general, Gödel's results on the trade-off between truth and provability, Turing's on the *Halting Problem for Turing Machines*, and the *Goodstein sequences*. However, his adherence to intuitionistic constructive mathematics, in *PCMC*, does *not* depend on any familiarity with these things.

[10] On the 20th of December, 1980, Sraffa wrote in my copy of *PCMC* (italics added):

> This book has the advantage of being *compact*.

Obviously, Sraffa does not mean compact in any mathematical sense!

It is equally obvious that *all* the mathematical economics literature of post-1930 era is replete with use and invoking of some form of the ***axiom of choice***, ***Zorn's Lemma*** and the ***well-ordering principle*** in their formulations of so-called theorems and proofs of them. Similarly, because almost all of the mathematical economics literature, whatever the ideology or school of thought—neo-Ricardian, neoclassical, Keynesian, post-Keynesian, new classical, Marxian, or whatever—invokes the ***tertium non datur***, especially in existence proofs, jointly with any form of the axiom of choice, their use is deeply anti-constructive (of any kind—especially the intuitionistic variant of Brouwer). Bishop's interpretation of the ***tertium non datur*** echoes Brouwer's fundamental objection to its use in non-finitary mathematics, but the former refers to it as the ***principle of omniscience*** (Bishop 1967, p. 9; italics in the original):

> This principle, which we shall call the *principle of omniscience*, lies at the root of most of the unconstructivities of classical mathematics. This is already true of the principle of omniscience in its simplest form: if $\{n_k\}$ is a sequence of integers, then either $n_k = 0$ for some $k$ or $n_k \neq 0$ for all $k$. We shall call this the *limited principle of omniscience*. Theorem after theorem of classical mathematics depends in an essential way on the limited principle of omniscience, and is therefore not valid constructively. …
>
> …
>
> Applications of the axiom of choice in classical mathematics either are irrelevant or are combined with a sweeping appeal to the principle of omniscience. The axiom of choice is used to extract elements from equivalence classes where they never should have been put in the first place. …; there is no need to drag in the equivalence classes. The proof that the real numbers can be well ordered is an instance of a proof of the sweeping use of the principle of omniscience is combined with an appeal to the axiom of choice. Such proofs offer little hope of constructivization.

I have quoted Bishop extensively because it was his enunciation of the principle omniscience and the axiom of choice, much more than Brouwer's intuitionistic constructivisation, that convinced me of the algorithmic, numerical, structure of *PCMC*—all those decades ago; more recently persuaded me that programming in Martin-Löf's (intuitionistic) type theory (Bauer 2017) just reinforced an earlier conviction.

Bishop, between the two quoted paragraphs, reinforces all the other points I have enumerated in the above list. However, I must confess that, till very recently, I was wont to use the excellent reasons given in Brouwer

1952, in being sceptical of the role of Sperner's Lemma in the proof of the Bolzano-Weierstrass theorem. I was also an early convert to Russian Constructivism, and as such the validity of the Specker sequences and the invalidity of the Heine-Borel theorem (compactness) was something I noted *PCMC* observed, although implicitly (Aberth 1980, § 13. 2).

*PCMC* is a *system*[11] *of propositions*, based on *definitions* and *assumptions* on the economic system, and *constructively proved* to exist by *rules of inference* and *computation rules* (and 'classic' considerations of 'thought experiments' of the Galileo or Einstein-Bohr kinds). The Sraffa *oeuvre* is noted for its concentration on the *deterministic laws* of economic propositions, and impeccable *constructive proofs*. *Laws*—for example, *laws of returns* (Sraffa 1926)—are 'contingent' and are similar, in engineering mathematics, to Newton's (mechanics), Carnot's (thermodynamics) or Maxwell's (electromagnetic) laws; or, if pure or applicable mathematics, the similarity is to Post's alternative to Church's conception of a *thesis* (Post 1936, p. 105).

As examples of these notions, primarily in (the paragraphs and pages of) *PCMC*, the following are almost paradigmatic:

- Definition: *viable economic system* (footnote 1, on p. 5);
- Assumption: 'wage is paid *post factum* as a share of the annual product' (§9, p. 10);
- Proposition: Construction of The Standard Commodity, as a 'balanced ratio' of deficit and surplus industries (§21, & chapter IV);
- *Constructive existence proofs*[12]: the tractor example of §81 (p. 68);
- Rules of inference: Elimination of non-basics in joint production systems (§61);
- Computation rules: On constructing 'sub-systems' (Appendix A);
- Thought experiment: as an 'imaginary experiment', §37, p. 26.

Sraffa, in *PCMC*, *never* uses *axioms*, the *axiomatic system*, *set membership relations* or even *sets*. I was tempted to interpret, at least the proofs,

[11] The notion of *system* in *PCMC* is clarified in Sraffa's interchange with Newman, reprinted in Bharadwaj (1970).

[12] The example of a mathematical economist of classical persuasions talking at cross purposes with a constructivist mathematical economists is illustrated in the above discussion between Newman and Sraffa; when the latter asks the former to *prove* (p. 426, in Bharadwaj, *op.cit.*), it is in the sense of *constructive existence proofs* that is meant (I think). Consequently, I don't think Newman understood Sraffa's admonishment at all.

the mathematical framework of *PCMC* in terms of *category theory*, especially because the set membership relation, ∈ (or ε) is superfluous in it (see Lawverre 1964, p. 1506), but I opted for Martin-Löf *type theory* (*intuitionistic type theory*[13]) due to the fact that one does not need to appeal to axioms in the theory. Martin-Löf (1998, p. 128; italics added) explicitly states:

> "I shall use the *notation*
>
> $$a \in A$$
>
> to express that
>
> *a* is an object of *type A*."

And (*ibid*; second & third italics, added):

> "A *proposition* is defined by prescribing *how* we are allowed to *prove* it."

Therefore, programming *PCMC* in Martin-Löf's *intuitionistic type theory*, makes it eminently *constructive*, *mathematically*[14] (as I have maintained at least since 1979, in the publicly available literature).

There is, therefore, a serious difference between intuitionistic, constructive, type theory of the Martin-Löf variety, with which *PCMC* can be analysed, and Sraffa's nihilism towards any kind of *set theory* (axioms, axiomatic systems, theorems and proofs). Just as an example:

$$X :- \{n \in N \mid \forall x, y, z, x^n + y^n \neq z^n\}$$

depicts a *set*, but *not* an intuitionistic *type*.

[13] It is Brouwer's *intuitionism* and Russell's *type theory*, along with Bishop's *constructive mathematics* that was the motivation for Martin-Löf's inspiration. Ramsey's revised type theory, developing the truth-table scheme of the early Wittgenstein, was decisive in the second edition of the *Principia Mathematica* by Whitehead and Russell. Russell's original motivation for introducing types in his logicism had been due to the logical paradox of too many or too large sets in Frege's naïve conception of sets and the discovery of the Burali-Forti paradox (see Russell 1937; van Heijenoort (edited), 2000, especially pp. 124–128). Brouwer's introduction of intuitionism, in opposition to Hilbert's formalism, was (partly, at least) due to what I think is Hilbert's equally naïve introduction of formalism, on the basis of his axiomatization of geometry. Sraffa did not subscribe to axiomatic formalism or any kind of logicism, in *PCMC*. Most—if not all—of the formalisers of mathematical economics, are followers of Hilbert.

[14] However, Nordström et al. (1990, chapter 2)), which I use for programming *PCMC* in Martin-Löf's intuitionistic type theory interpret the rule about proofs of propositions as one of many possibilities.

I conclude this section with some observations, along an intuitionistic constructive mathematical vision, very broadly conceived, given the indebtedness explicitly expressed in *PCMC* to the three mathematicians, Besicovitch, Ramsey and Watson, and the 'famous' acknowledgement to Sraffa expressed by Wittgenstein (1967, p. x/x$^{e}$).

The considerable danger of trivial 'repetition', let me quote *PCMC* on Sraffa's explicit acknowledgement to three mathematicians, and mathematics (pp. vi–vii; italics added):

> My greatest debt is to Professor *A. S. Besicovitch* for invaluable *mathematical* help over many years. I am also indebted for similar help at different periods to the late *Mr Frank Ramsey* and to *Mr Alister Watson*. It will be only too obvious that I have not always followed the expert advice that was given to me—particularly with regard to *the notation adopted*,[15] which I have insisted on retaining (although given to objections in some respects) as being easy to follow for the *non-mathematical* reader.

I'll discuss the three mathematicians in the reverse order in which they appear in *PCMC*. At the very outset, I would like to point out that the interpretation of the role of these mathematicians,[16] and mathematics, is on the basis of my intuitionistic constructive vision of the underlying structure of *PCMC*.

First of all, Watson's influential—at least in the Cambridge of the late 1930s and early 1940s—1938 *Mind* contribution was written as a contribution to an elucidation of Hilbert's *Entscheidungsproblem*, from the point of view of Turing's solution to it, with a Wittgensteinian intuitionistic foundations of mathematics vision (especially, p. 445). It is a vision to which Watson remained faithful all through his life, in particular during the time he helped Sraffa with his solutions to economically formalized problems.[17] Secondly, although intuitionistic constructive mathematics is

[15] If Sraffa, in *PCMC*, had 'adopted' matrix notation, the obvious claim by all adherents of classical mathematics, in economics (irrespective, again, of ideology), would have been to question why the *Perron-Frobenius theorem* had not been invoked—as if such a claim was not made by all and sundry, anyway! Note, also, that Leontief (1941) wrote out the equations—in 'long-hand'!—of inter-industry analysis without any appeal to matrix algebra.

[16] For these purposes, I add Wittgenstein and Turing to this list of three mathematicians, who were decisive in influencing the intuitionistic constructive mathematics of *PCMC*.

[17] Although I have the utmost respect for the Wittgenstein publications of McGuiness, I class the following observation about Alister Watson (McGuiness 2008, p. 8) in the same dubious class as the one about Sraffa (McGuiness 2002, p. x; italics added):

not explicitly discussed by Watson, in the way Russell's theory of types and the axioms in the *Principia Mathematica* are handled critically, makes it impossible to imagine that Sraffa was not aware of the content of the chapter. Thirdly, it is hardly conceivable that Watson, and Sraffa, were unaware of Wittgenstein's awareness of Brouwer's intuitionistic constructive stance; after all, Brouwer's two Vienna lectures on Intuitionism were instrumental in Wittgenstein's 'return' to formal philosophy.

Frank Ramsey, whose life was tragically cut short just before the age 27, in January, 1930, in the opinion of his friend (and editor), Braithwaite (1931) 'was converted to a *finitist*[18] view' of the philosophy of the foundations of mathematics. This is convincingly argued by Majer (1989), partly based in Ramsey's essays 53 & 54, in Galavotti (1991).[19] Ramsey, too, like his equally illustrious King's College successor, Alan Turing, worked on Hilbert's *Entscheidungsproblem* (Ramsey 1928), in the years he worked with Sraffa (who was, at this time, a member of King's College). Of course, the role of Ramsey in persuading Wittgenstein to return to academic philosophy, is only too well known. Ramsey may have helped Sraffa with finding explicit, constructive, solutions to economically formulated problems, but I contend that they would have been intuitively algorithmic (in the Turing sense), in an intuitionistic way. I think Watson remains correct in attributing to Ramsey the *extensional* logical view, whereas I hold that Sraffa was always an adherent of *intensional* logic. But to be a 'convert to intuitionism' meant it was only a question of time before Ramsey was also converted to *intensional* logic.

> Later Alister Watson, perhaps the only one of the charmed circle of young Apostles to remain with him, contributed in somewhat the same way presentations of Wittgensteinian ideas on the foundations (or lack of foundations) of mathematics.

Sraffa, whom Wittgenstein regarded as his severest critic, spent much of his time during the years of his closest association with the philosopher on the writing and rewriting of a devastating review of a *paper by Hayek* (who *survived cheerfully* enough).

First of all, it was a review of a *book* by Hayek; secondly, till at least 1974, Hayek did *not* 'survive cheerfully'—in fact he was a *bitter* man!

[18] See also Russell (1931, p. 477). Till Heyting and Kolmogorov in the early 1930s, eminent English mathematicians referred to intuitionists as finitists—as, for example, in Hardy (1929), p. 5: 'the finitists or intuitionists'.

[19] There is a curious error in Galavotti (*op.cit.*), where the references to Majer in footnote 37, on p. 22 and footnote 1, on p. 197, are to two different journals! It is, obviously, a trivial error.

That brings me to Besicovitch.[20] I shall concentrate on one of his many influential writings, which spans his interest in the kind of ***constructive mathematics*** that is of relevance here. More than a hundred years ago, in 1917, the Japanese mathematician, Soichi Kakeya posed the following problem (Besicovitch 1963,[21] p. 697; italics in the original):

> In the class of figures in which a segment of length 1 can be turned around through 360°, remaining always within the figure, which one has the smallest area?

In 1920, Besicovitch (ibid.; italics added), 'considered the [twin] problem':

> Given a function of two variables, ***Riemann-integrable*** on a plane domain, does there ***always exist*** a pair of mutually perpendicular directions such that the repeated ***simple integration*** along the two directions ***exists*** and gives the value of the integral over the domain?

Four remarks are warranted, here:

- First of all, they are both problems of ***computational geometry*** and the figures can be drawn on the plane with a straight edge and compass;
- The indicated integrations are both ***computable*** and ***constructive*** processes;
- ***Continuity*** requirements entail, in the case of the Besicovitch solution, '***Pál joins***',[22] very similar to the one-sided oscillator of nonlinear two-dimensional business cycle theory;
- The computed areas, or the calculated integrals, are ***approximate***[23]—although they can be ***exactly approximated***;

[20] Gibson (2010), p. 75, observes:

> Note that one of the Examiners was *A. S. Besicovitch*—Fellow of Trinity, advisor to Sraffa—***Wittgenstein sparring partner***, ...

'Sparring partner'—of philosophy, mathematics, philosophy of mathematics, or?

[21] The Mathematical Association of America, using a grant obtained from the National Science Foundation, established a Committee on Production of Films; the fourth of the films produced by this Committee, was the Besicovitch lecture on the geometric construction involved in his 1920s initial solution of the Kakeya problem (see Besicovitch, *op.cit.*, p. 697).

[22] See Pál (1921). I may mention that Pál developed Matsusaburo Fujiwara's formulation of Soichi Kakeya's problem.

[23] Besicovitch (op. cit., p. 697) gives the area of the circle answering the Kakeya question as *exactly* equal to.78 ($=\pi r^2$), whereas it can only be exactly approximated. Similarly, the area

Now, the computation and inference rules of *PCMC* result in *exact real values* that are also continuous.[24] So, even though Besicovitch's notation is not for approximation of areas, to solve the Kakeya problem, the notation is simply for convenience! For example, the general quintic cannot be solved exactly by radicals.[25] Besicovitch's computational geometry resulted in exact geometric figures, which were approximate answers to (various forms of) the Kakeya problem. For example, again, in Besicovitch (1963, pp. 705 & 698) he ended by wondering whether the Kakeya problem can be solved for simply connected regions, having quoted George Birkhoff bracketing Guthrie's four colour theorem with the Kakeya set. All papers, expository and technical, are replete with the preoccupation with geometric *constructions*. It is only a 'mild' step towards analytic constructive mathematics and computable constructions, of an approximate nature, based on exact real geometry of the Besicovitch kind, which was eminently consistent with the methods of deriving propositions, and proving them, in *PCMC*.

I can now, tentatively, draw the threads together. The Cambridge mathematicians, Besicovitch, Ramsey and Watson, together with Wittgenstein, were all close to Sraffa and he, in turn, utilized their expertise in solving the exact arithmetic problems that *PCMC* gave rise to. Turing was at King's College and Russell had been at Trinity College, both Cambridge Colleges with which Sraffa had deep associations.[26] Wittgenstein was Besicovitch's 'sparring partner', was also a foil for Turing, introduced by Watson, and Ramsey re-did the set theory and axioms of Principia Mathematica on the basis of truth table technique of the Tractatus, and a development of simple—that is, non-intuitionistic or non-finitist—type theory.

The publicly available literature, the only ones that I am familiar with, records that Sraffa was well acquainted with Besicovitch, Ramsey, Watson

of the equilateral triangle, with side length $4/\sqrt{3}$ is ≈ 0.58, not—as Besicovitch writes—exactly so.

[24] I was greatly helped, in this interpretation of *PCMC*, by a reading of Geuvers et al. (2007).

[25] Niels Henrik Abel's famous late 1870s result, which I conjecture one or the other of the many mathematicians around Sraffa, would have appraised him, or—versatile as he was—he may have independent knowledge of this significant exact unsolvability.

[26] In late 1979, when Sraffa walked with me in the 'backs', he stopped near King's College, and said:

> You know, I used to be a member of this College; but after the 'Book" [he was referring to Keynes' General Theory], it was impossible to live with Kahn! Robertson took me as a Fellow to Trinity.

and Wittgenstein[27]; I conjecture that it is eminently reasonable—also because Sraffa was very well acquainted with Maynard Keynes—that Sraffa was familiar with Russell and Turing. When combined with Russell's letter to Herbert Simon of 2nd November, 1956 (Simon 1996, p. 208):

> I am quite willing to believe that everything in deductive logic can be done by a machine.

It is quite natural for me to formulate the following conjecture (an 'early' formulation of it was given in the first section):

*Conjecture I*
*Every proposition and proof in PCMC is constructive.*

*Conjecture II*
*Every proposition, and therefore proof, in PCMC is programmable for machine implementation, using intuitionistic type theory.*

*Remark 1*
The *definitions* in *PCMC can* be considered *axioms* in the classical sense; they delineate *properties* of the system being studied.

*Remark 2*
Thus, *PCMC can* be considered an *axiomatic analysis* of an algorithmic theory of production—but not studied from any *topological* standpoint.

*Remark 3*
I refrain from doing so, because it is against the spirit and method of *PCMC*. This is partly, because I have not, as yet, formulated *informal rigour* from an *intuitive* point of view.

[27] If not for the difficulties of a personal relationship with Wittgenstein, I am sure Sraffa would have added the eminent philosopher's name to the three mathematicians he listed as being indebted to, for the results in *PCMC*. See Wittgenstein's mid-to-late 1940s letters to Sraffa republished in McGuiness (2008), in particular the sentence overheard by Smythies, of Sraffa saying to Wittgenstein (*op.cit.*, p. 339):

> I won't be bullied by you, Wittgenstein.

Incidentally, Sraffa and Wittgenstein were admitted to Trinity fellowships, on the same day, in 1938 (McGuiness, ibid., p. 288).

However, I am prepared to state as;

*Conjecture III*
*All rigour* is *informal*, defined on the basis of *intuitive mathematics*.[28]

*Remark 4*
*Conjectures* I and II can be *proved*, in the *constructive* sense; but it requires that one can enumerate *all* the *propositions* in *PCMC*. I am not sure, after about 45 years of 'struggling' with the understanding of a terse, aphoristic, *PCMC*, I have succeeded in identifying *all* of the propositions in the book. For the more 'standard' propositions, *existence* of the standard commodity, the 'tractor' example for fixed capital, the reduction to dated quantities of labour using the device of the 'subsystem', etc., it is easily (sic!) done (using Martin-Löf's identification, in intuitionistic type theory, between propositions and proofs).

*Remark 5*
The *algorithms* that *program* the *proof* of the propositions of *PCMC* (cf. *Conjecture* II) are the *finite initial segments* of (a possible) infinite *lawless computation*. This makes the algorithms *lawlike sequences* of *Brouwer's choice sequences*, and therefore, intrinsically, part of *intuitionistic constructive mathematics*. This is, of course, because algorithms are *finite* objects, since the programs that define them are of *finite* length. However, I do not think the intuitionistic type theory of Martin-Löf includes the Browuerian ideas of *lawless choice sequences*, *species*, *spreads* and the like, of the *constructive intuitive continuum*.[29]

The two first conjectures are the reasons for my claim that *PCMC* is an *algorithmic theory of production*. Why I have supplemented my initial idea of *PCMC* being constructive, even intuitionistically constructive, with *type theory* has a great deal to do with Bishop (1967), the review of the Bishop and Bridges (1985) by Mines (1988) and, of course Martin-Löf (1998). First of all, there was Bishop's observation (*op.cit.*, p. 360; italics added):

> Various mathematicians have devised methods for classifying (classical) real numbers according to their logical complexity—we mention *Russell's theory*

[28] Remark 3 and Conjecture III are inspired by Kreisel (1967).
[29] See the first footnote in Bickford et al. (2018).

> *of types* ... and the work of various modern logicians. Echoes of such *type-theoretic classifications* appear in *constructive analysis.*

Then, Mines noted (*op.cit.*, pp. 162–3; italics added):

> Many people have expressed the view that there should be close ties between the *theory of algorithms* and *constructive mathematics.* Perhaps *because algorithms are innately constructive.* ... Times have changed. ... [T]here are now pure existence proofs for computer programs that solve real problems and no hope of ever programming them.

Martin-Löf's intuitionistic type theory resurrected hope of being able to write programs for 'real-life' computers for *every proposition* and *proof* in *PCMC.*

## 3 Tentative Concluding Notes on the *Prelude* that is *PCMC*

> It is no wonder that [*PCMC]* took a long time to write. It will not be read quickly. Addicts of pure *economic logic* who find their craving ill satisfied by the wishy-washy products peddled in contemporary journals have here a double-distilled elixir that they can enjoy, drop by drop, for many a day.
>
> For some, indeed, *the logic* may be too pure. ...
>
> ...
>
> Presumably, it will be a little time before *the critique* to which this is *the prelude* will be published. (Robinson 1961, pp. 53, 58; italics added)
>
> [T]he *logic* which is developed [in Bourbaki] is, right from the start, completely *classical.* (Indeed, in the historical note—printed in a later volume—it is asserted that *Intuitionism* will subsist only as an historical curiosity). ... It is possible then that this book may itself soon have only historical interest. (Gandy 1959, p. 72; italics added)

I am not convinced that there is something identifiable called 'pure economic logic'; nor do I think any Bourbaki volume was even updated as to what was the then 'classical logic'. Moreover, I am of the belief, entirely substantiable, that any work by Bourbaki would not even have historical interest. Only Russell (of all those mentioned in the previous section), and that, too, the very early one, believed in an immutable logic to which any given mathematics can be 'reduced'.

None of the constructivists, whether they were adherents of intuitionism, any kind of type theory or even Hilbertian formalists, believed in a logic to which mathematics could be reduced. Of course, this was true of Brouwer, Heyting, Bishop and Martin-Löf; Kolmogorov used logic instrumentally.

I don't think Sraffa, certainly not in *PCMC*, believed in any immutable logic—particularly that to which a mathematics could be reduced. Ramsey, of the three for whom Sraffa was indebted for mathematical help, was alone in believing in a given logic, but he, too, 'was converted to a finitist view'—remembering, always, that it meant intuitionistic view. I will be wrong, but I am tempted to say, that there was much in common between Ramsey and Kolmogorov.

Brouwer, on the other hand, was convinced that mathematics was an autonomous discipline and he did not subscribe to any axioms or axiomatic formulations. His own proofs were based on inference and computation rules derived from the practice of mathematical intuitionism. He developed choice sequences and a system of ***counter-examples***, the latter depending on his deep and penetrating knowledge of mathematics, as a whole. He also construed the idea of the creative mathematician.

In all of these Brouwerian aspects, Sraffa was also a past master; I interpreted the method of *PCMC*, particularly the proofs of propositions, in a Brouwerian mode. But I now think that Sraffa was more than a follower, a defender or a practitioner of Brouwerian mathematics. This was why I felt able to suggest, in the form of provable conjectures, the two propositions of the previous section, that *PCMC* is an algorithmic theory of production. The discipline was in the use of inference and computation rules in proving the stated propositions.

Computation is, I am convinced, anything subject to the discipline of the Turing-Post theory of computability. It will take me too far off the subject matter of this particular approach to the mathematics of *PCMC* but I can do no better than refer the interested reader to Soare (2016) and Dowek (2015).

It is in extracting Sraffa's use of inference rules, in proofs of propositions, that I find some innovative possibilities. By constraining these inference rules to an algorithmic theory of production, I hope I can find the necessary 'trade-off' between the discipline provided by computability theory, for computation rules, and those that can be extracted by Sraffa's proofs in *PCMC*.

It is in this way that I interpret the *prelude to a critique of economic theory* that is *PCMC*. I am convinced that with *PCMC*, interpreted as an *algorithmic theory of production, economics enters a new age.*

## References


Aberth, Oliver (1980), *Computable Analysis*, McGraw-Hill International Book Company, New York.

Arrow, K. and F. Hahn (1971), *General Competitive Analysis.* Holden-Day, INC, San Francisco.

Bauer, Andrej (2017), *Five Stages of Accepting Constructive Mathematics, Bulletin (New Series) of the American Mathematical Society*, Vol. 54, Number 3, July, pp. 481–498.

Besicovitch, A. S. (1963), *The Kakeya Problem*, The American Mathematical Monthly, Vol. 70, No. 7, August-September, pp. 697–706

Bharadwaj, Krishna (1970), *On the Maximum Number of Switches Between Two Production Systems, Schweizerische Zeitschrift für Volkswirtschaft und Statistik*, Vol. 106, No. 4, December, pp. 409–429.

Bickford, Mark, Liron Cohen, Robert L. Constable & Vincent Rahli (2018), *Computability Beyond Church-Turing via Choice Sequences*, pp. 245–254, in: *LICS '18 33rd Annual ACM/IEEE Symposium on Logic in Computer Science*, ACM New York, NY.

Bishop, Errett (1967), *Foundations of Constructive Analysis*, McGraw-Hill Book Company, New York.

Bishop, Errett (1970), *Mathematics as a Numerical Language*, pp. 53–71, in: *Intuitionism and Proof Theory, Proceedings of the Summer Conference at Buffalo N.Y. 1968*, edited by A. King, J. Myhill and R. E. Vesley, North-Holland Publishing Company, Amsterdam.

Bishop, Errett (1985), *Errett Bishop: Reflection on Him and His Research*, edited by Murray Rosenblatt, *Contemporary Mathematics*, Vol. 39, American Mathematical Society, Providence, RI.

Bishop, Errett & Douglas Bridges (1985), **Constructive Analysis**, Springer-Verlag, Berlin.

Braithwaite, R. B. (1931, editor), *The Foundations of Mathematics—and other Logical Essays* by Frank Plumpton Ramsey, *Preface* by G E Moore, Routledge, London.

Bridges, Douglas S. (1982), *Recent Progress in Constructive Approximation Theory.* pp. 41–50, in: *The LEJ Brouwer Centenary Symposium*, edited by A. S. Troelstra and D van Dalen, North-Holland Publishing Company, Amsterdam.

Brouwer, L. E. J. (1952), *An Intuitionist Correction of the Fixed-Point Theorem on the Sphere, Proceedings of the Royal Society of London, Series A*, Vol. 213, No. 1112, June, pp. 1–2.

Brouwer, L. E. J. (1975), *Collected Works 1: Philosophy and Foundations of Mathematics*, edited by A. Heyting, North-Holland Publishing Company, Amsterdam.

Brouwer, L. E. J. (1976), *Collected Works 2: Geometry, Analysis, Topology and Mechanics*, edited by H. Freudenthal, North-Holland Publishing Company, Amsterdam.

Brouwer, L. E. J. (1981), *Brouwer's Cambridge Lectures on Intuitionism*. Edited by D. van Dalen, Cambridge University Press, Cambridge.

Dowek, Gilles (2015), *Computation, Proof, Machine: Mathematics Enters a New Age*, Cambridge University Press, Cambridge.

Galavotti, Maria Carla (1991; editor), *Notes on Philosophy, Probability and Mathematics*, Bibliopolis, Napoli.

Gandy, R. O. (1959), *Review* of *Théorie des ensembles*, *Journal of Symbolic Logic*, Vol. 24, No. 1, March, pp. 71–73.

Gandy, Robin (1980), *Church's Thesis and Principles for Mechanism*, pp. 123–148, in: *The Kleene Symposium* edited by J. Barwise, H. J. Keisler and K. Kunen, North-Holland Publishing Company, Amsterdam.

Gandy, Robin (1995), *The Confluence of Ideas in 1936*, pp. 51–102, in: *The Universal Turing Machine: A Half-Century Survey* (second edition), edited by Rolf Herken, Springer-Verlag, Wien.

Gantmacher, F. R. (1960), *The Theory of Matrices, Volume Two*, Chelsea Publishing Company, New York, NY.

Geuvers, Herman, Milad Niqui, Bas Spitters & Freek Wiedijk (2007), *Constructive Analysis, Types and Exact Real Numbers, Mathematical Structures in Computer Science*, Vol. 17, Issue 1, February, pp. 3–36.

Gibson, Arthur (2010), *The Wittgenstein Archive of Francis Skinner*, *Wittgenstein After his Nachless*, edited by Nuno Venturinha, Palgrave Macmillan, Hampshire, UK.

Gurevich, Yuri (2011), *What is an Algorithm?*, Technical Report MSR-TR-2011-116, Microsoft Research, July, Microsoft Corporation, WA.

Hardy, G. H. (1929), *Mathematical Proof*, *Mind*, *New Series*, Vol. 38, No. 149, January, pp. 1–25.

van Heijenoort, Jan (2000), *From Frege to Gödel: A Source Book in Mathematical Logic, 1879–1931*, toExcel, NE.

Knuth, Donald E. (1985), *Algorithmic Thinking and Mathematical Thinking*, *American Mathematical Monthly*, Vol. 92, No. 3, March, pp. 170–181.

Knuth, Donald E. (1996), *Foreward* to: *A = B*, by Marko Petkovšek, Herbert S. Wilf & Doron Zeilberger, A K Peters, Wellesley, MA.

Kreisel, Georg (1967), *Informal Rigour and Completeness Proofs*, pp. 138–186, in: *Problems in the Philosophy of Mathematics*, edited by Imre Lakatos, North-Holland, Amsterdam.

Lawverre, F. William (1964), *An Elementary Theory of the Category of Sets, Proceedings of the National Academy of Science of the United States of America*, Vol. 52, No. 6, December, 15, pp. 1506–1511.

Leontief, Wassily W. (1941), *The Structure of the American Economy, 1919–1929*, Harvard University Press, Cambridge, MA.

Majer, Ulrich (1989), *Ramsey's Conception of Theories: An Intuitionistic Approach, History of Philosophy Quarterly*, Vol. 6, No. 2, April, pp. 233–258.

Martin-Löf, Per (1984), *Intuitionistic Type Theory—Notes by Giovanni Sambin of a series of lectures given in Padua, June 1980*, Bibliopolis, Napoli.

Martin-Löf, Per (1998), *An Intuitionistic Theory of Types*, chapter 8, pp. 127–172, in: *Twenty-five Years of Constructive Type Theory—Proceedings of a Congress Held in Venice, October1995* edited by Giovanni Sambin and Jan M. Smith, Clarendon Press, Oxford.

McGuiness, Brian (2002), *Approaches to Wittgenstein: Collected Papers*, Routledge, London.

McGuiness, Brian (2008, edited), *Wittgenstein in Cambridge: Letters and Documents 1911–1951*, Blackwell Publishing, Oxford.

Metakides, George & Anil Nerode (1982), *The Introduction of Non-Recursive Methods Into Mathematics*, pp. 319–335, in: *The LEJ Brouwer Centenary Symposium*, edited by A. S. Troelstra and D van Dalen, North-Holland Publishing Company, Amsterdam.

Mines, Ray (1988), *Review* of Bishop & Bridges, 1985a, *The American Mathematical Monthly*, Vol. 95, No. 2, February, pp. 159–163.

Nikaido, Hukukane (1970), Introduction to Sets and Mappings in Modern Economics, translated by Kazuo Sato, North-Holland Publishing Company, Amsterdam.

Nordströn, Bengt, Kent Petersson & Jan M. Smith (1990), *Programming in Martin-Löf's Type Theory*, Clarendon Press, Oxford.

Pál, Julius (1921), *Ein Minimumproblem für Ovale, Mathematische Annalen*, Vol. 83, pp. 311–319.

Post, Emil L. (1936), *Finite Combinatory Processes—Formulation 1, The Journal of Symbolic Logic*, Vol. 1, No. 3, September, pp. 103–105.

Ramsey, F. P. (1928), *On a Problem of Formal Logic, Proceedings of the London Mathematical Society, Series 2*, Vol. 30, Part 4, pp. 338–384.

Robinson, Joan (1961), *Prelude to a Critique of Economic Theory, Oxford Economic Papers, New Series*, Vol. 13, No. 1, February, pp. 53–58.

Russell, Bertrand (1931), *Review* of *The Foundations of Mathematics and other Logical Essays.* by Frank Plumpton Ramsey, R. B. Braithwaite and G E Moore, Vol. 40, No. 160, October, pp. 476–482.

Russell, Bertrand (1937), *The Principles of Mathematics* (second edition), George Allen & Unwin Ltd., London.

Sen, Amartya (2003), *Sraffa, Wittgenstein and Gramsci, Journal of Economic Literature*, Vol. 41, No. 4, December, pp. 1240–1255.
Simon, Herbert A. (1996), *Models of My Life*, The MIT Press, Cambridge, MA.
Sinha, Ajit (2016), *A Revolution in Economic Theory—The Economics of Piero Sraffa*, Palgrave Macmillan, Springer International Publishing AG, Switzerland.
Soare, Robert I. (2016), *Turing Computability: Theory and Applications*, Springer-Verlag, Berlin.
Sraffa, Piero (1926), *The Laws of Returns under Competitive Conditions, The Economic Journal*, Vol. 36, No. 144, December, pp. 535–550.
Sraffa, Piero (1960), *Production of Commodities by Means of Commodities—Prelude to a Critique of Economic Theory*, Cambridge University Press, Cambridge.
Sreenivasan, K. R. (2005), *Comment on William C. Brainard and Herbert E. Scarf's 'How to Compute Equilibrium Prices in 1891'*, pp. 89–92, in: *Celebrating Irving Fisher—The Legacy of a Great Economist* edited by Robeert W. Dimand and John Geanakoplos, Blackwell Publishing Ltd., Oxford.
Watson, A. G. D. (1938), *Mathematics and its Foundations, Mind*, Vol. 47, No. 188, October, pp. 440–451.
Wittgenstein, Ludwig (1967), *Philosophical Investigations*, third edition, translated by G. E. M. Anscombe, Basil Blackwell, Oxford.

Comment

# A Comment on 'Definitions, Assumptions, Propositions and Proofs in Sraffa's *PCMC*'

*Stefano Zambelli*

Over the years Velupillai has made a strong and persuasive case in favour of embracing classical recursion theory and constructive-algorithmic mathematics in economics. He has highlighted the limitations of uncritical use of 'standard', non-constructive mathematics that is pervasive in economic theory. He has single-handedly defined the core elements of *Computable Economics.*[30]

For my own point of view, *Computable Economics* is and has always been an approach to economic theorising, where *computations* (and hence *algorithms*) are the natural constituents of any description of economic processes. Furthermore, computational and constructive aspects concerning theorising of economic behaviour, economic dynamics and decision making are closely investigated. Here *computations* and *algorithms* have to be understood in the deep, formal sense in which they were invoked by Gödel and Turing (and many other metamathematicians). In brief, for any algorithm there is a Turing machine equivalent and every computation is a Turing machine (program) that halts.[31]

As Velupillai has repeatedly emphasised, classical mathematics used by mathematical economists is often devoid of computational content and

[30] Note that this is not to be confused with *Computational Economics. Computable Economics* is much wider in scope and its foundations are firmly set based on results from metamathematics and *Computability Theory* (Velupillai 2000, 2010a).

[31] Clearly, this implicitly adhers to the Church-Turing Thesis. On computation and the Church-Turing thesis, see Velupillai (2010a) and/or Davis (1958).

presupposes that non-computable objects (for which an algorithm to generate or compute them cannot be devised) are in fact computable.

This presupposition is pervasive in mainstream and non-mainstream economic theory alike. For example, the existence proofs of competitive economic equilibrium presented in Debreu (1959) and Arrow and Hahn (1971) are provably uncomputable.[32] Unfortunately, this happens to be the case with many followers of Sraffa as well. Like the neo-Ricardians who *believe* that the rates of profits gravitate around a long term uniform rate of profits (Garegnani 1976, 1990; Kurz and Salvadori 1995), the faith in classical mathematics seems to be uncritical and scholars do not provide computable or constructive arguments.[33]

The requirement that theorising concerning economic behaviours, decisions, models, institutions and so on have to be algorithmic (or constructive) should be rather obvious. This point has been persuasively made by scholars like Velupillai and Herbert Simon.

[32] One should not be led into thinking that Walrasian General Equilibrium is computable because it is at times claimed that Scarf's algorithm can help compute. Scarf's Computable General Equilibrium is *"neither computable nor constructive in the strict mathematical senses"* (Velupillai 2006, p. 360). A list of non-computable or non-constructive results proved by Velupillai are to be found also in (Zambelli, 2010, pp. 37–38). For *"variations in the theme of conning in mathematical economics"* see Velupillai (2007) and for the *"unreasonable ineffectiveness of mathematics in economics"* see Velupillai (2005).

[33] One should not be conned into thinking that arguments expressed in English prose, as it is the case for many neo-Ricardians, cannot be formalized in the form of mathematical propositions or sentences. The logical structure of these arguments remain the same. In Sraffa's PCMC the propositions, whether presented in English or formulated in a mathematical language, have a definite computable algorithmic content. The same, however, is not true for the propositions made by many of Sraffa's followers. A good example would be the claims made by many that Sraffa's PCMC is dealing with a convergent (or with an already converged process towards) long-run equilibrium. Such claims are never supported by arguments that are constructive or algorithmic. Therefore, this interpretation of the Sraffa's schemes of being relative to a long-run equilibrium is postulated or assumed, rather than demonstrated. Surely, it is not something assumed by Sraffa himself.

Even Pasinetti, who in his work has made propositions that can in fact be reformulated constructively, has not followed Sraffa's method. As early as 1980, in his review of Pasinetti (1977) Velupillai wrote:

> There is a crucial distinction between the methodology followed by Pasinetti and that followed by Sraffa in proving the important propositions. [...] The distinction seems to be that Sraffa, whenever he gives an explicit proof, invariably gives us a constructive proof, whereas all the proofs Pasinetti (and almost everyone else who has attmpted to formalize and generalize Sraffa) gives, follows the method of formalist mathematicians (Velupillai 1980, pp. 64–65)

In the above chapter, Velupillai outlines a list of mathematical theorems, axioms and methods of proofs which are non-algorithmic and/or non-constructive. Though these feature ubiquitously in mathematical economics of today, they never feature in Sraffa's work. This corroborates the view that Sraffa's method was distinct and different from that of standard mathematical economics.

Velupillai conjectures that Sraffa was aware of the important notions, methods and results that form the core of what came to be known as *Computable Economics.* In essence, the arguments presented in the chapter suggest, based on the work and methodology of Sraffa, that he was aware of the algorithmic and constructive methods. Whether this conjecture is true or not is, as Velupillai himself states, unimportant: *"[Sraffa's] adherence to intuitionistic constructive mathematics, in PCMC, does not depend on any familiarity on these things"* (above, p. 5, fn.4).

Velupillai has made a convincing argument that Sraffa's method is constructive in several publications (Velupillai 1989, 2008, 2010b). It is worth drawing attention to two of the conjectures presented in the chapter:

- **Conjecture 1** *"Every proposition and proof in PCMC is constructive"* (p. 12)
- **Conjecture II** *"Every proposition, and therefore proof, in PCMC is programmable for machine implementation [...]"* (p. 12),

I find that these are not conjectures, but are facts. I claim them to be facts because in my own research I have worked on Sraffa's contributions and Sraffian schemes through an algorithmic lens. Following *Sraffa's English prose,* I have programmed all of Sraffa's propositions and/or computations for machine implementation (in MATLAB).[34]

From conjecture 2, I have omitted the final words *"using intuitionistic type theory"*. This is a novel interpretation proposed by Velupillai as a potential lens through which Sraffa's work can be viewed and developed. This part Velupillai's research work or agenda is unfortunately beyond my

[34] The computations of self-replacing prices (Zambelli 2018b), of the standard commodity and of the subsystems (Velupillai, Vela with Zambelli 1993), Velupillai-Fredholm-Zambelli algorithm (Zambelli et al. 2017, see Appendix) (which is an algorithmisation of chapter 12 of PCMC) and algorithms used in Zambelli (2018a) and in Venkatachalam and Zambelli (2021) can all be traced back to Sraffa's PCMC or explicitly based on it.

own expertise. The same goes for the remarks that concern several aspects of intuitionistic/constructive mathematics. But they do seem to promise an interesting research program within intuitionistic mathematics or logic.

The most crucial point that the chapter conveys is that Sraffa's method was intrinsically computational and constructive. Sraffa was concerned with the arithmetic of exchange and bookkeeping. Sraffa assumes that quantities and methods of production are given, and the whole book is about a *search* for self-replacing prices and the implication that these prices have for computation of capital and, most importantly, the distribution of the surplus. This is done using constructive methods without resorting to any non-constructive arguments. It is also worth noting that these quantities and methods of production are expressed in terms of *measurable* units. These cannot be non-computable or irrational numbers, but must be in terms of rational numbers. Note that infinitesimals are not measurable or observable economic magnitudes.

Exchanges in property rights are exchanges in magnitudes that are elements in the arithmetic of bookkeeping, that is, entries in the economic accounts and balance sheets. In reality, it would be impossible to have a bookkeeping entry, which is expressed as a non-computable real number or an infinitesimal. However, in modern mathematical economics, this distinction is often overlooked. The analysis is not restricted to deal with economically meaningful magnitudes.

*"The mathematics of [PCMC] is about formulating economic problems in the form of equations and finding methods to solve them"* (Velupillai 2008, p. 283). It is a well known fact that Sraffa worked intensively on the *"equations"* that became the core of PCMC during the second half of the 1920s. In order to solve the equations, Sraffa seem to have studied an Italian mathematics book (Chini 1923), which has the following subtitle (own translation): *To be used mainly by chemists and naturalists.* In the book, there are many annotations by Sraffa. He was clearly looking for methods to solve his system of equations. It is interesting to note that the book has many applications and that chemists of his time were mostly interested in solving problems concerning right proportions of inputs that would produce a certain output as the result of chemical reactions. These problems can be seen as being isomorphic with Sraffa's production schemes. The solutions are all algorithmic in nature.

The important result of PCMC from my point view is that the number of equations is less than the number of variables. Hence, the system is indeterminate. This extremely powerful result shows the impossibility of having *"[...] 'any' notion of capital as a measurable quantity independent of distribution and prices*" (PCMC, p. 38). No one, yet, has been able to refute Sraffa's result.

This has also the implication that, as Venkatachalam and I stress in our contribution to this volume, the so-called natural economic forces alone are insufficient for the determination of distribution and prices. If this is true, actions to change the distribution of physical surplus are not *'futile'* as pointed out by Sraffa.

I am convinced that PCMC is, as Velupillai suggests, an *algorithmic theory of production* and that attempts towards expanding Sraffa's work ought to be coherent with the constructive-algorithmic method. Velupillai's contributions on this topic provide foundations for this.

## References

Arrow, K. and F. Hahn (1971). *General Competitive Analysis*. San Francisco: Holden-Day, Inc.

Chini, M. (1923). *Corso speciale di matematiche. Con numerose applicazioni. Ad uso principalmente dei chimici e dei naturalisti — Sesta edizione con altre notevoli aggiunte*. (6th ed.). Raffaello Giusti, Livotno.

Davis, M. (1958). *Computability and Unsolvability*. New York: Dover (Paperback Edition 1982).

Debreu, G. (1959). *Theory of Value*. New York: John Wiley & Sons, Inc.

Garegnani, P. (1976). On the change in the notion of equilibrium in recent work on value and distribution: A comment to samuelson. In M. Brown, K. Sato, and P. Zarembra (Eds.), *Essays in Modern Capital Theory*. Amstredam: North Holland.

Garegnani, P. (1990). On some supposed obstacles to the tendency of market prices towards natural prices. *Political Economy 6*(1-2), 329–359.

Pasinetti, L. (1977). *Lectures on the Theory of Production*. New York: Columbia University Press.

Velupillai, K. V. (1980). Review of L.L. Pasinetti's book "Lectures on the Theory of Production". *Journal of Economic Studies (New Series) VII*, 64–65.

Velupillai, K. V. (1989). The existence of the standard system: Sraffa's constructive proof. *Political Economy Studies in the Surplus Approach 5* (1), 3–12.

Velupillai, K. V. (2000). *Computable Economics*. Oxford: Oxford University Press.

Velupillai, K. V. (2005). The unreasonable ineffectiveness of mathematics in economics. *Cambridge Journal of Economics 29* (6), 849–872.

Velupillai, K. V. (2006). Algorithmic foundations of *computable* general equilibrium theory. *Applied Mathematics and Computation 179* (1), 360–369.

Velupillai, K. V. (2007). Variations on the Theme of Conning in Mathematical Economics. *Journal of Economic Surveys 21* (3), 466–505.

Velupillai, K. V. (2008). Sraffa's economics in non-classical mathematical modes. In G. Chiodi and L. Ditta (Eds.), *Sraffa or Alternative Economics*, Chapter 14, pp. 275–294. Palgrave Macmillan.

Velupillai, K. V. (2010a). *Computable Foundations for Economics*. London: Routledge.

Velupillai, K. V. (2010b). A constructive interpretation of Sraffa's mathematical economics, in Velupillai, K. V. (2010a). *Computable Foundations for Economics*, pp. 285–308. London: Routledge.

Velupillai, K. Vela with Zambelli, S. (1993). The International Comparisons Programme and Production Based Indices of Economic Performances. mimeo, World Bank, Washington.

Venkatachalam, R. and S. Zambelli (2021). "Self-replacing prices with credit and debt", ASSRU Discussion Paper (DP-2021-3).

Zambelli, S. (2010). Computable and constructive economics, undecidable dynamics and algorithmic rationality: An essay in honour of Kumaraswamy (Vela) Velupillai. In S. Zambelli (Ed.), *Computable, Constructive and Behavioural Economic Dynamics: Essays in Honour of Kumaraswamy (Vela) Velupillai*, pp. 13–45. London: Routledge.

Zambelli, S. (2018a). The aggregate production function is NOT neoclassical. *Cambridge Journal of Economics 42* (2), 383–426.

Zambelli, S. (2018b). Production of Commodities by Means of Commodities and non-uniform rates of profits. *Metroeconomica 69* (4), 791–818.

Zambelli, S., T. Fredholm, and R. Venkatachalam (2017). Robust measurement of national technological progress. *Structural Change and Economic Dynamics 42*, 38–55.

CHAPTER 14

# Sraffa's Monetary Writings, Objectivism and the Cambridge Tradition

*Carlo Panico*

## 1 Introduction

During the 1980s and the 1990s the scientific literature circulated a set of contributions examining the links between Sraffa's writings on money and banking and those on value and distribution.[1] This literature disclosed the origins of Sraffa's theoretical interests, the homogeneity of his views on the factors affecting income distribution in earlier and later works, and the fruitful relation with Keynes.[2] The analysis of *Sraffa Papers* (SP), a large

I want to thank Neri Salvadori for helpful comments on a previous version of this essay.

[1] See Pasinetti (1979, 1985), Roncaglia (1983, 1984), Ginzburg (1986), Eatwell and Panico (1987), Potier (1987), Panico (1988a), De Cecco (1993, 1994), Davis (1998).

[2] See Roncaglia (1983, 1984, 1999), Ginzburg (1986), Eatwell and Panico (1987), Panico (1988a), De Cecco (1993, 1994), Ciocca and Rinaldi (1997), Davis (1998) and Bellofiore and Potier (1998). A different position is held by Skidelsky (1986), who describes the rela-

C. Panico (✉)
Facultad de Economia, Universidad Nacional Autonoma de México (UNAM), Mexico City, Mexico
e-mail: panico@unina.it



A. Sinha (ed.), *A Reflection on Sraffa's Revolution in Economic Theory*, Palgrave Studies in the History of Economic Thought,
https://doi.org/10.1007/978-3-030-47206-1_14

set of unpublished writings that became available in the Wren Library in Cambridge in the second half of the 1990s, confirmed those conclusions (see Panico 2001). It however disclosed a new aspect of his work, that is his intent to adopt a philosophical approach called "objectivism" 'to represent the production and circulation of commodities in material terms' (SP, D3/12/27: 11) in order to drive down the intrusion of ideological elements in the formation of economic theory (see Kurz and Salvadori 2005).

A thorough integration of this new aspect in the analysis of the relation between Sraffa's writings on monetary problems and on value and distribution is still lacking. The aim of this essay is to fill this gap by examining how Sraffa's "objectivist" approach can correlate with his analysis on the role of money in the theory of distribution.

The essay delineates how Sraffa linked theoretical work with that on history and policy, moving in accordance with the academic tradition, headed by Marshall, which dominated in Cambridge before WW2. It also discusses his interpretation of the objectivist approach and the implications for the study of the role of money in the theory of distribution. Finally, it compares the evolution of Keynes' and Sraffa's work on the role of money in the theory of value and distribution.

We underscore the existence of close cooperation with the work of Keynes on monetary problems. Like Keynes, Sraffa focussed on the concept of liquidity and ended up by developing a historical and conventional theory of the interest rate, highlighting the role of the institutional organization of financial markets and monetary policy. His writings emphasise, more than those of Keynes, that the pressures of the banking industry affect the formation of monetary legislation. They concentrate on the growth and the degree of concentration of this industry and on the ability of the largest corporations to influence social and political life.

These elements make his writings close to those of Marx. There are however significant differences with the analyses of Marx and of some Marxist circles. By placing attention to the application of an objectivist approach to the theory of value and distribution, Sraffa's conclusions on this subject diverge form those of Marx (see Kurz and Salvadori 2005).[3]

tion between Sraffa and Keynes as a case of non-communication. For a critique of Skidelsky's arguments, see Panico (1988a).

[3] The attention to the application of historical materialism to the analysis of the production and circulation of commodities led Sraffa to concentrate on 'the twin concepts of *physical*

By admitting that the conflicts within the capitalist class and the autonomous interests of political and administrative bodies can affect policy decisions, Sraffa's position differs from those expressed by some Marxist circles. We conclude this part of the inquiry by suggesting that Sraffa's historical materialism is helpful to interpret some features of the recent evolution of financial markets and of monetary legislation and policy, and the effects of these changes on income distribution.

The essay is organised as follows. Section 2 examines some applied monetary works that Sraffa wrote in the 1920s. Section 3 describes the relation between his theoretical works on Marshall's theory of competitive value and the approach that was dominant in Cambridge after Marshall's appointment to the Chair of Political Economy in December 1884. Section 4 illustrates Sraffa's adoption of an objectivist approach in the construction of a theory of value and distribution in terms of simultaneous determination of all prices. Section 5 compares the evolution of Sraffa's and Keynes' views on the role of money in the theory of value and distribution. Section 6 concludes.

## 2 Sraffa's Early Writings on Applied Monetary Problems in the 1920s

In the 1920s Sraffa wrote several essays on applied monetary problems. An account of these works can be found in Panico (1988a, 2001). This section summarises some of them, highlighting that the main issues there considered were the role of social conflicts in the formation of legislation and policy decisions, their effects on income distribution, and the determinants of the degree of liquidity of different assets.

Sraffa graduated at the University of Turin (Italy) in 1920 with a thesis titled *Monetary inflation in Italy*, here referred to as Sraffa (1920). Luigi Einaudi, an economist who preferred Ricardo, Mill and the pragmatism of the Marshall's school to the formalism of Walras and Pareto, supervised the thesis that dealt with some highly debated issues in the monetary literature of the time: the causes and consequences of inflation, the

*real cost* and *social surplus* within the framework of an analysis that conceived of production as a circular flow' (Kurz and Salvadori 2005, p. 414) and to criticise the view 'that a measure of labour can be elaborated that allows one to portray in a reliable way the material process of production and which therefore can be used in the theory of value, as Ricardo and Marx were inclined to think' (Kurz and Salvadori 2005, p. 418).

stabilisation of internal prices and exchange rates within an unstable international financial system, and the arguments for restoring the gold standard and revaluing the currency to the pre-war gold parity. These themes were at the centre of the International Financial Conference organised by the League of Nations in Brussels in 1920. As De Cecco (1993, p. 3, 1994, p. 158) clarifies, the thesis contains comments on the conclusions of the conference and on the arguments proposed by the major economists attending it.

Sraffa held the view, close to that of Cassel and Fisher in the conference, that the value of money tends to stabilise around the level that common opinion considers «normal». For him, the pre-war gold parity is 'no more «normal» than any other value of money' (Sraffa 1920, p. 42. English translation: 1993, p. 24). He wrote:

> The value of money is normal when forces which tend to make it change are absent. Thus, had the required time to let prices adjust to circulation elapsed, a new equilibrium would have established and the value reached by money would be perfectly normal. (Sraffa 1920, p. 42. English translation: 1993, p. 24)

In the thesis Sraffa considered monetary phenomena, like inflation and deflation, as part of the social conflicts shaping income distribution. He suggested that these conflicts affect State intervention and have a permanent influence on the level of distributive variables, a position that differed from that proposed by the dominant school of thought.[4] At the time he had not yet scrutinised the theoretical foundations of this view.

In 1922 Sraffa published two other works on applied monetary problems. They were promoted by his connection with Keynes. He met the British economist for the first time in Cambridge in August 1921. Impressed by Sraffa's expertise, Keynes asked him to write an article on the crisis of the Italian banking system for the Weekly Supplement of the *Manchester Guardian Commercial*, which dealt with the monetary and financial problems of the post war reconstruction in Europe. The article that he wrote (Sraffa 1922a) was too long for the *Manchester Guardian*, but suitable for the *Economic Journal*, where it was published in June

[4] For a comparison of Sraffa's (1920) and Keynes' (1923) analyses on this point, see Panico (2001, p. 286).

1922, while a shorter article (Sraffa 1922b) took its place in the 7 December issue of the *Manchester Guardian Commercial*.

By using the available evidence with accuracy and competence, Sraffa described the events that led to the crisis by illustrating the working of the "mixed" (or "universal", as it is named today) banking system, used in Italy since the last decade of the XIX century. He argued that the major problems of this system are not the risks related to the unbalanced assets-liabilities structure of the banks' balance sheet, but the embroiled relations between banks and industries and the consequent formation of large groups or "concentrations" able to control sections of the economy, of the media and of the political world and to attain monetary legislation and policies favourable to their own interests.

His reconstruction of the interactions among the industrial sector, the financial sector and the political world seems suitable for the interpretation of the recent evolution of financial regulation, of the related financial crisis, and of the change in income distribution (see Panico et al. 2012, 2016; Panico and Pinto 2018). It underlined that government interventions are part of the conflicts among different economic and social groups. His analysis, however, instead of being based on a rigid class division, acknowledged that the conflicts within the capitalist class and the autonomous interests of political and administrative bodies can affect policy decisions. Moreover, it underlined that the formation of large groups or "concentrations", able to disguise from the majority of the population the aims of their initiatives and to influence the exertion of power, can represent a danger for democracy.

> The general tendency seems to be towards the … formation of large 'groups' of companies of the most varied kinds concentrated around one or more banks, mutually related by the exchange of shares and by the appointments of Directors common to them. Within these "groups" the various interests are all equally subject to the interests of a few individuals who control the whole group. … Very little is known … about these groups. … What the public knows and feels … is the enormous financial and political power which they have and the frequent use they make of it to influence both the foreign and home policy of the government in favour of their own interests. Each group keeps several press organs which support its policy, and some of the accusations made against certain Ministries of being actuated by the interests not of a class, but of private concerns, and of favouring one financial group against another, have no doubt a basis of truth. (Sraffa 1922a, p. 196)

The analysis of how the autonomous interests of political and administrative bodies can affect policy decisions was again proposed in an exchange of correspondence with Angelo Tasca,[5] published in the monthly magazine of the Italian Communist Party *Stato Operaio* in the issue of November–December 1927 (referred to as Sraffa 1927), and in an unpublished essay titled "The Revaluation of the Lira" (SP, D2/3) that Sraffa wrote for a conference held in Cambridge at the Emmanuel Society on the 3rd of November 1927. Both documents comment on the causes of the revaluation of the Italian lira that occurred between August 1926 and March 1927.

The correspondence with Tasca disputed the view that the government adopted the revaluation policy to favour the *immediate* interests of the dominant economic groups. Sraffa disagreed with this position:

> In general, it seems a mistake to me—and a very dangerous one—to believe that every single action of the fascist government (and of any capitalist government) is *directly* dictated by the *immediate* interests of the banks and of the big industrialists. (Sraffa 1927, p. 1089; italics in the original; our translation)

His opposition to oversimplified interpretations of the events shows that his opinion on the role of the State differed from that held within Marxist political circles. According to Sraffa, in 1926 the Fascist government had turned in favour of the revaluation to gain the support of the middle and of part of the working class and so avoid the danger of the social isolation that it was facing. For him, the independent interests of political and administrative bodies can have a role in the formation of economic policy and the fact that the dominant groups made pressures on the government to avoid the revaluation confirms that it did not correspond to their *immediate* interests.

The unpublished essay titled "The Revaluation of the Lira" (SP, D2/3) deals with the same events showing Sraffa's ability at distinguishing between stability and revaluation of the exchange, between inflation of money, of credit and of prices, and between devaluation caused by price inflation and by speculative activity.

The essay also provides information on how speculative activity changed after the First World War, a theme that was at the centre of Keynes'

[5] Angelo Tasca was at the time a member of the Italian Communist Party.

concerns in those years. Sraffa pointed out that the funding of the war had caused a massive growth of financial markets and speculative activity, making it more difficult to control the exchange rates and to carry out gradual exchange revaluations.[6] Moreover, the substitution of professional for amateur speculators after the collapse of the German mark had reduced the seasonal regularities of the exchange markets and the opportunity to gain from arbitrage operations, another subject that attracted Keynes' and Sraffa's attention and saw cooperation among them (see de Cecco 2005, p. 354).

Finally, Sraffa's account of the revaluation of the lira confirmed his inclination to examine policy problems by focussing on the interests of the different groups and institutions participating in the decision-making process and by using the historical evidence in a rigorous and competent way.

Upon his arrival at the University of Cambridge in 1927, Sraffa was requested to teach some courses. Keynes asked him to lecture on "Continental Banking" focussing on the working of the German mixed banking system (see de Cecco 2005). A manuscript describing the content of these *Lectures* can be found in *Sraffa Papers* (SP, D2/5).

Sraffa argued that the German mixed or universal banks were effective in channelling financial resources to productive sectors. He also contended that, provided that they are actively supported by the central bank, they are not riskier than specialised banks as far as liquidity and solvability are concerned.

In the *Lectures* Sraffa further developed this point (see SP, D2/5/14) by focusing on the notion of liquidity in financial and commodities' markets (see de Cecco 2005).[7] He argued that the degree of liquidity attributed to the assets depends on the psychological perception that investors have of the stability of their prices. He cited several elements that can

[6] The upsurge of speculative activity produced negative effects on the growth of the British economy. These effects induced Lavington and Keynes to deepen the analysis of the speculative demand for money and to introduce this new element in the theory of interest and money.

[7] De Cecco (2005) claims that this part of Sraffa's *Lectures* is as an anticipation of the analysis he presented in the debate with Hayek, where he introduced the concept of "own interest rate", subsequently adopted by Keynes in chapter 17 of the *General Theory*. De Cecco points out that Sraffa had been working on these problems for a long time, providing Keynes with data on the Lira forward rates soon after they first met in August 1921 and that 'Keynes used these data in his exposition of the theory of forward rates in his famous Manchester Guardian Supplement article'.

influence this perception,[8] underscoring that the degree of liquidity depends more on the difficulty of selling the assets at the present price than on their maturity. For this reason, the existence of a large market and the availability of a big purchaser, as a central bank can be, to buy the securities at a given price increase the degree of liquidity of the assets.[9] The powers that legislation attributes to the authorities on the stabilisation of financial markets and the funding of credit institutions, he concluded, is the most important element deciding the degree of liquidity of the assets and the solvability of the banking system. Sraffa's reference to this element clarifies that in his analyses the role of the institutional organization of financial markets, which can be studied by observing its historical evolution, overshadows that of psychological perceptions on the degree of liquidity of the assets.

In the *Lectures* Sraffa recalled that, unlike what had happened in Britain, the configuration of the German banking system was the result of a deliberate choice to attribute to the banks the role of promoter of industrial development. The *Reichsbank*, i.e. the bank of issue that then became central bank, was given the power to establish closer co-operative relations with credit institutions than those existing in England. Since the German unification this form of institutional organisation had been allowing the mixed banks to fund the industrial development without running into problems of solvability related to their unbalanced assets-liabilities structure. For Sraffa, the *Reichsbank* worked as the leading entity of a planned economy in which the mixed or universal banks took a long-term concern in industrial firms by promoting their formation, providing the rating and the certification of their balance sheets, and favouring the sale of their

[8] Among these elements, Sraffa mentioned that the characteristics of the credit contracts and of the commodities used for speculative investment influence the perception of financial operators. Lenders, for instance, consider a credit given to a customer less liquid than a bond that can be sold in a secondary market. He also mentioned that that 'raw materials (e.g. raw cotton) are more liquid than manufactured products (e.g. cotton cloth of a given quality) since they have a wider market, since the number of people who buy products made out of raw cotton is greater than those who buy products made out of that particular quality of cloth. (Thus barley is more liquid than beer from the point of view of bankers)' (SP, D2/5/14). Moreover, he stated that the size of the funds demanded by a credit institution and the solvability pressures leading this entity to ask for them is considered relevant by the operators when they evaluate the degrees of liquidity of the assets (see SP, D2/5/14).

[9] As document (SP, D2/5/14) clarifies, the same applies to commodity markets. Gold and some raw materials, for instance, had a high degree of liquidity on account of the way their markets were organised.

shares in the market. While the banks funded production in strategic sectors of the economy, the *Reichsbank* assisted them with funds, which were systematically provided to stimulate development, rather than given as an emergency provision, as occurred in Britain. De Cecco (2005, pp. 355 and 358) points out that, until the WW1, the *Reichsbank* acted as "lender of first resort", rather than as "lender of last resort". This operation, however, had to change after the war.[10]

From this perspective too, Sraffa's analysis appears suitable for the interpretation of the recent evolution of monetary policy. According to de Cecco (1999), the dominance of universal banks in USA, after the 1980s, has induced the Federal Reserve to convert its operation from that of "lender of last resort" to that of "lender of first resort", trying to anticipate and avoid credit crunches by issuing base money.

Unlike what had happened in Germany before WW1, the transformation of the US banking system into a universal one was not the result of a process aiming at fortifying the industrial development. It was rather the result of a set of events, including the pressures of the financial industry, which led to a radical change in financial regulation and to the reduction of the discretionary powers of the authorities over the managers of that industry (see Panico et al. 2016). Owing to this change, the financial industry has been able to pursue its interests under reduced controls over their risky activities[11] and the Federal Reserve has been bound to systematically provide the markets with liquid means to prevent solvability problems.

To sum up, Sraffa's earlier writings on applied monetary problems submitted a view of the working of financial markets that focussed on the ability of large corporations to control relevant sections of the economy, of the media and of the political world in order to attain monetary legislation and policies favourable to their own interests. This analysis, instead of being based on a rigid class division, acknowledged that conflicts within the capitalist class and the autonomous interests of political and administrative bodies could affect policy decisions. This approach to monetary problems can still be used to elucidate how the interests of the different

[10] De Cecco (2005, pp. 356–58) points out that after WW1 the *Reichsbank* could not operate as before, owing to the distress of the German economy, the burden imposed by the Treaty of Versailles, the hyperinflation, and the restrictions on the working of the central bank set by the Dawes Plan.

[11] White (2009, p. 36) too recognises that the origin of the recent financial crisis can be found in the change of the approach used in financial regulation in USA since the 1970s.

groups have influenced the recent evolution of financial regulation and led to the financial crisis. Sraffa shared with Keynes a deep concern for the analysis of the determinants of the "degree of liquidity" of the assets, a subject that was at the centre of the literature of those years. His writings on this theme underscored that the institutional organisation of the markets takes the limelight to the psychological perceptions of financial operators in determining the "degree of liquidity" of the assets. As a consequence, the study of the evolution and organization of financial markets, which can be worked out by examining the available evidence according to the scientific method of the historians, can replace that of the psychological perceptions of financial operators, that is of a subjective and not directly observable element.

## 3 Sraffa's Work on Value and Distribution and the Cambridge Tradition

Probably aroused by what he had learnt from Einaudi at the University of Turin, Sraffa decided to give a course on Marshall's work when he started to teach at the University of Perugia in 1923.[12] The preparation of the course led to the development of his first theoretical works on Marshall's theory of competitive value (Sraffa 1925, 1926, 1930a, b). The content of these works reveals proximity to Marshall's methodological approach to theory, history and policy, an approach that became dominant in Cambridge after his election to the Chair of Political Economy in December 1884.[13]

Marshall described this approach in the Inaugural Lecture he gave in February 1885 (see Marshall 1885; Groenewegen 1995, 1997; Hodgson 2005). He stated that, like other sciences, political economy must identify general principles that 'will no doubt be presented as a perfectly pure or

[12] For a description of the events that occurred between the publication of the article in the *Manchester Guardian Commercial* in December 1922 and the start of his teaching activity at the University of Perugia in November 1923, see Pasinetti (1979, 1985), Roncaglia (1984), Naldi (1998) and Panico (1988a, 2001).

[13] Cunningham was Marshall's major opponent over the Chair and the establishment of this approach in Cambridge. He tried to stand out against it several times, presenting arguments similar to those of some exponents of German historical school, whom Menger (1883) criticised (see Cunningham 1892; Marshall 1892; Hodgson 2005, p. 124; Groenewegen 2007, p. 116).

abstract theory' (Marshall 1885, reprinted in Hodgson 2005, p. 129).[14] The validity of these theories has to be verified in terms of logical coherence. Alternative theories, that is, can be evaluated by identifying the assumptions that are required to make them logically coherent and by comparing the realism of these assumptions. The comparison allows one to evaluate the strength of the theories and the limits of their use in the interpretation of events and in the formulation of policies.

Marshall specified that the set of theories constituting the general principles of political economy

> is not a body of concrete truth, but an engine for the discovery of concrete truth, similar to, say, the theory of mechanics ... [that] supplies a universal engine, which will help in determining what strain any bridge will bear. But it has no universal dogmas by which this strain can be determined without observation of the particular facts of the case (Marshall 1885, reprinted in Hodgson 2005, p. 129).[15]

For him, the application of the principles of the economic discipline to any particular situation cannot ignore historical and empirical work.

The acceptance of this point did not lead the British economist to question abstract theoretical work. On the contrary, he opposed the standpoint of the economists contesting the value of abstract theories by claiming that facts do not speak for themselves:

> They are telling us to discard all theories, and to seek the solution of our economic difficulties in the direct teaching of facts. ... The answer is that facts by themselves are silent. Observation discovers nothing directly of the actions of causes, but only of sequences in time (Marshall 1885, reprinted in Hodgson 2005, p. 134).

Facts must be interpreted and their interpretation inevitably requires the use of theories:

[14] In the Introduction to the *Cambridge Economic Handbook Series*, Keynes stated that this approach characterised the "Cambridge School of Economics", whose members, according to him, did not hold unanimity of theoretical views (see Groenewegen 2007, p. 198).

[15] In the Introduction to the *Cambridge Economic Handbook Series*, Keynes presented this approach as 'an apparatus of mind' that does not furnish any 'settled conclusions immediately applicable to policy'. Sraffa (1926, p. 535) quoted this part of Keynes' Introduction.

They do not throw a direct light on particular economic problems of our age. They do not in any way help us to dispense with the use of the economic organon: but rather make use of its aid at every step. And those whose great achievements have made the school illustrious have never attempted to dispense with the aid of economic theory (Marshall 1885, reprinted in Hodgson 2005, p. 134).

He resolved that 'greedy then as the economist must be for facts, he must not be content with mere facts … He will thus work in the light of facts, but the light will not be thrown directly, it will be reflected and concentrated by science' (Marshall 1885, reprinted in Hodgson 2005, p. 137).

In line with what characterised the Cambridge Tradition, Sraffa's first theoretical works criticised Marshall's theory of competitive value by identifying the assumptions that are required to make it coherent and by evaluating the realism of these assumptions. He argued that Marshall introduced within the general principles of the economic discipline a version of the theory of competitive value in terms of partial equilibrium trying to show that this version, while favouring an intuitive appreciation of the working of the economic system, could achieve the same results as that in terms of general equilibrium.[16]

In his 1925 article and in the part of that of 1926 summarising the content of the former, Sraffa wrote that the partial equilibrium version of the theory of competitive value assumes independence among markets where there is actually interdependence among them. This approximation can be accepted if it can be proved that the errors implied by the results of the partial equilibrium version can be considered of secondary importance (see Panico 1991; Samuelson 1991; Panico and Salvadori 1994):

It is well known that such an assumption would not be illegitimate merely because the independence may not be absolutely perfect …; a slight degree of interdependence may be overlooked without disadvantage if it applies to quantities of the second order of smalls … But, of course, it is a very differ-

[16]According to Sraffa (1926, p. 541), Marshall's wanted to avoid having to rely only on the analysis of 'the conditions of simultaneous equilibrium in numerous industries: a well-known conception, whose complexity, however, prevents it from bearing fruit, at least in the present state of our knowledge, which does not permit of even much simpler schemata being applied to the study of real conditions'. Sinha (2016, p. 39) underlines that Sraffa (1926) used this argument to focus on the analysis of imperfect competition.

> ent matter, and the assumption becomes illegitimate, when a variation in the quantity produced by the industry under consideration sets up a force which acts directly, not merely upon its own costs, but also upon the costs of other industries; in such a case the conditions of the "particular equilibrium" which it was intended to isolate are upset, and it is no longer possible, without contradiction, to neglect collateral effects (Sraffa 1926, pp. 538–39).

To establish the adequacy of the partial equilibrium version it is thus necessary to define a criterion to evaluate up to which point mutual interdependence can be disregarded and the "ceteris paribus" assumption legitimately made.[17]

The criterion adopted by Sraffa distinguishes two ways in which mutual interdependence can operate when a change in the quantity produced by the isolated industry, call it $x_n$, occurs (see Panico 1991). The first way implies that variations in $x_n$ bring about non-constant returns operating *directly* on the technical coefficients (call them $a_{ij}$s with some $i \neq n$) of both the isolated and the other industries. This leads to variations in prices and distribution and to further feedback effects on the technical processes used and on demand. The second way implies that variations in $x_n$ bring about non-constant returns operating *directly* only on the technical coefficients of the isolated sector, call them $a_{nj}$s. This leads to variations in the price of the isolated good, which in turn can feedback on the price of other goods and cause changes in the technical processes used and in demand. According to Sraffa, mutual interdependence, operating *directly* on the technical coefficients of other industries too, cannot be neglected. On the contrary, that producing only *indirect* effects on other industries can be ignored.

By using this criterion Sraffa demonstrated that the partial equilibrium version of the theory of competitive value can be made coherent by introducing the assumption that the non-constant returns generated by a

[17]Viner too signified the relevance of establishing a criterion for the acceptance of Marshall's theory. He proposed to refer to the fact that the errors of the partial equilibrium version are quantitative, rather than qualitative: 'All partial equilibrium analysis ... rests on assumptions of the ceteris paribus order which posit independence where in fact there is some degree of dependence. For such logically invalid assumptions, there is the pragmatic defence that they permit of more detailed analysis of certain phases of economic interdependence than would be possible in their absence, and that ... it is reasonable to believe that the errors in the results obtained will be almost invariably quantitative rather than qualitative in character, and will generally be even quantitatively of minor importance' (Viner 1931, pp. 199–200).

change in the quantity produced by the isolated industry are "external to the firm and internal to the industry". Yet, Sraffa noticed, in *Industry and Trade*, the book where he 'wished to approach nearer to reality' (Sraffa 1926, p. 540), Marshall himself acknowledged that external economies 'can seldom be allocated exactly to any one industry: they are in great measure attached to groups, often large groups of correlated industries' (Marshall 1919, p. 188). That is, non-constant returns "external to the firm and internal to the industry" rarely occur in reality.

In the second part of his 1926 article, which does not summarise the content of that of 1925, after pointing out two possible solutions to the deficiencies of Marshall's theory, one in terms of imperfect competition and the other in terms of simultaneous determination of all prices (see Sraffa 1926, pp. 541–42), Sraffa focussed on the former. This choice however was soon abandoned to shift towards the solution in terms of simultaneous determination of all prices. The occurrence of this shift can be traced in the Preface of his 1960 book, where it is stated that its 'central propositions had taken shape in the late 1920s' (Sraffa 1960, p. vi). An early draft, already pursuing the solution in terms of simultaneous determination of all prices, was shown to Keynes soon after Sraffa's arrival in Cambridge for the academic year 1927–1928.[18]

The two short interventions published in the 1930 *Economic Journal* Symposium on "Increasing returns and the representative firm" confirm the dismissal of the partial equilibrium version of the theory of value. There, Sraffa completely ignored the imperfect competition solution and, dealing with Robertson's objections to his critique, concluded with a statement that highlights the links with the Marshallian tradition, dominant in Cambridge, on the relationship between theory and facts:

> I am trying to find what are the assumptions implicit in Marshall's theory; if Mr. Robertson regards them as extremely unreal, I sympathise with him. We seem to be agreed that the theory cannot be interpreted in a way which makes it logically self-consistent and, at the same time, reconciles it with the facts it sets out to explain. Mr. Robertson's remedy is to discard mathematics, and he suggests that my remedy is to discard the facts; perhaps I ought to have explained that, in the circumstances, I think it is Marshall's theory that should be discarded (Sraffa 1930a, p. 93).

[18] See Talamo (1976), Roncaglia (1978), Maneschi (1986), Schefold (1989), Garegnani (2005).

## 4 Sraffa's Theory of Value and Distribution and Objectivism

The abandon of Marshall's theory of competitive value and the construction of an alternative theory of value and distribution in terms of simultaneous determination of all prices raised additional methodological problems.

Sraffa had a deep interest in the literature focusing on the restraint of the influence of value judgements in the development of scientific knowledge. He examined how other sciences were dealing with this topic and read the contributions embracing an objectivist philosophy (Whitehead 1926, p. 124) adopted in different sciences by Hertz, Helmholtz, Poincaré, Picard, Simiand, Labriola, Heisenberg, Dirac, Bridgman, Whyte, and so on. According to this approach, scientific research must move from observable magnitudes and must look for "efficient causes", which relate present events to material elements or facts of the past, instead of "final causes", which relate present events to perceptions or foresight of future acts (see Kurz and Salvadori 2005). The material basis of the objectivist approach averts the formation of theories rooted in subjective and elusive factors that favour the intrusion of ideological elements.

When dealing with income distribution, however, the use of what Sraffa called 'an entirely objective point of view' raised a problem on which he had been working since 1927 (see Sinha 2016, pp. 69–72). He described this problem in a 1931 unpublished manuscript (SP, D3/12/7: 161) providing a solution that modified his view on the application of the objectivist approach to the theory of distribution (see Kurz and Salvadori 2005 and Davis 2012).

He noticed that in an objectivist approach, since any effect must have an efficient cause, there must be correspondence between the material inputs of production (the cause) and the objects produced (the effect) according to the rule 'production can transform objects, but not create them'. How then can any surplus come about from the production process and be distributed to different social groups? Sraffa's solution of this conundrum was to consider that the economic field is an "open system" within which "outside causes" operate. In the case of production the outside or external causes are the historical elements affecting the *level* of distributive variables, for instance the historical conditions imposing a real wage rate at subsistence level. These conditions, external to the production process, allow the identification of the surplus within the social product. They are facts of the past (efficient causes) that can be studied by observing the evolution of legislation, institutional organization and policy and can have a prominent influence on daily individual decisions. Even if the latter are related to psychological perceptions, they actually depend, as objectivism requires, on the observation of events arisen in the past.

The solution proposed by Sraffa allowed him to reconcile history with the materialism of the objectivist philosophy and to take into consideration the role of individual decisions. He was able 'to represent the production and circulation of commodities in material terms' (SP, D3/12/27: 11), developing it in such a way as to align social distributional factors alongside physical real costs (see Davis 2012). Moreover, it could contemplate how subjective phenomena operate in and influence physical real costs. According to Davis, the solution anticipated a philosophical position, known as "supervenience physicalism", that Davidson and Dennett developed in the 1970s and 1980s (see also Marion 2005). This view, today widely held, is 'central to understanding the nature and role of modern science' (Davis 2012, p. 1354) and can provide a solid scientific basis to examine the level of distributive variables and their evolution.

The integration of subjectivity and history into theory allowed Sraffa to take into account another element, underlined by the objectivist literature and related to the interaction between the behaviours of individuals and society. Instead of analysing the behaviour of society as the sum of those of the individuals composing it, the objectivist literature proposes to assume that individual behaviour depends on the legal and social frameworks in the sense that it is subject to the conventions that historically emerge from social interaction.[19] In this literature the formation of rules, legislation and institutions is an integral part of the analysis. It is the result of the conflicting relations among the different sectors of society, mirroring their ability to enact rules and agreements over the distribution of benefits and the future of society, which are functional to their interests.

[19] Kurz and Salvadori (2005, pp. 427–28) point out that, among Sraffa's reading of the literature on objectivism, a relevant one was Whitehead's book *Science and the Modern World*, read in April 1928. One theme that attracted Sraffa's attention was whether "the whole is constitutive of the part". He supported this view when he dealt with the interactions between the behaviour of the individual and that of society, submitting that individual behaviours depend on the legal and social frameworks. Kurz and Salvadori recall that Labriola (1922), which Sraffa read and annotated, and Keynes' 1933 essay on Edgeworth, held a similar position. Chapters 8 and 9 of the *General Theory* also appear to propose a theory of consumption where the individual behaviour depends on the legal and social framework.

The adoption of an objectivist approach, amended to take into account the intervention of historical elements, was the benchmark of Sraffa's work. It was in line with the view, expressed in the 1920s on the degree of liquidity of the assets, that the institutional organisation of the markets and the policy of the monetary authorities dominate the psychological perceptions of financial operators. This view was relevant, as we will argue, for his inquiry into the role of money in the theory of value and distribution.

## 5 Sraffa's and Keynes's Work on the Role of Money in the Theory of Distribution

When Sraffa arrived in Cambridge in 1927, Keynes was working on *A Treatise on Money*, which, like Volume 2 of Wicksell's *Lectures on Political Economy*, deals with the quantity theory of money and the fluctuations of the economic cycle.[20] Its analysis takes as given the relative prices, the distributive variables and the level of activity. These variables are determined, without considering the quantity of money in circulation, by what Keynes called the "real department of economic theory", which corresponds to Volume 1 of Wicksell's *Lectures* or to Marshall's *Principles of Political Economy.*

*A Treatise* expounds what Keynes called the "monetary department of economic theory". The natural interest rate, determined by the "real department", plays a crucial role in it. This equilibrium rate represents the reference point for monetary policy. When the market interest rate is equal to the natural rate, the economy does not suffer from inflation or deflation. 'Every departure of the market rate from the natural rate tends, on the other hand, to set up a disturbance of the price level' (Keynes 1930a, p. 139) and to generate fluctuations of the economy. To guarantee stability, monetary policy has to keep in line the market and the natural rates.

In *A Treatise* the fluctuations of the economy are analysed by considering that the level of activity, again determined by the "real department of economic theory", does not change. Any variation in aggregate demand causes similar variations in the price level. Variations in aggregate demand are produced by changes in the 'readiness to invest', due to changes either in the prospect of future returns or in the degree of uncertainty felt by

[20] Panico (1988b, pp. 109–21) provides primary and secondary evidence on this interpretation of *A Treatise*.

entrepreneurs in the market, and by variations in the market interest rate, caused by changes in the perceptions of financial operators as to the future prices of the assets.

The introduction of speculation on financial assets in the analysis of the demand for money was an important novelty of *A Treatise*. The massive expansion of the markets for government bonds, due to the funding of the First World War, had increased the size of this kind of speculation and made it a crucial element affecting the working of the economy. In previous work (including Keynes 1923), this element did not play a major role in the analysis of the demand for money and of the working of the economy.[21]

When Wall Street collapsed in October 1929, the galley proofs of *A Treatise* were already available. At the time the book was contained in one volume. Yet, the crisis convinced Keynes to postpone its publication and to add some new chapters. The final version, published in two volumes, contains an evaluation of the causes of the crisis. The additional chapters introduce some new elements, which however do not demolish the analytical framework of the book. The dichotomy between the real and the monetary departments of economic theory was maintained. The natural rate of interest was still there and its relationship with the market rate was used to interpret the origin of the crisis and to identify possible solutions to it. The novelties can be found in the recognition of the difficulty that the central bank can have in identifying the natural interest rate and in leading the market rate towards it. In the chapters on the 1929 crisis Keynes was more prudent on this point than in the previous parts of the book:[22]

> I have more sympathy today than I had a few years ago with some of the doubts and hesitations such as were express in 1927 by Governor Strong and other witnesses before the Committee of the United States Congress on Stabilisation. This committee was appointed to examine the wisdom of a proposed amendment to the Federal Reserve Act, the effect of which would have been to lay upon the Federal Reserve Board the duty of using all the powers at its disposal to "promote a stable price level for commodities in general". (Keynes 1930b, p. 305)

[21] Until then, the literature had focussed on speculation on inventories, rather than on financial assets.

[22] Lambert (1963, pp. 28~2, 1969, pp. 257–58) and Harrod (1951, pp. 413–23) also noticed the occurrence of this change of emphasis in Keynes' position.

What's more, Keynes (1930b, p. 344) recognised that the banking industry could unduly resist the reduction of the interest rate on long-term loan to cover its costs and profits.[23]

Keynes used the analytical framework of *A Treatise* to deal with the causes and solutions of the crisis in the essays written after the publication of this book until the end of 1932.

The literature has already highlighted Sraffa's participation in the debates raised by *A Treatise* before and after its publication (see Panico 1988a, 2001). The evidence shows, among other things, that he played a prominent role in the Circus, the group of Keynes' close colleagues debating *A Treatise* after its publication. This role is cherished by the expression 'To the Circus Master—May Term 1931' written by Joan Robinson on the cover of the abstract of her 'A parable in saving and investment' that she presented to Sraffa. This participation led the Italian economist to get more and more involved in the theoretical work on the relationship between money and the theory of value and distribution. Moreover, it led him to publish his first articles on this subject (Sraffa 1932a, b), a critical review of Hayek's 1931 book *Prices and Production*.[24]

The review suggests that, at the time, Sraffa's positions on money were close to those of Keynes.[25] Moreover, it indicates that his analyses on liquidity were highly developed. His usage of the concept of "own interest rate" inspired the content of chapter 17 of Keynes' *General Theory*. Finally, the review shows no sign that at the time Sraffa had become aware of the problems of the neoclassical theory, which he highlighted in *Production of Commodities*. Nor does it show sign that he was aware of the possibility to attribute to money an alternative role in the theory of distribution (see Panico 1988a and 2001).

On the latter point Keynes's work went through important changes by the end of 1932. The typed paper, from which he appears to have lectured in November of that year, is the first piece of evidence of this renovation.

[23] The reference to the costs and profits of the banking industry plays a central role in Keynes (1931, 1932) and was recalled in the *General Theory* (Keynes 1936, pp. 145–46 and 207–08). For an analysis of the role of the costs and profits of the banking industry in Keynes' thought, see Epstein (1995) and Panico (2008).

[24] Sinha (2016, p. ix) points out that Keynes asked Sraffa to write the review.

[25] The version of the quantity theory that Sraffa presented was similar to that of *A Treatise* (Sraffa 1932a, pp. 42 and 46–49). Moreover, like Keynes, he acknowledged the difficulty of the authorities to identify the natural interest rate and to keep the market rate in line (see Sraffa 1932a, pp. 49–51, b, p. 251).

The paper shows that Keynes wanted to deal with the long-period tendencies of the level of activity, rather than with its fluctuations, as he had done in *A Treatise*. While in this book the equilibrium levels of relative prices, distributive variables and output are independent of the quantity of money in circulation, in Keynes' new view they also depend upon monetary considerations. No separation is drawn between the "real" and the "monetary department of economic theory" and monetary factors are not seen as rigidities preventing the working of market mechanisms from leading the economy to a full employment position.

An impressive sequence of statements is introduced in this writing to clarify this point (see Panico 1988b, pp. 124–26). Here we recall one of them.

> The root of the objection, which I find to the theory under discussion, *if it is propounded as a long-period theory*, lies in the fact that ... it cannot be held that the position towards which the economy system is tending ... is entirely independent of the policy of the monetary authority. (Keynes 1979, pp. 54–55; italics added)

Similar claims can also be found in the introductory chapters of the *General Theory* (Keynes 1936, pp. 4–5) and in Chap. 18, which presents a summary of its arguments. Keynes referred to his new position as "monetary theory of production" and started to work on a novel book having as title this expression. The book was eventually published with the title *The General Theory of Employment, Interest and Prices.*

Keynes clearly perceived that his new theoretical approach implied the abandonment of a natural interest rate and of a neutral policy that the monetary authority is bound to pursue and devoted a large amount of time and energy to the elaboration of a different theory of the interest rate in the preparation of his new book. He considered the "natural interest rate" as an abstract concept leading the profession to a mistaken interpretation of the functioning of the economic system. Referring to *A Treatise*, he claimed that 'it was a mistake to speak of the natural rate of interest' (Keynes 1936, p. 242) and added that 'I am now no longer of the opinion that the concept of a «natural» rate of interest, which previously seemed to me a most promising idea, has anything very useful or significant to contribute to our analysis' (Keynes 1936, p. 243).

Several steps can be identified in the construction of his new theory. As in *A Treatise* he started by claiming that the market interest rate is determined at the level that makes the available supply of money equal to the demand for it and added that the demand for money for speculative purposes, which is the most likely cause of changes in the market interest rate on account of its high variability, reflects the individual perceptions of the future level of this rate. Then, he argued that the individual perceptions ultimately depends on what the common opinion considers a "safe" level of the interest rate and referred to the organization of the markets and to the policy of the monetary authorities as the main determinants of this safe level (Keynes 1936, p. 201). Like Sraffa's *Lectures* on the German mixed banking system, he acknowledged that the presence in the market of a big dealer, as a central bank can be, tends to stabilise the price of the assets and to establish a "durable" or "average" interest rate, around which the market rate fluctuates.

In *A Treatise*, in line with the traditional neoclassical theory, Keynes had stated that the central bank has to place the durable or average interest rate at its natural level. In the *General Theory* he rejected this view and claimed that the policy of the monetary authorities, rather than depending on an abstract and not directly observable rate, is the result of their evaluations of what is most convenient for the country under the current historical circumstances. The authorities take this decision by assessing several elements, like the relation between the national and the international financial systems, the state of the economy and of the balance of payments, the funding needs of the financial industry, of the other industries, and of the government sector, the level of systemic risk that the institutional organization of financial market can bring about.

Keynes (1936, p. 202) added that the average rate depends 'not only on the current policy of the monetary authority, but also on market expectations concerning its future policy'. For him, the average interest rate can be at any level that the authorities consider convenient, provided that financial operators consider their policy "credible":

> *Any* level of interest which is accepted with sufficient conviction as *likely* to be durable, *will* be durable; subject, of course, in a changing society, to fluctuations for all kinds of reasons round the expected norm. (Keynes 1936, p. 203; italics in the original)

The roles of the individual psychological perceptions and of monetary policy in Keynes' theory is highlighted by his statement on the dual character of this rate:

> The rate of interest is a highly psychological phenomenon ... It may be more accurate, perhaps, to say that the rate of interest is a highly conventional, rather than a highly psychological, phenomenon. For its actual value is largely governed by the prevailing view as to what its value is expected to be. (Keynes 1936, pp. 202–03)

This statement recalls Sraffa's claim in his *Lectures* that the role of monetary policy and of the institutional organization of the markets, which can be studied by observing their historical evolution, overshadows that of the operators' psychological perceptions.

Another important part of Keynes's work on the new theory of interest was the attempt to prove that the neoclassical theory lacked logical coherence. This part highlights the links with the Marshallian tradition, dominant in Cambridge, and the existence of another element of complementarity between Keynes' and Sraffa's writings.

The colleagues reading his work systematically demolished his critical attempts, which became more and more persistent as Keynes went into the final stages of preparation of the *General Theory* (see Panico 1988b, pp. 133–37). In the summer of 1934 he stated that this critique was necessary to obtain his objective. In the public broadcast in the series "Poverty in Plenty", reprinted in *The Listener* on 21 November 1934, he said that, in spite of the severity of the crisis, the "heretic" school had not been successful in convincing the profession to abandon the dominant school, since they had failed to recognise its strength, which was based on the acceptance of its theoretical principles.

> They [the heretics] believe that common observation is enough to show that facts do not conform to the orthodox reasoning.... But, they have made no impression on the citadel [the dominant school]. Indeed, many of them themselves accept the orthodox premises; and it is only because their flair is stronger than their logic that they do not accept its conclusions. (Keynes 1973, p. 489)

Keynes then proceeded by saying that, unlike the heretics, he was aware of the strength of the dominant school. He had been brought up in the "citadel" and knew that it was absolutely necessary to attack it on its own premises in order to ask the profession and the public to look at the economic problems in a radically new way. First, he claimed,

For me therefore, it is impossible to rest satisfied until I can put my finger on the flaw in that part of the orthodox reasoning which lends to the conclusions which for various reasons seem to me inacceptable. I believe that I am on my way to do so. There is, I am convinced, a fatal flaw in that part of the orthodox reasoning ... the flaw being largely due to the failure of the classical doctrine to develop a satisfactory theory of the rate of interest. (Keynes 1973, p. 489)

Then, he concluded,

If the basic system of thought on which Mr Henderson, Mr Brand and Professor Robbins rely is, in its essentials unassailable, then there is no escape from their broad conclusions, namely, that whilst there are increasingly perplexing problems and plenty of opportunities to make disastrous mistakes, yet nevertheless we must keep our heads and depend on the ultimate soundness of the traditional teaching—the proposal of the heretics, however plausible and even advantageous in the short-run, being essentially superficial and ultimately dangerous. Only if they are successfully attacked in the citadel can we reasonably ask them to look at the problem in a radically new way. (Keynes 1973, pp. 491–92)

Unfortunately, he failed to give his criticisms a satisfactory basis before the final version of the *General Theory* was published.

Keynes' insistence on the critique in terms of logical coherence, on the importance of the general principles of the discipline, and on the fact that empirical elements are unable to persuade the profession to abandon the neoclassical theory is evidence for his accord with the Marshallian tradition dominant in Cambridge. The critique in terms of logical coherence, on the other hand, establishes another link with Sraffa's work.

There is no sign that at the time of the publication of the *General Theory* Sraffa had reached the results presented in *Production of Commodities* on the critique of the neoclassical theory. The first documents showing awareness of this critique are dated 1942.[26] In the debate that followed the publication of *Production of Commodities*, the validity of this critique was recognised by outstanding economists like Samuelson. In the Summing up of the Symposium published in 1966 in the *Quarterly Journal of Economics* he warned the economic profession as to the relevance of Sraffa's critique as follows:

[26] For an analysis of this evidence, see Panico (2001, p. 300).

> If all this causes headaches for those nostalgic for the old time parables of neoclassical writing, we must remind ourselves that scholars are not born to live an easy existence. We must respect, and appraise, the facts of life. (Samuelson 1966, p. 583)

Sraffa's participation in the discussion on the *General Theory* before and after its publication has been documented by previous works, which have argued that he accepted Keynes' historical and conventional theory and took the interest rate as the independent variable in the analysis of production prices.

The documents of *Sraffa Papers* allow one to reconstruct how the Italian economist expounded this point.[27] He noticed that the English classical economists took the real wage rate as exogenously given. It was a subsistence wage determined by forces external to the production process. This practice, Sraffa said, is not suitable for the study of modern economies, in which wages appear to contain, beyond the element of subsistence, a part of the surplus value produced and the formation of the real wage has to be considered, like that of the rate of profits, as part of the historical social processes through which the surplus value is shared among different social classes. For this reason, any one of the two variables can be taken as given in the analysis of prices of production. As a matter of fact, Sraffa concluded, when the wage comes to be measured in terms of an abstract standard, it is preferable to take the rate of profits as the independent variable, assuming that the latter is determined by the money interest rates, that is again by forces which are external to the system of production under consideration:

> The last steps of the preceding argument have led us to reverse the practice followed from the outset of treating the wage, rather than the rate of profits, as the independent variable or 'given' quantity. The choice of the wage in the preliminary stages depended on its being assumed to consist of specified necessaries, determined by circumstances which (whether natural or social) were outside the system of production under examination. But the moment the possibility of variations in the division of the product was admitted, this consideration lost much of its force. And now that the wage comes to be 'given' in terms of a more or less abstract standard and does not acquire a definite meaning until the prices of commodities are determined, the position is reversed. The rate of profit, as a ratio, has a significance which is

[27] For an analysis of this evidence, see Panico (2001, pp. 300–02).

> independent of any prices, and can well be 'given' before the prices are fixed. It is accordingly susceptible of being determined from outside of production in particular by the level of the money rates of interest. In the following sections the rate of profits will therefore be treated as the independent variable. (SP, D3/12/80, 8–9; dated 29/3/57)

On the determination of the interest rate the documents of *Sraffa Papers* provide some new elements with respect to what is contained in paragraph 44 of *Production of Commodities*. For Sraffa, the forces external to the production process are the institutional organization of the market, the policy of the monetary authorities, and the way in which financial markets view this policy:

> The point of view, however, of variation depending on the pull and push for the distribution of the surplus is more suited to regarding the rate of profit as being the independent variable and as being itself a reflection of the rate of interest which is determined by the operation of the Stock Exchange or regulated by banking policy. If however, as is more convenient with the sharing of surplus, we regard r as determined by the money rate, and this by Bank or Stock Exchange, the problem is much simplified. (SP, D3/12/68, 2)

The reference to monetary policy and to the prevailing views of financial markets is frequent in the preparatory documents of *Production of Commodities* and in those subsequent to its publication. In a correspondence with Garegnani after the publication of his book, Sraffa first underlined the link between the hint in paragraph 44 and the objective of his work, that is the elaboration of a theory of distribution independent of the relative scarcity of the factors of production. Then, he clarified his intention to pursue this objective by using a historical and conventional theory of the interest rate. Finally, he underlined the provisional character of his proposal, which was susceptible of being further elaborated and modified:

> I am convinced that the maintenance of the interest rate by the bank and (or) the stock exchange has had its part in the determination of income distribution among social classes ... I did not want to commit myself much, and in general I only wanted to signal something in order to avoid the belief that the system is presented as 'foundation' for a theory of the relative supplies of capital and labour! It is what is denied that seems important to me: as to what is affirmatively claimed, I have no intention to put forward another mechanical theory which, in one form or another, states again that

> income distribution is determined by natural, or technical or even accidental, circumstances, which in any case are such that they make any action taken by either part, in order to modify it, futile. … I do not see any difficulty in the determination of the rate of profit through a controlled or conventional interest rate, provided that the rate of profit will not be assumed to be determined by external unchangeable circumstances. (SP, D3/12/111; stress in the original; our translation)

In conclusion, the available evidence shows that Sraffa, in line with Keynes's position in *The General Theory*, wanted to give an active role to monetary policy in the determination of the average interest rate. This implies that the decisions of the monetary authorities play an active role in the determination of distributive variables and are part of the historical and institutional factors affecting the division of the surplus value produced.

## 6 Conclusions

The documents of *Sraffa Papers* can clarify how the monetary theory of distribution can be integrated within the objectivist approach. Although the psychological perceptions of financial operators is central for the evaluation of the degree of liquidity of the assets and for the determination of the market interest rate, the institutional organization of financial market and the policy of the monetary authorities, which can be studied by observing their historical evolution, are dominant in the formation of the what the common opinion considers the "safe" interest rate and, therefore, in the determination of what Keynes called the average or durable rate. Keynes' historical and conventional theory of the interest rate refers to these elements. Sraffa's hint in paragraph 44 of *Production of Commodities* is in line with this standpoint.

Although it can appear that Sraffa was more deeply involved than Keynes in using the objectivist approach, there is evidence showing that the British economist also used this approach in some of his writings (e.g., the critique of Edgeworth's *Mathematical Psychics* in Keynes 1933).

The documents of *Sraffa Papers* indicate that Sraffa and Keynes were closely cooperating since they first met in August 1921. Sraffa inspected in detail the concept of liquidity, interchanging with Keynes on several aspect of this notion. His expertise on the theme led him, probably on Keynes' request, to publish a review of Hayek's *Prices and Production*. There,

Sraffa focussed on the concept of "own interest rate" that Keynes used in the *General Theory*, recognising its debt to the Italian economist (see footnote 1 of Chap. 17).

Keynes parted with the neoclassical theory and adopted a monetary theory of production in November 1932. The evidence shows no sign that, at the time, Sraffa was aware of the possibility to attribute to money a different role in the theory of distribution. What's more, Keynes tried to develop his new view on this point by focussing on the theory of interest. He proposed a historical and conventional determination of the interest rate and unsuccessfully tried to identify the existence of logical flaws in the neoclassical theory. There is reason to assert that Sraffa accepted Keynes' historical and conventional determination of the interest rate and that his critique of the neoclassical theory of distribution could have been used to make successful Keynes' attempts to identify the existence of logical flaws in this theory.

Contrary to Skidelsky's (1986) view, the relationship between Sraffa and Keynes can be considered a case of beneficial communication and cooperation.

## References

Bellofiore R., Potier J.P., 1998, Piero Sraffa: nuovi elementi sulla biografia e sulla ricezione di *Produzione di Merci* in Italia, *Il Pensiero Economico Italiano*, VI (1), 55–103.

Ciocca P., Rinaldi R., 1997, L'inflazione in Italia, 1914–20. Considerazioni a margine della tesi di laurea di Piero Sraffa, *Rivista di Storia Economica*, 13, 3–40.

Cunningham W., 1892, The relativity of economic doctrines, *Economic Journal*, 2 (1), 1–16.

Davis J.B., 1998, Sraffa and Keynes: differences and shared perceptions, *Il Pensiero Economico Italiano*, VI (1), 133–54.

Davis J.B., 2012, The change in Sraffa's philosophical thinking, *Cambridge Journal of Economics*, 36 (6), November, 1341–56.

De Cecco M., 1993, Piero Sraffa's "Monetary inflation in Italy during and after the war": An introduction, *Cambridge Journal of Economics*, 17 (1), March, 1–5.

De Cecco M., 1994, Introduzione a Piero Sraffa, «L'inflazione monetaria in Italia durante e dopo la guerra», *Economia Politica*, XI (2), Agosto, 155–62.

De Cecco M., 1999, The lender of last resort, *Economic Notes*, 28 (1), 1–14.

De Cecco M., 2005, Sraffa's lectures on continental banking: A preliminary appraisal, *Review of Political Economy, Special Issue: Piero Sraffa 1898–1983*, 17 (3), 174–88.

Eatwell J.L., Panico C., 1987, Piero Sraffa, in P. Newman, M. Milgate and J.L. Eatwell (eds.), *The New Palgrave Dictionary of Economics*, vol. 4, London: Macmillan, 445–52.

Epstein G., 1995, The illiquidity trap, *Eastern Economics Journal*, 21 (3), Summer, 309–18.

Garegnani P., 2005, On a turning point in Sraffa's theoretical and interpretive position in the late 1920s, *European Journal of the History of Economic Thought*, 12 (3): 453–92; reprinted in Kurz H., Pasinetti L.L., Salvadori N., eds., *Piero Sraffa: The Man and the Scholar*, London: Routledge, 2008: 79–118.

Ginzburg A., 1986, Sraffa e Keynes su inflazione e deflazione, in R. Bellofiore (a cura di), *Piero Sraffa: tra Teoria Economica e Grande Cultura Europea*, Milano: Angeli, 59–73.

Groenewegen P.D., 1995, *A Soaring Eagle: Alfred Marshall 1842–1924*, Aldershot: Edward Elgar.

Groenewegen P.D., ed., 1997, *Alfred Marshall. Collected Essays 1872–1917*, Bristol: Overstone Press, 2 volumes.

Groenewegen P.D., 2007, *Alfred Marshall, Economist: 1842–1924*, Basingstoke: Palgrave Macmillan.

Harrod R.F., 1951, *The Life of John Maynard Keynes*, London: Macmillan.

Hayek, F.A. (1931) *Prices and Production*, London: Routledge.

Hodgson G.M., 2005, The present position of economics by Alfred Marshall, *Journal of Institutional Economics*, 1 (1), 121–37.

Keynes J.M., 1923, *A Tract on Monetary Reform*, in Moggridge D.E., ed., *The Collected Writings of J.M. Keynes*, vol. IV, London: Macmillan (1971).

Keynes J.M., 1930a, *A Treatise on Money*, in Moggridge D.E., ed., *The Collected Writings of J.M. Keynes*, vol. V, London: Macmillan (1971).

Keynes J.M., 1930b, *A Treatise on Money*, in Moggridge D.E., ed., *The Collected Writings of J.M. Keynes*, vol. VI, London: Macmillan (1971).

Keynes J.M., 1931, Unemployment as a world problem: Reports of Round Tables, vol. II, Harris Foundation Institute Round Tables, in Moggridge D.E., *The Collected Writings of J.M. Keynes, Activities 1931–1939: World Crises and Policies in Britain and America*, vol. XXI, London: Macmillan and Cambridge University Press for the Royal Economic Society, 1982.

Keynes J.M., 1932, A note on the long-term rate of interest in relation to the conversion scheme, *Economic Journal*, September, reprinted in Moggridge D.E., *The Collected Writings of J.M. Keynes, Activities 1931–1939: World Crises and Policies in Britain and America*, vol. XXI, London: Macmillan and Cambridge University Press for the Royal Economic Society, 1982.

Keynes J.M., 1933, Francio Ysidro Edgeworth, *Essays in Biography*, in Moggridge D.E., ed., *The Collected Writings of John Maynard Keynes*, vol. X, London: Macmillan for the Royal Economic Society, 251–66.

Keynes J.M., 1936, *The General Theory of Employment, Interest and Money*, in Moggridge D.E., ed., *The Collected Writings of J.M. Keynes*, vol. VII, London: Macmillan (1973).

Keynes J.M., 1973, *The General Theory and After, Part I: Preparation*, in Moggridge D.E., ed., *The Collected Writings of J.M. Keynes*, vol. XIII, London: Macmillan.

Keynes J.M., 1979, *The General Theory and After: A Supplement*, in Moggridge D.E., ed., *The Collected Writings of J.M. Keynes*, vol. XXIX, London: Macmillan.

Kurz H.D., Salvadori N., 2005, Representing the production and circulation of commodities: On Sraffa's objectivism, *Review of Political Economy*, 17 (3), 414–41.

Labriola A., 1922, *Il valore della scienza economica: introduzione a una critica dell'economia politica*, Napoli: Alberto Morano.

Lambert P., 1963, *L'oeuvre de John Maynard Keynes*, Le Haye: Nijhoff.

Lambert P., 1969, The evolution of Keynes' thought from the Treatise on money to the general theory, *Annals of Public and Cooperative Economy*, 40, 243–63.

Maneschi A., 1986, A comparative evaluation of Sraffa's "The laws of returns under competitive conditions" and its Italian precursor, *Cambridge Journal of Economics*, 10 (1), 1–12.

Marion M., 2005, Sraffa and Wittgenstein: Physicalism and constructivism, *Review of Political Economy*, 17 (3): 381–406; reprinted in Kurz H.D., Pasinetti L.L., and Salvadori N., eds., *Piero Sraffa: The Man and the Scholar*, London: Routledge, 2008: 217–42.

Marshall A., 1885, *The Present Position of Economics, An Inaugural Lecture at Cambridge University*, February. A large part of this article was reprinted in Hodgson, 2005.

Marshall A., 1892, The perversion of economic history: A reply, *Economic Journal*, 2 (7), September, 507–19.

Marshall A., 1919, *Industry and Trade*, London: Macmillan.

Menger C., 1883, *Untersuchungen über die Methode der Sozialwissenschaften und der politischen Ökonomie insbesondere*, Tübingen: J.C.B. Mohr. Published in English as *Investigations into the Method of the Social Sciences with Special Reference to Economics*, New York: New York University Press, 1985.

Naldi N., 1998, Some notes on Piero Sraffa's biography: 1917–1927, *Review of Political Economy*, 10, 493–515.

Panico C., 1988a, Sraffa on money and banking, *Cambridge Journal of Economics*, 12 (1), March, 7–28.

Panico C., 1988b, *Interest and Profit in the Theories of Value and Distribution*, London: Macmillan; New York: St. Martin Press.

Panico C., 1991, Some notes on Marshallian supply functions, *Economic Journal*, 101 (406), May, 557–69.

Panico C., 2001, Monetary analysis in Sraffa's writings, in Cozzi T. and Marchionatti R. (eds.), *Piero Sraffa's Political Economy: A Centenary Estimate*, London: Routledge, 285–310.

Panico C., 2008, Keynes on the control of the money supply and the interest rate, in Forstater M. and Wray R., *Keynes and Macroeconomics After 70 Years: Critical Assessments of the 'General Theory'*, Aldershot: Elgar, 157–75.

Panico C., Pinto A., 2018, Income inequality and the financial industry, *Metroeconomica*, 69 (1), February, 39–59.

Panico C., Pinto A., Puchet Anyul M., 2012, Income distribution and the size of the financial sector: A Sraffian analysis, *Cambridge Journal of Economics*, 36 (6), November, 1455–77.

Panico C., Pinto A., Puchet Anyul M., Vázquez Suárez M., 2016, A Sraffian approach to financial regulation, in Freni G., Kurz H.D., Lavezzi M. and Signorino R., eds., *Economic Theory and Its History: Essays in Honour of Neri Salvadori*, London: Routledge, 249–70.

Panico C., Salvadori N., 1994, Sraffa on returns, *The European Journal of the History of Economic Thought*, 1 (2), 323–43.

Pasinetti L.L., 1979, Sraffa Piero, in *International Encyclopaedia of Social Sciences, Biographical Supplement*, vol. 18, New York: Macmillan and The Free Press, 736–38.

Pasinetti L.L., 1985, In memoria of Piero Sraffa: economista italiano a Cambridge, *Economia Politica*, II (3), Dicembre, 315–32.

Potier J.P., 1987, *Un Economiste non Conformiste: Piero Sraffa (1898–1983)—Essay Biographique*, Lyon: Presses Universitaires.

Robbins L. (1932), *An Essay on the Nature and Significance of Economic Science*, London: Macmillan and Co. Ltd, 2nd ed., 1935.

Roncaglia A., 1978, *Sraffa and the Theory of Prices*, Chichester: John Wiley & Sons.

Roncaglia A., 1983, Piero Sraffa: una bibliografia ragionata, *Studi Economici*, XXXVIII (21), 137–66.

Roncaglia A., 1984, Sraffa e le banche, *Rivista Milanese di Economia*, 10, Aprile–Giugno, 104–12.

Roncaglia A., 1999, *Sraffa: la Biografia, l'Opera, le Scuole*, Bari: Laterza.

Samuelson P.A., 1966, A summing up, *Quarterly Journal of Economics*, 80 (4), November, 568–83.

Samuelson P.A., 1991, Sraffa's other leg, *Economic Journal*, 101 (406), May, 570–74.

Schefold B., 1989, Comment on Samuelson, in Bharadwaj K. and Schefold B., *Essays in Honor of Piero Sraffa: Critical Perspective on the Revival of Classical Theory*, London: Unwin and Hayman.

Sinha A., 2016, *A Revolution in Economic Theory: The Economics of Piero Sraffa*, London: Palgrave Macmillan.

Skidelsky R., 1986, Keynes e Sraffa: un caso of non-comunicazione, in R. Bellofiore (a cura of), *Piero Sraffa: tra Teoria Economica e Grande Cultura Europea*, Milano: Angeli, 73–84.

Sraffa P., 1920, *L'Inflazione Monetaria in Italia*, University of Turin. Published in the same year in Milan with the Premiata Scuola Tipografica Salesiana. Translated in English by W.J. Harcourth and C. Sardoni and published in 1993 as Piero Sraffa: 'Monetary inflation in Italy during and after the war', *Cambridge Journal of Economics*, 17 (1), March, 7–26.

Sraffa P., 1922a, The bank crisis in Italy, *Economic Journal*, XXXII (126), June, 178–97.

Sraffa P., 1922b, Italian banking to-day, *The Manchester Guardian Commercial, Reconstruction in Europe, Supplement*, 7 December, 675–76.

Sraffa P., 1925, Sulle relazioni tra costo e quantità prodotta, *Annali of Economia*, 2, 277–328. English translation in L.L. Pasinetti (ed.), 1998, *Italian Economic Papers*, Bologna: Il Mulino; Oxford: Oxford University Press, 323–63.

Sraffa P., 1926, The laws of return under competitive conditions, *Economic Journal*, XXXVI, December, 535–50.

Sraffa P., 1927, Politica monetaria. Due lettere sull'articolo di Angelo Tasca «La rivalutazione della lira e la crisi dell'economia italiana», *Lo Stato Operaio*, Anno I, no. 9–10, Novembre–Dicembre, 1089–95.

Sraffa P., 1930a, A criticism' and 'Rejoinder, contributions to the Symposium on «Increasing Returns and the Representative Firm», *Economic Journal*, XL, March, 89–93.

Sraffa P., 1930b, An alleged correction of Ricardo, *Quarterly Journal of Economics*, XLIV, May, 539–44.

Sraffa P., 1932a, Dr. Hayek on money and capital, *Economic Journal*, XLII (165), March, 42–53.

Sraffa P., 1932b, A rejoinder, *Economic Journal*, XLII (166), June, 249–51.

Sraffa P., 1960, *Production of Commodities by Means of Commodities*, Cambridge: Cambridge University Press.

Talamo O., 1976, Piero Sraffa: su alcuni problemi di interpretazione, in Marcuzzo M.C., ed., *Problemi di Storia delle Teorie Economiche*, Milano: Mazzotta.

Viner J., 1931, Cost curves and supply curves, *Zeitschrift fur Nationalökonomie*, 3, September, 23–46, reprinted in Stigler G.J. and Boulding K.E., eds., *Readings in Price Theory*, 1953, London: Allen and Unwin.

White E.N., 2009, Lessons from the history of bank examination and supervision in the United States, 1863–2008, in Gigliobianco A. and Toniolo G., eds., *Financial Market Regulation in the Wake of Financial Crises: The Historical Experience*, Workshops and Conferences, November, 15–44.

Whitehead A.N., 1926, *Science and the Modern World*, Cambridge: Cambridge University Press.

Wicksell K., 1901, *Föreläsningar i Nationalekonomi*, Första delen: Teoritisk Nationalekonomi, Lund. English edition: *Lectures on Political Economy*, translated from the Swedish by E. Classen and edited with an Introduction by Lionel Robbins, vol. I, The Macmillan Company, New York, 1935.

Comment

# A Comment on 'Sraffa's Monetary Writings, Objectivism and the Cambridge Tradition'

*Guglielmo Chiodi*

Panico's paper is a remarkable reconstruction of the outstanding scientific works which Sraffa has undertaken since his undergraduate thesis of 1920 until his seminal, corner-stone 1960 book.

The remarkable feature of Panico's paper is that he succeeded in presenting even the most intricate aspects of Sraffa's published and unpublished papers with extreme rigor and precision by making use of a plain and crystal-clear language.

In taking into account practically the entire scientific Sraffian works, he meticulously reconstructs in parallel the other path-breaking Keynesian works, specifically *A Treatise on Money* and *The General Theory*, by picking up the most salient intersections with the Italian economist's works.

The ultimate result is really amazing, for in a few pages no one of the most characterizing features of both Sraffa's and Keynes's works seem to have been lost.

Panico brings to the fore at the very beginning a strand of philosophical thought called 'objectivism', which Sraffa seemed to have adopted 'in order to drive down the intrusion of ideological elements in the formation of economic theory' in order 'to represent—using Sraffa's own words of an unpublished paper, quoted by Panico—the production and circulation of commodities in material terms'. This is actually a crucial and a very delicate aspect of Sraffa's methodological approach, which Panico aims to 'correlate with his [Sraffa's] analysis on the role of money in the theory of distribution'.

More generally, it might be said that Sraffa's methodological choice seems to have been dictated by the necessity of taking due distance from the psychological-based subjective neoclassical theory and—at the same time—also from some unnecessary parts of Marxian and Marxist thought.

The rejection of the ideological elements, which Sraffa is supposed to have made by embracing 'objectivism', should, however, not to be intended in any dogmatic way. The role attributed by Sraffa to 'objectivism' seems to be merely functional in representing data and in making calculations on them which *any* observer can make, without any prejudice or biased perspective whatsoever.

Then 'ideology' should not necessarily be identified with having a 'political vision', when observing, representing and interpreting a given society. This is because the economic discipline is a *social* discipline, thus fully immersed in the complex functioning of communities of human beings, and no scholar can put herself or himself completely outside the frame. Thus, 'objectivism' does not mean having no 'political vision' of the society or, to put it other way, no 'pre-analytical vision' at all—to borrow Schumpeter's words.

Sraffa wanted in fact to radically depart from the traditional point of view until then maintained by the dominant (neoclassical) economic theory, in which quantities *and* prices—in the stereotyped general equilibrium framework—are simultaneously determined by the anonymous market forces of demand and supply. He instead explicitly took the Classical economists' standpoint, which meant immediately separating quantities from prices, and representing the system by means of *objectively given* quantities of commodities used and produced in a given period, in a given society—which implies that a *value-judgment* is *ipso fact* expressed, and that a 'political vision' of the society *must* preliminarily be maintained.

From this point of view, it is not quite easy to swallow Pasinetti's 'separation theorem', Pasinetti (2007), according to which in '*Production of Commodities* he [Sraffa] does not rely on any institutional set-up, he does not make reference to any historical context', p. 275. The 'theorem', interpreted to the letter, might perilously signify a sort of 'neutrality' of the economic discipline introduced by Sraffa like a Trojan horse, whereas he simply does not rely on any *specific* institutional set-up or historical context. However, the 'extremely simple society' contemplated by Sraffa in chapter I of his book reflects indeed a *general* historical context, and the introduction of 'the rate of profits' and the class of 'capitalists' in chapter II and the 'money rates of interest' in chapter V, all reflect an institutional set-up which contemplates the recognition of property rights, class division, and financial institutions.

With this 'political' key-reading, I think it would be far easier to follow Panico's admirable reconstruction of Sraffa's research path from his early writings on monetary problems in the 1920s until the value and distribution theoretical problems of his 1960 book. Panico, in fact, rightly emphasizes the fundamental role played throughout Sraffa's writings by the historical context, by the evolution and organization of financial markets and by the asymmetric *powers* of different institutions—which all reflects his 'political vision', the latter patently in sharp contrast to the neoclassical one, so naively and badly hidden by a supposed 'scientific neutrality' of the theory, along the lines first explicitly stated by Walras (1874, p. 52) and outlined afterwards by Robbins (1932).

The other interesting feature characterizing Panico's paper is his punctual reference to Keynes's main works on issues which cross the Sraffian ones, particularly those related to monetary problems.

Keynes' crucial transition from *A Treatise* to *The General Theory* is remarkably testified by the facts that monetary and the real parts of the economy are considered indissolubly intertwined in the 1936 work, and it is perhaps one of the most important 'bridge' linking the 'monetary' treatment of Sraffa to that of Keynes, which is admirably outlined by Panico. Due emphasis is given to the institutional and conventional character attributed by Keynes, and afterwards by Sraffa, to the money rates of interest, which in fact represents the most significant break with that tradition. The latter supposes the existence of a 'natural' rate of interest, still alive in *A Treatise*, so much characterized by some Wicksellian 'flavour'—although, to be honest, Wicksell was not so dogmatic in his conception of the 'natural' rate of interest, and in any case he was quite ahead with respect to other neoclassical economists in the years to come, such as Modigliani and Patinkin. (It is also worth noticing, at this juncture, that Wicksell (1901) was the most lucid economist to denounce at the beginning of the last century the intricacies surrounding the notion of 'capital' and the impossibility of its definition without arguing in a circle.)

The profound and highly significant intersections between Sraffa and Keynes highlight the 'revolutionary' common feature of their works.

It is a pity that one of the most outstanding Sraffian-inspired economists such as Garegnani (1964, 1965), has contributed to leave Keynes' *General Theory* within the uncomfortable boundaries of the neoclassical citadel, for having appreciated exclusively Keynes's idea of 'effective demand', while judging his monetary theory irremediably neoclassical. Panico's paper is highly welcome in this respect, for having done full justice to Keynes' masterpiece.

## References

Garegnani P., 1964, Note su consumi, investimenti e domanda effettiva, Parte Prima, *Economia Internazionale*, Novembre, pp. 591–631; abridged version reprinted in *Cambridge Journal of Economics*, December 1978, pp. 335–353.

Garegnani P., 1965, Note su consumi, investimenti e domanda effettiva, Parte Seconda, *Economia Internazionale*, Novembre, pp. 575–611; abridged version reprinted in *Cambridge Journal of Economics*, March 1979, pp. 63–82.

Pasinetti L.L., 2007, *Keynes and the Cambridge Keynesians. A 'Revolution in Economics' to be Accomplished*, Cambridge: Cambridge University Press.

Walras L., 1874, *Éléments d'économie politique pure*, Corbaz, Lausanne, translated by William Jaffé, Edition Définitive 1926, George Alen and Unwin Ltd, London, 1954.

Whitehead A.N., 1926, *Science and the Modern World*, Cambridge: Cambridge University Press.

Wicksell K., 1901, *Föreläsningar i Nationalekonomi*, Första delen: Teoritisk Nationalekonomi, Lund. English edition: *Lectures on Political Economy*, translated from the Swedish by E. Classen and edited with an Introduction by Lionel Robbins, vol. I, The Macmillan Company, New York, 1935.

CHAPTER 15

# Sraffa, Money and Distribution

*Ragupathy Venkatachalam and Stefano Zambelli*

## 1 Introduction

Sraffa's contributions to economics can be broadly divided into two parts, which may at first seem disconnected. The first part of his research was dominated by themes that are closely related to monetary economics and banking theory (Sraffa 1920, 1922a, 1932a). The second concerns themes related to the production, measurement and distribution of physical surplus in an economy (Sraffa 1925, 1926, 1951, 1960). The second part concerning value and distribution has received quite a lot of (well-deserved) attention from scholars, while his early writings on monetary themes have received comparatively little attention. The connection between these two parts and how one may have influenced the other have often been overlooked.

R. Venkatachalam
Institute of Management studies, Goldsmiths, University of London; New Cross, London, UK
e-mail: r.venkatachalam@gold.ac.uk

S. Zambelli (✉)
Department of Economics and Management, University of Trento; via Inama, Trento, Italy
e-mail: stefano.zambelli@unifctn.it



A. Sinha (ed.), *A Reflection on Sraffa's Revolution in Economic Theory*, Palgrave Studies in the History of Economic Thought,
https://doi.org/10.1007/978-3-030-47206-1_15

The absence of money in Sraffa's magnum opus, *Production of Commodities by Means of Commodities* (Sraffa 1960, from now on PCMC), has led many to conclude that money is not an important element in his theoretical system. Sraffa has been explicit in stating that his system is indeterminate. In other words, it is not possible to simultaneously determine the self-replacing prices, wage rate and profit rate, with the knowledge of the methods of production alone (see the Appendix).

Bharadwaj (1963, p. 1450) highlights this point explicitly:

> Distribution in Sraffa's system is not endogenously generated through production relations. [This is a significant departure from the widely prevalent practice of obtaining distributive shares from the production function applying the marginal method]. In conjunction with prices, production relations determine only the net surplus that is to be distributed. [Rather prices and the rate of surplus are determined **simultaneously** given the production relations]. No theory of distribution is offered in the book. His wage rate is a variable that could be considered to vary with the same ease as the profit rate. Wage is assumed to be paid post factum, so that profits and wages are surplus sharing entities. (emphasis added, text within parenthesis are footnotes in the original version)

Sraffa takes one of the prices (i.e., the uniform rate of profits) to be an independent (i.e., exogenous) variable, thereby making his system determinate. It is important to note that this assumption concerning choice of the uniform rate of profits is in itself not crucial. One could equally work with choosing wage rate or a commodity price as the independent variable. Hence, the important point concerns the indeterminacy in the Sraffian system, which implies that there is a crucial role for forces outside the economic system in the determination of self-replacing prices.

Sraffa remarks that the money rates of interest, which may be determined from outside the system of production, might influence the rate of profits. As we show below, it would be incorrect to interpret Sraffa as suggesting that the system would be determined when the money interest rates are given. Quite the contrary! Sraffa in fact makes a reference to the money rates of interest to reinforce the idea that economic forces alone are not sufficient to determine exchange-values and distribution.[1]

[1] In the correspondence with Garegnani, Sraffa makes this point quite forcefully, see Sect. 3.1

Therefore, when one attempts to reconstruct economic theory with Sraffian schemes and Sraffa's work as the starting point, it would be essential to consider institutional factors and the banking/financial sector. More generally, one ought to consider the power relations that underpin production activities in an economy that could decisively shape the distribution of physical surplus.

The money rates of interest is one of the tools that can be used to control financial costs and hence profits of production units. A selective use of financial funds favouring some sectors with respect to others would obviously influence the exchange values as well. However, it is not the only way in which the power of certain groups may influence the exchange values and hence distribution. The recognition that the system is indeterminate and allocations of funds may be controlled by powerful groups opens up space for investigating a monetary explanation of distribution. Motivated by this, we argue that money (seen as deferred means of payment), offers a fruitful (and necessary) generalisation of the Sraffian schemes.

In Sect. 2, we provide a brief overview of Sraffa's contributions. We argue that money was very much a part of his research agenda. We offer a classification of his contributions and argue that they have to be viewed as a coherent whole. In Sect. 3, we focus on the explicit and implicit role of money and distribution in PCMC and, more broadly, in Sraffa's work. We identify and emphasise a common theme in his work that concerns the insufficiency of the natural forces of the market in determining the prices, rates of profit, wages and the distribution. In Sect. 4, we outline the motivations for generalising Sraffian schemes by incorporating money, viewed as a deferred means of payment. Appendix A provides a formal treatment of Sraffian schemes for the case of single product industries extended to the general case of non-uniform rates of profits. Appendix B provides a concise outline of how credit and debt can be incorporated inside the traditional Sraffian framework.

## 2 An Overview of Sraffa's Contributions

Sraffa's first scholarly publication was his graduate thesis in 1920, titled *L'inflazione monetaria in Italia durante e dopo la guerra* ('Monetary Inflation in Italy during and after the war', Sraffa (1920)) written under the supervision of Luigi Einaudi. In this, Sraffa focused on the themes related to money and monetary policy in the Italian context of that time.

It investigated the causes of inflation, analysed issues concerning stabilisation of domestic prices, exchange rates and distributional aspects. This work, as argued in Panico (1988a), is quite applied in character and bears many similarities with A *Tract on Monetary Reform* by Keynes.

Apart from his lucid and incisive political commentaries in *L'Ordine Nuovo* (Sraffa 1921), his next significant scholarly publication was also centred around the Italian monetary affairs. Keynes, who was then the editor of *Manchester Guardian Supplement on Reconstruction in Europe* had commissioned Sraffa to write an article for the magazine. This led to two articles—one in the *Manchester Guardian* and other was published in the *Economic Journal*. His analysis of the case of Banca Italiana di Sconto which went bankrupt by the end of 1921 was particularly insightful.

These articles (Sraffa 1922a, b) focused on the crisis of the Italian banking sector. They showcase his remarkable familiarity and understanding concerning the operations of the banking sector, the regulatory framework, its relationship with the industries, the role of interest groups and the myriad of issues they face.[2] Thus, characterisations of Sraffa's body of work merely as being an exercise in theory with no practical relevance cannot be justified. This is evident even from a cursory look at the explicit themes that characterise his contributions on money, even without delving into the implicit practical, empirical and policy implications of his theoretical work.

Sraffa (1925) put forward an important critique of Marshallian theory of firm and its relation to the equilibrium position of an industry. This was published in the Italian journal *Annali di economia,* titled *Sulle relazioni fra costo quantitá prodotta.*[3] This contribution, which systematically analysed the relationship between cost and quantity, is to be seen in the context of a raging debate between different scholars during the 1920s on the 'law of returns' in the Marshallian theoretical apparatus. A summary of this article was published in the *Economic Journal* (Sraffa 1926).

Sraffa's review of Hayek's *Prices and Production* (Sraffa 1932a, b) published in the *Economic Journal* deconstructs a theoretical system put forward by Hayek and highlights some inconsistencies in

[2] It is also important to note in this context that Sraffa has spent time learning the practical workings of the wheels of banking from the inside as an apprentice at the provincial bank *Banca di Legnano e Busto Arsizio* post his graduation.

[3] Sraffa's Italian translation of *A Tract on Monetary Reform* was also published in 1925. Sraffa also published brief book reviews in the *Giornale degli Economisti* that concerned monetary and banking issues.

Hayek's framework. Sraffa's review also highlights his nuanced understanding of money, monetary theory and stabilisation policy.[4] In particular, his views on the nature of monetary interest rate, attempts to compare it to the 'natural' rate of interest, and the impossibility of having a unique equilibrium interest rate were important insights.

Sraffa undertook the task of editing the collected works and correspondences of David Ricardo in the 1930s. This monumental, meticulous work took well over two decades and was published the early 1950s. In the first volume—*On the Principles of Political Economy and Taxation*—Sraffa offered an impressive introduction to Ricardo's works and resurrected the classical surplus approach to political economy, where circular flow takes a centre stage. Many scholars regard his introduction as par excellence in offering a balanced, informative and interpretive judgement on a great scholar's body of work.

Sraffa's arguably best known work, *Production of Commodities by Means of Commodities,* was published in 1960. In this, Sraffa builds up a system following the classical tradition. Given the methods of production and taking one of the distributive variables as being exogenously given, Sraffa resolves some important issues concerning value and distribution. In particular, he has shown that values (i.e., self-replacing prices) and distribution cannot be computed independently of each other. There is a unique association between prices and the distribution of the physical surplus. Therefore, given a distribution, we can calculate the prices or vice-versa. This was and is a devastating result for (Neo-) Walrasian general equilibrium and Austrian value theory and for any attempt to measure aggregate capital and aggregate national product independent of the distribution.[5]

In Sraffa's framework, distribution becomes the key aspect in determining the relative prices. It is a penetrating (prelude) critique of neoclassical theory and its marginal foundations. This slender volume breathed a new

[4] Although the title of Hayek's book is *Prices and Production*, it is really about a monetary theory of the business cycle. The title of Sraffa's review in fact is *Dr. Hayek on Money and Capital.* One of Sraffa's points was that a disequilibrium scenario requires the use of deferred means of payments—i.e., money or credit-debt contracts.

[5] This point made by Sraffa is often misunderstood or misinterpreted. Zambelli (2018b) takes the numerical values used by Sraffa and demonstrates that the value of capital and the industry capital/output ratios are all functions of distribution (prices). The values of aggregate capital, or the value of capital per industry do not remain constant as the distribution changes. Furthermore, their changes are not monotonic with respect to changes in distribution.

life into the classical, surplus based approach and offered it as a credible alternative to the marginal approach. Finally, there is also his important interjection in the Corfu conference on Capital theory, which in our opinion provides us a hint to find a persuasive way to interpret his body of work.

### 2.1 Sraffa's Contributions: A Coherent Whole

Sraffa's early work, as we have outlined previously, is highly applied in character. It focuses on monetary and banking issues. Around the same time, he dealt with some thorny issues in pure theory. The later part of his work focused mostly on theoretical issues concerning value, distribution and capital, with relatively less applied elements. This has often led scholars to conclude that his work on the two areas—money and capital, for the convenience of exposition—as disconnected, distinct spheres of Sraffa's scholarly contributions.

Instead, we argue that his early work on money, his critique of marginal theory and capital are not isolated exercises. Following Panico (1988b), we claim that Sraffa's research endeavours in these two parts are coherent with respect to a focused research agenda.[6] First, it is important to bear in mind that Sraffa's research agenda had been set fairly concretely during the 1920s and early 1930s. We now know that essential aspects of his theoretical system were already in place by then. In fact, his subsequent research work was mainly about furthering this research agenda, which only saw marginally modifications over time. Second, from his early writings on money, it is evident that he had a concrete stance about the role of power, class, interest groups and non-market forces in the determination

[6]In his paper discussing Sraffa's early work Sraffa (1920, 1922a) on money and banking, Panico (1988b, p. 7) underlines that:

> Sraffa started his research working on applied monetary and banking problems, making some relevant contributions to the area. The theoretical problems that made his fame as an outstanding theoretician came to dominate his interests only at a later stage.

Panico also argues that:

> … there is a close link between [Sraffa's] earlier writings on applied problems and his later theoretical works, a link which clarifies the origins of his theoretical interests, showing the error of considering his contributions as 'merely abstract exercise in pure theory'.

of the money rates of interest. There is no evidence that we can find to believe that his stance had changed in any substantial manner throughout his academic life. Third, we argue that this position can be seen to later manifest itself in the form of Sraffa having an *exogenous* distribution. This exogenous distribution variable (profit or wage rate), as argued by many scholars (Bharadwaj 1963), is a crucial element in determining the relative prices in PCMC.

Even if one accepts the lack of complete disconnect between Sraffa's early and later writings, we are still left wanting for a coherent narrative to understand his body of work. To this end, we propose that Sraffa's research may be characterised as one that is geared towards achieving precision in theory. In order to make this point clear, we construct a spectrum in which contributions range from being applied in nature on one end to being purely theoretical on the other end (see Fig. 15.1).

- Works closer to the applied end of the spectrum take existing theory largely as a given and explore innovative applications and interpretations, without drastically challenging or reworking the theoretical premises.
- Contributions closer to the right-hand side of the spectrum, on the other hand, constitute fundamental contributions that challenge, construct and clarify core theoretical categories.

The former, given their applied nature, necessarily deal with and even rely on imprecise measurements. No such imprecision is allowed when it comes to purely theoretical concepts and definitions involving economic measures. This attitude is well encapsulated in his intervention at the 1958 conference of the International Economic Association at Corfu on the '*Theory of Capital*'.

> [O]ne should emphasize the distinction between two types of measurement. First, there was the one in which the statisticians were mainly interested. Second there was measurement in theory. The statisticians' measures were only approximate and provided a suitable field for work in solving index number problems. **The theoretical measures required absolute precision. Any imperfections in these theoretical measures were not merely upsetting, but knocked down the whole theoretical basis.** [...] The work of J. B. Clark, Böhm-Bawerk and others was intended to produce pure defini-

tions of capital, as required by their theories, not as a guide to actual measurement. If we found contradictions, then these pointed to defects in the theory, and an inability to define measures of capital accurately. It was on this—the chief failing of capital theory—that we should concentrate, rather than on problems of measurement.—Piero Sraffa (Lutz and Hague 1961, pp. 305–306, emphasis added)

We can place Sraffa's scholarly contributions to the two different spheres (money and capital) on the above mentioned spectrum in Fig. 15.1.

First, on Sraffa's contributions relating to monetary matters:

(a) On the left hand side of the segment, we place his work on the causes of inflation and issues concerning monetary stabilisation; institutional characteristics and regulatory framework that characterised the Italian banking system in the early 1920s and their consequences; his analysis of the relationship between industrial and banking sectors; the role of interest groups (Sraffa 1920, 1922a, b).[7]

(b) On the right hand side (theoretical extreme), we place his review of Hayek's 'Prices and Production', where his critique strives to high-

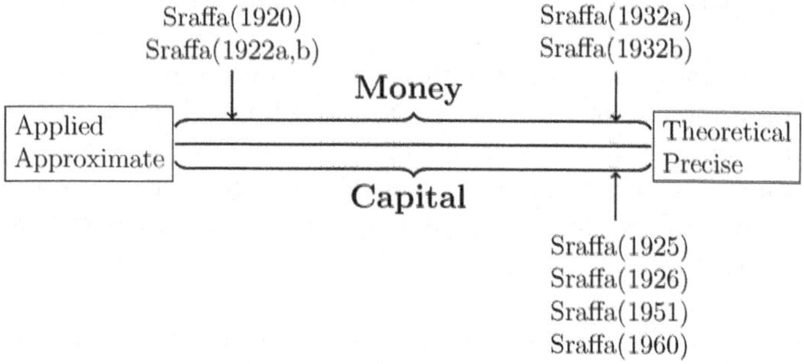


**Fig. 15.1** A classification of Sraffa's contributions

[7] In his thesis, for instance, he takes the quantity theory of money as a given and applied it to monetary issues of his time (Panico 1988b, p. 10).

light and remove imprecisions from pure theory that concern the role of money in business cycles.[8]

Second, moving on to Sraffa's contributions pertaining to value, distribution and measurement of capital (non-monetary issues, loosely speaking, which we characterise under capital for the ease of exposition): all of them invariably fall on the right-hand extreme of the spectrum—involving theoretical contributions which require definitions and measurements with utmost precision. Thus, in the right end of the spectrum, we have the problems of:

- measurement of capital and returns at the firm level (Sraffa 1925, 1926)
- value and the measurement of capital at the aggregate level (Sraffa 1951, 1960)

In each of these cases, Sraffa is uncompromising in his theoretical rigour, be it in fleshing out inconsistencies in Marshallian economics, clarifying the conceptual issues in Ricardo's theoretical apparatus or constructing a theoretical system of his own. It certainly won't be an exaggeration to claim that precision in pure theory has been one of the major concerns of Sraffa throughout his academic life. For instance, other than the works mentioned above, his interventions in the famous 'Symposium on increasing returns and the representative firm' (Robertson, Sraffa and Shove 1930, pp. 89–92, 93) bear evidence to his penchant for theoretical precision. Equally, there are also several instances in his archives and correspondences that further support this. Sraffa was critical of the use of marginal productivity theory to determine aggregate wage shares by Hicks in his

[8] His critique of Hayek's book was made alongside Keynes' own critique of Hayek (Keynes 1931; Hayek 1931b, 1932b; Sraffa 1932a, b). There is evidence of Sraffa's involvement in the corrections of 'A Treatise on Money' (Keynes 1930) as well. It is worth noting that Sraffa translated (Keynes 1923, 'A Tract on Monetary Reform') into Italian and was involved in the discussions surrounding the writing of Keynes' 'General Theory'. In this context, it is also important to bear in mind that Chapter 17 of the *General Theory*, (Keynes 1936, Ch: The Essential Properties of Interest and Money) may be interpreted as addressing the relation between interest and money using some of the arguments put forward by Sraffa in his review of Hayek. In fact, Keynes does acknowledge that the idea of the own rate of interest was taken from Sraffa's critique (Sraffa 1932a) of Hayek's 'Prices and Production' (Keynes 1936, p. 223, fn. 1). All these attest to the interpretation that money and monetary issues were not in anyway insignificant in Sraffa's academic interests.

*Theory of Wages*. In his correspondence with Hicks (Sraffa papers, C133, undated), Sraffa objects to certain assumptions concerning the elasticity of substitution. These astute remarks may be seen as an anticipation of capital theory debates of the 1960s.

Although we have placed Sraffa's works exclusively on one of the two spheres, we need to exercise caution while interpreting such a characterisation. For example, *PCMC* (Sraffa 1960) certainly has elements which are clearly related with money and credit. This can be seen from the importance that he places, explicitly or implicitly, on the relevance of money interest rates in influencing the exogenous distributive variable. One may argue that these reference points found in PCMC are simply 'marginal'. They may be so for a *critique* (or a prelude to a critique) of certain kinds of economic theory. However, they are, at least in our view, central starting points for a rigorous generalisation of Sraffian schemes. We discuss more about money and distribution in Sraffa in the next section.

## 3 Sraffa, PCMC, Money and Distribution

We have outlined earlier that money was very much a part of Sraffa's research interest right from his early days as an economist. The lack of detailed discussion of money in PCMC has led many to believe that money perhaps was not important in his theoretical system or that there is no place for it. We argue that money (seen as credit and debt), far from being irrelevant, offers a fruitful generalisation of the Sraffian schemes. Second, even within the context PCMC, there are two important points where the money rates of interest become particularly relevant.

### *3.1 Importance of Money Rates of Interest: Explicit Reference*

The explicit point concerning the importance of money rates of interest in PCMC is in the well known quotation:

> The rate of profits [...] is accordingly susceptible of being determined from outside the system of production, in particular by the level of the money rates of interest. (PCMC, p. 33)

The relationship between the rate of profits and the money rates of interest is one of the central themes of economic theory. In particular, a crucial issue is whether there exists a (unique) *natural* rate of profits which is related to the productivity of physical capital. Additionally, whether money rates of interest converge towards a natural rate of profits and whether the sectoral rates of profits converge towards a unique money rate of interest is also relevant. Fisher (1906, 1907, 1911) and Wicksell (1936) have made important contributions to the debate centred on these issues.

If the money rates of interest determine the rate of profits of commodities, and this rate of profits, in turn, determines the distribution of the physical surplus, it reasonably follows that the very forces that determine the money rates of interest implicitly determine the distribution as well.[9]

If the money rates of interest do indeed regulate the rate of profits, then it could be argued that specifying this functional relationship in some unique manner can help ameliorate the issue of exogeneity, making the system determinate. However, Sraffa does not find merit in this. That the money rates of interest are important factors (but not the only one) for the determination of the distribution of the surplus is stated in a response to Garegnani (Sraffa Papers D3_12_111_144–146), who was seeking clarifications with respect to the above quotation.

In 1962, Garegnani was preparing a review of PCMC to be published in the Italian journal *Moneta e Credito*. In a correspondence with Sraffa, Garegnani enquired about the relation between the profit rate and the money rates of interest. Garegnani seemed to be in doubt on whether he should address the issue of the profit rate as determined by the money rates of interest.

> [...] you opt for the rate of profit "determined from outside the system ... in particular by the level of the money rates of interest" [...] I am reluctant to venture, even if only with hints, on the terrain of a theory of profit determined by the money rates of interest, with the risk of attributing to you opinions that you do not have.
>
> This for the review. The question however interests me a lot [...] Expressed in very rough terms the way I see the problem is as follows. From

[9] This was an important point already put forward in Sraffa (1922a) and subsequently in his debate with Hayek (Sraffa 1932a). Chapter 17 of Keynes (1936) may be interpreted to be addressing this problem using arguments put forward by Sraffa.

Ricardo to Wicksell the idea was that **the rate of interest** would **adapt** to the **rate of profit** obtained from production.

The **justification for that theory** lies, it seems to me, in the idea of a **demand for loans with almost infinite elasticity** (given the time required to formulate the investment plans) around the levels of interest corresponding to the rate of profit obtained in production. If such an elasticity in fact would exist, the **market interest rate** should always tend to its **"natural"** level (any attempt by the monetary authority to fix a different rate by creating or destroying credit would fail, then, because the resulting inflation process or deflation would force to return to the "natural" rate). [Sraffa Papers, D3/12/144–145, our translation, emphasis added][10]

To this question, Sraffa replies in his letter as below:

From the point of view of your research it seems to me that it is difficult to unify the views of Ricardo and Wicksell, even though they both make the **rate of interest** depend on the rate of profit. But as for the **"rate of profit"** they only have the words in common: for the former it is the excess above the wages given, for the latter the marginal product of capital! How can you

[10] Text in the original:

[…] Lei opta […] per il saggio del profitto "determinato da influenza esterne del sistema …e particolarmente dal livello dei tassi monetari dell'interesse" […] sono restio ad avventurarmi, anche solo per accenni, sul terreno di una teoria del profitto determinato dai saggi monetari dell'interesse, col rischio di attribuire a lei opinioni che non ha.

Questo per la recensione. La questione peró mi interessa molto […] Espresso in termini molto rozzi il modo con cui vedo il problema è questo. Da Ricardo a Wicksell l'idea è stata che **il saggio di interesse si adeguasse al saggio del profitto** ottenuto nella produzione.

**La giustificazione di quella teoria** sta, mi sembra, nell'idea di una **domanda di prestiti di elasticità pressoché infinita** (dato il tempo richiesto per formulare i piani di investimento) intorno ai livelli dell'interesse corrispondenti al saggio di profitto ottenuto nella produzione. Se infatti tale elasticità esistesse, il saggio **di interesse di mercato** dovrebbe sempre tendere al suo livello **"naturale"** (l'eventuale tentative della autorità monetarie di fissare un diverso saggio creando o distruggendo credito fallirebbe, allora, perché il risultante processo di inflazione o deflazione costringerebbe a ritornare al saggio "naturale") [Sraffa Papers, D3/12/111/150, emphasis added]. To this question, Sraffa replies in his letter as below:

> put them together? I would like to know the sense in which it can be said that "the justification for that theory" lies in the "demand for loans with almost infinite elasticity at an interest rate equal to the rate of profit". **Justification my foot**! I call it an "identical repetition in difficult words", a serious tautology.
>
> I do not see the difficulty of determining the rate of profit by means of a rate of controlled or conventional interest, **provided that the rate of profit is not assumed to be determined by external unavoidable circumstances**. [Sraffa Papers, D3/12/111/150, our translation, emphasis added][11]

But Sraffa goes on to stress his disapproval of any mechanical theory of determining distribution in the economic system.

> … I am convinced that the maintenance of **the rate of interest** by the bank or the stock exchange has played its part in determining the distribution of income among the social classes: because it is an **obligatory passage** for those who give and those who borrow loans.
>
> … I did not mean anything very binding, and in general I only wanted to send out some signals **to avoid anyone to think** that the system [of PCMC] is presented [by me] as a "foundation" for a theory of relative supplies of capital and labour! It is the <u>negation</u> that seems important to me: as to the <u>affirmative</u> I have no intention of putting forward yet **another mechanical theory** which, in one form or another, reinforces the idea that distribution is determined by natural, or technical, or perhaps even accidental circumstances, but such as to render **futile** whatever **action**, from one side or the

[11] Text in the original:

> Dal punto di vista dei suoi studi mi pare che sia difficile unificare i punti di vista di Ricardo e Wicksell, per quanto entrambi facciano **dipendere** il **saggio dell'interesse** dal **saggio del profitto**. Ma quanto al "saggio del profitto" hanno in comune solo il nome: per il primo è l'eccedente al si sopra del salario dato, per il secondo il prodotto marginale del capitale! Come si fa a metterli in pariglia? Io vorrei sapere in che senso si possa dire che "la giustificazione di quella teoria" sta nella "domanda di prestiti ad elasticità infinita a saggio di interesse uguale al saggio di profitto". **Giustificazione un corno!** Io la chiamo "ripetizione identica in parole difficili", grave tautologia.
>
> Non vedo la difficoltà della determinazione del saggio del profitto mediante un saggio dell'interesse controllato o convenzionale, a **condizione che non si presupponga il saggio del profitto determinato da circostanze ineluttabili esterne**. [Sraffa Papers, D3/12/111/150, our translation, emphasis added]

other, aimed to **modify the distribution**. (Sraffa Papers, D3/12/111/149, our translation, in bold emphasis added)[12]

Clearly, Sraffa views banks and the stock market (which are relevant for setting of the money rates of interest) as being important in the determination of distribution. But he makes a very subtle point. In PCMC, Sraffa does show that the determination of prices, and hence distribution, cannot be determined *exclusively* within the economic system. In his system, there is *at least* one degree of freedom and he picks the rate of profits to be the variable which is determined from outside the economic system (see Appendix A). Once the rate of profits is given, the distribution of the surplus is simultaneously determined. This is the *negation* that Sraffa seems to emphasise, that is the fact that the system as a whole is indeterminate (see A.6). But, on the contrary, suppose that there was a theory that would *mechanically* explain the determination of the rate of profits through the money rates of interest. This, in turn, would make the system determinate by the mechanical forces that shape the functioning of the banking system or the stock market. Even in such a case, Sraffa seems to point out that any attempt to modify the distribution would be futile.

The point here is that the allocation of new loans could depend partly on the power structure and decisions of banks that provide the *necessary financial backing* to powerful companies that may be controlled by a few individuals. Obviously, strictly mechanical forces of the economic system are not adequate to exactly capture such tendencies. This seems to be the point that Sraffa is making when he writes "*I did not mean anything very*

[12] The text in Italian:

> … sono convinto che il mantenimento del **saggio d'interesse** da parte della banca o della borsa abbia avuto la sua parte nel determinare la distribuzione del reddito fra le classi sociali: perché è un **passaggio obbligato** per chi dà e per chi prende a prestito.
>
> […] io non ho inteso dir niente di molto impegnativo, e in generale ho solo voluto metter fuori qualche segnale **per evitare che si creda** che il sistema viene presentato come "fondamenta" per una teoria delle offerte relative di capitale e lavoro! E' la negazione che mi sembra importante: quanto alla affermativa non ho nessuna intenzione di mettere avanti **un'altra teoria meccanica** che, in una forma o nell'altra, ribadisca l'idea che la distribuzione sia determinata da circostanze naturali, o tecniche, o magari accidentali ma comunque tali da rendere **futile** qualsiasi **azione**, da una parte o dall'altra, per **modificarla** (Sraffa Papers, D3/12/111/149, in bold emphasis added).

*binding* (impegnativo)*, and in general I only wanted to send out some signals to avoid anyone to think that the system [of PCMC] is presented [by me] as a "foundation" for a theory of relative supplies of capital and labour!*[13]

Sraffa seems to take a similar position as early as 1922 Sraffa, where he writes:

> The large industries are stimulated on their part to make themselves independent by acquiring control of a bank so as to obtain from it, without undergoing heavy impositions, **the necessary financial backing**. As a result of this opposition, however, it cannot be said, generally speaking, that either of the two opposite tendencies has the absolute upper hand over the other. The general tendency seems to be towards the elimination of this opposition by the formation of large "groups" of companies of the most varied kinds concentrated round one or more banks, mutually related by the exchange of shares and by the appointment of Directors common to them. Within these "groups" the various interests are all equally **subject to the interests of a few individuals** who control the whole group, possessing on their own only a very few shares of the various companies. Very little is known and very little can be generalised about these groups, on account of the undetermined state of their structure, of their **unofficial character**, of the variety of the various groups, and of the continual shifting of the elements which compose them. What the public knows and feels-not only when disasters take place, fatal to the existence to some of them, or when hostilities break out between one group and another-is **the enormous financial and political power which they have and the frequent use they make of it to influence both the foreign and home policy of the Government in favour of their own interests**. Each group keeps several press organs which support its policy, and some of the accusations made against certain Ministries of being actuated by the interests not of a class, but of private concerns, and of favouring one financial group against another, have no doubt a basis of truth. (Sraffa 1922a, p. 196, emphasis added)

### 3.2 Implicit Influence of the Money Rate of Interest on the Measurement of Capital and Distribution

The implicit influence of the money rates of interest in PCMC relates to the critique that Sraffa develops about the possibility of measuring

[13] If our interpretation is correct, it is puzzling that many Sraffians have placed a strong emphasis on the mechanics of the economic determination of the long term uniform profit rate.

aggregate capital. In particular, the use of the Austrian notion of 'roundaboutness of capital' or the 'period of production'. The interest rate in the Austrian framework has an influence on the roundaboutness of capital or the length of production processes (Samuelson 1966, pp. 568–569).

Sraffa shows in PCMC that the length of 'the period of production' (Ch.VI 'Reduction to dated quantities of labour') cannot be independent of distribution or the rate of profits. The view put forward by authors like Jevons and Böhm-Bawerk was that as the money rate of interest falls, the 'period of production' (roundaboutness of capital) would increase.[14] Sraffa writes:

> (The reduction to dated labour terms has some bearing on the attempts that have been made to find in the **'period of production'** an independent measure of the quantity of capital which could be used, without arguing in a circle, for the determination of prices and of the shares in distribution. [What demonstrated in this book] seems to be **conclusive** in showing the impossibility of aggregating the **'periods'** belonging to the several quantities of labour into a single magnitude which could be regarded as representing the quantity of capital. The reversal in the direction of the movement of relative prices, [...], cannot be reconciled with **any notion** of capital as a measurable quantity **independent** of **distribution and prices**)—PCMC, p. 38, emphasis added.

The critique that Sraffa advances has to do with the lack of theoretical foundation concerning the monotonic inverse relation between the 'period of production' (i.e., capital) and the interest rate.

Sraffa's intervention at the Corfu conference is related to the discussion of Hicks' article, titled *The Measurement of Capital in Relation to the Measurement of other Economic Aggregates* (Hicks 1961). The concluding phrase in this article is: *'The marginal productivity of capital is the marginal productivity of roundaboutness [period of production], after all.'* (Hicks 1961, p. 31).

It is here that Hicks' economic theory clashes directly against Sraffa's critique. Sraffa was in fact correct that a monotonic relationship between the value of capital (period of production) and the rate of profit cannot be

[14] On the notion of the *period of production* and on Sraffa's personal notes on it, see Sinha (2016, Chap. 5, pp. 111–151). On Sraffa's critique of the period of production as being independent of the rate of profits, see PCMC, Chapter 6, and a rather harsh comment of Sraffa (1962) on the review of PCMC by Harrod (1961).

established unambiguously.[15] In our parallel work incorporating money inside the Sraffian schemes, we show that a monotonic relation between value of capital and the monetary interest rate is seldom observed.

We have so far drawn attention to Sraffa's view on the crucial role of distribution, money rates of interest and his explicit dismissal to engage in building a mechanical theory, where distribution is determined by the natural forces of the market. It is also instructive to contrast Hicks' view and Sraffa's view on the matter.

One of Sraffa's objectives is to determine the

> *... set of exchange-values which if adopted by the market restores the original distribution of the products and makes it possible for the process to be repeated; (PCMC, p. 3, emphasis added)*

so that the system is in a *self-replacing state*.

Sraffian self-replacing prices are market clearing prices. Sraffian schemes can be interpreted as budget constraints of the overall economic system. It is a system where producers have to exchange their products (endowments) at the end of the production cycle. They do so in order to buy the means of production necessary for production to be replicated during the next production cycle.[16]

*Value and Capital* (Hicks 1939) is recognized as a foundation of Neo-Walrasian General Equilibrium. There are several similarities between the production structure that one observes in Hicks (1939) and in Sraffa's PCMC. Hicks separates the production cycle (the week) and the market day (Monday) in the same way in which Sraffa separates the production cycle (the year) with the annual market day (occurring after the harvest). Hicks also points to his discussions with Sraffa in the *Preface*: "I have had some very useful criticism from Mr. Sraffa" (Hicks 1939, p. vi).[17] In this

[15] See also Zambelli (2018a) for an empirical demonstration of this point.

[16] Hahn (1982), Hahn and Petri (2002), Fratini (2018), Garegnani (2000), Mandler (1999a, b, 2002a, b, 2005), Negishi (2014, p. 7), Parrinello (2008), Schefold (2005, 2008), Sinha and Dupertuis (2009) are exam- ples where Sraffian schemes are studied in relation to Walrasian (intertemporal) general equilibrium. In these contributions there is a clear acknowledgment that Sraffian schemes are equivalent to the aggregate budget constraints of the producers, workers and consumers that compose the economic system.

[17] There is evidence from Sraffa's diaries that they met frequently while Hicks was in Cambridge (from 1935 to 1938). There is also the draft of the letter from Sraffa to Hicks in which Sraffa makes a precise quote in relation to Hicks' "The Theory of Wages", first Edition, Hicks (1932). In a personal conversation, Massimo Di Matteo at the University of

book, Hicks also aimed at providing firm foundations to the theory of monetary interest rates:

> [...] it is evident that any treatment which pretends to deal with the economic system as a whole [...] cannot possibly regard the rate of interest in isolation. It is a price, like other prices, and must be determined with them as part of a mutually interdependent system. The problem is not one of determining the rate of interest *in vacuo*, but is really the general problem of price-determination in an economy where borrowing and lending are practised, and in which the rate of interest is therefore a constituent part of the general price-system. (Hicks 1939, Ch.XII, *The Determination of the Rate of Interest*, p. 154)

When we consider Sraffa's response to Garegnani's query (see Sect. 3.1), in conjunction with the content found in PCMC, this quote is quite illuminating. At first reading, it seems that Hicks and Sraffa are saying the same thing. But there is a very crucial and fundamental difference.

Sraffa considers that the rate of profits and/or the money rates of interest may not be determined mechanically by the working of economic forces alone. Therefore, from his standpoint, one can claim that there is a scope for attempts to determine the distribution in a society through attempts to change the power structure or through policy.

On the contrary, in Hicks's world, any attempt to change the distribution of the surplus produced would be *futile* because of the very mechanics of the working of the economic system, which solely determine distribution at full employment. For Sraffa, General Equilibrium prices are not determined by the mechanical forces of the markets alone.[18]

Siena, who has been a student of Hicks, told the second author that he had indeed asked Sraffa about what the 'useful criticisms' were. He recalls Hicks' answer that they discussed matters related with Walrasian General Equilibrium and Pareto.

[18] See also Panico (1988b, p. 26) in support of this interpretation:

> A well-defined thread, going throughout his writings up to Production of Commodities by Means of Commodities, can thus be found in Sraffa's work. His theoretical interests were enhanced by the need to provide firm analytical foundations for his original view of the working of the economic system. This view emphasised the role of state intervention in determining income distribution and the equilibrium position of the economy, and the fact that policy decisions are influenced by the pres-

## 4 Towards a Sraffian Monetary Theory of Production and Distribution

We have argued in the previous sections that monetary issues are an important subject of Sraffa's research agenda and that it could explain some of the theoretical positions that were adopted in his later contributions. We also outlined the explicit and implicit role that money plays in PCMC. In order to build a monetary theory of production and distribution in the Sraffian vein, we first need to present the motivations for introducing money inside the Sraffian schemes. Further, we need to clarify what we mean by 'money' in this context and outline how the structure presented in PCMC can be suitably modified for this purpose.

It is well known that PCMC, as the subtitle indicates, was a *Prelude to a Critique of Economic Theory.* This critique is based on the idea of a self-replacing system and the associated notions of self-replacing prices, wages and profit rates. Self-replacing is the situation where production can continue to take place as it did during the previous production cycle. This condition is important because it allows one to focus on prices and distribution of the surplus in a situation where produced physical quantities are not changing. Sraffa has shown that the *natural prices* of the commodities in such a situation are not unique and they simultaneously determine the distribution of the physical surplus. Additionally, the values of gross or net output and the means of production (the value of capital) are themselves a function of prices and hence of distribution.

There are several directions in which this framework can be and have been extended in the literature, ranging from the introduction demand in to this framework, exploring compatibility with the Keynesian system, to devising a complete general equilibrium system. In the next subsection, we outline the motivations for generalising this framework to acommodate a form of money.

sures of the dominant groups, and are not the result of purely technical considerations. His earlier analyses on the influence of monetary policies on income distribution were closer to the classical and the Marxist tradition than to the neoclassical one. The study of these analyses, together with that of his contacts with Keynes, thus provides useful insights in the evolution of Sraffa's thought, underlining that his later work is not 'merely abstract exercise in pure theory'.

### 4.1 *Money in the Sraffian Schemes: Some Motivations*

Right from the outset, we would like to acknowledge that there have been attempts in the past to introduce money inside the Sraffian schemes (Panico 1988a; Pivetti 1991; Ciccarone 1998). However, to the best of our knowledge, the possibility of generalising this framework by the introduction of money is far from being realised. There are several motivations behind the need for introducing money inside the Sraffian schemes.

First, money is an essential aspect of the capitalistic mode of production. This point hardly seems to need any elaboration and this case is eloquently made by Sraffa himself in his review of Hayek's book. Second, the fact that money may not have been necessary for Sraffa's critique of marginalist theory, which in fact was the main thrust of PCMC, does not automatically imply that there is no place for money in his entire theoretical apparatus. It seems useful to extend this scheme—not necessarily to get a complete theory—in order to see its potential, yet remaining faithful to the spirit in which Sraffa constructs his theoretical system. Third, if we view money not as a commodity like other commodities in the economic system, then the role it plays in the facilitating production needs to be studied. Fourth, there is both a need and scope for money—viewed as credit or a deferred means of payment—within Sraffian schemes.

Prices in Walrasian general equilibrium are those that are associated with equilibrium exchanges: no trade is allowed outside equilibrium.[19] Within Sraffian schemes the endowments are quantities in the possession of agents after the harvest and before the beginning of the market day. Evidently, the self-replacing prices are market clearing prices. Here the production (and consumption) plans are implemented due to the fact that producers and workers have the necessary purchasing power to buy the necessary means of production and the surplus. That is, if deferred means of payments do not exist or are not generated, during the market day the prices must be such that the revenues of producers and workers must

[19] After twelve chapters, defining and proving theorems about Walrasian general equilibrium, Arrow and Hahn (1971) state this clearly:

> Of course, our model is in no shape to give a satisfactory formal account of the role of money. In particular it would be hard to "explain" the holding of money or why it mediates in most acts of exchange. (p. 338)

exactly match their expenditures (i.e., the excess supply and excess demand functions are zero).

The necessity of money for the exchanges is well explained by Arrow and Hahn (1971):

> The terms in which contracts are made matter. In particular, if money is the good in terms of which contracts are made, then the prices of goods in terms of money are of special significance. This is not the case if we consider an economy without the past and the future.[...] **If a serious monetary theory comes to be written, the fact that contracts are indeed made in terms of money will be of considerable importance**. (Arrow and Hahn 1971, pp. 365–7, emphasis added)

As prelude to a critique, the assumption of uniform rate of profits and the absence of monetary means of exchange (which imply accounting equilibrium prices) are more than justified. The relevance of Sraffa's results have been questioned by Hahn (1982), among others, on the ground that the uniform rate of profits assumption present in PCMC reduces Sraffian schemes to be just a special case of the Walrasian General Equilibrium. In a recent paper (Zambelli 2018b), the assumption of the uniform rate of profits has been removed and the properties of the Sraffian system have been studied under more general conditions (See Appendix A).

Building on this, here we argue for an extension where there is a possibility of having exchanges of real goods or services against deferred means of payments (money, credit and debt). In order to introduce money in the Sraffian scheme, we can consider cases in which prices alone are not enough to allow for self-replacing. There is a distinct possibility that commodities may be exchanged due to the issue of new loans (credit and debt) or through the transfers of fiat money. The introduction of deferred means of payments inside Sraffian schemes provides a good starting point for introducing money (See Appendix B for a brief sketch of how this can be accomplished).

### 4.2 *Marx, M–C–M and the Capitalistic Mode of Production*

Although Sraffa's work has been widely interpreted as a critique of marginalism, it is also seen by some as a critique of the embodied labour theory of value. In Chapter VI, *Reduction to Dated Quantities of Labour* Sraffa is very clear about this. Sraffa demonstrates that the embodied labour terms[20] when summed up cannot be used as an invariable measure of capital and this has been seen as a critique of Ricardo's labour theory of value and its use by Marx. Most importantly, he shows that it cannot be reconciled with 'period of production' as an independent measure of the quantity of capital.[21] Sraffian schemes are sometimes viewed not just as a critique but also as a solution of Ricardo's embodied labour value problem and as a way to address the Marxian transformation problem (Marx 1894, Capital, volume III, Ch. XII).[22]

Marx characterizes economic systems with two circular processes: the Commodity- Money-Commodity, *C–M–C*, and the Money-Commodity-Money, *M–C–M*.

> In the circuit *C–M–C*, the money is in the end converted into a commodity, that serves as a use-value; it is spent once for all. In the inverted form, *M–C–M*, on the contrary, the buyer lays out money in order that, as a seller, he may recover money. ... He lets the money go, but only with the sly intention of getting it back again. The money, therefore, is not spent, it is merely advanced (Marx 1867, vol. 1, p. 148)

There are those that interpret Sraffa's contribution as holding an *'antagonistic'* view with respect to Marx's work (Pilling 1972; Lebowitz 1973; de Brunhoff 1975, 1990; Rowthorn 1974; Nicholas 2014) and others who see it as a *'harmonious'* one (Dobb 1975; Eatwell 1974; Laibman 1975; Garegnani 1978; Lippi 1979; Hodgson 1982; Steedman 1977).

[20] The quotation we refer to in Sect. 3.2 is made at the end of section 48 in PCMC. Figure 2 in page 36 of Sraffa's book shows the value of different 'labour terms' which when summed up make the value in terms of 'embodied labour' of capital, which varies as the profit rate (i.e. distribution of the surplus) varies. This result is also stressed with the aid of Fig. 3, which is found in the same page as the above quotation.

[21] There is a possibility here for a misleading conclusion that this critique of economic theory that is based on the impossibility of measuring capital, or the values in general, is restricted to the adherence of a labour theory of value. This would be incorrect. Zambelli (2018b), for example, has shown that the values, even when the prices and values are determined by taking the surplus as the *numéraire,* are a function of distribution. Furthermore, when the agricultural sector (or any other sector) is taken to be the *numéraire,* there is no simple, unique measure of capital that is independent of distribution. This is a 'modern' critique and does not depend on the labour theory of value and the Austrian notion of the 'period of production' (Zambelli 2018b).

[22] For a contemporary discussion of the transformation problem see also, among others, Dobb (1967), Laibman (1973), Samuelson (1971), Seton (1957), Shaikh (1984), Steedman (1977), Winternitz (1948).

The *'antagonists'* do recognize that Marx had some *unsolved* issues in his analysis (i.e., the relation between commodity, money and capital when defined as social necessities).[23] However, they consider the Sraffian solution to the transformation problem (and value theory) strictly as a solution to Ricardo's search for an *'invariable measure of value'*, but not as a solution to Marx's value problem.

The major reason for this would be that commodity prices, money and capital can perhaps be studied in the form in which they may be observed from a pure economic theory point of view (as in Ricardo). However, when discussing value, according to Marx, one has to consider the inescapable role of money in commodity circulation. In the capitalistic mode of production, the change from the *use value* of the commodities into their *exchange value* means a change in focus from the *C–M–C* circuit to the capitalistic circuit *M–C–M*. Hence Sraffian schemes as they stand may shed light on the *C–M–C* circuit, but cannot yet shed light on the *M–C–M* circuit. However, according to Marx, the latter characterizes the capitalistic mode of production, because *it is money which is the form that the commodities take.*[24] A formal treatment of money in the context of Sraffian

[23] Pilling (1972); de Brunhoff (1973, 1975) discuss the difference between Ricardo (and a Ricardian interpretation of the work of Sraffa made by the neo-Ricardians) and Marxian theory of value as it is dealt with in Marx's Capital. *"For Marx did not attempt to construct a pure economie theory, his field of problems and his point of departure is quite different from that of Ricardo and that of pure economics"* (de Brunhoff 1973, p. 423). Crucial points are (1) Marx's distinction between 'labour-time', 'abstract labour', 'labour force power' and 'socially determined labour' and (2) the different, unique meaning given to money.

[24] Marx's critique to Ricardo on this point may be summarized by the following:

> With him [Ricardo], however, wage labour and capital are again conceived as a natural, not as a historically specific social form *[Gesellschaftsform]* for the creation of wealth as use value; i.e. their form as such, precisely because it is natural, is *irrelevant,* and is not conceived in its *specific* relation to the form of wealth, just as wealth itself, in its exchange- value form, appears as a merely formal mediation of its material composition; thus the specific character of bourgeois wealth is not grasped precisely because it appears there as the adequate form of wealth as such, and thus, although *exchange value* is the point of departure, the specific economic forms of exchange themselves play no role at all in his economics. Instead, he always speaks about distribution of the general product of labour and of the soil among the three classes, as if the form of wealth based on exchange value were concerned only with use value, and as if exchange value were merely a ceremonial form, which vanishes in Ricardo just as money as medium of circulation vanishes in exchange. Therefore, in order to bring out the true laws of economics, he likes to refer to this relation of money as a merely formal one. This does not mean that commodities do not have *use values, but it is*

schemes may be necessary if one aims at shedding light on the mechanics of the economic aspects of both C–M–C and M–C–M circuits and of the Marxian value theory in general.

We conclude with the suggestion by Nuti (1971, p. 33) reported in the introduction of Panico (1988a, p. 7):

> The most appropriate way of approaching the theory of distribution, reintroducing the reality of class struggle into this important branch of Political Economy, seems therefore that of combining the Sraffian relation between wage and profit rates with the little we know—not least from Marx—about the interaction of real and monetary phenomena.

We undertake first steps in this direction in our companion piece (Venkatachalam and Zambelli 2021) and a concise summary is presented in Appendix B. More importantly, we hope to have convinced the reader that this is a journey that seems worth undertaking and that it promises to offer interesting insights.

## Appendix A: Sraffian Schemes and Distribution

### *Production and Methods of Production*

In PCMC (p. 3, §1 and p. 10, §9) Sraffa assumes that there is *'an annual cycle of production with an annual market'*.

At the end of the production cycle there are *n* produced commodities:

$$\mathbf{b} = [b_1, b_2, \ldots b_i, \ldots, b_n]^T \qquad (15.1)$$
$$i = 1, 2, \ldots, n.$$

The method of production (PMCM, p. 3, §1 and p. 6 §4) producing $b_i$ is a linear combination of means of production and labour:

$$a_i^1, a_i^2, \ldots, a_i^j, \ldots, a_i^n, l_i \rightarrow b_i \qquad (15.2)$$

*money which is the form that the commodities take*. Hence also his weakness in the doctrine of money proper. (de Brunhoff 1973, p. 429, author translation from Marx's *Grundrisse*)

where $a_i^j$ denotes the means of production produced by industry $j$ used in the production of commodity $i$ and $l_i$ the labour used in the production of $b_i$.

In a more compact form, the method of production $i$ may be written as $\mathbf{a}_i, l_i \rightarrow b_i$.

### *Surplus to Be Distributed*

*The 'economy produces more than the minimum necessary for replacement and there is a surplus to be distributed' (PCMC, p. 6, §4).*

That is,

$$\begin{array}{ccccc} s_1 & = & b_1 & - & \sum_{i=1}^{n} a_i^1 \\ s_2 & = & b_2 & - & \sum_{i=1}^{n} a_i^2 \\ \vdots & = & \vdots & - & \vdots \\ s_i & = & b_j & - & \sum_{i=1}^{n} a_i^j \\ \vdots & = & \vdots & - & \vdots \\ s_n & = & b_n & - & \sum_{i=1}^{n} a_i^n \end{array} \tag{15.3}$$

where $s_i$ is the surplus of commodity $i$ available for distribution after the quantities $\left\{a_i^j\right\}$ have been put aside for the next year's production. Alternatively, it is the quantity produced in the previous period which is left over, once the inputs used in production have been removed.

In compact matrix notation, we have

$$\mathbf{S} = (\mathbf{B} - \mathbf{A})^T \mathbf{e} \tag{15.4}$$

where: $\mathbf{e}$ is the $n \times 1$ unit or summation vector (each element is 1); T is the transpose operator; $\mathbf{S}$ is the $n \times 1$ Physical Surplus vector or Physical *Net National Product*; $\mathbf{B}$ is the $n \times n$ diagonal matrix composed of gross production $\mathbf{b}$ as its diagonal elements.

### *Self-Replacing Prices*

Sraffa also assumes, for most of his book, that wage is paid *post factum* (PCMC, 9–10, §89). Given the knowledge of the methods of production used during the previous annual production cycle (and assuming that the same methods will be used), Sraffa searches for uniform prices that would allow the system to replicate these commodities during the next production cycles. These prices, following the classical tradition, could have different names (PCMC, pp. 7–8, §8). Here we refer to them as self-replacing prices.

If this is our problem setup, the accounting balance would require that

$$\begin{aligned}(1+r_1)\mathbf{a}_1\mathbf{p}+l_1 w &= b_1 p_1\\(1+r_2)\mathbf{a}_2\mathbf{p}+l_2 w &= b_2 p_2\\ \ldots &= \ldots\\(1+r_i)\mathbf{a}_i\mathbf{p}+l_i w &= b_i p_i\\ \ldots &= \ldots\\(1+r_n)\mathbf{a}_n\mathbf{p}+l_n w &= b_n p_n\end{aligned} \tag{15.5}$$

where: $\mathbf{p} = [p_1, p_2, \ldots, p_n]^T$ is the price vector, $r_1, r_2, \ldots, r_n$ are the sectoral profit rates and w is the uniform wage rate.

In matrix notation, Eq. 15.5 become

$$(\mathbf{I}+\mathbf{R})\mathbf{Ap}+\mathbf{L}w=\mathbf{Bp} \tag{15.6}$$

where: **I** is the identity $n \times n$ matrix; $\mathbf{R} = diag(\mathbf{r})$ is the diagonal matrix, whose diagonal elements are the rate of profits in each single industry, $r_1$, $r_2, \ldots, r_i, \ldots r_n$ (vector **r**); **A** is the $n \times n$ matrix denotes the means of production $\{a_i^j\}$; **L** is the $n \times 1$ vector whose elements $\{l_i\}$ are the labour used in production.

### *Number of Equations and Number of Variables*

The system of Eqs. 15.5 (or 15.6) is indeterminate. There are at least two cases that we can consider.

(1) **Non-uniform rates of profits:** in this case, there are $2 \times n + 1$ variables ($n$ prices, $n$ profit rates and the wage rate $w$) and $n$ equations. This is a more general case and it is examined in Zambelli (2018b).

(2) **Uniform rate of profits:** this is case that Sraffa examined in PCMC. Sraffa simplifies the problem by assuming that the rates of profits are uniform. In that case, the n rates of profits are assumed to be equal to a single rate of profits: $r = r_1 = r_2 = \cdots = r_n$ (PMCM, p. 11, §11).[25] Here there are $n$ equations and the number of variables are reduced to $n + 2$ (i.e., $n$ prices, a wage rate $w$ and a uniform rate of profits $r$). Hence the system is still indeterminate.

## *Net National Income in a Self-replacing System*

### The Share between Workers and Producers of the Surplus in Value Terms

*'The national income of a system in a self-replacing state consists of the set of commodities which are left over when from the gross national product we have removed item by item the articles which go to replace the means of production used up in all industries [the surplus vector S]. The value of this set of commodities [...] we make equal to unity' (PCMC, p. 11, §12).*

Formally, the value of the Net National Product ( $Y^{\text{NNP}}$ ) is:

$$Y^{\text{NNP}} = \mathbf{S}^{\mathbf{T}}\mathbf{p} = 1 \tag{15.7}$$

If the value of the Net National Product is unity, this implies that $\mathbf{S}^{\mathbf{T}}\mathbf{p} = 1$ is the *numéraire*.

In value terms, the value of the surplus that goes to producers is $\sum_{i=1}^{n}(r_i\mathbf{a}_i\mathbf{p}) = \mathrm{e}_{n\times1}^{T}\mathbf{RAp}$. Similarly, the value of the surplus that goes to the workers is $\sum_{i=1}^{n}(wl_i) = \mathrm{e}_{n\times1}^{T}w\mathbf{L}$ :[26]

$$Y^{\text{NNP}} = \overbrace{\mathrm{e}_{n\times1}^{T}\mathbf{RAp}}^{\substack{\text{Share to}\\ \text{Producers (value)}}} + \overbrace{\mathrm{e}_{n\times1}^{T}w\mathbf{L}}^{\substack{\text{Share to}\\ \text{Workers (value)}}} \tag{15.8}$$

[25] Zambelli (2018b) discusses this assumption. The uniform rate of profits is obviously a special case of Eqs. 15.5. This assumption is justified in PCMC, which is meant as a *Prelude to a Critique of Economic Theory*.

[26] $\mathrm{e}_{n\times1}$ is the summation vector.

For the special case in which the rate of profits is uniform, as in PCMC, we have:

$$Y^{\text{NNP}} = \overbrace{\mathbf{e}_{n\times 1}^{T} r\mathbf{A}\mathbf{p}}^{\substack{\text{Share to} \\ \text{Producers (value)}}} + \overbrace{\mathbf{e}_{n\times 1}^{T} w\mathbf{L}}^{\substack{\text{Share to} \\ \text{Workers (value)}}} \tag{15.9}$$

## Distribution of the Surplus to Industries and Workers: Value and Physical Terms

The value of the Net National Product, $\mathbf{S}^{\mathrm{T}}\mathbf{p}$ is distributed to the *n* industries and workers:

- $d_1 S^T\mathbf{p}$ is the share going to industry 1, $d_2\mathbf{S}^{\mathrm{T}}\mathbf{p}$ is the share to industry 2, …, $d_n\mathbf{S}^{\mathrm{T}}\mathbf{p}$ is the share to industry *n*
- $d_w\mathbf{S}^{\mathrm{T}}\mathbf{p}$ is the share that goes to the workers

Clearly $\sum_i^n d_i + d_w = 1$.

$$\mathbf{d}S^T\mathbf{p} = \overbrace{\begin{bmatrix} d_1 S^T\mathbf{p} \\ d_1 S^T\mathbf{p} \\ \vdots \\ d_1 S^T\mathbf{p} \\ d_1 S^T\mathbf{p} \end{bmatrix}}^{\substack{\text{Distribution of} \\ \text{Surplus or NNP}}} = \overbrace{\begin{bmatrix} b_1 p_1 - \mathbf{a}_1\mathbf{p} - l_1 w \\ b_1 p_1 - \mathbf{a}_2\mathbf{p} - l_2 w \\ \vdots \\ b_n p_n - \mathbf{a}_n\mathbf{p} - l_n w \\ \mathrm{e}^T\mathbf{L}w \end{bmatrix}}^{\substack{\text{Purchasing Capacity} \\ \text{Income}}} = \overbrace{\begin{bmatrix} r_1\mathbf{a}_1\mathbf{p} \\ r_2\mathbf{a}_2\mathbf{p} \\ \vdots \\ r_n\mathbf{a}_n\mathbf{p} \\ \mathrm{e}^T\mathbf{L}w \end{bmatrix}}^{\substack{\text{Purchasing Capacity} \\ \text{Income}}} \tag{15.10}$$

Given that $\mathbf{S}^{\mathrm{T}}\mathbf{p} = 1$, the vector $\mathbf{d} = [d_1, d_2, \ldots, d_n, d_w]^T$ denotes distribution of the surplus **S** both in value and physical terms.

Recall that $\mathbf{S}^{\mathrm{T}}\mathbf{p} = 1$ (see Eq. 15.7), the above accounting identity, Eq. 15.10, may be written in a compact form as

$$\mathrm{d} = \begin{bmatrix} d_{n\times 1} \\ d_w \end{bmatrix} - \begin{bmatrix} (\mathrm{B}-\mathrm{A}) & -\mathbf{L} \\ \mathbf{0}_{1\times n} & \mathrm{e}^T\mathbf{L} \end{bmatrix}\begin{bmatrix} \mathbf{p} \\ w \end{bmatrix} = \begin{bmatrix} \mathbf{RA} & \mathbf{0}_{n\times 1} \\ \mathbf{0}_{1\times n} & \mathrm{e}^T\mathbf{L} \end{bmatrix}\begin{bmatrix} \mathbf{p} \\ w \end{bmatrix} \tag{15.11}$$

Clearly, when the self-replacing prices **p** and wage rate *w* are given, the distribution is uniquely determined and the profit rates **r** are determined as well. Also, when the distribution **d** is given, prices **p** and the wage rates $w$ are also determined. In this case we have

$$\begin{bmatrix} \mathbf{p} \\ w \end{bmatrix} = \begin{bmatrix} (\mathbf{B}-\mathbf{A}) & -\mathbf{L} \\ \mathbf{0}_{1\times n} & \mathrm{e}^T\mathbf{L} \end{bmatrix}^{-1} \mathrm{d} = \begin{bmatrix} (\mathbf{RA})^{-1} & \mathbf{0}_{n\times 1} \\ \mathbf{0}_{1\times n} & (\mathrm{e}^T\mathbf{L})^{-1} \end{bmatrix} \mathrm{d} \qquad (15.12)$$

From Eqs. 15.11 and 15.12, we see that distribution and prices are determined **simultaneously**.

The vector **d** is the physical distribution of the surplus **S** to producers and workers, but it is also the distribution in value terms, that is, the share in value of the net national product shown in Eq. 15.8. The share going to the producers is given by $\mathbf{e}^T\mathbf{RAp}$ and that which goes to the workers is given by $\mathbf{e}^T\mathbf{L}w$.

Furthermore, Zambelli (2018b, Appendix A) shows an equivalence between the actual physical distribution and the equivalent distribution in terms of a proportion of the surplus vector **S**. In other words, if $d_i > d_j$, it implies that $d_i\mathbf{S}^T\mathbf{p} > d_j\mathbf{S}^T\mathbf{p}$ and it as if $d_i\mathbf{S} > d_j\mathbf{S}$. This allows for a physical comparison of the distribution of the surplus.

### *Indeterminacy or Exogeneity*

In both cases (1) and (2) mentioned earlier, the system can be closed only by adding some additional exogenous elements or additional theories. Here we have chosen the physical surplus **S** as the *numéraire*. The economic system is described in terms of $n$ equations (Eqs. 15.5) and the *numéraire* equation (Eq. 15.7).

This implies that when considering a more general case with non-uniform rates of profits, we have $n + 1$ equations and $2n + 1$ variables. Therefore, there are $n$ degrees of freedom: that is, all the self-replacing prices, rates of profits and wage rate cannot be computed by the knowledge of the quantities and the methods of production alone. In order to solve the system of equations, $n$ variables among $\{p_1, p_2, \ldots, p_n, r_1, r_2, \ldots, r_n, w\}$ have to be given exogenously, that is, from outside the economic system.

When we consider Sraffa's special case with uniform rate of profits (case ii), we have $n + 1$ equations (i.e., Eqs. 15.5 is restricted to the case in which the rates of profits are uniform and the *numeraire* Eq. 15.7) and

$n + 2$ variables $\{p_1, p_2, \ldots, p_n, r, w\}$. In this case, there is only one degree of freedom and one of the $n + 2$ variables have to be exogenously given.

## Appendix B: Incorporating Credit and Debt Inside a Sraffian Framework: An Outline

We now outline the way in which we can introduce money, credit and debt, inside the Sraffian schemes (for details, see our companion paper Venkatachalam and Zambelli (2021)).

When we discuss an economic system, whether real or virtual, we often deal with a world of bilateral or multilateral exchanges. In such a world, when someone buys a good, there is always a counterpart who sells. In the traditional Sraffian schemes, the possibility of deferred means of payment have not been adequately explored. For the aims and arguments developed in PCMC, the introduction of deferred means of payments are not strictly necessary and their omission may be understandable. But as claim in Sect. 3, money and deferred means of payments are indeed important and that there is a clear scope to extend the framework presented in PCMC by incorporating money, credit and debt.

In Appendix A, we provided a formal treatment of generalized Sraffian schemes, where the rates of profits are non-uniform (system presented in PCMC with uniform rate of profits is a special case). As we have discussed above, when the system is indeterminate, there is ample space in theory for including elements that are not typically considered as belonging to the realm of economics.

Consider the more general case where the rates of profits are allowed to be nonuniform. We have $n + 1$ equations and $(2n + 1)$ the variables (see Appendix A.6). If we now assume that the commodity prices $\bar{\mathbf{p}}$ and wage rate $\bar{w}$ are exogenously given we have the following inequalities:

$$
\begin{array}{ccc}
\overbrace{\phantom{a_1^1\bar{p}_1+\ldots+a_1^j\bar{p}_j+\ldots+a_1^n\bar{p}_n+l_1\bar{w}}}^{\substack{\text{Virtual}\\ \text{Expenditures}}} & & \overbrace{\phantom{b_1\bar{p}_1}}^{\substack{\text{Virtual}\\ \text{Revenues}}} \\
a_1^1\bar{p}_1+\ldots+a_1^j\bar{p}_j+\ldots+a_1^n\bar{p}_n+l_1\bar{w} & \substack{<\\=\\>} & b_1\bar{p}_1 \\
a_2^1\bar{p}_1+\ldots+a_2^j\bar{p}_j+\ldots+a_2^n\bar{p}_n+l_2\bar{w} & \substack{<\\=\\>} & b_2\bar{p}_2 \\
\vdots+\vdots+\vdots+\vdots+\vdots+\vdots & \substack{<\\=\\>} & \vdots \\
a_i^1\bar{p}_1+\ldots+a_i^j\bar{p}_j+\ldots+a_i^n\bar{p}_n+l_i\bar{w} & \substack{<\\=\\>} & b_i\bar{p}_i \\
\vdots+\vdots+\vdots+\vdots+\vdots+\vdots & \substack{<\\=\\>} & \vdots \\
a_n^1\bar{p}_1+\ldots+a_1^j\bar{p}_j+\ldots+a_n^n\bar{p}_n+l_n\bar{w} & \substack{<\\=\\>} & b_n\bar{p}_n
\end{array}
\qquad (15.13)
$$

In matrix notation, this could be written as:

$$
\overbrace{\mathbf{A}\bar{\mathbf{p}}+\mathbf{L}\bar{w}}^{\substack{\text{Virtual}\\ \text{Expenditures}}} \lesseqgtr \overbrace{\mathbf{B}\bar{\mathbf{p}}}^{\substack{\text{Virtual}\\ \text{Revenues}}} \qquad (15.14)
$$

The prices and the wage rate in the above equations are to be seen as virtual or bookkeeping prices, which may or may not be the actual exchange prices. In fact, they cannot be the actual exchange prices when there is at least one industry for whom the bookkeeping expenditures (left-hand side) would be greater than bookkeeping revenues (right-hand side). In this case, the rate of profits of this industry is obviously negative. This is a situation where there would not be enough purchasing power for the exchanges to take place and some industries would be left with unsold commodities. This, in turn, would mean that the economic system as a whole would not be in the self-replacing condition.

Clearly, the industries with virtual expenditures higher than their virtual revenues would not have the necessary purchasing power to buy the means required to replicate production of the previous period. These industries, therefore, are in a condition where there is a potential financial deficit. Concurrently, there would be industries which would not be able to sell all of their product. The industries in 'financial deficit' would be able to purchase the necessary means of production only by agreeing to a deferred payment to take place during the years to follow. At the same time, the industries in potential 'financial surplus' would be able to sell all of their product by agreeing to deferred payments by the borrowers. The traditional Sraffian schemes outlined in Appendix A can be extended or generalised to accomodate this situation.

We can conveniently view these deferred means of payments as a form of an I Owe You (IOUs). This can be seen to include all forms of financial contracts, where there is a promise to return goods in the future. These obligations may be with an explicit delivery date (as in the case of forward contracts associated with real goods), or with a relatively loose delivery date (as in the case of standard means of exchange such as cash, checks, debt and credit accounts, bonds etc.).

We can consider a situation wherein a commodity is exchanged for a promise to pay back at a future point in time (deferred payment). The entity selling this commodity would see its credit increasing and correspondingly, the counterpart buying this commodity would have its debt increase. This might take place by the writing off of means of payments previously generated or by issuing new means of payments. Furthermore, this could take place through an institution, like the banking system, or through direct contracts. A full-fledged treatment of Sraffian schemes with deferred means of payments can be found in our companion piece Venkatachalam and Zambelli (2021).

Given this structure, we can ask several interesting questions and perform thought experiments. To start with, we can explore the set of prices and possible financial conditions that would allow the system to reproduce itself, if one allows for transfer of purchasing power in the form of credit and debt. In other words, identifying the set of prices that would allow self-replacing. Note that IOUs may be generated and transferred from one period to another, which in turn might influence the set of self-replacing prices in the subsequent periods. These may offer us insights concerning the potential paths that are available for the system to evolve in the future. To answer questions of this nature, it is also important to study the effects that already existing future promises to pay may have on determining the set of self-replacing prices. All of this require keeping a close track of the credit and debt structure with careful accounting built in to the framework.

We also note that in our attempts to introduce money, we follow Sraffa's approach in PCMC and do not provide a theory of prices and distribution. Instead, the objective would be to determine, for given past methods of production, usage of labour and output, those set of prices and the possible financial conditions that would allow the system to reproduce itself. We hope that this provides some pointers to the direction in which Sraffian schemes can be generalised.

## References

Arrow, K. and F. Hahn (1971). *General Competitive Analysis.* San Francisco: Holden-Day.

Bharadwaj, K. (1963). Value through exogenous distribution. *The Economic Weekly,* 1450–54.

Ciccarone, G. (1998). Prices and distribution in a Sraffian credit economy. *Review of Political Economy 10* (4), 399–413.

de Brunhoff, S. (1973). Marx as an a-Ricardian: value, money and price at the beginning of Capital. *Economy and Society 2,* 421–430.

de Brunhoff, S. (1974/1975). Controversies in the theory of surplus value: A reply to John Eatwell. *Science & Society 38*(4), 478–482.

de Brunhoff, S. (1990). Reflections on Marx and Sraffa. In K. Bharadwaj and B. Schefold (Eds.), *Essays on Piero Sraffa. Critical Perspectives on the Revival of Classical Theory.* London: Unwin Hyman.

Dobb, M. (1967). Marx's 'Capital' and its place in economic thought. *Science & Society 31*(4), 527–540.

Dobb, M. (1975). A note on the Ricardo-Marx-Sraffa discussion. *Science & Society 39*(4), 468–470.

Eatwell, J. (1974). Controversies in the theory of surplus value: Old and new. *Science & Society 38*(3), 281–303.

Fisher, I. (1906). *The Nature of Capital and Income.* London: Macmillan.

Fisher, I. (1907). *The Rate of Interest.* New York: Macmillan.

Fisher, I. (1911). *The Purchasing Power of Money. Its Determination and Relation to Credit Interest and Crises (Second Revised Edition, 1920th ed.).* New York: Macmillan.

Fratini, S. (2018). On the second stage of the Cambridge capital controversy. *Working Papers—Centro Sraffa* (30).

Garegnani, P. (1978). Sraffa's revival of Marxist economic theory. *New Left Review 112,* 71–75.

Garegnani, P. (2000). Savings, investment and capital in a system of general intertemporal equilibrium. In H. Kurz (Ed.), *Critical Essays on Piero Sraffa's Legacy in Economics.* Cambridge: Cambridge University Press.

Hahn, F. (1982). The neo-Ricardians. *Cambridge Journal of Economics 6*(4), 353–374.

Hahn, F. and F. Petri (Eds.) (2002). *General Equilibrium: Problems and Prospects.* London: Routledge.

Harrod, R. (1961). Review: Production of commodities by means of commodities. by Piero Sraffa. *Economic Journal 71* (284), 783–789.

Hayek, F. A. (1931a). Reflections on the pure theory of money of Mr. J. M. Keynes. *Economica 11* (33), 270–295.

Hayek, F. A. (1931b). A rejoinder to Mr. Keynes. *Economica 34,* 398–403.

Hayek, F. A. (1932a). Money and capital: A reply. *The Economic Journal 42*(166), 237–249.

Hayek, F. A. (1932b). Reflections on the pure theory of money of Mr. J. M. Keynes (continued). *Economica 35*, 22–44.
Hicks, J. (1932). *The Theory of Wages*. London: Macmillan.
Hicks, J. (1939). *Value and Capital*. Oxford: Clarendon Press.
Hicks, J. (1961). The measurement of capital in relation to the measurement of other economic aggregates. In F. Lutz and D. Hague (Eds.), *The Theory of Capital*, Chapter 2, pp. 18–31. London: Macmillan.
Hodgson, G. (1982). *Capitalism, Value and Exploitation. A Radical theory*. Oxford: Martin Robertson.
Keynes, J. M. (1923). *A Tract on Monetary Reform*. London: Macmillan.
Keynes, J. M. (1930). *A Treatise On Money*. London: Macmillan.
Keynes, J. M. (1931). The pure theory of money. A reply to Dr. Hayek. *Economica 34*, 387–397.
Laibman, D. (1973). Values and prices of production: The political economy of the transformation problem. *Science & Society 37*(4), 404–436.
Laibman, D. (1974/1975). Controversies in the theory of surplus value: A comment. *Science & Society 38* (4), 482–487.
Lebowitz, M. A. (1973). The current crisis of economic theory. *Science & Society 37*(4), 385–403.
Lippi, M. (1979). *Value and Naturalism in Marx*. London: New Left Books.
Mandler, M. (1999a). *Dilemmas in Economic Theory. Persisting Foundational Problems of Microeconomics*. Oxford: Oxford University Press.
Mandler, M. (1999b). Sraffian indeterminacy in General Equilibrium. *Review of Economic Studies 66* (3), 693–711.
Mandler, M. (2002a). Classical and neoclassical indeterminacy in one-shot versus ongoing equilibria. *Metroeconomica 53*(3), 203–222.
Mandler, M. (2002b). The sequential indeterminacy problem. In F. Hahn and F. Petri (Eds.), *General Equilibrium: Problems and Prospects*. London: Routledge.
Mandler, M. (2005). Well-behaved production economies. *Metroeconomica 56*(4), 477494.
Marx, K. (1965 [1867]). *Capital*, Volume I. Moscow: Progress Publishers.
Marx, K. (1965 [1894]). *Capital*, Volume III. Moscow: Progress Publishers.
Negishi, T. (2014). *Elements of Neo-Walrasian Economics. A survey*. Japan: Springer Verlag.
Nicholas, H. (2014). Whither Sraffa's theory of price? *World Review of Political Economy 5*(3), 315–330.
Nuti, D. M. (1971). "Vulgar Economy" in the theory of income distribution. *Science & Society 35* (1), 27–33.
Panico, Carlo (1988a), *Sraffa on money and banking*, *Cambridge Journal of Economics 12*(1), pp. 7–28.

Panico, C. (1988b). *Interest and Profit in the Theories of Value and Distribution*. New York: St. Marin's Press.

Parrinello, S. (2008). The stability of general intertemporal equilibrium: A note on Schefold. *Metroeconomica 59*(2), 305–3012.

Pilling, G. (1972). The law of value in Ricardo and Marx. *Economy and Society 1*(3), 281–307.

Pivetti, M. (1991). *An Essay on Money and Distribution*. London: Macmillan.

Robertson, D., P. Sraffa, and G. Shove (1930). Symposium on increasing returns and the representative firm. *Economic Journal 40*(157), 79–116.

Rowthorn, B. (1974). Neo-Classicism, Neo-Ricardianism and Marxism. *New Left Review 86*, 63–87.

Samuelson, P. A. (1966). A summing up. *Quarterly Journal of Economics 88* (4), 568–583.

Samuelson, P. A. (1971). Understanding the Marxian notion of exploitation: A summary of the so-called transformation problem between Marxian values and competitive prices. *Journal of Economic Literature 9*(2), 399–431.

Schefold, B. (2005). Zero wages—no problem? A reply to Mandler. *Metroeconomica 56* (4), 503–513.

Schefold, B. (2008). Savings, investment and capital in a system of general intertemporal equilibrium—an extended comment on Garegnagni with a note on Parrinello. In G. Chiodi and L. Ditta (Eds.), *Sraffa or an Alternative Economics*. London: Palgrave Macmillan.

Seton, F. (1957). The 'transformation problem'. *Review of Economic Studies 24* (3), 149–160.

Shaikh, A. (1984). The transformation from Marx to Sraffa. In E. Mandel and A. Freeman (Eds.), *Ricardo, Marx, Sraffa—The Langston Memorial Volume*. New York: Verso.

Sinha, A. (2016). *A revolution in economic theory: the economics of Piero Sraffa*. London: Palgrave Macmillan.

Sinha, A. and M. Dupertuis (2009). Sraffa and the question of equilibrium. *Cahiers d'economie politique 56*(1), 91–100.

Sraffa, P. (1920). *L'inflazione monetaria in Italia: durante e dopo la guerra*. Premiata Scuola Tipografica Salesiana.

Sraffa, P. (1921). Industriali e governo inglese contro i lavoratori. *L'Ordine Nuovo, 24 July*.

Sraffa, P. (1922a). The bank crisis in Italy. *The Economic Journal 32*(126), 178–197.

Sraffa, P. (1922b). *Italian banking today. The Manchester Guardian Supplement, Reconstruction in Europe, Supplement, 7 December.*

Sraffa, P. (1925). Sulle relazioni tra costo e quantita prodotta. *Annali di Economia II*, 277–328.

Sraffa, P. (1926). Laws of returns under competitive conditions. *Economic Journal 36*(144), 535–550.

Sraffa, Piero (1932a), *Dr. Hayek on Money and Capital, The Economic Journal 42*(165), March, pp. 42–53.
Sraffa, P. (1932b). Money and capital: A rejoinder. *The Economic Journal 42*(166), 249–251.
Sraffa, P. (1951). *Works and Correspondence of David Ricardo (ed.), Vols I-X, 1951–1955, and Vol. XI (index), 1973*. Cambridge: Cambridge University Press.
Sraffa, P. (1960). *Production of Commodities by Means of Commodities*. Cambridge: Cambridge University Press.
Sraffa, P. (1962). Production of commodities: A comment. *Economic Journal 72*(286), 477–479.
Steedman, I. (1977). *Marx after Sraffa*. Oxford: Unwin Brothers Limited.
Venkatachalam, R. and S. Zambelli (2021), Self-replacing prices with credit and debt, ASSRU Discussion Paper (DP-3-2021).
Wicksell, K. (1936). *Interest and Prices*. London: Macmillan, Trans. by R. Kahn [First published in 1898, Jena: Gustav Fischer].
Winternitz, J. (1948). Values and prices: A solution of the so-called transformation problem. *The Economic Journal 58* (230), 276–280.
Zambelli, S. (2018a). The aggregate production function is NOT neoclassical. *Cambridge Journal of Economics 42*(2), 383–426.
Zambelli, S. (2018b). Production of Commodities by Means of Commodities and nonuniform rates of profits. *Metroeconomica 69*(4), 791–818.

Comment

# A Comment on 'Sraffa, Money and Distribution'

*K. Vela Velupillai*

It is a daunting prospect, to envisage comments from an old dilettante, but who nevertheless thinks very highly of Sraffa's contribution to economics, in general; it is especially daunting to be asked to provide—what will turn out to be innocuous and inexpert—remarks on the comprehensive studies of Sraffa's published and unpublished material and its resultant outcome in a highly competent paper on **Sraffa, Money and Distribution** by two scholars of integrity and originality. Moreover, the senior author has studied the unpublished writings and notes of Sraffa, deposited in the Wren Library, at Trinity College, Cambridge. All this makes my comments take on a perspective that strengthens my regrettable reputation of being a glorified dilettante! However, my respect for the editor of this volume means that I have to give it a try.

*Sraffa, Money and Distribution* is a competently written, impeccably researched—in a paper of 17 pages of text, with two appendices of 6 dense pages, the references numbering 77, are spread over 4 pages—closely and, to the extent I can gauge, correctly reasoned, fine contribution. My remarks, confined to a reading of the 17 pages of main text, except for two, very minor dissenting points, on the main arguments in the two appendices (even although a substantiation of the theme in Appendix B is said to be 'work in progress', by the two distinguished authors).

As I read this fine paper, (partly) based on the extensive reading (by the senior author) of the unpublished material of the Sraffa papers, primarily at the Wren Library of Trinity College, Cambridge—and ruminations by both authors—I understood the main theme to be that Sraffa's earlier work in money, banking, returns-to-scale, all under some kind of equilibrium condition, is one with the propositions in **PCMC**. This main theme is due to his—Sraffa's—increasing awareness of 'the insufficiency of market forces to determine *values*.'

As the authors state, already in the *Abstract*, but strengthened in the main text:

> In contrast, we suggest that Sraffa's contributions *have to be viewed as a coherent whole* .... We point to the *unifying thread* between his early and later work, which concerns *the insufficiency of* economic mechanisms or *market forces* to exclusively determine values (prices, profit rates and wage rates). (Italics added)

The authors seem to accept, to substantiate the above theme, that 'Sraffa's contributions to economics can be broadly divided into *two parts*' (p. 2, italics added)—one being monetary and banking theory and the other, *for simplicity*, can be referred to as capital (or value) and distribution theory (although the authors are more careful in their nomenclature).

They seem to want to buttress their stance on the 'two part' theory by appealing to, for example, Panico (1988a).[27] I don't think they need this support, particularly because it is dubious in its theoretical underpinnings. I, *as a non-reader of Sraffa's unpublished works*, do *not* subscribe to this *monolithic*[28] story. For one, there was the returns-to-scale issue, strongly dependent on various determinate and indeterminate concepts of *equilibria*; the other is, in Adarkar's felicitous words, *Mr. Sraffa's Commodity Rate* (chapter VII, in Adarkar 1935)—also known as the *own rates of interest*. This, in turn, via Sraffa's Keynes inspired critique of Hayek (Sraffa 1932a, b), leads to the dichotomy between the *short-run* and *long-run*,[29] on the one hand, and what came to be the construction of the *standard commodity*.

When he constructed the arguments, the economics and the mathematics of *PCMC*, he gradually divested himself of all dependence on money, banking, returns-to-scale, competitive vs. non-competitive, short-run versus long-run and, above all, any notion of equilibrium. PCMC is a

[27] I possess a copy of Keynes (1923) and both of Panico's references to it, on p. 9 of his article, are incorrect (at least with regard to page numbers). Moreover, the quotation on the quantity theory of money is not entirely faithful to the original, in addition to the fact that neither Keynes, nor Sraffa accepted 'the form dominant at that time', which was Irving Fisher's version of the quantity theory; as a matter of fact Keynes went into great trouble, in chapter III, § 1, of the same book, to give details of the version he (and Sraffa) worked with.

[28] I shall explain, below, what is meant by a *monolithic* story.

[29] This may be one reason for Sraffa's disapproval of the famous chapterm17 of Keynes's GT.

non-equilibrium, eternal-run, theory of production and distribution which cannot determine, uniquely or otherwise, values—even after taking into consideration the exogenously determined profit rate (§44, *PCMC*).

I think, largely based on Professors Chiodi and Zambelli's reporting to me of the contents of the unpublished material in Sraffa's works, there is no mileage in working with a *monolithic* story; I think scholars *evolve*,[30] and sweep aside what they think are irrelevant accoutrements to a developing theory of a system.

I come, now, to the two dissenting points, that I wish to state:

- There is no justification for Nuti to state, what the authors report on p. 18, as being quoted by Panico; in particular, no one can determine what is the 'most appropriate way' for anything.
- I have difficulties of giving up the Sraffian assumption of uniform rates of profit; this is why I refer, now, to *PCMC* dealing with an economic system in the *eternal-run* of *non-equilibrium* configurations.

*PCMC* is, if anything, a manifesto of *David*—without *Ricardo*'s equilibrium, long-run, accoutrements. That is why Sraffa was able to write Keynes, on 20 December, 1932:

> [D]on't treat too ill my David.

## References

Adarkar, Bhalchandra P. (1935), *The Theory of Monetary Policy*, P. S. Kong & Son, Lyd., London.

Panico, Carlo (1988), *Sraffa on money and banking*, *Cambridge Journal of Economics*, Vol.12, No. 1, pp. 7–28.

Sraffa, Piero (1932a), *Dr. Hayek on Money and Capital*, *The Economic Journal*, Vol. 42, No. 165, March, pp. 42–53.

Sraffa, P. (1932b). Money and capital: A rejoinder. *The Economic Journal 42*(166), 249–251.

[30] This does not mean I subscribe to a Darwinian-Mendelian theory of evolution. Even without reading Sraffa's unpublished writings, I am sure he did *not* subscribe, either, to a conventional theory of evolution.

Reply
# A Reply to Velupillai

*Ragupathy Venkatachalam and Stefano Zambelli*

Vela Velupillai has provided thoughtful comments on our piece and this allows us to provide some clarifications that may be necessary. Velupillai writes:

> I understood the main theme to be that Sraffa's earlier work in money, banking, returns–to–scale, all under some kind of equilibrium condition, is one with the propositions in **PCMC**. This main theme is due to his—Sraffa's—increasing awareness of 'the insufficiency of market forces to determine *values*'.

Based on this, he seems to interpret our position as one that characterises Sraffa's research program as being *monolithic*. Such an interpretation would be incorrect and hence we want to clarify this further to avoid confusion. First, we argue that Sraffa's research contributions are coherent and that his views on money are not necessarily incompatible with that of PCMC. We do not, however, make a much stronger claim that it is *monolithic*. Second, the only equilibrium notion that we attribute to Sraffa (from his early work onwards) is the idea of accounting or bookkeeping equilibrium.[31] The continuing thread that we trace across his body of work does not have any reliance on equilibrium. We do not imply at any point

[31] In our remarks concerning Keynes and Sraffa's works during the early 1920s on the monetary theory of their time, we do not imply that they approved Fisher's (or even that of Wicksell, or Hayek) version of the quantity theory of money. We emphasise that they both did, inside that framework, use the quantity theory to speculate on the relationship, for example, between prices and distribution.

that Sraffa assumed ***some kind of equilibrium condition***, except for the accounting equilibrium. Third, Sraffa's production prices or self-replacing prices are accounting equilibrium prices, that is, prices for which the quantities sold multiplied by the prices (Revenues) are equal to the quantities bought multiplied by the prices (Expenditures) plus what is left or residual (Profits or Losses, which are Revenues minus Expenditures).

The prices considered by Sraffa, in PCMC and in the preparatory work,[32] are never actual or market prices. Instead, they are virtual (uniform) prices which, if realized during the annual market, would allow the system to be in a self-replacing state. This seems to be the ***thought/imaginary experiment*** that Sraffa conducts for a thorough critique of economic theory. However, it is not Sraffa's theory of distribution, as Bharadwaj (1963, p. 1450, quoted in our chapter) has elegantly summarised.

When we attempt to relax or modify some of the assumptions of PCMC, we have not used any notion of equilibrium, except that of the bookkeeping or accounting equilibrium and search for the prices that would guarantee it. In the Appendix B and in Venkatachalam and Zambelli (2021), we conduct a thought experiment where we search for the prices and the nature of credit and debt contracts that would allow the system to remain in the self-replacing state. We believe that this is coherent with Sraffa's method of investigation, although our goals may be different.

PCMC, as we have stressed in our contribution, is a *Prelude to a Critique of Economic Theory*. This is often acknowledged by the scholars working with PCMC, but very few take it as such. The major critique, which is much more than a '*Prelude*', is the impossibility of computing the self-replacing prices based solely on information concerning the methods of production.

Once the major critique concerning ***indeterminacy*** has been put forward, Sraffa continues by considering the imaginary case in which one of

[32]We both, as well as many others, have consulted Sraffa's archives, which are now available online. The material found in the archives confirms Sraffa's statement that

> [...] the central propositions [of PCMC] had taken shape in the late 1920's [...] (PCMC, p. vi)

This and the archival material are evidence in support of our view that Sraffa's research is to be seen as a coherent (although surely not monolithic!) whole.

the distributive variables $r$ (or the wage rate $w$) is exogenously given and prices are computed as a function of the distributive variable. This critique is strong and it is based on the fact that as the profit rate changes, prices may not change monotonically (PCMC, p. 38, Fig. 3)[33] and this allows him to conclude that

> The reversal in the direction of the movement of relative prices, in the face of unchanged methods of production, cannot be reconciled with *any* notion of capital as a measurable quantity independent of distribution and prices. (PCMC, p. 38)

Supported by textual evidence, we think that Sraffa's view or research program, from the beginning of the 1920s to 1960, is a coherent whole. It *concerns the insufficiency of economic mechanisms or market forces to exclusively determine values (prices, profit rates and wage rates)*. His contributions are all critical of the standard notions of equilibrium and we wholeheartedly embrace this.

As we have emphasised in our contribution (see. p. 15), non-uniform rate of profits, money or notions of credit or debt are not *necessary* for

[33] In passing, we conjecture that this might have been the reason why Sraffa did not like or approve of Ch. 17 of the General Theory. Keynes (1936, Ch. 17, §II and §III ) develops an argument that there would be a 'natural' tendency for the own rates of interest of commodities to converge towards the own rate of interest of money:

> Thus with other commodities left to themselves, 'natural forces,' i.e. the ordinary forces of the market, would tend to bring their rate of interest down until the emergence of full employment had brought about for commodities generally the inelasticity of supply which we have postulated as a normal characteristic of money (Keynes 1936, Ch. 17, §III, p. 235)

According to Keynes, full employment would not be reached, and hence the 'natural forces' of the markets would not fully operate, because money has a higher liquidity premium with respect to the carrying costs than any other commodity. Keynes does not challange or consider the direction of motion of prices. On the contrary, Sraffa is clearly concerned with this and argues that when the distribution changes the prices could change in all different directions (Lutz and Hague 1961, see Sraffa's remark, p. 325). Even when the system is made determinate by assuming uniform rate of profits (or the wage rate), prices could move in different directions and consequently, the commodity own rates of interst could also do so.

The similarity of arguments in Garegnani's letter to Sraffa (see our chapter), with those of the converging mechanisms described by Keynes in Ch. 17 is striking. By analogy and induction, it may be reasonable to suppose that Sraffa's response to Keynes' Ch. 17 would be similar to that of his repsonse for Garegnani's justification.

Sraffa's prelude to a critique in PCMC. However, this does *not* imply that there is no place for them in Sraffa's system or that it is illegitimate to introduce them. We argue that there is such a scope and that this will open up new avenues for research.

We take note of Velupillai's difficulties in giving up the Sraffian assumption of the uniform rates of profit. However, we have difficulties in understanding the connection between the uniform rates of profits assumption and Velupillai's view that *"PCMC [is] dealing with an economic system in the eternal–run of non-equilibrium configurations"*. We are sure that he is not suggesting that the rules of accounting are broken, because this would dismantle PCMC. Sraffa has shown that the system of equations in PCMC is indeterminate and this could be seen as a non-equilibrium configuration. If this is what Velupillai means, we are in complete agreement, however, with the difference that we do not find it difficult to give up the assumption of uniform profit rates. The system remains indeterminate even when the uniform rate of profits assumption is dropped.

We are indeed sympathetic to the idea that an economic system ought to be studied without the shackles of equilibrium and as *non-equilibrium configurations*. In fact, we believe that our work is a research agenda in this direction. We are also convinced that the constructive and algorithmic method, one that Velupillai has advanced over the years, is a promising way to achieve this. We attempt to do so by combining constructive methods, Sraffa's core insights and by removing simplifying assumptions which we do not consider to be essential in PCMC and by adding other elements.

## References


Bharadwaj, K. (1963). Value through exogenous distribution. *The Economic Weekly*, 1450–54.

Keynes, J. M. (1936). *The General Theory of Employment, Interest, and Money.* London: Macmillan.

Lutz, F. and D. Hague (Eds.) (1961). *The Theory of Capital.* London: Macmillan.

Venkatachalam, R. and S. Zambelli (2021). Self replacing prices with credit and debt, ASSRU Discussion paper (DP-2021-3).

CHAPTER 16

# From Ricardo to Sraffa: A Quest for a Modern Classical Standpoint on Money

*Ghislain Deleplace*

## 1 Introduction

It is generally acknowledged that the theory of prices built by Sraffa in *Production of Commodities by Means of Commodities* (Sraffa 1960) is a development of Ricardo's theory of natural prices in *On the Principles of Political Economy, and Taxation* (Ricardo 1817–1821) and that it allows establishing on a firmer ground the main static results obtained by Ricardo, namely a determination of the price system consistent with the assumption of a uniform rate of profit on all units of capital and an inverse relationship between this rate of profit and the wage share in the aggregate income. However, although Ricardo insisted that all prices—whether natural or market prices—were expressed in money and he devoted many pages to the currency, in *Principles* as before and after, there is no theory of money in *Production of Commodities* which could complement the price system. It is thus surprising that, in comparison with the vast literature on the Sraffa system of prices, very few works have explored the possibility of its

G. Deleplace (✉)
University of Paris 8 Vincennes-Saint-Denis, Saint-Denis, France
e-mail: ghislain.deleplace@univ-paris8.fr



A. Sinha (ed.), *A Reflection on Sraffa's Revolution in Economic Theory*, Palgrave Studies in the History of Economic Thought,
https://doi.org/10.1007/978-3-030-47206-1_16

consistency with Ricardo's theory of money or with any other theory of money. The object of the present chapter is to discuss some directions in which such consistency has been looked for, and to focus on a specific one, based on my understanding of Ricardo's theory of money (in Deleplace 2017) and on some writings by Sraffa prior to *Production of Commodities*. As such, it aims at contributing to a quest for a modern Classical standpoint on money, if by "modern Classical" one means faithful to Ricardo—the greatest Classical author on money—and consistent with a Sraffa system of prices.

The famous § 44 of *Production of Commodities*, in which Sraffa writes that the rate of profits is "susceptible of being determined from outside the system of production, in particular by the level of the money rates of interest" (Sraffa 1960, p. 33), has been often interpreted as a way to introduce money into Sraffa's theory of price. This does not seem appropriate for a money *à la* Ricardo for two reasons. First Ricardo contended that the quantity of money did not affect permanently the market rate of interest, which was regulated by the rate of profit, not the other way round. Second, the store-of-value function of money was absent from Ricardo's theory—and, also, as we will see, from some early writings by Sraffa on money.[1] For Ricardo money was both the unit of account in which the prices of commodities are expressed and the circulating medium which purchases these commodities; for a monetary system to be "sound", there must exist a standard of money. This is what should be made consistent with a Sraffa system of prices.

The present chapter is organised as follows. In Sect. 2 I contend that the adoption of Sraffa prices by a monetary economy, in accordance with Ricardo's theory of money, requires a specific working of the market for the commodity selected as standard of money, so that this commodity is "outside the system of production" (to borrow Sraffa's expression about the rate of profit). In Sect. 3 I suggest that the extension of Ricardo's analysis of gold to a modern economy calls for the adoption of a public bond as standard of money, in line with Sraffa's reflections in the 1930s about the various standards in terms of which to fix one's debt.

[1] For a more detailed critique of the introduction of money through the rate of interest, see Deleplace (2017, pp. 276–8) about Ricardo and Deleplace (2014, pp. 141–6) about Sraffa.

## 2 Sraffa's Theory of Price and Ricardo's Theory of Money

Sraffa explicitly assumed that the prices which reflect the physical constraints of production and the social distribution of the product—in Ricardo's terms, the natural prices—are "adopted by the market" (Sraffa 1960, p. 3). One should then inquire about the relationship between money and market prices. When this question is raised in *any* monetary theory, the focus is on the quantity of money. To sum up: how does a change in the quantity of money affect the market prices of commodities? The most widely accepted answer to this question in the literature is the Quantity Theory of Money: a change in the quantity of money causes a *uniform* change in the market prices of *all* commodities in the same direction, which preserves their relative prices (money is neutral). Contrary to a widespread opinion, it has been advocated by some scholars that this does not need to be the case from a Classical standpoint, including Ricardo's. Two recent reconstructions illustrate in a different way this rejection of the Quantity Theory of Money to analyse a monetary system endowed with a standard (e.g. gold). For the one, a change in the quantity of money causes a *non-uniform* change in the market prices of all commodities in the same direction, *except* for the standard of money (Sect. 2.1.). For the other, a change in the quantity of money causes an adjustment in the market price of the standard of money which is *specific* to it (Sect. 2.2.). This raises a question which will be dealt with in Sect. 3 below: what is precisely the role of the monetary standard in a Classical approach to money?

### *2.1 Money, Market Prices, and the "Cantillon rule"*

A monetary reconstruction has been recently proposed for Part I of Sraffa (1960), that is, a system of production of commodities by means of commodities with circulating capital only and one technique of production for each commodity. The purpose of the exercise is to offer a representation of the economy that takes as given data the monetary flows circulating between its various component parts rather than the physical quantities of commodities exchanged between them. Two studies illustrate such a reconstruction: Benetti et al. (2014), and Cartelier (2018, pp. 215–7). The difference between these two studies is that in the latter the monetary flows taken as starting point of the analysis are equilibrium ones (they are consistent with one another), while in the former they are arbitrary, the equilibrium monetary system being deduced from the initial (disequilibrium) one. The important conclusion, common to both studies, is that all the salient results obtained by Sraffa on the basis of a matrix of physical quantities may be obtained on the basis of a matrix of monetary flows. This is in particular true for the inverse relationship between the rate of profit and the wage share in the aggregate income.

According to their authors, one advantage of such exercise is to get rid of the questionable objectivity and observability of physical goods, which have to be assumed in the "real" analysis. The data required to build a monetary representation of the economy are objective and observable, since they are the monetary flows written in the accounts of the economic units. It is therefore noticeable that results considered as landmarks of the Classical tradition reconstructed by Sraffa may be freed from the difficulties associated with the assumed physical objectivity of commodities. There is, however, a restriction to this monetary duplication of Sraffa's results, and it is the obvious consequence of an exercise that aims at dispensing with the assumed knowledge of physical quantities: the very notion of price per unit of commodity vanishes.

It is then worthwhile inquiring whether this notion of price may be reintroduced when the monetary flows taken as the data of the system are supposed to be spent in the purchase of given quantities of commodities. The matrix of money flows is then not only considered as alternative to the matrix of physical quantities exchanged at natural prices but also as a description of actual transactions at market prices. An immediate consequence is that money is not only a unit of account—something sufficient to duplicate the Sraffa system of (natural) prices—but also the circulating medium with which commodities are purchased in the market. The theory of market prices consistent with such analysis is the "Cantillon rule", in which the market price of a commodity is the ratio of the purchasing power in money spent in its market to the quantity actually offered of the commodity.

Benetti and Cartelier (2020) combines a Sraffa system of natural prices, the "Cantillon rule" of determination of market prices, and a monetary system exclusively composed of notes convertible into a standard (gold) at a fixed legal price.[2] The authors' purpose is to explore a theory of a

[2] There are other studies which involve, jointly or separately, the authors already mentioned and develop what may be called a monetary approach; see for example Benetti et al.

standard-based money which can be made consistent with the Classical theory of prices, meaning by that a theory of natural prices as developed in Sraffa (1960) and a theory of market prices according to the "Cantillon rule". The main result obtained by the authors is that, due to this rule, a change in the note issue causes a *non-uniform* change in the market prices of commodities in the same direction. This result is obviously at odds with the Quantity Theory of Money, in which the market prices of all commodities vary uniformly with the quantity of money, a condition to ensure the neutrality of money in respect to relative prices.

In Benetti and Cartelier (2020), there is, however, one exception to the (non-uniform) positive effect of a change in the quantity of money on the market prices, and it concerns the monetary standard itself (gold). Because of the existence of a fixed legal price of gold, the authors maintain that its market price cannot depart from this level: if it did, the absence of one side of the bullion market would rule out any actual transaction. If the market price of gold were below the legal price, no one would offer it in the market because it would be more advantageous to part from it at the bank of issue for notes. If the market price of gold were above the legal price, no one would demand it in the market because it would be more advantageous to obtain it from the bank against notes. The absence of transactions on gold out of equilibrium is, according to the authors, a necessary property of a monetary system in which notes are convertible both ways into the standard at a legal price. In other words, the market for gold may only exist in equilibrium: transactions can only occur at a market price equal to the legal price. This statement has an important consequence: since the market price of gold cannot be affected by the quantity of notes issued, the regulation of this quantity does not result from a price adjustment in the market for gold but it is to be explained by the (non-uniform) price adjustment in all other markets.

Such an exclusion of the market price of the standard from the regulation of the quantity of money as a consequence of convertibility would imply that the general emphasis in England—not only Ricardo's—on "the high price of bullion" (as compared with the mint price) was characteristic of a period (from 1797 to 1819) when the convertibility of the Bank of England note had been suspended. In a monetary system with convertible notes, the practical existence of a spread between the market price and the

(2015) and the volume Cartelier (2018). I restrict here myself to those studies which explicitly aim at making the Sraffa system and the existence of money mutually consistent.

legal price of the standard, and the theoretical interest for the relationship between this spread and the quantity of money, could only be linked to the coexistence of notes and coins, but they would both disappear with a circulation exclusively composed of notes convertible into the standard at no cost.

This conclusion might be somewhat surprising to scholars familiar with Ricardo's Ingot Plan where such paper money was advocated and the note issue was regulated through varying it inversely with the sign of the spread between the observed market price and the legal price of gold—a rule of monetary policy which implied that such a spread might exist even in this system. Even if one takes at face value Benetti and Cartelier's warning that they assume to be unfaithful to Ricardo's monetary thought, the exclusion of the market price of gold from the regulation of the quantity of money contradicts an important argument taken by Ricardo from Thornton (who nevertheless originated a distinct Classical approach to money from Ricardo's) and applying to *any* system of convertible notes. This argument is what I have called "the Penelope effect" (Deleplace 2017, pp. 259–62). In short, when the metallic reserve of the bank of issue falls below a certain limit, the bank is forced to purchase bullion in the market, although by doing so it has to pay a higher price than the legal price at which it gives gold against its notes. The bank thus provides the missing side of the market for bullion, in the absence of any demand by all other agents who may obtain gold from the bank at the legal price.

It is thus erroneous to maintain that the market price of gold cannot depart from the legal price under convertibility because no transaction might take place in this circumstance. Indeed, such transactions are not between individual agents who all behave like arbitragers and are only on one side of the market for gold. On the other side, the bank of issue is not arbitraging because its behaviour is constrained by convertibility. The fact that this side of the market is the monetary institution does not prevent a transaction on gold from being a market one (at a market price), distinct from the convertibility relationship between the same individuals and the same monetary institution (at a legal price). In fact, the market and the legal relationships between the bank and the arbitragers are complementary: both sides are "living on the same spot, and giving constant employment to each other", as Ricardo quoted from Thornton (Ricardo 1810–1811, p. 59). Convertibility does not imply that the market price of gold is necessarily equal to its legal price; in the contrary, this supposition would amount to assuming away convertibility: why would arbitragers return their notes for gold to the bank at the legal price if they could not sell this gold in the market at a higher price, where it is purchased *by the bank itself*? This arbitrage is central in the Classical analysis of the regulation of the quantity of money by the market price of the standard, as we will now see.

### 2.2 *The Quantity of Money and the Market for Gold Bullion*

I have contended elsewhere (Deleplace 2017, pp. 143–6) that Ricardo's theory of money may be summed-up by a Money-Standard Equation, which states that the rate of change in the value of money is equal to the rate of change in the value of the standard minus the rate of change in its market price. The first determinant reflects the change in the conditions of production of the commodity selected as standard of money (e.g. gold) while the second one reflects the discrepancy between the change in the actual quantity of money and the change in the "conformable" quantity of money, that is, the quantity required by "the wants of commerce". When there is no such discrepancy, the value of money only varies with the value of the standard (ibid., pp. 245–7). Contrary to the Quantity Theory of Money in which a change in the quantity of money *directly* causes a uniform change in the market prices of all commodities in the same direction (hence a change in the value of money in the opposite direction), in Ricardo a change in the quantity of money causes an opposite change in its value *indirectly*, through its effect on the market price of the standard. The monetary standard is therefore central in Ricardo's theory of money, and it is of the utmost importance to analyse how its market price adjusts when either cause of change in the value of money—a change in the conditions of production of the standard or a change in the actual quantity of money relatively to its conformable level—operates.

It may be shown that this adjustment process is different for each cause. When, for whatever reason (the discovery of a new highly productive mine or of a new technique of production, or the necessity to work less productive mines to answer a new demand for monetary purpose) the conditions of production of gold change, its market price adjusts in accordance with Ricardo's analysis of the formation of the natural price of a competitively produced commodity (ibid.: Chap. 5). When the cause of change is monetary (e.g. when the issuing of notes is greater than what is required by "the wants of commerce", to accommodate government budgetary needs or because the bank of issue "forces" its notes by discounting at a lower rate than the market for loans), an arbitrage is set in motion between the market for bullion and the bank of issue which eventually eliminates the discrepancy between the actual and the conformable quantities of money and restores the equality between the market price and the legal price of gold (ibid.: Chap. 7).

This adjustment of the quantity of money is a consequence of the arbitrage emphasised above between the market for the standard (gold) and the monetary institution (the bank of issue). Speaking of arbitrage implies that: (a) the market price of gold may diverge from the legal price; and (b) this divergence will generate correcting forces which will eliminate it. In Ricardo, the divergence between the market price and the legal price of gold simply reflected the discrepancy between the actual and the conformable quantities of money. As for the forces which eliminated this discrepancy, hence the divergence of prices, they resulted from "the Penelope effect" mentioned above. Since the bank of issue purchased bullion in the market (to replenish its metallic reserve) above the price at which it was legally compelled to give gold against its notes, it made losses. Sooner or later, the bank had to change its behaviour and it contracted the issue, in the hope that, issuing fewer notes through discounting, fewer notes would be returned to it to be converted into gold: "The Bank would be obliged therefore ultimately to adopt the only remedy in their power to put a stop to the demand for guineas. They would withdraw part of their notes from circulation, till they should have increased the value of the remainder to that of gold bullion." (Ricardo 1810–1811, p. 59)[3]

Even if the reduction of the excess quantity of notes eventually occurred, this adjustment might take quite a long time, depending on the lag with which the bank first purchased bullion in the market, and second contracted the issue to stop its losses. During this interval, the market price of gold bullion remained above the legal price: money was depreciated. According to Ricardo, this was one of the reasons why the monetary system could be improved by his Ingot Plan that replaced the mixed circulation of coins and notes by an exclusive circulation of notes

[3]Contrary to what is usually ascribed to Ricardo, the exportation of bullion, which under inconvertibility adjusted the aggregate quantity of money when it was in excess, only played a role at the beginning of the adjustment process under convertibility. As soon as the bank of issue purchased bullion to replenish its metallic reserve, the exportation of bullion was replaced by a domestic adjustment: the reduction of the bank issue under the pressure of the Penelope effect. For the international part of the adjustment, see Deleplace (2017: Chapter 8).

(convertible into bullion), the quantity of which was to be immediately adjusted when the market price of gold diverged from the legal price.

The real and the monetary adjustment processes have obviously opposite results for the market price of gold. In the first, this price was set at a new level, which corresponded to the permanent new conditions of production (the new natural price of gold). In the second, after it had changed, the market price of gold sooner or later returned to the legal level, illustrating that the monetary cause could only be temporary: the inadequacy of the actual quantity of money was eventually corrected. Consequently, there seems to be a contradiction between these two features of the economic system, both of them emphasised by Ricardo. According to his theory of value and distribution, the market price of gold must change with the conditions of production of this commodity (which determine what Ricardo calls its value) since it is competitively produced. According to his theory of money, the market price of gold should adjust to the legal price since gold is the standard of money. In other words, the market price of gold should be "regulated" by its ever-changing value (as for any competitively produced commodity) and by its fixed legal price (because gold is the standard of money). As Ricardo emphatically declared in 1819: "In a sound state of the currency, the value of gold may vary, but its price cannot"; Ricardo 1951–1973, V, p. 392).

I contend that there is no contradiction if it is assumed as a necessary condition of a standard-based monetary system that the commodity selected as standard of money in a given economy should be produced outside this economy, that is, in another economy where *another* standard of money is used.[4] In *Principles*, the historical example of gold being used as standard of money in England and produced "in Spanish America" (under a silver standard) was a metaphor of a theoretical requisite: for a monetary system to be "sound", the standard of money should be outside the system of production of the commodities circulated by this money.

This assumption was not explicitly made by Ricardo, even if in Chapter III "On the rent of mines" of *Principles* he observed that "by the discovery of America and the rich mines in which it abounds, a great effect was produced on the natural price of the precious metals" (Ricardo 1817–1821, p. 86) and in Chapter XIII "Taxes on gold" he mentioned that for this

[4] Gold should also be a non-basic product, so that it does not affect the distribution of income, hence the ranking of the mines by their physical productivity; see Deleplace (2017, pp. 182–3).

metal "its market value in Europe is ultimately regulated by its natural value in Spanish America" (ibid., p. 195). This assumption is obviously contradictory with the well-known supposition in *Reply to Bosanquet* of "the gold mine to be actually the property of the Bank [of England], even to be situated on their own premises". (Ricardo 1811, p. 216) But it should be noted that in the Bullion Essays of 1810–1811 Ricardo was not yet equipped with his later theory of value and distribution. What I contend is that the exteriority of the standard of money from the system of production of commodities is a necessary condition for applying Ricardo's theory of value and distribution to a monetary economy.

## 3 From Ricardo's Standard of Money to Sraffa's Standard in Which to Fix a Debt

### 3.1 *Ricardo's and Sraffa's "standard commodity"*

The expression "standard commodity" was used by Ricardo in his *Proposals for an Economical and Secure Currency* (Ricardo 1816, p. 58) to emphasise the necessity of having a commodity as standard "to regulate the quantity, and by the quantity the value of the currency" (ibid., p. 59). On this occasion, he not only discarded the "idea of a currency without a specific standard" (ibid.) but also the opinion "that we might judge of its value [of the currency] by its relation, not to one, but to the mass of commodities" (ibid.), which means that he excluded that the standard of money could be a composite commodity.[5] In *Principles*, after having analysed the effect of differences in the durability of capital on the relative values of competitively produced commodities, he supposed that gold could be considered as the standard commodity.[6]

[5] The exclusion by Ricardo of "the mass of commodities" as standard of money is not contradictory with his definition of the value of money by its purchasing power over all commodities except the standard. See Deleplace (2017, pp. 89–93; 115–16)

[6] See Ricardo (1817–1821, pp. 45–6). In his *Notes on Malthus*, Ricardo justified this supposition: "It was never contended that gold under the present circumstances was a good measure of value, it was only hypothetically, and for the purpose of illustrating a principle, supposed that all the known causes of the variability of gold, were removed. [...] I beg you [Malthus] to suppose all causes of variation removed, that we may speak about the variations of other things in an unvarying measure without confusion." (Ricardo 1951–1973, II, pp. 82–3).

As is well-known, Sraffa's "standard commodity" is a composite one which is constructed by the combination of the basic commodities in appropriate proportions. In *Production of Commodities*, the link with Ricardo's standard of value used as standard of money was suggested by Sraffa himself in an indirect way: "The conception of a standard measure of value as a medium between two extremes (§ 17 ff.) also belongs to Ricardo[1] and [...] the Standard commodity [...] has been evolved from it." (Sraffa 1960, p. 94) Footnote 1 attached to "Ricardo" directs the reader to Sraffa's own introduction to his edition of *Principles*, where he wrote: "In edition 3 [of *Principles*], therefore, the standard adopted was money 'produced with such proportions of the two kinds of capital as approach nearest to the average quantity employed in the production of most commodities'." (Sraffa 1951, p. xliv) One observes that Sraffa refers to "money" when Ricardo's quoted excerpt was attached to "gold considered as a commodity" (Ricardo 1817–1821, p. 45); this was consistent with Ricardo's expression in a later sentence: "To facilitate, then, the object of this enquiry, although I fully allow that *money made of gold* is subject to most of the variations of other things, I shall suppose it to be invariable, and therefore all alterations in price to be occasioned by some alteration in the value of the commodity of which I may be speaking" (ibid., p. 46; my emphasis). According to Sraffa, his "standard commodity" thus descended from Ricardo's standard of value supposed to be also the standard of money (gold).

One should, however, resist the temptation of linking Sraffa's standard commodity and Ricardo's standard of money.[7] As seen above, a monetary theory *à la* Ricardo presupposes the existence of arbitrage between the market for the standard of money and the monetary institution.[8] This rules out any conception of the standard of money as a commodity constructed on the basis of the actual commodities that belong to the system of production. Of course, such a composite commodity may substitute for gold as numéraire to measure prices, and it might even be imposed legally (as in the mechanism of indexation of the minimum wage in some countries). But to become the standard of money (and not simply an arbitrary

[7] For another discussion of the relation between the standard commodity and the standard of money, see Marcuzzo and Rosselli (1994).

[8] The emphasis on an arbitrage implying the market for gold—whether international arbitrage with its foreign markets, as in Marcuzzo and Rosselli (1991), or domestic arbitrage with the monetary institution, as in Deleplace (2017)—is a distinctive aspect of the unorthodox understanding of Ricardo's monetary theory; see Deleplace (2020).

numéraire), this composite commodity should fulfil two conditions: (a) convertibility both ways into the currency, which means a legal price at which the monetary institution guarantees this convertibility; and (b) the existence of a market in which this commodity is exchanged for the currency at a variable price, and not only the existence of a market for each of the commodities entering the composition of the standard. In the absence of these two conditions, arbitrage *à la* Ricardo is impossible. If modern financial markets where contracts on indexes are traded offer an example of fulfilment of the second condition, it is difficult to imagine that a market determination of the prices of commodities might coexist with a legal fixation of the price of their constructed aggregate. In addition, I have contended above that, for a monetary system to be "sound" in Ricardo's sense, the standard of money should be outside the system of production of commodities. This excludes the selection of Sraffa's standard commodity as standard of money.

Discarding Sraffa's standard commodity as standard of money in Ricardo does not mean that Sraffa's earlier views on the question of the standard cannot be useful to build a modern Classical standpoint on money. Since such a standpoint puts the standard of money centre stage it is first necessary to mention a possible extension of its analysis by Ricardo to modern times. In Deleplace (2017, pp. 388–9) I have suggested that Ricardo's conditions for "a sound state of the currency" in no way require the standard to be metallic: any marketable asset that is legally convertible both ways into money at a fixed price may play the same role, provided that it does not belong to the system of production of commodities. This exteriority of Ricardo's gold standard implied the supposition that gold was produced in a foreign economy, where it was not the standard of money. Another kind of exteriority can be imagined for a financial asset if it is a public one, issued outside the privately operated system of production of commodities. The standard of money would then be a public bond accepted and given against money by a central bank at a fixed legal price and traded at a variable price on a secondary market. The market price of such a public-debt standard would be determined, as for any financial asset, by the anticipation of its future returns. Here the introduction of a time horizon makes room for Sraffa's writings before *Production of Commodities*.[9]

[9] For a detailed study of money in Sraffa from the late 1920s to *Production of Commodities*, see Deleplace (2014).

### 3.2 Money, Gold, or Fresh Fish: In Which Standard to Fix One's Debt?

In his manuscript critique of Keynes's own-rate of interest (inserted in his own copy of *General Theory*), Sraffa made a comparison between gold and fresh fish: "From the whole treatment of this section [in Chapter XVII of *General Theory*], from the examples he gives (houses—use; money—liq.[uidity] pref.[erence]; wheat—carry cost) it is obvious that K.[eynes] has in the back of his mind two wrong notions, which have entirely misled him: (a) that commodities are borrowed for holding them till the end of the loan; (b) that only durable articles can therefore be borrowed—But in fact it is as convenient to make a loan of fresh fish for 100 years, as it is to make it of gold." (*Sraffa Papers*: I 100/9)

The role which fresh fish might play as conveniently as gold was explained two pages later: "P. 228 last para.[graph of *General Theory*] The idea that the advantages of possessing a given article have something to do with the own rate of *that* article, leads to this extraordinary paragraph. To understand it, notice that the underlying assumption is that people borrow an article in order to *keep it* and enjoy its advantages (liquidity for money, use for house, carrying costs for wheat) and therefore he [Keynes] has in mind permanent assets. But in fact people borrow money for parting from it, and buying things: the thing they borrow is, not what they want to use, but the standard in which they fix their debt: thus they might borrow fresh fish for 100 years, although it has neither liquid.[ity] pref.[erence], nor use at so much per annum—and it would have almost infinite carrying costs." (ibid.: I 100/11; Sraffa's italics) For the individuals, fresh fish or gold was thus "the standard in which they fix their debt" and not an asset in which they own their wealth.

One might think that each of these two standards being "as convenient", selecting one or the other would be a matter of individual choice. However, since borrowing such an "article" to make it a standard implies purchasing it spot with money today and selling it forward for money later,[10] the way it allows to "fix [one's] debt" depends on the role of money as a means of spot and deferred payment.

[10] "Loans are currently made in the present world in terms of every commodity for which there is a forward market. When a cotton spinner borrows a sum of money for three months and uses the proceeds to purchase spot, a quantity of raw cotton which he simultaneously sells three months forward, he is actually 'borrowing cotton' for that period." (Sraffa 1932, pp. 49–50)

In preparatory notes for his article against Hayek (Sraffa 1932), Sraffa had observed that the existence of money supposed the absence of a *complete* system of forward markets: "If money did not exist, all effects would be identical as if there were perfect forward markets for all commodities. In this case, money would not be stand.[ard] of d.[eferred] p.[ayments], because everybody would hedge." (*Sraffa Papers*: D3/9/44) In the contrary, sequential trading was a characteristic of a monetary economy, and consequently the means of exchange was also "standard of deferred payments": "The use of money as a 'medium of exchange' cannot go without its being 'a standard of deferred payments' or a 'store of value', two attributes which are included in the above: this is obvious if money transactions succeed each other in time; and if they are simultaneous, they must be cleared against each other and no money is required" (ibid.: D3/9/104).[11] Money was thus for Sraffa a "standard of deferred payments."

In the 1932 paper and the preparatory notes for it, another aspect characterised a monetary economy: "the essential feature of money [is] the singleness of the standard," in contrast with "non-monetary economies, in which different transactions are fixed in terms of different standards." (Sraffa 1932, p. 51) This singleness was a necessary attribute of "the standard in terms of which debts, and other legal obligations, habits, opinions, conventions, in short all kinds of relations between men, are more or less rigidly fixed," (ibid., p. 43) what in his preparatory notes Sraffa called "monetary constants" (*Sraffa Papers*: D3/9/49).

It should be observed that in all these quotations, the word "standard" used by Sraffa refers to money in which "all kinds of relations between men are more or less rigidly fixed," *not* to the standard of money in the sense of Ricardo, that is, something in which money itself is fixed. When money is used to borrow fresh fish or gold so as to fix a debt in terms of one or the other rather than in money, these articles become *a* standard of money. But so doing they do not necessarily retain the social character attached by Sraffa to money when he spoke of "monetary constants", and they certainly do not acquire the legal character attached by Ricardo to *the* standard of money. In a non-monetary economy where "different

[11] In this extract Sraffa mentioned as "attribute" of money "a store of value" (see also Sraffa 1932, p. 43). As seen above, in his manuscript notes on *General Theory* (four years later), he discarded Keynes's "underlying assumption […] that people borrow an article in order to *keep it* and enjoy its advantages." It is possible that Keynes's faulty use (in Sraffa's view) of the notion of "commodity-rate of interest" which Sraffa had introduced in his 1932 article led him to stress that fixing one's debt in terms of an article was different from storing it.

transactions are fixed in terms of different standards," fresh fish might be one of these standards, and in a monetary economy it could only be a private standard, chosen by some individuals rather than the single common one (money). By contrast, gold was during centuries the social standard of money, thanks to legal convertibility both ways into money. Fixing debts in terms of money amounted to fix them in terms of the standard of money (gold), although it was always possible, for matter of individual convenience,[12] to fix them in terms of a private standard (e.g. fresh fish), by buying spot (with borrowed money) and selling forward this particular article. As suggested above, a public bond might be substituted for gold as the legal standard of money in a modern economy, leaving open the possibility for all kinds of future contracts to be used as private standards in which to fix one's debt and so fulfil particular needs.

A last point may be worth mentioning. In a *Notebook* dated November 1927, Sraffa drew about the notion of cost a distinction between "natural" and "institutional economics": "Cost in the sense of ***means*** belong to ***natural economics***, i.e. they are equal in all forms of society, and are independent of institutions. Cost in the sense of ***inducements*** belong to ***institutional economics***, they vary according to 'social standards'. [...] (*Means* are habitual necessaries, as Ricardo says, i.e. physical since that habit is physical; ***not*** conventional necessaries, as Marshall says—these are psychological and therefore are part of inducement, not of possibility)." (ibid.: D3/12/11/98; Sraffa's emphasis) It is ironical that, in the case of the standard of money *à la* Ricardo (whether gold or a public bond), the necessary requisite of being outside the system of production echoes Sraffa's project to separate, in the determination of the system of prices, what is "natural" from what is "institutional".

## References

Benetti, C., Cartelier, J. (2020) "From Ricardo to Sraffa: Gold as Monetary Standard in a Classical Theory of Money," in M.C. Marcuzzo, G. Deleplace, P. Paesani (eds.) *New Perspectives on Political Economy and Its History*, Cham: Palgrave Macmillan: 241–57.

[12] "Now we might ask, whether the 'monetary constants' referred to above are in terms of this money, or not; and if yes, how can it be indifferent for a debtor, when the money price of wheat falls, whether his debt is fixed in money or wheat." (*Sraffa Papers*: D3/9/50)

Benetti, C., Klimovsky, E., Rebeyrol, A. (2014) "Une reconstruction monétaire du modèle de Sraffa," in F. Tricou et D. Leeman (eds.) *Economie, mathématique et histoire. Hommage à Christian Bidard*, Paris: Presses Universitaires de Paris-Nanterre: 75–88.

Benetti, C., Bidard, C., Klimovsky, E., Rebeyrol, A. (2015) "Temporary disequilibrium and money in a Classical approach," *Cahiers d'économie politique*, 69: 159–184.

Cartelier, J. (2018) *Money, Markets and Capital. The Case for a Monetary Analysis*, Abingdon: Routledge.

Deleplace, G. (2014) "The Essentiality of Money in the *Sraffa Papers*," in R. Bellofiore, S. Carter (eds.), *Towards a New Understanding of Sraffa. Insights from Archival Research*, Houndmills: Palgrave Macmillan: 139–166.

Deleplace, G. (2017) *Ricardo on Money. A Reappraisal*, Abingdon: Routledge.

Deleplace, G. (2020) "Orthodox *versus* Unorthodox Views on Ricardo's Theory of Money," *The European Journal of the History of Economic Thought*, 27 (6), December: 819–36.

Marcuzzo, M.C., Rosselli, A. (1991) *Ricardo and the Gold Standard. The Foundations of the International Monetary Order*, London: Macmillan.

Marcuzzo, M.C., Rosselli, A. (1994) "The standard commodity and the standard of money," *Cahiers d'économie politique*, n° 23: 19–31.

Ricardo, D. (1810–1811) *The High Price of Bullion, A Proof of the Depreciation of Bank Notes*, in Ricardo (1951–1973), vol. III, 1951: 47–127.

Ricardo, D. (1811) Reply to Mr. Bosanquet's 'Practical Observations on the Report of the Bullion Committee'. in Ricardo (1951–73), vol. III, 1851: 155–256.

Ricardo, D. (1816) *Proposals for an Economical and Secure Currency*, in Ricardo (1951–1973), vol. IV, 1951: 49–141.

Ricardo, D. (1817–1821) *On the Principles of Political Economy, and Taxation*, 1st edition: 1817, 2nd edition: 1819, 3rd edition: 1821, in Ricardo (1951–1973), vol. I, 1951.

Ricardo, D. (1951–73) *The Works and Correspondence of David Ricardo*, edited by P. Sraffa with the collaboration of M.H. Dobb, Cambridge: Cambridge University Press.

Smith, A. (1776) *An Inquiry into the Nature and Causes of the Wealth of Nations*, in R.H. Campbell, A.S. Skinner (eds.) *The Glasgow Edition of the Works and Correspondence of Adam Smith*, Oxford: Clarendon Press, 1976.

Sraffa, P. (1932) "Dr. Hayek on Money and Capital," *The Economic Journal*, Vol. 42, March: 42–53.

Sraffa, P. (1951) "Introduction," in Ricardo (1951–1973), vol. I: xiii–lxiv.

Sraffa, P. (1960) *Production of Commodities by Means of Commodities. Prelude to a Critique of Economic Theory*, Cambridge: Cambridge University Press.

*Sraffa Papers*, Cambridge: Trinity College, Wren Library.

Comment

# A Comment on 'A Quest for a Modern Classical Standpoint on Money'

*Carlo Benetti and Jean Cartelier*

The subtitle of GD's chapter is explicit: "A quest for a modern Classical standpoint on money"; while wishing to be "faithful to Ricardo", GD's aim is to be "compatible with Sraffa's price system". Benetti and Cartelier (2020) chose to favour the second objective at the cost of the first one. The coherence of Sraffa's price system with Ricardo's monetary theory, as interpreted by GD, raises difficult problems.

GD's central statement is that "the adoption of Sraffa's prices by a monetary economy, in accordance with Ricardo's theory of the currency, requires a specific working of the market for the commodity selected as standard of money." The recent, excellent book published by GD *Ricardo on Money* reinforces this thesis. However, two points need to be raised: the specificity of the gold market and its relation to Sraffa's natural prices.

It is true that, according to Ricardo, the market for the standard of money works in a specific way. But the real distinctive feature of the gold standard is its legal price in terms of £, the consequence of which is that, in a closed economy, the market price of gold cannot diverge from the legal price. On the other hand, in GD's construction of a "modern Classical standpoint on money" gold is assumed to be produced outside the economy. As a result, it becomes problematic to link gold with the domestic production system determining natural prices and the rate of profit. The project of being "consistent with a Sraffa system of prices" is seriously impacted.

### *The Specificity of the Gold Market*

GD states that, in the Classical monetary theory (i.e. inspired by Ricardo), the suitable amount of convertible notes issued by the Bank is regulated by the arbitrage between the market price and the legal price of gold. This proposition will be discussed after a brief summary of GD's analysis.

#### *Smith-Thornton's Arbitrage*

In Benetti and Cartelier (2020, § 2) we showed that in an (abstract) economy with convertible banknotes, the price of gold in the domestic market cannot differ from the legal price and any possible gap between these prices quickly vanishes.[13] GD's main critique consists in an argument taken by Ricardo from Thornton who, in turn, was inspired by Smith (1776). The reference text is found in *High Price of Bullion*, p. 58–9 where Ricardo presents Thornton's analysis as a possible alternative to the regulation of the quantity of money by means of the exportation of gold which he has just analysed in detail (1810–1811, pp. 54–8, see also our paper, pp. 252–3). According to Thornton, in order to increase an insufficient stock of gold, the Bank of issue[14] purchases gold bullion in the domestic market, although the market price is higher than the legal price at which individuals obtain gold against their notes. Individual gains are the Bank's losses. The whole argument is taken literally by Ricardo as showed by Thornton's long quotation (1810–1811, pp. 58–9), and endorsed by GD: "The bank thus provides the missing side of the market for bullion, in the absence of any demand by all other agents who may obtain gold from the bank at the legal price" (p. 504).

In the gold market, as Thornton notices, "The one party [the individuals] will be melting and selling while the other [the Bank] is buying and coining". The objects traded on this market are also heterogeneous. Individuals sell a commodity against £. The Bank does not buy a luxury good. It buys gold as the material of coins and the basis of issuing banknotes.

[13] See also Ricardo's clarification about the position of the Bullion Committee: "Their principle, when fairly stated, is, not that gold as a commodity may not rise above its value as coin, but that *it cannot continue so* (ital. added), because the convertibility of coin into bullion would *soon* (ital. added) equalize their value." (Ricardo 1811, p. 187)

[14] It is the only function of the Bank we deal with, see D. Ricardo (1824, p. 277): there is "no necessary connection" between the issuance of "paper currency, as a substitute for a metallic one" and the credit operations.

The Bank inevitably suffers losses. This is why Thornton speaks about a "very unequal war" so that the Bank "will be likely to be tired sooner than its adversaries". It is Ricardo who sets out the true solution (mentioned by GD p. 506): the Bank "would be obliged ultimately [...to...] withdraw part of their notes from circulation, till they should have increased the value of the remainder to that of gold bullion". As in the case of the exportation of gold, it is the only sensible measure to reach the desired value of the ratio between the quantity of banknotes and the Bank's stock of gold (we dubbed $\mu$).

*Comments*

It should be emphasized that our only aim is to clarify the notion of arbitrage between two prices when one of them is fixed by law.[15] The main theoretical weakness of Smith-Thornton's argument is that the losses of the Bank result from its odd monetary policy. For the same amount of £ received by the Bank against gold and spent to redeem gold in the market, the quantity of gold in the Bank's vault decreases; for the same quantity of gold in the vault of the Bank, the amount of circulating banknotes increases. In both cases, multiplier $\mu$ increases. The "Penelope effect" as called by GD,[16] implies that Penelope undid in the night *more* than what she did in the day. It is easy to guess the end of such a tale.

The difficulty does not lie in the Bank's losses as such. Depending on circumstances they could be justified. The problem is that these losses are the direct consequence of the Bank monetary policy which is diametrically opposed to the very aim of the Bank i.e. to decrease the ratio $\mu$. Why does the Bank provide "the missing side of the market for bullion" (p. 504) whereas this action worsens the problem that motivated its decision to act? GD states that "when the metallic reserve of the bank of issue falls below a certain limit, the bank is *forced* (ital. added) to purchase bullion in the

[15] This issue is independent of "*facts*", for instance the "high price of bullion" when the convertibility had been suspended (see GD p. 503), the monetary policy at the time of Ricardo, the evolution of the gold price and its possible difference from the legal price during the convertibility which is considered in the Ingot Plan, and erroneously interpreted by GD as a critique of our position (see GD p. 504). Actually, we adopt Rousseau's famous methodological precept: " Commençons par écarter tous les faits, car ils ne touchent point à la question " (1754, p. 132).

[16] GD takes this metaphor from Smith. See A. Smith (1776, book IV, Chap. VI, pp. 516–17): the Bank's gold demand is mainly due to the fact that "the greater part of the current coins is almost always more or less worn". See also book II, ch. II, p. 286.

market, although by doing so it has to pay a higher price than the legal price at which it gives gold against its notes". Such an assertion is not acceptable: the insufficient gold reserve cannot "force" the Bank to apply a monetary policy that reduces it further. The constraint on the Bank is quite different. As stated by Ricardo, the Bank ends up being "obliged" (1810–1811, p. 59) to reduce the amount of circulating notes, which is the only way to bring $\mu$ down.

Finally, why does not the Bank adopt this measure *from the beginning*? We do not see how to justify the idea that the initial reserve of gold is both too low—so that the Bank is motivated to buy gold at a price higher than the legal price—but not low enough to force the Bank to adopt the only rational action that is the withdrawal of "part of their notes from circulation" as stated by Ricardo. Why must the initial stock of gold be reduced further?

In our paper we argued that there is no arbitrage between two prices when one of them is legally fixed. We now complete our argument: the supply of gold in the domestic market is positive only if the market price is higher than the legal price; but in such a case the demand for gold is also positive only if there is at least one agent whose economic behaviour is incoherent. As rightly stated by GD, our position "contradicts" the "Penelope effect" he qualifies as "an important argument." We have just showed that such a contradiction is in favour of our position and *not* against it as GD asserts.[17]

Besides, it seems to us that Ricardo clearly favours the regulation of the quantity of money through the export of gold by individuals. Gold is bought by foreigners and the role of the Bank would be to regulate the issuance of money and not to act in the market instead of individuals, as stated by GD. A deep doctrinal reason is suggested by Ricardo (1810–1811, pp. 55–6): "The exportation of the specie may at all times be safely left to the discretion of individuals; it will not be exported more than any other commodity, unless its exportation should be advantageous to the country. If it be advantageous to export it, no laws can effectually prevent its exportation. Happily in this case, as well as in most others in commerce where

[17] We put aside GD's critique according to which our position "would amount to assuming away convertibility: why would arbitragers return their notes for gold to the bank at the legal price if they could not sell this gold in the market at a higher price, where it is purchased *by the bank itself*?." This assertion is obviously incorrect: "arbitragers [would] return their notes for gold to the bank at the legal price" simply in order to export gold. (See Ricardo 1810–1811, p. 58).

there is free competition, the interests of the individual and that of the community are never at variance."

### *What About a Relation with Sraffa's Natural Prices?*

Two points will be made thereafter. GD's requirement that gold be produced outside the economy (i) is not well-founded since there is in fact no contradiction between the legal price of gold (in £) and the natural price of gold (in a single or composite commodity) when gold is domestically produced; and (ii) is not appropriate for an integration, intended by GD, of Sraffa's natural prices in Ricardo's monetary theory.

*No Contradiction to be Found When Gold is Domestically Produced*
In Benetti and Cartelier (2020), we also attempt at exploring the spiny question of a possible link between Sraffa's system of prices, on the one hand, and the Classical thought about money and market prices, on the other. What follows is tentative and is put forth for the sake of discussion. GD maintains that compatibility between Sraffa's theory of natural prices and Classical theory of money requires that gold (the standard of money) must be produced outside the domestic system of production. GD's assertion is surprising. Not as a general principle since anthropologists have shown that in many societies money effectively comes from abroad, but as a specific feature of Ricardo's views. A straight consequence of that requirement is that *the natural price of gold is no longer determined by the conditions of production of the economy under consideration.*

Why does the integration of money in Sraffa's system of natural prices require the expulsion of gold production away from the economy? This is in fact related to a wall separating money and prices. According to GD, the market price of gold is subject to "two apparently contradictory adjustment processes", gravitation around Sraffa's natural price, on the one hand, regulation through the legal price, on the other. "A conclusion emerges: this double regulation of the market price of gold bullion may only be non-contradictory if one assumes that *gold bullion is produced outside of the country in which it is used as the standard of money*" (Deleplace 2017, pp. 272 and 273; Deleplace's emphasis).

In Benetti and Cartelier (2020) the same problem occurs but we have not found any contradiction. Let us start from the following system of natural prices determination *à la* Sraffa (system (2*) in Benetti and Cartelier 2020):

$$\begin{aligned}&(1+r)A'p = p\\&(1+r)\lambda\sum_{i\neq g} a_{ig}p_i = \lambda p_g \qquad (2^*)\\&p_g = 1\end{aligned}$$

where $r$ is the rate of profit, $A'$ the transpose of $A$ the square matrix of the $a_{ij}$'s, $a_{ij}$ the quantity of commodity $i$ used for the production of commodity $j$, and $p$ the vector of natural prices.

The second line of (2*), which concerns gold (subscript $g$), is formally identical to the others except for parameter $\lambda$ which applies to the natural quantity of gold (equal to one as for any other commodity). Gold is supposed to be a non-basic commodity in accordance with GD's position, which means that its conditions of production do not determine the rate of profit and the natural prices of basic commodities. Once the system of production of basic commodities has been solved (first line of (2*)) and when market prices have converged toward natural prices, as GD seems to take for granted, it is easy to get money (natural and market) prices by using the legal definition of the £, that is:

$$p^{£} = p^{g}\frac{1}{\beta}£ \qquad (3^*)$$

where $\beta$ is the quantity of gold which defines the £ and $p^g$ is the vector of natural prices—solution of (2*)—expressed in gold.

If "a mine of gold were discovered" (Ricardo 1810–1811, p. 54) and if the technique did not change, parameter $\lambda$ would increase but not the "wants of commerce" (see Benetti and Cartelier 2020). Therefore no variation of the quantity of means of payment would take place. The additional quantity of gold is used as a commodity by contrast with what happens according to Ricardo who considers that the total quantity of gold is coined.

If the relative value of gold decreased as an effect of a better production technique, vector $p^g$ would be multiplied by a scalar greater than 1. The same scalar would also apply to the prices expressed in £, which are nothing but natural prices in gold multiplied by constant $\frac{1}{\beta}$. The "wants of commerce" increase and are met by the issuance of additional means of payment as a counterpart of additional loans (see Benetti and Cartelier 2020, p. 247)

*Some Puzzling Consequences of Gold Being Produced Abroad*
Requirement that gold has to be produced abroad does not seem to affect GD's reasoning. This is the consequence of his neglect of the *general* problem of gravitation of market prices around natural prices. His focus is on the *specific* question of the gold market only. However, elaborating a coherent theory embracing both Sraffa's natural price system and Classical theory of money cannot avoid considering market prices in general. This is what GD implicitly comply with in his book where market prices are present under the disguise of the so-called "wants of commerce". It is not the proper place for developing this difficult point. We shall content ourselves with insisting on the necessity to inquire into the relation between natural and market price in the specific case of gold (the market price of gold being its legal price).

Two formal expressions of gold being produced abroad may be considered.

In the first one, the second line of system (2*) is cancelled which amounts to eliminating the natural price of gold. However, the natural price of gold (produced abroad) enters his equation (5.8) (in Deleplace 2017, p. 187), which gives the condition for importing gold. Gold imports are subject to a comparison between the cost of buying gold abroad (including the domestic natural rate of profit) and its legal price. He determines the cost of importing gold as its natural price abroad (assumed to be equal to its market price) corrected by the rate of exchange *times* $(1 + r)$ given by (2*). That importing gold from abroad makes sense only if the cost of it is inferior or equal to its legal price is pretty sure. Pretty sure also is the fact that the natural price of gold does not play any role here the domestic natural price of gold has disappeared with the cancellation of the second line of (2*) and the natural price abroad is replaced by its market price as a consequence of gravitation. Note that GD also requires that gold is not the standard of the monetary system abroad. This last consideration shows that GD cannot get rid of the embarrassing question of gravitation, at least for gold produced abroad, gold not being the standard there. Why should the market price of gold be equal to its natural price if no gravitation mechanism is considered?

Notwithstanding a further point. GD's equation (5.8) has a clear general signification but we have to wonder who is concerned by it. Does it mean that a special branch or industry runs gold external trade? If not so, why invoke the domestic natural rate of profit? Any industry (corn, iron, etc.) could get an extra profit (in addition to its natural one) by importing gold if condition alluded above (equation 5.8) is met even if this extra profit is small.

A second interpretation has to be explored in order to escape from the critique above. Let us consider the complete system (2*) with its second line. Instead of production *stricto sensu*, that second line would show the gold external trade as a special branch. Such a view seems perfectly compatible with GD's requirement that gold be a non-basic commodity produced abroad. Inputs $a_{ig}$ would be interpreted as the elements of the vector of commodities bartered against gold, imported gold being viewed as the specific output of that branch subject to the rule of the uniform rate of profit. Then, it could be possible to get the (domestic) natural price of gold given by (2*). It would be the result of inserting in the second line of (2*) $a_{ig}$'s natural prices and the rate of profit which are solutions of the first line of (2*). The last line of (2*) allows getting natural prices expressed in gold.

Now we face two natural prices (domestic and foreign) of gold. But they do not play any role in GD's equation (5.8) since only the foreign market price of gold and the domestic legal price of gold are relevant (with the rate of exchange).

It is not possible to go further unless we would be obliged to embark upon a theory of international natural prices which is beyond the scope of these limited comments. It is enough to be aware of the numerous and difficult questions that any Classical theoretician of money has to overcome in order to make that theory consistent with a Sraffa system of prices.

Another problem comes from the *market price* of gold. In a closed economy, the market price of gold is necessarily equal to its legal price (Benetti and Cartelier 2020). What can be said about it when gold is imported?

For Classical economists, since Smith's *Wealth of Nations* chapter vii, "Cantillon's rule" determines market prices in general. That rule applies here in a specific way. Let us imagine that gold external trade branch, as a special industry, borrows a given amount of \$ from the Bank abroad. Thanks to these \$, it buys gold abroad at market price. Assume that quantity of imported gold, when brought to the Bank of England, allows that industry getting a sufficient amount of £ and allows paying back the foreign Bank after conversion in \$ (we abstain from any consideration about exchange rate for the sake of simplicity). The foreign market price of gold is given, according to Cantillon's rule, by the ratio of the total of means of payments (\$ and £ made homogenous thanks to the rate of exchange) divided by the quantity of gold brought to market. That quantity and the total amount of means of payment are not known. Even if this does not affect the condition alluded by GD (equation 5.8), it shows how long is the way toward an integration of a Classical theory of money—not separable from market prices determination—in Sraffa's natural prices theory.

## References


Benetti, C., Cartelier, J. (2020) "From Ricardo to Sraffa: Gold as Monetary Standard in a Classical Theory of Money," in M.C. Marcuzzo, G. Deleplace, P. Paesani (eds.) *New Perspectives on Political Economy and Its History*, Cham: Palgrave Macmillan: 241–57.

Deleplace, G. (2017) *Ricardo on Money. A Reappraisal*, Abingdon: Routledge.

Ricardo, D. (1810–1811) *The High Price of Bullion, A Proof of the Depreciation of Bank Notes*, in Ricardo (1951–1973), vol. III, 1951: 47–127.

Ricardo, D. (1811) *Reply to Mr. Bosanquet's 'Practical Observations on the Report of the Bullion Committee'*, in Ricardo (1951–1973), vol. III, 1951: 155–256.

Ricardo, D. (1824) *Plan for the Establishment of a National Bank*, in Ricardo (1951–1973), vol. IV, 1951: 271–300.

Ricardo, D. (1951–1973) *The Works and Correspondence of David Ricardo*, edited by P. Sraffa with the collaboration of M.H. Dobb, Cambridge: Cambridge University Press, 11 vols.

Rousseau, J.-J. (1754) *Discours sur l'origine et les fondements de l'inégalité parmi les hommes*, Paris: La Pléiade, 1964.

Reply

# A Reply to Benetti and Cartelier

*Ghislain Deleplace*

## *The Compatibility between Sraffa's System of Prices and Ricardo's Theory of Money*

The common issue under discussion in my chapter, in Benetti and Cartelier's comments on it, and in the present reply, is the compatibility between Sraffa's system of prices and Ricardo's theory of money. Consequently what are left aside pertain to Sraffa's own monetary views before *Production of Commodities by Means of Commodities*—tackled in Sect. 3 of the chapter—and to Ricardo's monetary views not directly linked to his theory of prices—such as the detail of his stipulations about the best monetary system. This discussion, however, cannot be freed from two other references by the same participants, which have a distinct object in their own right. One is my book *Ricardo on Money. A Reappraisal*: although it claims that it may contribute to the integration of money in the modern (Sraffa's) theory of Classical prices (Deleplace 2017, pp. 384–6), it pursues another goal, which is to provide a coherent reconstruction of Ricardo's theory of money. The second reference is a recent paper by Benetti and Cartelier entitled "From Ricardo to Sraffa: Gold as Monetary Standard in a Classical Theory of Money": the authors explicitly warn that "to establish 'what Ricardo has really said' is not our topic. We have preferred to inquire into what could or should be a Classical theory of money and prices sixty years after Sraffa's *Production of Commodities by Means of Commodities* publication." (Benetti and Cartelier 2020, p. 241). One difficulty of the discussion is that some of the arguments exchanged cannot be fully understood without referring to these two other publications.

An interesting aspect of these two references is that they both offer a more or less detailed formalization. This should help locating precisely on which points there is a disagreement between studies which have in common to praise Sraffa's contribution to the Classical theory of prices and to reject the usual understanding of Ricardo's monetary theory in terms of the Quantity Theory of Money. The model of Ricardo's theory of money developed in Deleplace (2017) leads to three main results:[18] (a) The value of money varies positively with the natural price of the standard and negatively with its market price (what I call "the Money-Standard Equation"); (b) The standard (e.g. gold) is necessarily exterior to the system of production that determines the natural prices of the commodities in the economy where the standard-based currency circulates; and (c) When the quantity of money is in excess of what is required by "the wants of commerce", it is endogenously adjusted through a contraction of the note issue that is forced by the unfortunate consequences for the issuing bank of the rise in the market price of the standard (what I call "the Penelope effect").

In their comments, Benetti and Cartelier target result (b): "GD's requirement that gold be produced outside the economy (i) is not well-founded since there is in fact no contradiction between gold legal price (in £) and gold natural price (in a single or composite commodity) when gold is domestically produced; and (ii) is not appropriate for an integration, intended by GD, of Sraffa's natural prices in Ricardo's monetary theory" (p. 519). They also target result (c): "As rightly stated by GD, our position 'contradicts' the 'Penelope effect' he qualifies as 'an important argument'" (p. 518). Benetti and Cartelier's assumption that under convertibility the market price of gold cannot diverge from the legal price also amounts to rejecting result (a) since the value of money does not vary with a market price of the standard that they assume to be fixed. This assumption of a fixed market price of gold is central in the alternative Classical model of a standard-based money in a Sraffian system of prices which they propose in Benetti and Cartelier (2020).

[18] These results support the two main conclusions of the book for the history of economic thought: the existence of an evolution in Ricardo's monetary theory from his initial views (at the time of the *Bullion Report*) to his mature views (in relation to his genuine theory of value and distribution developed in *Principles)*, and the rejection of the usual understanding of this theory as an incoherent mix of a quantity-theory of money in the short run and a commodity-theory of money in the long run.

In the same order as they appear in Benetti and Cartelier's comments, I will discuss below their critiques of my results (c) and (b), before emphasising that their fixity assumption is inappropriate in Ricardo's theory of money, hence in any attempt at making it compatible with Sraffa's system of prices. But before that, a caveat may help the reader. In accordance with their commitment in their 2020 paper to leave aside "what Ricardo has really said", Benetti and Cartelier do not suggest that my reconstruction of Ricardo's theory of money might be unfaithful. Consequently, the critiques they raise at me should be considered as applying to Ricardo himself. Besides, my object has been to reconstruct Ricardo's theory of money in a way that makes it: (1) consistent with the whole of Ricardo's writings on money; (2) internally coherent; and (3) compatible with Sraffa's system of prices. It has *not* been to contend that this reconstruction is the only (and even the best) way to analyse a monetary economy—an assertion which would be beyond the point.

### *The Divergence of the Market Price of the Standard from its Legal Price*

Two questions may be asked about this divergence: can it exist, and if yes, what happens to it? I have shown elsewhere (see Deleplace 2020) that, even if they may disagree on the answer to the second question, the (minority) unorthodox scholars on Ricardo's theory of money all answer to the first question in the positive and consider this divergence as a central aspect of Ricardo's theory.[19] It should be noted that for these scholars it is so whether Ricardo dealt with a monetary system exclusively composed of inconvertible notes (as testified by the title of his famous first monetary pamphlet *The High Price of Bullion, A Proof of the Depreciation of Bank Notes*) or a system containing (partly or exclusively) convertible notes. By contrast, Benetti and Cartelier boldly assert that "the real distinctive feature of the gold standard is its legal price in terms of £, the consequence of which is that, in a closed economy, the market price of gold cannot diverge from the legal price" (p. 515), and this confirms what they claim

[19] Contrary to the (majority) orthodox scholars who consider that the market price of gold is only in Ricardo a proxy of the general price level, in the absence of a solution to the index-number problems. In such an interpretation, the divergence between the market price of gold and its legal price is obviously deprived of any significance.

in the opening lines of their comments: they accept to bear "the cost" of not being "faithful to Ricardo".

*Can There be a Divergence?*

In response to this question, the statement that, in a closed economy (the one, according to Benetti and Cartelier, for which theoretical propositions ought to be demonstrated), the market price of gold is fixed (provided its legal price remains itself fixed) may come as a surprise since in their numerous writings they have (rightly) advocated for long that a theory of prices should not be restricted to equilibrium but account for disequilibrium situations. They maintained in these writings that in a decentralised economy the equilibrium situation cannot be known *ex ante* by the individual agents, so that a theory of price adjustment is required, which should not violate the decentralised ("market") character of the economy. In accordance with this *motto*, they now claim that, to analyse a monetary economy, the non-uniform adjustment of all prices in response to a change in the quantity of money should be made explicit. *All except one*: the price of gold, which is supposed to be unaffected by such a change.

The reason for this exception is for Benetti and Cartelier crystal-clear: being the standard of money, gold has a legal price, contrary to all other commodities. The market price of gold cannot diverge from the legal price at which gold may be brought to or obtained from the Bank: there cannot be a market for gold because it is impossible that sellers and buyers of it coexist. If the market price of gold were above the legal price, potential buyers would find more advantageous to go to the Bank and they would ignore the market, as potential sellers would do in the symmetrical case of the market price being below the legal price. No individuals face one another about gold: the Bank is the only counterpart. But its relationship with individuals (two-way convertibility) is not a market one, since the Bank is compelled to give or to accept gold at a legal price.

Leaving aside the consequences of such an assumption for the very notion of standard of money (see below Section "Sraffa's Theory of Prices and Classical Theory of Money"), it is easy to see why it clashes with Ricardo's understanding of a monetary economy. Ricardo repeats over and over that the divergence between the market price and the legal price of gold reflects the divergence between the actual quantity of money and that which satisfies "the wants of commerce". No individual—and not more the issuing bank itself—may know *ex ante* the latter. Benetti and Cartelier are right when they say that "the wants of commerce" depend on what happens in the markets for all other commodities, and it is precisely why—according to their own *motto*—they cannot be known *ex ante*. If one assumes (something which is not denied by them) that the Bank can issue more or less liberally the notes demanded by individuals, the decentralised nature of the economy makes it impossible for the Bank to issue an aggregate quantity of notes that would spontaneously equalise with what is required by "the wants of commerce". It is thus impossible to reconstruct Ricardo's monetary theory or to use it as starting point of any Classical theory of money by presupposing, as Benetti and Cartelier do, that "in a closed economy, the market price of gold cannot diverge from the legal price", if this "closed economy" is a decentralised one.

*What Happens in Case of Divergence?*

Supposing that the market price of gold is above the legal price (in Ricardo's words, the currency depreciates), what happens next? Benetti and Cartelier's answer is: nothing. Nothing between individuals, and I agree. As I wrote in my chapter (on p. 504): "The bank thus provides the missing side of the market for bullion, in the absence of any demand by all other agents who may obtain gold from the bank at the legal price." But here comes their second critique concerning the market price of gold: nothing will happen between individuals and the Bank either, because this would be assuming that the Bank implements an "odd monetary policy" (p. 517). In other words, its "economic behaviour is incoherent." (p. 518) This is a dead blow to a reconstruction that emphasises the coherence of Ricardo's theory of money.

To see exactly at which level this critique operates, let me again sum up how Ricardo describes the regulation of the issuing of convertible notes by the Bank of England in case of its excess (for details see Deleplace 2017, Chapter 7). The adjustment process is in four steps: (1) Individuals obtain gold for notes at the Bank and export it. (2) If the Bank reissues the notes which are returned to it, there comes a time when its metallic reserve will be viewed by it as too low to guarantee convertibility. It then starts to purchase gold in the market with its notes, in spite of the loss it suffers by buying gold at a higher (market) price than the (legal) price at which notes are returned to it. (3) Arbitragers stop exporting the gold since it is now more advantageous to sell it to the Bank in the market. (4) This goes on and on ("Penelope effect") until the Bank, to stop its losses, contracts its note issue. This contraction is implemented as long as notes are returned to the Bank, that is, as long as the market price of gold is above the legal price. In the end both prices equalise.

Benetti and Cartelier do not object to (1); indeed it is what they retain from Ricardo: "It seems to us that Ricardo clearly favours the regulation of the quantity of money through the export of gold by individuals. Gold is bought by foreigners and the role of the Bank would be to regulate the issuance of money and not to act in the market instead of individuals, as stated by GD." (p. 518) An ambiguity lies in the word "favours". If it means that Ricardo contended that money in excess generates an export of gold, no one will deny it; I devoted the whole Chap. 8 of my book to the analysis of how this occurs. If "favours" means that for Ricardo the export of gold should be the preferred means of regulating the note issue, why is it that "the role of the Bank would be to regulate the issuance of money", as they say? If the export of gold does the job, the sole "role" of the Bank is to give gold passively against its notes when it is required to do so under convertibility. Indeed Ricardo never considered that the Bank should buy and sell gold "instead of individuals, as stated by GD" (I nowhere stated that), but he did consider that the implementation of a specific rule (see below) by the Bank was preferable to the export of gold by individuals as means of regulating the note issue.[20]

Since the export of gold implies a cost (as for any commodity that must be brought to the market), Benetti and Cartelier cannot either object to (3), that is, the substitution of domestic arbitrage (at no cost, as they suppose) for the international one. Hence their critique does not concern the behaviour of the arbitragers but that of the Bank. They object to (2): "Why does the Bank provide 'the missing side of the market for bullion' ([GD] p. 504) whereas this action worsens the problem that motivated its decision to act? [...] The insufficient gold reserve cannot 'force' the Bank to apply a monetary policy that reduces it further." (pp. 517–8) They also object to (4): "Why does not the Bank adopt this measure [the contraction of the note issue] *from the beginning*? [...] The demand for gold is positive only if there is at least one agent [the Bank] whose economic behaviour is incoherent." (p. 518; BC's emphasis)

[20] The long quotation from Ricardo given by Benetti and Cartelier had another object: to stress that, *if* this regulation was done through international arbitrage, freedom of gold export should be granted—a sensible stipulation since it reduced the limits between which the exchange rate might fluctuate (the gold points).

Benetti and Cartelier's critique is entirely right. It just misses its target: their "whys" do not question my reconstruction of Ricardo's argument but they are in fact addressed by Ricardo himself to the Bank of England. According to him, the Bank behaved stupidly, *from its own point of view*, as testified by the fact that it was eventually obliged to adopt the opposite behaviour (contracting the issue rather than expanding it to purchase gold).[21] But mistaking the real target they fail to see that, although the Bank behaves stupidly in the first place, this does not prevent the regulation of the note issue from operating (that is, the divergence between the market and the legal prices of gold will sooner or later disappear, contrary to what happens under inconvertibility), precisely because the "Penelope effect" obliges the Bank to stop behaving stupidly. In other words, the "incoherent behaviour" of the Bank is indeed "not acceptable", as Benetti and Cartelier write (p. 518). But it is so, not as a condition of my reconstruction of Ricardo's argument (which logically would condemn this reconstruction) but, according to Ricardo himself, as the explanation of the poor functioning of the monetary system (poor because the delay with which the Bank changes its behaviour has adverse effects on the economy).

Benetti and Cartelier are also right when they ask, as quoted above: "Why does not the Bank adopt this measure [the contraction of the note issue] *from the beginning*?" This is exactly Ricardo's point, and there is no contradiction between my reconstruction of what happens when it is not so and the contention that all would be better if it were so. On the contrary, this reconstruction justifies Ricardo's advocacy of his monetary policy rule: the varying of the note issue inversely with the sign of the spread between the market price of gold and its legal price. However, Benetti and Cartelier also fail to see that such adoption, on the one hand ruled out, not only domestic arbitrage (the "Penelope effect") but international arbitrage as well (because, in the case of excess note issue, its contraction would occur *before* the export of gold became profitable), and on the other hand it would dispense the Bank from bothering about the ratio between its metallic reserve and the stock of notes in circulation (because,

[21] Benetti and Cartelier make fun of the Penelope "metaphor" (which I do not "take from Smith" as they say but borrow from Ricardo himself who takes it from Smith; see Ricardo 1810–1811, p. 58) on the motive that it "implies that Penelope undid in the night *more* than what she did in the day. It is easy to guess the end of such a tale." (p. 517; BC's emphasis). Contrary to what they write ("the difficulty does not lie in the Bank's losses as such"; ibid.), the loss experienced by the Bank and the expansion of its note issue to purchase gold in the market are one and the same thing: the latter is the exact measure of the former.

although returning one's notes to the Bank was always possible, no one would find advantageous to do it). Ricardo was very clear on these two aspects in his *Proposals for an Economical and Secure Currency*: "The most perfect liberty should be given, at the same time [as the proposed note convertibility into bullion], to export or import every description of bullion. *These transactions in bullion would be very few in number*, if the Bank regulated their loans and issues of paper by the criterion which I have so often mentioned, namely, the price of standard bullion, *without attending to the absolute quantity of paper in circulation*." (Ricardo 1816, p. 67; reprinted in the 2nd and 3rd editions of *Principles*, Ricardo 1821, pp. 357–8; emphasis added) Ricardo could stress that he had "so often mentioned" this issuing rule: in the 4th edition of *High Price* he had already stated: "An unfavourable exchange can ultimately be corrected only by an exportation of goods, by the transmission of bullion,—or *by a reduction in the amount of the paper circulation*." (Ricardo 1810–1811, p. 125; Ricardo's emphasis).[22]

At the end of the day, my disagreement with Benetti and Cartelier may be summed-up as follows: one thing is to assume that under convertibility the market price of gold is necessarily equal to its legal price (their position), another thing is to account for the adjustment process through which the market price of gold equalises with its legal price. According to me, this was the specific object of Ricardo's theory of money, which can thus be called a theory of monetary disequilibrium (meaning by that a situation in which the quantity of money does not "conform" to "the wants of commerce").[23]

[22] By insisting on the export of gold as the "favoured" regulating mechanism in Ricardo and by giving to the reserve coefficient a prominent role in their model, Benetti and Cartelier land unexpectedly in the realm of the Currency principle.

[23] It is interesting to note that the link between monetary theory and disequilibrium was made by Sraffa in his preparatory notes for his 1932 article against Hayek: "The dividing line, which appears to assert itself more and more definitely, is another one [than Hayek's]. The non monetary theory studies a state of equilibrium, and the conditions which determine it: it goes as far as comparing two, or more states of equilibrium, and measuring the differences in their conditions—but goes no further. Here begins the field of monetary theory: or rather, jumping over the study of the path followed in the transition from one position to another, it sets to study states of disequilibrium." (*Sraffa Papers*, D3/9[181])

### *The Necessary Exteriority of the Standard*

Benetti and Cartelier also criticize my proposition that the standard (e.g. gold) is necessarily exterior to the system of production that determines the natural prices of the commodities in the economy where the standard-based currency circulates. According to them, this proposition is both unwarranted and inappropriate.

Quoting my book, they rightly observe that this exteriority of the standard is supposed to result from the contradiction between a determination of the market price of gold by its natural price (if gold belongs to the system of production) and its regulation by a legal price (when gold is the standard of money). However, referring to their own model (in Benetti and Cartelier 2020), they see no such contradiction so that my proposition is unwarranted. It does not come as a surprise that this contradiction does not show up in *their* model: as testified by their equation (3*) on p. 520, the money price of any commodity other than gold is equal to the price of this commodity in terms of gold (solution of the Sraffa system in which the numéraire chosen is gold) multiplied by the money price of gold, *which they suppose constant* since it is the reciprocal of "the quantity of gold which defines the £" (p. 520).

As emphasised in the previous section, Benetti and Cartelier's assumption that the market price of gold cannot be but equal to its legal price is at odds with Ricardo's view that it may diverge from it, and that this divergence is caused by the discrepancy between the actual quantity of money and the quantity required by "the wants of commerce". Substituting the definitional equality of the market price of gold to its legal price for the regulation of the former by the latter, Benetti and Cartelier boldly eliminate one of the terms of the contradiction, hence the contradiction itself.

After having magically made my exteriority proposition disappear, they set themselves to the task of showing that it is inappropriate for the making of a Classical theory of money consistent with the Sraffa system of natural prices. It is so because of my "neglect of the *general* problem of gravitation of market prices around natural prices. His [GD's] focus is on the *specific* question of the gold market only." (p. 521; BC's emphasis) So doing I cannot, however, get rid of this gravitation problem, which hides itself in the quantitative determination of "the wants of commerce" and in the adjustment of the market price of gold in the foreign mines ("Why should the gold market price be equal to its natural price if no gravitation mechanism is considered?"; p. 522).

Now Benetti and Cartelier do not like gravitation as a theory of the determination of market prices, and they prefer the "Cantillon rule" according to which the market price of a commodity is determined by the ratio of the quantity of money spent on its market to the quantity of the commodity present on it and entirely sold at that price. This is a position which merits respect but again it misses its target. It is true that "the wants of commerce" must be expressed at market prices, but I never claimed that my reconstruction of Ricardo's theory of money accounted for them: my purpose was to elucidate what, following Ricardo, happens when the actual quantity of money diverges from the quantity required by "the wants of commerce" (unknown by the agents and by the Bank), not to determine the latter, which is to be considered as an independent variable. In the terms used in the previous section, I contend that Ricardo's theory of money is concerned with monetary disequilibrium (when the actual quantity of money diverges from the level required by "the wants of commerce"), not with market disequilibrium (when the market prices of commodities diverge from their natural level).

As for the equalization of the foreign market price of gold with its natural price, it may be amusing to note that, in the model I propose of the adjustment in foreign mines (see Deleplace 2017, Chapter 5), the specificity of gold as a commodity demanded for monetary purposes—namely that its demand expressed in money terms should be considered as given, independently of the adjustment process—explains that its market price is determined by what amounts to a "Cantillon rule" (see equations (5.3) to (5.5) in Deleplace 2017, p. 184). In the gold-producing country as in the gold-importing one, the adjustment of the market price of gold is entirely the consequence of its role in a standard-based monetary economy.[24]

[24] The existence in the importing country (where gold is used as standard of money) of a sector specialized in the gold trade explains that there is also a natural price of gold in this country. However, it is not determined by the conditions of production of gold in this economy (since it is not produced there) but by the natural price in the gold-producing country (to which the market price there is equated), plus the cost of transportation and the profit at the general rate in the importing country (see Deleplace 2017, pp. 187–91). The equalisation of this domestic natural price of gold with the legal price (thanks to the adjustment of the quantity of money) determines the exchange rate between the currencies of the two countries. This exchange rate is thus solely determined by the conditions prevailing in the market for gold in both countries and not by the state of the balance of payments.

### *Sraffa's Theory of Prices and Classical Theory of Money*

Benetti and Cartelier's comments—substantiated by their 2020 paper—and my reply suggest that the issue of linking Sraffa's theory of prices with a Classical theory of a standard-based money may be explored along two alternative routes.

The one I have proposed thanks to a reconstruction of Ricardo's theory of money focuses on the relationship between the issuing of money and the market price of the standard (gold), which diverges from its legal price when the quantity of money issued is not at the level required by "the wants of commerce". Sraffa's theory of (natural) prices may then apply at two necessarily distinct levels: in the country producing the commodity used abroad as standard of money (but not in this country) and in the country importing this commodity to use it as standard of money (but which does not produce it). At both levels, a general theory of market prices is unnecessary—this is not the object of the theory of money—and, although the standard of money is a commodity subject (in the gold-producing country) to the Sraffa system, it is outside this system in the country where the currency is based on it.

The other road, suggested by Benetti and Cartelier, assumes that the market price of the standard cannot be but equal to its legal price—in that, although Classical, this road moves away from Ricardo—and focuses on the relationship between the issuing of money and the market prices of all commodities except the one chosen as standard (gold), thanks to the adoption of the "Cantillon rule". Sraffa's theory of (natural) prices may apply to all commodities produced in a supposedly closed economy, whether the commodity chosen as standard or all other commodities. The theory of money cannot be separated from the theory of (monetary) market prices.

As I see them, the critiques addressed by Benetti and Cartelier to the first road are not conclusive. By contrast, the two main points they contend—the necessary equality between the market and the legal prices of the standard of money and the assumption that gold is produced in the economy which uses it as standard—raise two serious difficulties along the road they advocate. On the one hand, stating that its market price is fixed independently of the quantity of money issued amounts to considering gold as a simple numéraire, and not as a standard regulating the monetary system. Contrary to the title of their 2020 paper, Benetti and Cartelier's object is not to build a Classical theory of money in which the quantity of money is regulated by a standard but a theory in which it is regulated through the adjustment of the money prices of all commodities except the standard (reduced to a numéraire). On the other hand, assuming that the "Cantillon rule" applies to all commodities except the one chosen as standard of money (because its price is fixed) questions the statement that this standard does belong to a system of production of commodities subject to this rule.

More generally, my feeling is that Benetti and Cartelier's exercise is flawed by the fact that they are trying to pin a Classical-but-not-Ricardian theory of (market) price adjustment in response to a change in the quantity of money on a Ricardo-Sraffa theory of natural prices. As testified by many others of their writings, published jointly or separately, what they really believe in general is that a Classical theory of prices should not be built on a Ricardo-Sraffa basis but rather along Torrens's lines, in which the use of the surplus does matter to determine prices (contrary to what happens in *Production of Commodities by Means of Commodities*). The "Cantillon rule" that determines market prices is consistent with this view. But in their comments as in their 2020 paper, the exercise is based on Sraffa prices and there lies the root of their ambiguity as regards Ricardo. The question is therefore: can a non-Ricardian Classical theory of money reconcile the "Cantillon rule" of determination of market prices—which belongs to a non-Sraffian natural habitat—and Sraffa prices?

## References


Benetti, C., Cartelier, J. (2020) "From Ricardo to Sraffa: Gold as Monetary Standard in a Classical Theory of Money," in M.C. Marcuzzo, G. Deleplace, P. Paesani (eds.) *New Perspectives on Political Economy and Its History*, Cham: Palgrave Macmillan: 241–57.

Deleplace, G. (2017) *Ricardo on Money. A Reappraisal*, Abingdon: Routledge.

Deleplace, G. (2020) "Orthodox versus Unorthodox Views on Ricardo's Theory of Money," *The European Journal of the History of Economic Thought, 27* (6), December: 819–36.

Ricardo, D. (1810–1811) *The High Price of Bullion, A Proof of the Depreciation of Bank Notes*, in Ricardo (1951–1973), vol. III, 1951: 47–127.

Ricardo, D. (1816) *Proposals for an Economical and Secure Currency*, in Ricardo (1951–1973), vol. IV, 1951: 49–141.

Ricardo, D. (1821) *On the Principles of Political Economy, and Taxation*, 3rd edition, in Ricardo (1951–1973), vol. I, 1951.

Ricardo, D. (1951–1973) *The Works and Correspondence of David Ricardo*, edited by P. Sraffa with the collaboration of M.H. Dobb, Cambridge: Cambridge University Press, 11 vols.
Sraffa, P. (1960) *Production of Commodities by Means of Commodities. Prelude to a Critique of Economic Theory*, Cambridge: Cambridge University Press.
*Sraffa Papers*, Cambridge: Trinity College, Wren Library.

CHAPTER 17

# Sraffa and the Revenue of the Owner of Non-Renewable Natural Resources: A Note on the Literature

*Yoann Verger*

## 1 Introduction

From Parrinello (1983) to Huang (2018), a seemingly never-ending debate runs in the Sraffian literature about the way to handle non-renewable natural resource in the context of classical theory. Sraffa scarcely talks about natural resources, and about the revenue that you can get by owning these resources, in his book. However, it is interesting to note that the three basic commodities that are introduced as an example of an industrial system in chapter IV of his book are iron, coal, and wheat, that is three commodities that needs a direct exploitation of three natural resources to be produced: one renewable resource, land, for wheat, and

Y. Verger (✉)
UMR ECOSYS, INRA, AgroParisTech, Université Paris-Saclay, Thiverval-Grignon, France
e-mail: yoann.verger@uvsq.fr



A. Sinha (ed.), *A Reflection on Sraffa's Revolution in Economic Theory*, Palgrave Studies in the History of Economic Thought,
https://doi.org/10.1007/978-3-030-47206-1_17

two non-renewable resources, iron and coal deposits, for iron and coal.[1] Natural resources are presented by Sraffa as key components for the reproduction of the economic system, hence the question of their exhaustion appears truly important.

This article presents how the revenue of the owner of non-renewable natural resources is calculated in a Sraffian framework. The difference between renewable and non-renewable resources is actually not straightforward; and one can assume that there should be no difference in kind between the revenue of the owner of a renewable resource and the revenue of the owner of a non-renewable resource (Sect. 2). However, natural resources that are non-renewable may be viewed as wasting assets, and the question of the depreciation of this asset arises (Sects. 3 and 4). Finally, it appears from empirical facts that the revenue of the owner can be presented as a share of the net product of the company mandated for the extraction of the natural resource (Sect. 5).

## 2 Differential Rent

In chapter XI, Sraffa states that if the natural resource is scarce, its owner can get a differential rent. The fact that the natural resource is scarce can only be seen by looking at the conditions of production (the equations of production). If one commodity is produced by two distinct processes that use a natural resource, then a differential rent will arise that may be earned by the owner of the natural resource. The two distinct processes may use different qualities of the natural resource, or they may use the same quality, but with different proportions of the means of production and labour. The first case gives rise to the extensive differential rent, the second case to the intensive differential rent.

Sraffa only presents an example with a renewable natural resource, land, but in the beginning of the chapter he writes that the owner of a non-renewable natural resource can also earn this kind of revenue[2]:

> Natural resources which are used in production, such as land and mineral deposits, and which being in short supply enable their owners to obtain a rent, can be said to occupy among means of production a position equivalent to that of 'non-basics' among products. (Sraffa 1960, § 85)

[1] He previously introduced the commodity "pig" in Chap. 1.
[2] This idea is also defended by Parrinello (2001, 2004).

Thus, starting with the example presented by Sraffa in his chapter IV, and then adding two processes in order to express the scarcity of the non-renewable natural resources only, we could get the following system:

$$
\begin{array}{lllllllll}
45t.iron & + & 60t.iron & + & 30qr.wheat & + & \frac{3}{32}labour & \rightarrow & 85t.iron \\
45t.iron & + & 60t.iron & + & 30qr.wheat & + & \frac{3}{32}labour & \rightarrow & 95t.iron \\
30t.iron & + & 95t.iron & + & 60qr.wheat & + & \frac{6}{32}labour & \rightarrow & 225t.iron \\
20t.iron & + & 30t.iron & + & 90qr.wheat & + & \frac{4}{32}labour & \rightarrow & 225t.iron \\
40t.iron & + & 40t.iron & + & 200qr.wheat & + & \frac{16}{32}labour & \rightarrow & 480qr.wheat \\
\overline{180} & & \overline{285} & & \overline{410} & & \overline{1} & & Totals
\end{array}
\qquad (17.1)
$$

The case of the iron industry will lead to the rise of an extensive differential rent, as the technique of production is the same in both processes. The case of the coal industry will lead to the rise of an intensive differential rent, as the techniques of production are different in both processes. In both cases, the rent may be sized by the owner of the non-renewable resources, coal and iron deposits.

In order to calculate the rents, the price system may be represented as given below in equation system (17.2) (knowing that for the intensive rent, the process which is more productive per unit of natural resource should need more capital and labour per unit of output, otherwise the rent will be negative):

$$\left(45p_{iron}+60p_{coal}+30p_{wheat}\right)(1+\pi)+\frac{3}{32}w+\rho_{iron\,ore1}=85p_{iron}$$

$$\left(45p_{iron}+60p_{coal}+30p_{wheat}\right)(1+\pi)+\frac{3}{32}w+\rho_{iron\,ore2}=95p_{iron}$$

$$\left(30p_{iron}+95p_{coal}+90p_{wheat}\right)(1+\pi)+\frac{6}{32}w+\rho_{iron\,ore}=225p_{coal} \quad (17.2)$$

$$\left(20p_{iron}+30p_{coal}+60p_{wheat}\right)(1+\pi)+\frac{4}{32}w+30\rho_{coal\,ore}=225p_{coal}$$

$$\left(40p_{iron}+40p_{coal}+200p_{wheat}\right)(1+\pi)+\frac{16}{32}w=480p_{wheat}$$

As for the rent on iron ore, the first process appears to be less profitable than the second one: thus the first process shall not earn a differential rent, that is $\rho_{iron\ ore}\ 1 = 0$ (in this particular example, this can be asserted without knowing the prices, but in the general case, a comparison between the level of the different wages if the rate of profits is fixed from outside, or the different rates of profits if the wages are fixed from outside, will help to find the most profitable process).

## 3 Depreciation of Wasting Assets

Kurz and Salvadori (2015) report that Sraffa, in a draft of *Production of Commodities by Means of Commodities* written in 1958, had an extra passage at the end of Chapter XI on Land, § 91. In this passage Sraffa writes that wasting assets such as mineral deposits should make allowance for depreciation in the same way as "live" capital such as machine. Below I reproduce the draft with the first paragraph that appears in the final publication of the book along with the second paragraph that was eventually deleted:

> Machines of an obsolete type which are still in use are similar to land in so far as they are employed as means of production, although not currently produced. The quasi-rent (if we may apply Marshall's term in a more restricted sense than he gave it) which is received for those fixed capital items which, having been in active use in the past, have now been superseded but are worth employing for what they can get, is determined precisely in the same way as the rent on land. And like land such obsolete

> instruments have the properties of non-basics and are excluded from the composition of the Standard product. {…}
>
> On the other hand, as in the case of other wasting assets (such as mineral deposits) the annual depletion must be taken into account, which gives rise to as many separate processes as are the years of the asset's prospective residual life, on the same principle as was done in Chapter X for 'live' fixed capital. (Kurz et al. 2015, quotation from Sraffa 1960, § 91 and from unpublished Sraffa Papers, folder D3/12/96)

The principle of depreciating fixed capital is that, depending on the initial price, the expected life-span and the decreasing productivity of the asset, an annual charge is paid for interest and depreciation during all the expected life-span (instead of accounting the price of the fixed capital only for the year of the purchase). The method devised in Chapter X to calculate this annual charge is to consider that at each period of production during the life-span of the asset an older fixed capital is jointly produced, with a different price, and reused the next year. At the last period, nothing (or only scrap) is jointly produced. Thus a price appears for each dated fixed capital, and the subtraction of the price of the fixed capital jointly produced from the price of the fixed capital used during the period of production gives the annual charge to be paid to the producer of the fixed capital.

In order to know the initial price of a "live" capital, it is sufficient to know the means of production and labor required to produce this asset. One reason that could explain why Sraffa did not finally insert the above passage in his book is that, contrary to fixed capital, the initial price of a wasting asset such as exhaustible resource deposits is not a straightforward concept since assets are not produced by any human industry. Gibson (1984) suggests, however, that the discovery cost should be taken into account. For instance, if the owner of the natural resource has invested some means of production and labour to find the new mine, he can then sell the right to exploit the mine to an extracting company and the annual charge to be paid to him can be calculated on the same principle as for 'live' fixed capital.

## 4 The Hotelling Rule

The kind of revenue that most frequently appears in the neo-Ricardian literature is also linked to the depreciation of the wasting asset: the Hotelling rent (Parrinello 1983; Schefold 1989; Kurz and Salvadori 1997, 2000, 2001, 2009, 2011, 2015; Bidard and Erreygers 2001a, 2001b; Huang 2018). The Hotelling rule states that, given long-term equilibrium conditions (i.e. equal rate of profits for all industries), the rentier sells units of his natural exhaustible resource at a price which increases at each period following the rate of profits, so that he can earn the uniform rate of profits of the economy on his "conservation" activity. That is, if he decides during one period of production to sell nothing, the value of his stock of natural resource will increase at a rate equal to the ruling rate of profits from this period of production to the next one.[3]

In this regard, two points must be stressed. First, this kind of revenue is a logical consequence of the assumption that we consider a long-period equilibrium state of the economy where it is not possible for an economic actor to earn less (or more) than the ruling rate of profits. Second, to calculate the price of the natural resource and consequently the revenue of its owner at the present period of production, it is not enough to know the present conditions of production. A back-stop technology that will replace the existing wasting asset must be anticipated in order to calculate the exact level of the price at any given period of time.

For instance, Bidard and Erreygers (2001a, 2001b) show that, at the period of exhaustion of the wasting asset, the revenue of its owner will equal the differential rent that will exist between the value of the commodities produced with the help of its natural resource, and the value of the commodities produced by the back-stop technology. The revenue of the owner of the natural resource at the present period can be known by going backward from the period of exhaustion, knowing the Hotelling rule (with $y_t$ the price of the natural resource at period $t$ and $r$ the uniform rate of profits):

[3] If he earns less than the ruling rate of profits on its conservation industry, he would have an incentive to change its activity and to sell everything in the present period of production. If he earns more than the rate of profits, he would have an incentive to sell nothing in present period of production. Thus, the Hotelling theory states that he must earn exactly the ruling rate of profits on its conservation activity so that the economy can be in an equilibrium state.

$$y_t = (1+r)\,y_{t+1} \tag{17.3}$$

Then it is clear that to calculate this kind of revenue, a "snapshot" of the market at one point of time is not enough. And it is clear that, if the uniform rate of profits is not assumed to be the equilibrium rate of profits of an economy in a long-term position, but rather a "conventional" uniform rate of profits, reflecting the class conflict about the distribution of the surplus (see for instance for such a point of view Hicks 1985; Sinha 2012, 2016), then the logic behind the Hotelling rule falls into pieces. These two points may explain why Sraffa did not introduce this kind of revenue into his book.

## 5 Rent as a Share of the Product

In a recent paper, Fratini (2016) investigates rent as a share of the gross product. He shows how, in the Sraffian framework, this view of rent can help to better understand the institutions governing the distribution of the national product (in the line of Adam Smith, Ricardo and Marx), instead of only considering rent per unit of land. In his numerical examples, he does not relate the existence of the rent with the existence of a differential rent. Then the rent as a share of the gross product appears in Fratini as a tax (or a tithe) that could ensure a revenue to the owner of the natural resource even if no differential rent exists for the company extracting the natural resource.

Much before that, Ravagnani (2008) showed that in the context of oil market in the United States of America, the revenue earned by the owner of the oil deposits was much more related to conventions and relations of power with the company that extracts the deposit than with consideration of the Hotelling rule:

> We therefore submit for discussion, as a constructive contribution to the current debate, the view that royalties on exhaustible resources could appropriately be encompassed within the logical structure of the classical theory of value and distribution as separately determined 'independent variables'. The conventional arrangements based on percentage royalties could be accommodated with no difficulty in the normal positions defined by Sraffa's equations by taking the share of the resource price attributed to landowners as a 'given' coefficient reflecting the (persistent) share paid on average in the actual economy. (Ravagnani 2008)

While Ravagnani asserts that the introduction of rent as a share of the gross product into Sraffa's equations would be of no difficulty, he did not attempt to make a formalisation of it: this is why we will in the following make a step further and present an example of such formalisation.

The study of empirical data (IMF 2012) on the fiscal regimes imposed on extractive companies by governments owning exhaustible natural resources shows that parameters that determine the conventional arrangements on percentage royalties are: the expected differential rent, uncertainty (on prices, geology, input costs, change in fiscal or political regime, exhaustion of the resource), asymmetric information between the extractive company and the government, and monopolies (either on the extraction of the resource or on the ownership of the resource). Governments have then several possibilities to collect their share of the product. The different mechanisms listed in IMF (2012) are: signature and production bonus, royalties (rent as a share of product value), corporate or variable income tax (tax on profits and differential rents), resource rent taxes (tax on differential rents only, that is income less the average rate of profits), production sharing (rent as a share of the physical product), state participation (rent received as dividends, as any shareholder), and social investments from the extractive company (the company pays the rent through the construction of a road or a school, for instance).

No matter the fiscal regime chosen by the government, the tax rate can be reduced to a share of the net product of the extractive industry, the latter being defined as the sum of the profits and the differential rent earned by the extractive industry. Simulation methods (FARI model, Luca and Puyo 2016) and analysis of accounting data shows that the tax rate on the net product $t$ is between 65 and 85 per cent for oil and gas industries, and between 45 and 65 per cent for other mining industries (IMF 2012).

The introduction of the share of the net product as a "given" coefficient change the price system. In the case of a private company that extracts a natural resource owned by a State, the net product may be shared between the State and the private company. If the sharing occurs after the product has been sold, or if the company does not try to change prices once the tax is announced, the taxation of the net product has no impact on prices. The price system may be represented in this way, with $t$ the tax rate on the net product:

$$
\begin{aligned}
&45p_{iron}+60p_{coal}+30p_{wheat}+\frac{3}{32}w+t_{iron1}\left(45p_{iron}+60p_{coal}+30p_{wheat}\right)\pi\\
&\quad+\left(1-t_{ironore1}\right)\left(45p_{iron}+60p_{coal}+30p_{wheat}\right)\pi=85p_{iron}\\
&45p_{iron}+60p_{coal}+30p_{wheat}+\frac{3}{32}w+t_{iron2}\left[\left(45p_{iron}+60p_{coal}+30p_{wheat}\right)\pi+\rho_{ironore2}\right]\\
&\quad+\left(1-t_{iron2}\right)\left[\left(45p_{iron}+60p_{coal}+30p_{wheat}\right)\pi+\rho_{ironore2}\right]=95p_{iron}\\
&30p_{iron}+95p_{coal}+90p_{wheat}+\frac{6}{32}w+t_{coal1}\left[\left(30p_{iron}+95p_{coal}+90p_{wheat}\right)\pi+\rho_{coalore}\right]\\
&\quad+\left(1-t_{coal1}\right)\left[\left(30p_{iron}+95p_{coal}+90p_{wheat}\right)\pi+\rho_{coalore}\right]=225p_{coal}\\
&20p_{iron}+30p_{coal}+60p_{wheat}+\frac{4}{32}w+t_{coal2}\left[20p_{iron}+30p_{coal}+60p_{wheat})\pi+30\rho_{coalore}\right]\\
&\quad+\left(1-t_{coal2}\right)\left[\left(20p_{iron}+30p_{coal}+60p_{wheat}\right)\pi+30\rho_{coalore}\right]=225p_{coal}\\
&\left(40p_{iron}+40p_{coal}+200p_{wheat}\right)\left(1+\pi\right)+\frac{16}{32}w=480p_{wheat}
\end{aligned}
\tag{17.4}
$$

If the net product is taxed before the product is sold, or if the private company tries to avoid the effect of tax by increasing the prices (to see an example of this concerning a tax on profits, see Dome 1992), prices could be impacted. The price system may be represented in this way[4]:

[4] Only a part of what is effectively produced by the private company is brought to the market. Let us say that $T$ is the share of the physical gross product that is taken by the government. The price of the non-renewable resource will then be of the following form, if the company tries to earn the uniform rate of profits $\pi$ and to distribute the uniform wage $w$:

$$Kp_k\left(1+\pi\right)+lw+\rho=Xp_x\left(1-T\right)$$

With $K$ the quantity of the input (assuming that there is only one input), $X$ the quantity of the output (physical gross product), and $\rho$ the (possible) differential rent. The net product being defined as the value of the gross product less the value of the input and the value of labour, the share of the net product that goes to the government is defined as:

$$t=\frac{TXp_x}{Kp_k\pi+\rho+TXp_x}$$

Then the price system can also be written in this way:

$$45p_{iron} + 60p_{coal} + 30p_{wheat} + \frac{3}{32}w +$$
$$\left(1 + \frac{t_{iron\,ore1}}{1 - t_{iron\,ore1}}\right)(45p_{iron} + 60p_{coal} + 30p_{wheat})\pi = 85p_{iron}$$

$$45p_{iron} + 60p_{coal} + 30p_{wheat} + \frac{3}{32}w +$$
$$\left(1 + \frac{t_{iron2}}{1 - t_{iron2}}\right)\left[(45p_{iron} + 60p_{coal} + 30p_{wheat})\pi + \rho_{iron\,ore2}\right] = 95p_{iron}$$

$$30p_{iron} + 95p_{coal} + 90p_{wheat} + \frac{6}{32}w + \tag{17.5}$$
$$\left(1 + \frac{t_{coal1}}{1 - t_{coal1}}\right)\left[(30p_{iron} + 95p_{coal} + 90p_{wheat})\pi + \rho_{coal\,ore}\right] = 225p_{coal}$$

$$20p_{iron} + 30p_{coal} + 60p_{wheat} + \frac{4}{32}w +$$
$$\left(1 + \frac{t_{coal2}}{1 - t_{coal2}}\right)\left[(20p_{iron} + 30p_{coal} + 60p_{wheat})\pi + 30\rho_{coal\,ore}\right] = 225p_{coal}$$

$$(40p_{iron} + 40p_{coal} + 200p_{wheat})(1 + \pi) + \frac{16}{32}w = 480p_{wheat}$$

It is interesting to notice that in the last system of equations, the differential rent is distributed entirely to the extractive company. As we are considering only basic commodities (for a definition of basic and non-basic commodity, see Sraffa 1960, § 60), as soon as the tax rate on the net

$$Kp_k(1+\pi) + lw + \rho + \frac{t}{1-t}(Kp_k\pi + \rho) = Xp_x$$

Or in this way:

$$Kp_k + lw + \left(1 + \frac{t}{1-t}\right)(Kp_k\pi + \rho) = Xp_x$$

product t of any company increases, all prices and rents will change and the maximum rate of profits will decrease. The case of the second iron company is an exception, as the increase of the tax will only lower the extensive differential rent earned by this company.[5]

## 6 Conclusion

This article clarifies the different concepts that we must consider when we calculate the revenue of the owner of a non-renewable natural resource in a Sraffian framework. For a snapshot point of view, what matters in calculating prices and revenues are the conditions of production of the commodities that need a non-renewable natural resource to be produced, and the distribution of the net product of the companies that produced these commodities. From a dynamic point of view, what matters is the stocks of the natural resource, the evolution of demand and technologies, the possibility of back-stop technologies, and the change of the institutional framework that ensure the distribution of the net product. From a policy point of view, it is important to understand the real cost of production of the commodities that use non-renewable natural resources, so that taxes can be adequately designed depending on the policy chosen (either recovery of maximum revenue, or incentive for private prospecting and exploitation, or protection of the natural resource, etc.).

The debate about non-renewable resources in the context of classical theory largely focused on the Hotelling rule, and more specifically on how to calculate the Hotelling rent. Only recently another line of thought started to emerge, where the revenue of the owner of the natural resource

[5] It is important to notice that the real cost of production is not changed by the tax, only the distribution is changed. This point is touched upon by Sraffa in one of his unpublished papers:

Pantaleoni, Saggi, I, 146 n.

(quoted by Jannaccone, 342 n)

Is real cost of production increased by an increase in a the imposition of a specific tax on product? Yes, says Pantaleoni: the cost per unit of wheat is increased in exactly the same way if 11 hours instead of 10 are required to produce it, or if, having the government taken 10% of the produce, 10 hours now produce only 9/10 of a unit. The reply to Pantaleoni is—is then the wheat obtained by the Government produced without cost? (Unpublished Sraffa Papers, folder D1/13: 6 (17); B.III.12.xxv)

depends on the institutions governing the relations between the owner of the natural resource and the company that extracts the natural resource (Ravagnani 2008; Fratini 2016). A revenue above uniform profits exists for the extractive company as soon as there is differential rent, and this revenue may be seized by the owner of the natural resource. But empirical data show that in many cases a part of the differential rent is kept by the extractive company. Hence, from the point of view taken in Sect. 5, there exists no "natural law" about who should earn the revenue coming from the extraction of a non-renewable resource, as well as there exists no "natural" property regime for such resources (they could be considered as a common good of humanity, or as a State property, a community property, etc.).

## References


Bidard, C., and Guido Erreygers. "Further reflections on the corn-guano model", *Metroeconomica*, pp. 254–268, 2001a.

Bidard, C. and Guido Erreygers. 2001b. "The Corn-Guano Model". *Metroeconomica*, 52, 243–253.

Dome, Takuo. "Ricardo's theory of tax incidence: a Sraffian re-interpretation", *Cambridge Journal of Economics*, pp. 43–53, 1992.

Fratini, Saverio M. "Rent as a share of product and Sraffa's price equations", *Cambridge Journal of Economics*, pp. 599–613, 2016.

Gibson, Bill. "Profit and Rent in a Classical Theory of Exhaustible and Renewable Resources", *Journal of Economics*, pp. 131–149, 1984.

Hicks, J. (1985) 'Sraffa and Ricardo: A Critical View', in G. A. Caravale (ed.), *The Legacy of Ricardo*. New York: Basil Blackwell, pp. 305–319.

Huang, B. 2018. An Exhaustible Resources Model in a Dynamic Input-Output Framework: A Possible Reconciliation Between Ricardo and Hotelling, *Journal of Economic Structures*, 7:8, 1–24.

IMF. Fiscal Regimes for Extractive Industries: Design and Implementation, 2012. Retrieved from www.imf.org/external/np/pp/eng/2012/081512.pdf.

Kurz, Heinz D., Neri Salvadori: "Exhaustible resources in a dynamic input-output model with 'Classical' features", *Economic Systems Research*, pp. 235–252, 1997.

Kurz, Heinz D., Neri Salvadori: "Economic dynamics in a simple model with exhaustible resources and a given real wage rate ", *Structural Change and Economic Dynamics*, pp. 167–179, 2000.

Kurz, H.D. and Salvadori, N. 2001. "Classical Economics and the Problem of Exhaustible Resources", *Metroeconomica*, 52, 282–296.

Kurz, Heinz D., and Neri Salvadori. "Ricardo on exhaustible resources, and the Hotelling Rule." in Aiko Ikeo, Heinz D. Kurz (Eds.) *A History of Economic Theory. Essays in honour of Takashi Negishi*. London: Routledge (2009): 68–79.

Kurz, Heinz, and Neri Salvadori. "Exhaustible resources: rents, profits, royalties and prices." in Caspari, Volker (Ed.), *The Evolution of Economic Theory. Essays in honour of Bertram Schefold*, London, Routledge, pp. 39–52, 2011.

Kurz, Heinz D., Neri, Salvadori. "The 'classical' approach to exhaustible resources: Parrinello and the others", in Heinz, D. Kurz; Neri, Salvadori (Eds.), *Revisiting Classical Economics: Studies in Long-period Analysis*, Routledge: Abingdon, pp. 304–316, 2015.

Luca, Oana, and Diego Mesa Puyo. Fiscal Analysis of Resource Industries: (FARI) Methodology, 2016.

Parrinello, Sergio. "Exhaustible natural resources and the classical method of long-period equilibrium". In: Kregel, Jan Allen (Ed.), *Distribution, effective demand and international economic relations*. Palgrave Macmillan, London, 1983. pp. 186–199.

Parrinello, Sergio. "The price of exhaustible resources", *Metroeconomica*, pp. 310–15, 2001.

Parrinello, Sergio. "The Notion of Effectual Supply and the Theory of Normal Prices with Exhaustible Natural Resources", *Economic Systems Research*, pp. 311–22, 2004.

Ravagnani, Fabio. "Classical Theory and Exhaustible Natural Resources: Notes on the Current Debate", *Review of Political Economy*, pp. 79–93, 2008.

Schefold, B. 1989. *Mr Sraffa on Joint Production and Other Essays*. Unwin Hyman, London.

Sinha, Ajit. "Listen to Sraffa's Silences: A New Interpretation of Sraffa's Production of Commodities", *Cambridge Journal of Economics*, pp. 1323–1339, 2012.

Sinha, Ajit. *A revolution in economic theory. The economics of Piero Sraffa*. Palgrave Macmillan, 2016.

Sraffa, P. *Production of Commodities by Means of Commodities*. K. K. Vora, Vora & Co, 1960.

Comment

# A Comment on 'Sraffa and the Revenue of the Owner of Non Renewable Natural Resources: A Note on the Literature'

*Murray G. Patterson*

Given the impacts of climate change, unprecedented levels of biodiversity loss and other crucial environmental problems facing humanity, Verger is correct to bring to our attention that natural resources and by extension ecological processes should be part of the Sraffian analytical framework. No longer can economists of any persuasion, ignore environmental problems, or merely treat them as "externalities". These problems need to be central to any welfare calculus. Verger's approach is to draw on ideas from both classical and neoclassical economics, and link them and in some cases 'integrate' them into the Sraffian framework. Firstly, Verger addresses how the Sraffa framework takes account of "extensive differential rent" and "intensive differential rent", and he suggests that this rent "may be seized by the owner" (emphasis added) which leaves open the question of the role of government, power relations, property rights and other factors that might impinge on how this rent/surplus does (and should) be captured in contemporary situations where the relationship between the archetype of "tenant" and "owner" is less straightforward.

Next Verger argues that land (and presumably more generally natural capital) can be treated in the same way as manufactured capital such as machines. That is, to use the method devised by Sraffa (1960, in Chapter X) to calculate an annual charge for capital during its lifespan, for where each period the "wasting" capital is considered to be a joint product, until the last period where nothing (or only scrap) is produced. As an ecological economist I can only comment that using a framework (which Sraffa uses to routinely analyse manufactured capital) to then use it to analyse natural capital is more difficult and challenging. This is because natural capital is far more difficult to measure or even conceptualise, given its very diffuse, multiscale and organic character; and the fact that it involves a whole complexity of interacting ecological processes which aren't easily conflated into one unit of analysis, as simply is the case for a machine. Furthermore, as an ecological economist I need to point out that the whole idea of "wasting" is inextricably connected to the idea of entropy, the inevitability of energy/mass dissipation and other consequences of the second law of thermodynamics (Bisson and Proops 2002; Lozda 2017). All of these ideas stemming from ecological economics (or more narrowly biophysical economics) should be areas that Verger could incorporate into any future analysis of "wasting" assets or capital.

The third area that Verger focuses on is to connect the Hotelling Rule to the Sraffa framework and more broadly the neo-Ricardian literature. This I consider to be a more contentious part of Verger's work as he attempts to bring a rule which is seemingly "neoclassical" into the Sraffa framework which is arguably antithetical to the neoclassical approach. That said, as Verger points out, he is not alone amongst neo-Ricardians in linking Hotelling rent to the Sraffa approach [for example: Heinz et al. 2000; Huang 2018; Kurz and Salvadori 2001; Schefold 1989]. Furthermore, Verger argues like others such as Bidard and Erreygers (2001b) that the Hotelling rule is neither neoclassical or classical but just a logical consequence of long period equilibrium (where all industries have the same profit rate). My reservations with Verger's approach of linking the Hotelling rule to Sraffa are several: (1) as I pointed out below, Sraffa's equations appear to generate equal profits as this is an artefact of the Eigensolution method that Sraffa implicitly uses. A more assumption-free solution method such as Singular Value Decomposition actually shows that in most cases there are uneven profit rates, which is what you would expect in systems where there are inconsistent equations (or what Sraffa refers to as self-contradictory equations). Therefore, once it is admitted that uneven profit rates are a possibility, the Hotelling Rule which is a logical consequence of even profit rates, is open to criticism on these grounds alone; (2) the Hotelling Rule as pointed out by Kronenberg (2008) and very many others, does not play out in the real world as evidenced by price trends for many natural resources such as crude oil, zinc, iron ore and copper that show a slight decline in price over time, whereas the Hotelling Rule predicts price increases.

The fourth area Verger examines is considering "rent" from natural resources to be part of the gross product of the economy. This of course is an argument that would be supported by many ecological economists and policy advisers, now that it is becoming more and more important that natural resources should be part of national macroeconomic accounts, and indeed this is the central focus of endeavours such as the United Nations SEEA approach to bring environmental factors into national accounts (United Nations 2014). Verger however correctly points out that including rent in the gross product is not a simple task, as in the real world there are challenging factors such as uncertainty, asymmetric information and the influence of monopolies that complicate the calculation of the price and value of natural resources.

Verger outlines a system of overdetermined (5 processes, 4 quantities/commodities[6]) homogeneous equations that he doesn't actually solve. Like almost all overdetermined systems these equations are inconsistent. These equations could be manipulated[7] to form a square system and be 'solved' using the Sraffa method that Pasinetti (1977) showed was an Eigensolution method. Instead of adopting this Eigensolution method used by Sraffa (1960), these overdetermined equations could be solved by Singular Value Decomposition, a method that would not have been available to Sraffa even in the late 1950s when he was completing 'Production of Commodities by Means of Commodities'. The advantages of using the Singular Value Decomposition method are twofold: (1) Singular Value Decomposition can be applied to any real matrix including overdetermined matrices or for example matrices that translate to inconsistent equations. Whereas the Eigensolution method that Sraffa implicitly uses only applies to certain classes of square matrices, which Patterson (2012) shows is far more limiting in empirical analyses of complicated real-world systems; (2) in the Singular Value Decomposition (SVD) approach there is no *a priori* assumption that all industries have the same profit rate. Indeed, the SVD approach lets the data "speak for itself" and explicitly tests the hypothesis that all industries have the same profit rate, and in this regard

[6] This includes: coal, wheat and iron, as well as labour which is considered to be the fourth quantity.

[7] To convert an over-determined system to a determined (square) system, several approaches could be used: (1) aggregate rows that have the same single product, but this becomes a more ambiguous process where there is a complicated system that has very many joint products as for example encountered by Patterson (2002); (2) deleting rows that are seen to be 'unprofitable'—however, once the observer starts to remove rows deemed to be unprofitable, the observer is now altering the system which arguably contradicts Sraffa's intention of purely taking a 'snapshot' of the system as it exists.

Patterson et al. (2006) and Patterson (2014) argue that the occurrence of even profit rates[8] are the exception to the rule.

Before we can solve Verger simultaneous equations using Singular Value Decomposition: (1) for illustrative purposes land must be added to Verger's equations; (2) a 'residual' is added to each equation to allow for the possibility of uneven profits for each industry. Thus, Verger's simultaneous equations are modified for illustrative purposes:

$$(45\,\text{t iron}) * \text{p1} + (60\,\text{t coal}) * \text{p2} + (30\text{qr wheat}) * \text{p3} + (3/32\,\text{Labour}) \\ * \text{p4} + (1\,\text{acre Land}) * \text{p5} + \text{e1} \rightarrow (185\,\text{t iron}) * \text{p1} \tag{1.1}$$

$$(45\,\text{t iron}) * \text{p1} + (60\,\text{t coal}) * \text{p2} + (30\text{qr wheat}) * \text{p3} + (3/32\,\text{Labour}) \\ * \text{p4} + (5\,\text{acre Land}) * \text{p5} + \text{e2} \rightarrow (145\,\text{t iron}) * \text{p1} \tag{1.2}$$

$$(30\,\text{t iron}) * \text{p1} + (95\,\text{t coal}) * \text{p2} + (60\text{qr wheat}) * \text{p3} + (6/32\,\text{Labour}) \\ * \text{p4} + (2\,\text{acre Land}) * \text{p5} + \text{e3} \rightarrow (232\,\text{t coal}) * \text{p2}. \tag{1.3}$$

$$(20\,\text{t iron}) * \text{p1} + (30\,\text{t coal}) * \text{p2} + (90\text{qr wheat}) * \text{p3} + (4/32\,\text{Labour}) \\ * \text{p4} + (6\,\text{acre Land}) * \text{p5} + \text{e4} \rightarrow (245\,\text{t coal}) * \text{p2}. \tag{1.4}$$

$$(40\,\text{t iron}) * \text{p1} + (40\,\text{t coal}) * \text{p2} + (200\text{qr wheat}) * \text{p3} + (16/32\,\text{Labour}) \\ * \text{p4} + (10\,\text{acre Land}) * \text{p5} + \text{e5} \rightarrow (505\text{qr wheat}) * \text{p3} \tag{1.5}$$

Where:
p1 … p5 = prices of the respective quantities.
e1 … e5 = residuals of each process.
These equations can now be more compactly written:

$$\mathbf{Ap} + \mathbf{e} = \mathbf{Bp} \tag{2}$$

[8] Although Patterson (2006, 2012) uses 'Emergy Efficiency of a Process', this is directly related to the 'Profit Rate of a Process'. That is: Profit Rate = Emergy Efficiency − 1
Where:
Profit Rate = (Value of Process Outputs − Value of Process Inputs)/(Value of Process Inputs)
Emergy Efficiency = Value of Process Outputs / Value of Process Inputs
Value of Process Inputs = Sum of all Input Quantities times their Respective Prices
Value of Process Outputs = Sum of all Output Quantities times their Respective Prices

Where:

**A** = inputs (*m x n*) matrix of *m* processes (rows) and *n* quantities (columns).

**B** = outputs (*m x n*) matrix of *m* processes (rows) and *n* quantities (columns).

**e** = residuals vector (*m x 1*).

By letting **Bp** – **Ap** = **X**, the prices can therefore be determined by solving:

$$\mathbf{Xp} + \mathbf{e} = \mathbf{0} \tag{3}$$

Subject to the constraint **p** ≠ **0** that we need in order to obtain a non-trivial solution.

The prices **p** are then determined by the Singular Value Decomposition that consists of the factorisation of **X**:

$$\mathbf{X} = \mathbf{UDL}^{\mathrm{T}} \tag{4}$$

Where: **U** = orthogonal matrix *(n x n)* consisting of 'left hand singular' vectors; $\mathbf{L}^{\mathrm{T}}$ = orthogonal matrix *(m x m)* consisting of 'right hand singular' vectors; **D** = non-negative diagonal matrix *(n x m)* of singular values. The singular values are the non-zero square roots of the eigenvalues of both $\mathbf{XX}^{\mathrm{T}}$ and $\mathbf{X}^{\mathrm{T}}\mathbf{X}$. The eigenvectors of $\mathbf{XX}^{\mathrm{T}}$ are called the left hand singular vectors (i.e. the columns of **U**); and the eigenvectors of $\mathbf{X}^{\mathrm{T}}\mathbf{X}$ are called the right hand singular vectors (i.e. the columns of $\mathbf{L}^{\mathrm{T}}$). From the point of view of solving these homogenous simultaneous linear equations (**Xp** - **e** = **0**) in order to quantify the prices, the least squares solution to these equations is of primary interest. The least squares solution is given by the column of $\mathbf{L}^{\mathrm{T}}$ which corresponds to the smallest diagonal entity of **D** which is usually the last column entry since it is conventional to order the diagonal elements of **D** in descending order.

Once the prices are determined by factorisation of Eq. (4), then the profit rates can be determined. First, the "value of the inputs" $\boldsymbol{\gamma}$ and "value of the outputs" $\boldsymbol{\delta}$ for each process is determined:

$$\mathbf{Ap} = \gamma \text{ and } \mathbf{Bp} = \delta \tag{5}$$

Second, the elements in the vector $\boldsymbol{\delta}$ are divided by the corresponding elements in vector $\boldsymbol{\gamma}$, to give the vector $\boldsymbol{\eta}$. The profit rate of each process $\boldsymbol{\Phi}$, as a proportion is:

$$\eta - \mathbf{1} = \Phi \tag{6}$$

Where **1** is a column vector (*m x 1*) in which each element is of value unity (=1).

By using this Singular Value Decomposition method simultaneous equations outlined by (1.1) to (1.5) can be solved. The prices thus obtained for iron, coal, wheat, labour (wages) and land (rent), as well as the profit rates for each process and the entire system/economy, can be found in the last column of Table 17.1. This result is interesting, as it measures the marginal productivity of labour and land, which is not the case in conventional Sraffian analysis as such an approach can be seen not to be compatible with Sraffa's approach.[9] That is the price of land (rent) and labour (wages) compares the productivity of these two factors of

**Table 17.1** Solution of modified Verger equations by Singular Value Decomposition

| *Variables determined by solution of equations* | *Surplus to profits*[a] | *Surplus to labour*[b] | *Surplus to land*[c] | *Surplus to labour and land*[d] |
|---|---|---|---|---|
| Prices | | | | |
| Iron | 1.000 | 1.000 | 1.000 | 1.000 |
| Coal | 0.740 | 0.900 | 0.863 | 0.839 |
| Wheat | 0.537 | 0.969 | 0.783 | 0.929 |
| Labour | 0.000 | 428.000 | 0.000 | 328.000 |
| Land | 0.000 | 0.000 | 15.400 | 3.900 |
| Process profit rates (%): | | | | |
| Iron production 1 | 75 | 10 | 36 | 15 |
| Iron production 2 | 37 | -14 | -26 | -18 |
| Coal production 1 | 30 | -18 | 6 | -14 |
| Coal production 2 | 100 | 18 | 1 | 12 |
| Wheat production | 53 | 1 | 3 | 1 |
| System profit rate (%) | 56 | 0 | 0 | 0 |

[a]Land and labour set to zero (respective columns deleted in **X**) before solution
[b]Land set to zero (land column deleted in **X**) before solution
[c]Labour set to zero (labour column deleted in **X**) before solution
[d]No deletions in **X** matrix before solution

[9] It could be seen that marginal analysis requires a 'change' which arguably is not possible in the Sraffa framework as Sraffa equations represent a snapshot in time. However, 'changes' as in the above analysis can be considered in a cross-sectional sense, and not just necessarily in a longitudinal case which requires changes across time.

production, in terms of how much output they produce from respectively one extra input of each of these factors of production.

Another approach to solving these equations by Singular Value Decomposition is to examine three different "extreme cases" where: (1) all of the surplus is allocated to profit; (2) all of the surplus is allocated to labour (wages); and (3) all of the surplus is allocated to land (rent)—for results of these solutions refer to the first three columns of Table 17.1. By interpolating, between these three 'extreme cases' a triangular surface[10] can be constructed that describes 'intermediate cases' all of which represent feasible allocations of the surplus between profits, wages and rent. This is reminiscent of Sraffa's (1960) linear trade-off function between wages and rate of profits, where at one 'extreme' all of the surplus is allocated to wages and at the other 'extreme' all surplus is allocated to profits, with 'intermediate cases' located between of each of these extremes. From a political economy point of view it can be argue that any of the solutions on this triangular surface (expressing how surplus can be allocated between profit, wages and rent) are feasible, and exactly where society locates itself on this triangular surface is a reflection of the power relations in society and how they manifest themselves.

## References

Bisson, K. and Proops, J.L.R. 2002. *Waste in Ecological Economics*. Edward Elgar Publishing, Northampton, MA.

Heinz, D., Kurz, H.D., Salvadori, N. 2000. Economic Dynamics in a Simple Model with Exhaustible Resources and a Given Real Wage Rate. *Structural Change and Economic Dynamics*, 11, 167–179.

Huang, B. 2018. An Exhaustible Resources Model in a Dynamic Input-Output Framework: A Possible Reconciliation Between Ricardo and Hotelling, *Journal of Economic Structures*, 7:8, 1–24.

Kronenburg, T. 2008. Should We Worry About the Failure of the Hoteling Rule. *Journal of Economic Surveys*, 22:4, 774–793.

Kurz, Heinz D., Neri Salvadori: "Economic dynamics in a Simple Model with Exhaustible Resources and a Given Real Wage Rate", *Structural Change and Economic Dynamics*, pp. 167–179, 2000.

[10] The triangular surface referred to has 100% allocation of the surplus to profits at one corner point, 100% allocation of the surplus to wages at another corner point, and 100% allocation of surplus to rent at yet another corner point.

Kurz, H.D. and Salvadori, N. 2001. Classical Economics and the Problem of Exhaustible Resources, *Metroeconomica*, 52, 282–296.

Lozda, G.A. 2017. The Hotelling Rule for Entropy-Constrained Economic Growth. *Ecological Economics*, 133, 35–41.

Pasinetti, L.L. 1977. *Lectures on the Theory of Production*. Columbia University Press, New York.

Patterson, M.G. 2002. Ecological Production Based Pricing of Biosphere Processes. *Ecological Economics*, *41*(3), 457–478.

Patterson, M.G. 2012. Are All Processes Equally Efficient From An Emergy Perspective? Analysis of Ecological and Economic Networks Using Matrix Algebra Methods. *Ecological Modelling*, 226, 77–91.

Patterson, M.G. 2014. Evaluation of Matrix Algebra Methods for Calculating Transformities from Ecological and Economic Network Data. *Ecological Modelling*, 271, 72–82.

Patterson, M.G., Wake, G., McKibbin, R., Cole, A. 2006. Ecological Pricing and Transformity: A Solution Method for Systems Rarely at General Equilibrium, *Ecological Economics*, 56:3, 412–423.

Schefold, B. 1989. *Mr Sraffa on Joint Production and Other Essays*. Unwin Hyman, London.

Sraffa, P. 1960. *Production of Commodities by Means of Commodities*. Prelude to a Critique of Ecnomic Theory, Cambridge University Press, Cambridge.

United Nations. 2014. *System of Environmental-Economic Accounting 2012—Central Framework*. United Nations, New York. ISBN: 987-92-1-161563-0.

Reply

# A Response to 'Sraffa and the Revenue of the Owner of Non Renewable Natural Resources: A Note on the Literature'

*Yoann Verger*

I thank Professor Patterson (2021) for offering his critical comment on my paper. This gives me an opportunity to clarify an important point. Rent is looked at as two different things in the literature: first as a technical thing and second as the revenue of the landlord, or more generally as the revenue of the owner of a natural resource. These two ways of looking at rent are summed up in the two approaches given by Patterson to solve the system of equations (1) in his paper. The first approach allows him to measure the marginal productivity of land and capital, whereas the second examines three cases where the distribution of the surplus is given from outside.

The first approach directly leads from the classical standpoint to the neoclassical one. If a rent on a natural resource which is used to produce a given commodity can be calculated by the addition of a second process using the same natural resource and producing the same commodity, then the method can be extended to labour and capital. The level of rent, wages and profits thus becomes a technical requirement in order to solve the system of prices. To call them the measure of marginal productivity of capital, labour and land is just another way to say that their level is fixed by the system (Pasinetti 1999).

The second approach leads to the analysis of the socio-economic factors governing the distribution of the surplus. Prices are the result of a given distribution of the surplus, and rent is part of the distribution. Of course, in real economic systems, rent appears because there are several processes operating with different rates of profits, so the apparition of a rent is a technical requirement. In the same way, the surplus appears because of the technical requirements for reproduction. But the way rent is distributed between capitalists, landlords and workers cannot be given by the system. Rent being a part of the surplus, the questions are on what grounds the landlords are able to gain a part of the surplus? And similarly for the capitalists. These questions are raised if we follow the second approach, and this was the goal of Marx, and, I believe, of Sraffa.

## References

Pasinetti, L.L. 1999. 'Economic theory and technical progress', Economic Issues Lecture, Royal Economic Society Annual Conference, Nottingham, March 31.

Patterson, M.G. 2021. 'Reflection on "Sraffa and the revenue of the owner of non renewable natural resources: a note on the literature"' in A. Sinha (ed.), *A Reflection on Sraffa's Revolution in Economic Theory*, London: Palgrave Macmillan.

CHAPTER 18

# Antonio Gramsci's Letters that Piero Sraffa did not Forward to the Italian Communist Party

*Nerio Naldi*

I would like to thank Giancarlo de Vivo, Donatella Ianelli, Eleonora Lattanzi, Luisa Righi, Alessandro Roncaglia and Shoshanna Zuckerman. If not otherwise stated, letters of Antonio Gramsci, Piero Sraffa, Tatiana Schucht and of other members of the Gramsci-Schucht family mentioned in the paper are conserved in Rome at Fondazione Gramsci (until 1982 it was called *Istituto Gramsci*). The Piero Sraffa Papers are conserved at the Wren Library, Trinity College, Cambridge and are available on line at https://janus.lib.cam.ac.uk/db/node.xsp?id=EAD%2FGBR%2F0016%2FSRAFFA. Translations of Gramsci's letters have been taken from Gramsci (1994); other translations are mine. The following abbreviations will be employed: SP (Piero Sraffa Papers), FG (Fondazione Gramsci), PCI (Partito Comunista d'Italia). Most of the letters referred to have been published in Sraffa (1991), Gramsci (1992, 1994), Gramsci and Schucht (1997). Additional correspondence between Tatiana Schucht and Piero Sraffa and letters exchanged with other correspondents mentioned in this paper are due to be published as part of the *Edizione nazionale degli scritti di Antonio Gramsci*.

N. Naldi (✉)
Dipartimento di Scienze statistiche, Università di Roma, La Sapienza, Rome, Italy
e-mail: nerio.naldi@uniroma1.it



A. Sinha (ed.), *A Reflection on Sraffa's Revolution in Economic Theory*, Palgrave Studies in the History of Economic Thought,
https://doi.org/10.1007/978-3-030-47206-1_18

## 1 Piero Sraffa and Antonio Gramsci

Even though his position as an elected Member of Parliament should have meant he was protected by parliamentary immunity, Antonio Gramsci, the foremost leader of the PCI at the time, was arrested on 8 November 1926 and shortly afterwards sent to the island of Ustica in internal exile. In the days immediately following his arrival, he wrote to Sraffa (11 December 1926) asking him to send him "some books", and in particular "a good treatise on economics and finance". A few days later (21 December 1926), evidently upon receiving news from Sraffa that he had sent or was about to send him books and open an account for him in a Milan bookshop, Gramsci answered thanking him and telling him that he had already put in "quite a large order, certain that I wasn't being indiscreet, for I know your kindness" (Gramsci 1994, pp. 44, 52–3). From Ustica, Gramsci wrote at least four letters to Sraffa between 11 December 1926 and 2 January 1927 (Gramsci 1994, pp. 44, 45, 52–4, 57–9). None of Sraffa's replies have come down to us, but it emerges quite clearly from Gramsci's letters that Sraffa's immediate concern was to satisfy Gramsci's explicit or implicit requests: besides sending books, Sraffa also set out to supply Gramsci with both material and financial support. Sraffa must have realised that these efforts were not without some risk. We know that there were consequences to his sending Gramsci, on 10 January 1927, a postal order for a thousand lire:[1] the Palermo Police Department contacted their colleagues in Milan, and Sraffa was interrogated. Evidence of this is to be found in the police file dedicated to Sraffa in the *Casellario Politico Centrale* (*Central Political Records*, Archivio Centrale dello Stato), and it was on this occasion that the police began keeping a record of Sraffa's doings (see Lattanzi and Naldi 2018, pp. 74–5).

As far as we can judge from his letters, the period that Gramsci spent in Ustica was relatively trouble-free, but all too soon, on 20 January 1928,

[1]On 21 December 1926 Gramsci had written to Sraffa: "Our financial situation is still good: they give us, the political prisoners, ten lire a day: the *mazzetta* of the common detainees at Ustica amounts to four lire a day [...] six of us [...] live in a small house that costs us ninenty lire a month each, all services included" (Gramsci 1994, p. 53). On 2 January 1927 he had come again on the same subject: "I've received the books that you mentioned in your penultimate letter and a first batch of the ones I ordered. So I have plenty to read for some time. I thank you for your great kindness, but I would not want to abuse it. Yet I assure you that quite frankly I will turn to you whenever I am in need of something. As you can imagine, there is no opportunity to spend much here, just the opposite; sometimes one can't spend one's money even when the purchase is necessary"(Gramsci 1994, p. 57).

an arrest warrant arrived and he was transferred to Milan, where he was to be interned in the prison of San Vittore pending conclusion of the preliminary investigation that would lead to his trial before the *Tribunale Speciale per la Difesa dello Stato* (*Special Tribunal for the Defence of the State*) with the accusation of having organised armed insurrection. When he got wind of it, Gramsci sent telegrams to Tatiana Schucht and Sraffa announcing the imminent unexpected transfer (Gramsci and Schucht 1997, p. 42 n.1; Sraffa 1991, p. LI n.20), representing further evidence of just how crucial Gramsci the prisoner saw these contacts to be.

Some light on Sraffa's position in these circumstances is cast by a letter and some observations by Camilla Ravera (Sraffa 1991, pp. XXIV–XXV, LI; Ravera 1973, p. 267), who was then in charge of the underground office (based in Genoa) of the PCI.[2] From a letter by Camilla Ravera to Ruggero Grieco and Palmiro Togliatti (Togliatti virtually became the leader of the PCI from the moment of Gramsci's arrest; Grieco was another prominent figure in the PCI) dated 29 January 1927 and from the memoirs of Ravera herself we learn that just prior to that date, therefore only a few days after Gramsci's telegram about his transfer, Sraffa, who was not acquainted with her, went to see her accompanied by Felice Platone, another leader of the PCI. Thus it seems to have been Sraffa himself who informed the PCI underground office of Gramsci's transfer. So it was that, at a time when the PCI found itself going underground and suffering a great many arrests, Sraffa—even though in a letter to Gramsci in 1924, shortly before their acquaintance evolved into a close friendship, had described himself as *an undisciplined communist*—[3] assumed the role of liaison between Gramsci and the PCI, doing so on the strength of certain pre-established contacts and by making further contacts with members of the by now clandestine party. In the meantime the party and its members were being subjected to new legislative measures involving, among other things, capital punishment for assassination attempts against the highest representatives of the State, attempted organisation of insurrection and crimes of espionage. It was against this background that Camilla Ravera wrote that Sraffa "was sharing in our troubles and worries, and wanted to know if it was possible to do anything that might help Gramsci to endure the hardships of imprisonment" (Sraffa 1991, p. XXV).

[2] See also note 8 below.

[3] See Sraffa (1924, p. 4) (English translation in Naldi 2000, pp. 106–7).

For a better understanding of the climate at the time and the position taken by Sraffa, let us also recall that in February 1926, while he was still teaching at the University of Perugia, he came under fire from the local fascist newspaper (*L'assalto*) subsequent to having suggested adopting a book by Francesco Saverio Nitti (former Minister of the Treasury and Prime Minister who had fled Italy to escape fascist persecution) as a textbook for his course on Public Finance. In the spring of 1927, Sraffa came under further attack in the fascist journal *Libro e Moschetto*, intent on preventing his nomination to a professorship in Genoa. But in the same period Keynes, whom he had met in 1921 thanks to the good offices of Gaetano Salvemini and Mary Berenson (Naldi 1998a), suggested he move to Cambridge for a few years. Sraffa took up the proposal and at the beginning of June the University of Cambridge appointed him to a teaching post for a period of three years.[4]

Subsequent to contact with the above-mentioned underground office of the PCI, the first evidence we have of Sraffa taking steps in support of Gramsci was the visit he was allowed to make to Gramsci in the San Vittore prison, as mentioned in a letter written by Gramsci himself to Tatiana Schucht dated 14 July 1929 (Gramsci 1994, p. 277). The decision to apply for this visit was most probably made upon agreement between Sraffa and Camilla Ravera, or other exponents of the PCI, and we can assume that it took place in the first three weeks of June or the early days of July (Naldi 2000). Not long before Sraffa had made two journeys to Switzerland in April, and one to Tunis in May (SP A1/6). From a letter in which the historian Domenico Zucaro summarises the issues dealt with in a meeting with Sraffa, we can deduce that the purpose of the visit to San Vittore was to discuss the line in defence to be taken by Gramsci and other communist leaders awaiting trial (letter from Domenico Zucaro to Sraffa, 8 November, 1962, SP C343).

Setting out from Italy at the beginning of July bound for Cambridge, Sraffa stopped off in Paris for a few days, where he met the communist leader Angelo Tasca. With respect to Sraffa's efforts to support Gramsci, this meeting was to lead to publication in the daily *Manchester Guardian*,

[4] Sraffa's position had already been extremely dangerous and difficult in December 1922, when Mussolini (who had been nominated Prime Minister less than two months before) tried to force him to retract the content of a short article on the state of Italian banks he had published in a supplement to the *Manchester Guardian Commercial* edited by Keynes (Naldi 1998a, b; Lattanzi and Naldi 2018, pp. 69–70).

on 24 October 1927, of a letter entitled *The methods of fascism. The Case of Antonio Gramsci* (see Naldi 1998a). The idea of appealing to British public opinion to raise awareness of the case of Gramsci had been advanced to Sraffa by Tasca in the month of September (Angelo Tasca to Sraffa, 21 September, 1927, SP C309). The letter had been written by Tasca himself and translated into English with the help of Maurice Dobb (SP F1/9/1-5), Sraffa's colleague at Cambridge and friend since his first stay in England, in 1921–1922, and a member of the Communist Party of Great Britain. Over and above the letter in itself, however, a point worth stressing is that, subsequent to its publication in the *Manchester Guardian*, Sraffa found himself in the uncomfortable position of being threatened with application of the Italian *special laws* enacted in 1926 as engaged in activities that might *harm the prestige* of the country, facing, among other measures, a ban on returning to Italy, on penalty of immediate arrest. In fact, due to an editorial error, the letter which was meant to be published anonymously appeared in a form that implicated association with Sraffa. The letter was signed "An Italian in England", but Sraffa's name appeared on an index page where also the letters were listed (Naldi 1998a).[5]

Sraffa does not appear to have suffered as a result of the editorial error made by the *Manchester Guardian*, but this was most probably a matter of sheer luck. The letter had not escaped the attention of the Ministry of Foreign Affairs, which sifted through foreign press in search of *anti-Italian* publications. Italian officers (as shown by documents from the Italian ministries for foreign and internal affairs dated to October–November 1927 and conserved in the Archivio Centrale dello Stato and in the Archivio Storico Diplomatico—see Lattanzi and Naldi 2018, pp. 77–8) had hit on the letter as early as 25 October; yet they evidently failed to make the connection between the letter and the name published on another page of the newspaper and any further investigations that might have been made do not appear to have led to any consequences. Prudently, Sraffa neither left England nor returned to Italy during the Christmas holidays of 1927. When he eventually did so, in the Spring of 1928, he did not come up against any particular difficulties.

In 1928 Sraffa returned to Italy for the Easter holidays, and a note in his diary (SP E1) suggests that on 29 March he met a lawyer (Giovanni Ariis) who had been in touch with both Gramsci and Tatiana Schucht for some time. Sraffa's visit to Ariis might have had to do with developments

[5] For further details, see Naldi (2012, p. 1405) and Lattanzi and Naldi (2018, pp. 77–8).

in the trial against Gramsci, who had designated Ariis as his defence lawyer, but it is also possible that Sraffa was simply trying to make contact with Tatiana Schucht. Be that as it may, on the evidence of various documents we can place the first meeting between Sraffa and Tatiana Schucht since Gramsci had been arrested in the second half of September 1928, and more precisely between the 17 and 27 September (SP A1/6/1; letters by Tatiana Schucht to Gramsci, 15 and 17 September, 1928 and 5 and 6 October, 1928—Gramsci and Schucht 1997, pp. 259–62, 267–68). Gramsci made no response to Tatiana Schucht's mention in her October 6th letter to him of the meeting she had had with Sraffa. He may have avoided mentioning Sraffa as a matter of prudence.

## 2 Piero Sraffa, Tatiana Schucht and Gramsci's Letters

Gramsci's wife, Julia Schucht, had been living in Russia since 1917, when her exiled family had left Italy to return to their home country and join the Soviet revolution. Tatiana Schucht was the only member of the Schucht family who had decided to postpone moving to Russia and remain in Italy. While it was in the Soviet Union, in 1922, that Julia Schucht met Gramsci, Tatiana met her brother-in-law in Rome, most likely in late January 1925.[6] After Gramsci's arrest, Tatiana immediately began providing him assistance, first during his imprisonment in Rome, then during his internal exile on the island of Ustica, continuing during his imprisonment in Milan, his trial in Rome, his impisonment in Turi and eventually when he was moved to a clinic.

The purpose of the September 1928 meeting between Sraffa and Tatiana Schucht was almost certainly focused on establishing an efficient liason (which had clearly been lacking) between Gramsci—already serving a 20-year sentence—and his party.[7] Indeed, the following winter Sraffa began to liase regularly between Tatiana Schucht, who visited and

[6] See the letter from Antonio Gramsci to Giulia Schucht of 2 February 1925 (Gramsci 1992, pp. 412–13).

[7] See Sraffa to Angelo Tasca (not to Togliatti, as stated by the editor), 26 December, 1928, in Spriano (1979, pp. 169–71).

corresponded with Gramsci in prison, and the *centro estero* (*external centre*) of the Italian Communist Party (PCI).[8]

As far as we know, Tatiana Schucht prepared a copy of each letter she received from Gramsci and sent it or handed it to Sraffa. He, in turn, forwarded it to the *centro estero*. After having prepared another copy of each letter, Tatiana sent the original to the other family members Gramsci had addressed it to, or, if the letter had been addressed to herself, to her sister Julia. Indeed, we know that this method was extremely effective, in terms of both the conservation and the transmission of Gramsci's letters. However we also know that on some occasions originals and copies of the letters were not forwarded to the usual receivers. In particular, as highlighted by Giancarlo de Vivo through a comparison of lists of Gramsci's letters prepared in various years as well as from other documents (de Vivo 2017, pp. 38–52), it is quite clear that between the end of 1932 and the end of 1933 Sraffa refrained from forwarding a significant number of copies of letters which he normally would have sent to the *centro estero*. First among these, following explicit requests formulated by Gramsci, were copies of two crucial letters that Gramsci had sent to Tatiana Schucht, one on 5 December 1932 and the other on 27 February 1933 (Gramsci 1994, pp. 236–9, 274–8). Then, a few months later, Sraffa withheld copies of almost three-quarters of the 30 letters (including a telegram) from Gramsci to Tatiana which he likely received between mid-April and early December 1933. As far as we know, no specific request had been formulated by Gramsci in this case. Indeed, in 1974 those copies were still in Sraffa's hands.[9]

His decision to withhold so many of these letters was certainly intentional and important. The list of Gramsci's letters prepared by Togliatti between 1937 and 1941 shows that copies of virtually every letter Gramsci

[8] After it had been declared illegal the PCI was forced to move its main headquarters abroad. These offices (first based between Switzerland and France, then in Paris) were conventionally called *centro estero*.

[9] In 1962, writing to Camilla Ravera, Sraffa hinted at the copies of Gramsci's letters he still had with himself (SP C342/2). de Vivo's analysis was substantially based on a list of Gramsci's letters prepared by Palmiro Togliatti between May 1937 and January 1941. Togliatti presumably studied the copies of the letters forwarded by Sraffa to *centro estero* as well as letters held by Gramsci's wife in Moscow (Daniele and Vacca 2005, pp. 233–88). de Vivo also drew on the list of photocopies of documents relating to Gramsci that Sraffa gave to Giorgio Napolitano in 1972 (SP C115/5/25/1-7), and on the lists of copies of Gramsci's letters that Sraffa gave to Elsa Fubini in 1974 (FG, "Gramsci dopo la morte 1970–1977" and SP C115/5/13/1-4).

had written to Tatiana Schucht between December 1928 and November 1932—126 letters out of 128[10]—had reached Togliatti, quite likely having been forwarded by Sraffa to the *centro estero*.[11] The available documents further suggest that Sraffa continued giving *centro estero* copies of all Gramsci's letters for some months also after December 1932 and February 1933.

In the next Section we will provide information concerning the content of the two letters which Gramsci had intended exclusively for Tatiana and Piero. Section 4 will put forward an explanation of why Sraffa may have modified the well-established praxis of over four years and, beginning in April 1933, withheld a substantial number of Gramsci's letters from the *centro estero*. Focusing on those addressed directly to Tatiana Schucht, we will identify two sets of unforwarded letters among those sent by Gramsci after 27 February, 1933. The first set comprises Gramsci's letters written between mid-April and mid-July 1933. The second set comprises those from the end of September to mid-November 1933.[12] As we shall see, the

[10] The exceptions, justified by their unique characteristics, are the letters dated January 4 and November 6, 1932 (Gramsci 1994, pp. 123–4, 224–5). In the first of these, in answer to a question posed by Sraffa himself based on a query from a friend, Gramsci discussed the transcription of Russian names into Italian. It is therefore possible that upon receiving his copy Sraffa forwarded it just to this friend, and not to the *centro estero*, and that Tatiana Schucht did not forward it to Giulia. On the other hand, it is likely that the 6 November letter (formed by two parts, respectively addressed to Tatiana and to Julia) was never sent to Tatiana Schucht. Indeed, the subsequent letter of 9 November opens with a phrase suggesting that Gramsci had asked prison authorities to re-write the letter he had written few days before: "Dearest Tania, on Sunday I had already written a letter for you with a section for Giulia, but I have asked [prison authorities] to please let me do the letter over again because it was too much under the influence of a telegram sent me by Carlo that gave credibility to the wild rumors circulating among the prisoners [on potentially upcoming release]" (Gramsci 1994, p. 226—an amnesty law had actually been enacted, but the sentence Gramsci had to serve was only marginally reduced).

[11] Strictly speaking, any letter included by Togliatti in his list could have been forwarded by Sraffa or shown him by Julia Schucht. Therefore, that list can only tell us which letters Sraffa did not forward (possibly underestimating their number); it cannot tell us which letters Sraffa did forward. However, additional evidence allows us to extend our assessments, with a certain degree of confidence, also to the latter set (see, for instance, Sraffa to Togliatti 4 May 1932, where Sraffa explicitly mentioned letters he was forwarding to Togliatti (Sraffa 1991, pp. 224–5).

[12] Over the course of the years following 1933, when he was no longer in jail but in a clinic, Gramsci penned only two letters to Tatiana Schucht, on 22 July and 11 August, 1935 (Gramsci 1994, pp. 348–9, 350–1).

reasons why Sraffa did not forward the letters written in those two periods may be connected to what he himself described as "two first class disasters".

## 3 The Letters of 5 December 1932 and 27 February 1933

In the letter dated 5 December 1932, the first of Gramsci's letters written between December 1932 and December 1933 which were not on Togliatti's list (hence, *a fortiori*, not received by the *centro estero*), Gramsci commented harshly on a letter he had received while incarcerated in Milan in February 1928 from Ruggero Grieco. Gramsci had already touched on Grieco's letter writing to his wife on 30 April, 1928 (a letter which was included in Togliatti's list), and had spoken about it with Tatiana Schucht.[13] In June 1930, he also mentioned it to his brother Gennaro, who had been visiting him in Turi, describing it as extremely harmful to his position.[14] But on 5 December, 1932, Gramsci's remarks were even stronger than before. He wrote that the letter had been presented to him by the pretrial judge, accompanied by these disquieting words: "you have friends who undoubtedly want you to remain in prison for quite some time" (Gramsci 1994, p. 237). And he added:

> Was this a wicked act or an act of irresponsible superficiality? It is hard to say which. Perhaps both at the same time; perhaps the person who wrote it was only irresponsibly stupid and someone else, less stupid, induced him to write. But there is no point in racking one's brain over such questions. There remains the objective fact that has its significance. (Gramsci 1994, p. 238)

Why did Gramsci believe that Grieco's letter warranted such a strong reaction? From what Tatiana Schucht sent to her relatives in Moscow on 12–14 May, 1928 (Gramsci 1994, pp. 201–2), we may infer that Gramsci

[13] See Gramsci to Giulia Schucht, 30 April, 1928 (Gramsci 1994, pp. 201–2) and Tatiana Schucht to Giulia Schucht 12–14 May, 1928 (Rossi and Vacca 2007, p. 82; Vacca 2012, p. 55).

[14] See the notes Gennaro submitted to the *centro estero* a few weeks later (Rossi and Vacca 2007, pp. 209–17). The original of Grieco's letter is not extant, but a picture taken by Italian police in 1928 is conserved at the Archivio Centrale dello Stato (Direzione Generale di Pubblica Sicurezza, 1929, b.196); the latter has been reproduced (with minor imprecisions) and translated in Spriano (1979, pp. 151–3) (see also de Vivo 2009).

believed it provided Italian authorities with information which could be used against him. We know from Gennaro Gramsci's account that in the summer of 1930 Antonio had told him that Greico's letter became the main *capo d'accusa* at the trial where he was sentenced to more than twenty years. But while the first hypothesis is supported by the fact that Italian political police photographed the letter, Gramsci's trial's court report offers no evidence to support the latter statement. In January 1933, however, Gramsci also maintained that this letter had been instrumental in effectively obstructing a Soviet-coordinated effort towards his release—this effort involved approaching the Italian government by means of communication with the Vatican.[15] Gramsci most certainly reiterated this assessment to Tatiana Schucht, and likely to Piero Sraffa as well, in their meetings during his stay in the clinics, when he was permitted visitors, albeit still under strict police surveillance.[16]

In the second of the letters of the period in question which were not on Togliatti's list, that of 27 February 1933, Gramsci once again made reference to Grieco's February 1928 letter, but extended the blame:[17]

[15] See the letter from Tatiana Schucht to Piero Sraffa written on 11 February 1933 (Sraffa 1991, p. 228). That Gramsci did not communicate this view before 1933 does not necessarily mean that he had not conceived of it earlier. The restrictions he had to face in his communication (his letters were read by prison censors and his conversations with relatives always took place under strict surveillance by prison agents) certainly made it extremely difficult for him to express the notion that Ruggero Grieco's letter had significantly contributed to the failure of a Soviet government attempt to obtain his release. Recent research has argued that the view Gramsci had adopted is contradicted by documents kept in both Russian and Vatican archives (Fabre 2015, pp. 115–17). However, it has also been argued that in Gramsci's view his release was a dossier constantly open between Soviet and Italian governments, but that it would have had no chance of positive development if the PCI presented it as its own initiative—which could be read in Grieco's February 1928 letter (Vacca 2012, pp. 239).

[16] See Tatiana Schucht to Piero Sraffa 16 and 28 September 1937, and Piero Sraffa to Tatiana Schucht 18 September 1937—forthcoming (see also Tatiana Schucht to Giulia Schucht 5 July 1934—Edizione Nazionale, forthcoming, and Sraffa 1991, pp. 187–8).

[17] In this letter Gramsci referred to his wife with the nickname *Iulca*. It has been suggested that Gramsci may have been using this variation of her name as a code word, with *Giulia* actually referring to his wife and *Iulca* to indicate the PCI. Although close examination of Gramsci's letters does not reveal evidence to support such an hypothesis, it is nevertheless noteworthy that when Tatiana Schucht sent Sraffa the February 27 letter, she described it as "an Aesopian masterpiece" (Tatiana Schucht to Sraffa, 9 March, 1933—Gramsci and Schucht 1997, p. 1228) and that on the following day she added: "I wrote to him to let him know that I understood perfectly what he meant by referring to either Iulca or Giulia, (naturally using carefully selected wording)" (Tatiana Schucht to Sraffa, 10 March 1933, Edizione

it is certain that during all these years I've always thought about certain events (in the specific instance about a series of events that can be symbolically summarized in the famous letter mentioned to me by the pretrial judge in Milan and of which I talked to you recently), but it is also certain that in the last few months these thoughts have, let me say, intensified, perhaps because there dwindled away in me any thought of being able to clarify them personally, to deal with them «philologically», to go back to the sources and arrive at a plausible explanation. What I want to tell you today is this: that I connect the external manifestations of my relationship with Julca with this series of events. That is, that to this series of preoccupations are linked certain letters that I wrote to you a long time ago and that perhaps you have not forgotten, all the way down to the last that you sometimes refer to as «notorious» and that is not very distant in time.[18] In any case, I am convinced to this day that in my relations with Julca there exists a certain equivocation, a false bottom, an ambiguity that prevents us from seeing clearly and being completely frank; my impression is that I am set aside, that I represent so to speak «a bureaucratic dossier» to be annotated and nothing more. I assure you that I am the first to believe that I have made mistakes, but my impression is that these are not the errors involved but other things that escape me and that I am unable to identify with precision. On the other hand, as you can imagine, although I live in jail, isolated from all sources of communication, direct or indirect, you mustn't think that elements of judgement and reflection don't still reach me. [...] I know how to select, distinguish, tone down intentional exaggerations, integrate,

Nazionale, forthcoming; see also Tatiana Schucht to Gramsci, 10 March, 1933—first letter—Gramsci and Schucht 1997, p. 1227). It is our opinion that Gramsci was not employing a code in his February 27, 1933 letter, but was superimposing considerations about his wife with those about the PCI, thus expressing himself on both a private and a political level—and one can easily discern a reference to Party leaders, with no need to resort to any prearranged code, while the reference to his wife can also be understood in the light of other letters written by Gramsci from prison which we shall not go into at this time (see the letters from Gramsci to Tatiana Schucht on 3 October, 1932 and from Tatiana Schucht to Gramsci October 11, 1932—Gramsci 1994, pp. 214–5; Gramsci and Schucht 1997, p. 1093; see also Gramsci to Tatiana Schucht 6 March, 1933 and Tatiana Schucht, to Gramsci, 10 March, 1933—Gramsci 1994, pp. 278–81; Gramsci and Schucht 1997, p. 1227; see also Gramsci to Giulia Schucht, 16 June, 1936—Gramsci 1994, pp. 357–8).

[18]The *last letter* Gramsci was referring to is most certainly his letter to Tatiana Schucht written on 14 November, 1932 (Gramsci 1994, pp. 228–30). We are unable to identify a group of letters akin to that one written in earlier years. Even though Gramsci's prison corrispondence contains several critical analyses of the relationship between his wife and himself, and on several occasions he stated he would have not written again to his wife unless she had first sent him some letters, no letter is equally dramatic to the one written on 14 November, 1932.

etc. There obviously must be a few mistakes here and there, I'm ready to admit it, but they are not decisive, not such as to impart a different direction to the course of my thoughts. Besides, there are other things that I don't consider appropriate to write to you. You know my way of thinking: what is written acquires a «moral» and practical value that far transcends the mere fact of being written, which is still a purely material thing ... The conclusion, to put it summarily, is this: I was sentenced on June 4, 1928 by the Special Tribunal, that is, by a specific collegium of men, which could nominally be indicated by name, address, and profession in civilian life. But this is a mistake. Those who sentenced me belong to a much vaster organism, of which the Special Tribunal was only the external and material expression, which compiled the legal documents for the sentence. I must say that among the «sentencers» there was also Julca, I believe, indeed I'm firmly convinced she was there unconsciously, and then there is a series of less unconscious people. This at least is my conviction, by now irremovably anchored in me because it is the only one that explains a series of facts that are successive and congruent with one another. (Gramsci 1994, pp. 275–6)

We know that the original of the first letter—that of 5 December, 1932—is still preserved in a file housed at the RGASPI Archives, in Moscow,[19] because it had been sent to the Komintern as part of the documentation accusing Togliatti of orchestrating attempts to sabotage negotiations for Gramsci's liberation. This accusation arrived into the hands of a member of the Secretariat of the Communist Party of the Soviet Union from the Schucht sisters, presumably in the first months of 1939, shortly after Tatiana's re-entry into the Soviet Union.[20] This same letter, not on Elsa Fubini's January 1964 list of Gramsci's letters held by Istituto Gramsci,[21] was in fact first published in 1965. The publication reproduced a typed copy that Tatiana Schucht had evidently prepared and kept for herself, while sending Sraffa a carbon copy of the same typescript (the carbon copy was still in Sraffa's possession in 1974).[22] Gramsci's 27

[19] See Canali (2013, pp. 174, 251–4).

[20] See Pons (2004, pp. 92–5 and 106–10), Daniele and Vacca (2005, pp. 19–20), Vacca, pp. 348–9 and Canali (2013, pp. 174–5).

[21] See Daniele and Vacca (2005, pp. 191–8).

[22] Today, both copies are housed in the Fondazione Gramsci; a photocopy of the one belonging to Sraffa, with his handwritten note identifying it as "second photocopy," is preserved in SP C115/1/167/1-2. We have no information regarding any "first photocopy," but we may speculate that it was sent to the Gramsci Institute in a dossier entrusted to Alessandro Roncaglia in 1973–1974, before July 1974 (see Naldi 2020, pp. 258–66).

February, 1933 letter to Tatiana Schucht, on the contrary, was already on the list drafted by Maria Teresa Lanza de Laurentiis in 1956,[23] and was published in the book *2000 pagine di Gramsci* in 1964. Tatiana Schucht's typewritten copy of the letter of December 5, 1932 (presumably already in Togliatti's possession or in the archives of the PCI for a number of years) was probably not delivered to the curators of the editions of Gramsci's letters until 1964. The decision to publish it was presumably made by Togliatti himself, or by his successor at the helm of the PCI, between 1964 and 1965. While Togliatti's concern about the letter's content could have been a major factor in hindering its publication, the fact that the original manuscript was not available to corroborate its authenticity may have also discouraged an earlier disclosure.

This body of information regarding the content of these two letters and the circuitous route they travelled before publication can be supplemented with some additional facts and conjectures.

First of all, we can identify some indications that Tatiana Schucht did not send to Sraffa the December 5, 1932 letter as soon as she received it. We base this affirmation on the inference that Sraffa on 19 December, writing after receiving Gramsci's December 12 letter,[24] was making no reference to the contents of that of the 5th. He does seem to reference it, however, in a subsequent letter to Tatiana on 5 January, 1933, expressing alarm over her missed visits to Turi: "Unfortunately, we are now realizing how disastrous it was that you did not visit him in 1932!" (Sraffa 1991, p. 105) This was a point in Gramsci's 5 December, 1932 letter and we may surmise that Tatiana did not give Sraffa this letter until he had already returned to Italy.[25] It is plausible that she did not think it prudent to forward it to England. Perhaps she feared that Sraffa would have delivered it along with those from preceding months to the *centro estero* as he was passing through France, as there was no indication in it that he should not have.

But we may also note that Sraffa's 5 January 1933 letter, even though explicitly lamenting that Tatiana Schucht had not visited Gramsci in 1932, made no mention to the other themes in the 5 December letter. We may then presume that the sentence from Sraffa's 5 January letter quoted

[23] This list, first examined by de Vivo (2017, p. 46), is kept in SP C115/1/182/1-10.

[24] Compare Sraffa's 19 December 1932 letter to Tatiana Schucht (Sraffa 1991, pp. 100–1) with Gramsci's 12 December 1932 letter to Tatiana Schucht (Gramsci 1994, pp. 239–41).

[25] Sraffa left England for the 1932–1933 winter holidays around 15 December.

above was the effect of his reading of Gramsci's letter of 19 December (Gramsci 1994, pp. 242–4), where the lack of any visit by Tatiana Schucht in 1932 was also stressed. We may then assume that on 5 January 1933 Sraffa was still unaware of Gramsci's letter of 5 December 1932, and that it was handed in to Sraffa only when he met Tatiana Schucht in Rome, between 8 and 11 January 1933 (we know that he positively received it from references in his letter to Tatiana of 7 February 1933 and in Tatiana's letter to Sraffa of 11 February 1933, Sraffa 1991, pp. 108, 228).

It is also interesting to note that Gramsci gave no specific indication regarding who should have received the 5 December letter. It appears like that Schucht and Sraffa decided to ask Gramsci explicitly whether the *centro estero* should also receive a copy. Sraffa got his answer in a letter from Schucht dated 11 February, 1933, where she disclosed what Gramsci told her when she visited him in prison on 19 January: "When I remarked that we believed Nino's letter should not be passed on, he answered, 'Of course not.'" (11 February, 1933 letter from Tatiana Schucht to Piero Sraffa—Sraffa 1991, p. 228—"we believed" may certainly be understood to mean *Sraffa and I were of the opinion that*...). This leads us to assume that Gramsci trusted Tatiana and Sraffa to evaluate which of his letters were to be forwarded to the PCI, but that they generally considered this relay to be standard practice, as per his instructions. On the other hand, perhaps as a result of this request for clarification, Gramsci specifically indicated to whom his 27 February, 1933 letter was directed in the body of the letter itself: "What I am writing here is meant for you only [Tatiana Schucht] and for the attorney who is handling my affairs [Piero Sraffa]" (Gramsci 1994, p. 277).

Both Sraffa and Tatiana Schucht adhered to these requests. However, the information available suggests that Sraffa did not respect Gramsci's request to keep his Italian comrades in the dark regarding his petition that Soviet government tried to secure his release by negotiating with the Italian government. This can be deduced from information suggesting that, in addition to the copies of Gramsci's letters to Tatiana Schucht, Sraffa also usually transmitted the letters he himself was receiving from her to the *centro estero*.[26] While he did not forward to *centro estero* Gramsci's 27 February, 1933 letter, he seems to have submitted Schucht's 5 March,

[26] This question is considered in detail in Naldi (2020, pp. 258–66).

1933 letter, which allowed Party leaders to get wind of the potential negotiations pressed for by Gramsci (Sraffa 1991, pp. 442–5), and also to understand how Gramsci's 6 March 1933 letter to Tatiana Schucht (Gramsci 1994, pp. 278–81) hinted at that crucial theme. This, we may presume, may have been a result of the understanding that it would have been impossible for PCI leaders to be kept in the dark on such matters.[27] But we may also note that it would have happened before Sraffa's trust for PCI leaders was shaken by the "two first class disasters" which led him to withhold a large number of Gramsci's letters.

## 4 Two Disasters

As mentioned in Sect. 2, Sraffa not only complied with Gramsci's explicit request that his December 5, 1932 and February 27, 1933 letters to Tatiana be not forwarded to the *centro estero*, but he apparently also withheld a far greater number of Gramsci's letters to her composed between mid-April and early December, 1933—and no specific instructions about those letters have come to light. There were about 22 or 23 of these unforwarded letters[28] compared to the 29 or 30 that Tatiana had copied and forwarded to Sraffa. Similarly, from the list of documents of Gramsci-related material which Sraffa gave Giorgio Napolitano in 1972 and from Fubini's lists of the documents she had received from Sraffa, we may gather that in the 1970s Sraffa still had in his possession the copies of at least 18 (more likely, 22)[29] of the letters coming from the periods between mid-April and mid-July, and from the end of September to November-December 1933. Conversely, an analysis of the available lists suggests that up to mid-April 1933, notwithstanding his retainment of the December 5 and February 27 letters, Sraffa continued his usual practice of providing the *centro estero* with copies of all Gramsci's letters. In fact, only one of Gramsci's letters to Tatiana is lacking from Togliatti's list for the period between the end of December 1932 and mid-April 1933: the one dated 21 March 1933,[30] in which Gramsci provided further details about the

[27] A similar point was to be raised by the Soviet ambassador in Italy (V.P.Potëmkin) in summer 1933 (see letter from Tatiana Schucht to Sraffa 27 August 1933—Gramsci and Schucht (1997, p. 1343).

[28] See Table 18.1.

[29] See Table 18.1.

[30] See note 35 below.

state of his health following the serious episode he had suffered on the 7 of March and after his medical examination by Dr Uberto Arcangeli. This is further corroborated by Fubini's lists, from which we may surmise that Sraffa had none of those letters in his possession in 1974. In particular, we may assume that Sraffa had even sent the *centro estero* the letter dated March 6 which not only dramatically compared his situation to a shipwreck and refered to his own transformation on a "molecular" level, but also somewhat vaguely hinted at notions that Gramsci himself had labeled "secret" (*riservata*) which, taken together with the contents of the already mentioned Tatiana's March 5 letter to Sraffa, revealed Gramsci's desire for the Soviet government to intervene on behalf of his liberation by negotiating with the Italian government.

A reconstruction of events which may explain why the two aforementioned blocks of letters were not forwarded to the *centro estero*, while those prior to and between those two periods were duly submitted may be connected to what Sraffa himself described as "two first class disasters".[31] The first of these referred to the publication in the newspaper of the French Communist Party—l'*Humanité*—of the report Dr Arcangeli had filed on Gramsci's medical condition after visiting him in prison. The second was the Fascist police seizure of a PCI directive outlining how incarcerated communist militants should proceed when applying for parole.

Both episodes had dire consequences. The Fascist regime saw them as challenge to its prestige and power. Consequently, Fascist repression of political prisoners escalated. The effects on Gramsci's life were particularly disastrous, as they obstructed Sraffa's efforts to get his sentence reduced. In March of 1933, at Sraffa's request, attorney Saverio Castellett had filed an appeal with the Special Tribunal. In order to apply the amnesty law enacted 5 November 1932, each criminal conviction of a prisoner had to be reviewed individually. Castellett applied for one of Gramsci's sentences to be overturned on the grounds that the article of the 1889 Penal Law Code for which Gramsci had been sentenced to serve three years and four months (art.134) was not included in the new, 1931 Penal Law Code. This appeal, if granted, would have significantly reduced Gramsci's

[31] "Due disastri di prim'ordine" are the exact words used by Sraffa in a letter to Paolo Spriano on 18 December, 1969 (Sraffa 1991, pp. 271–2); similar words can be found in a couple of Sraffa's notes cited by de Vivo (2009, pp. 90–1)—the original documents can be found in SP E44 and J13.

sentence in addition to the benefits of the amnesty law, and would have allowed him to apply for parole within a few months. Initially, Sraffa had been confident that the appeal could be won and that parole could be granted,[32] but after the "two first class disasters" it was rejected and, as a matter of fact, no communist militants were ever granted parole.

### 4.1 *The First Disaster and Gramsci's Letters from mid-April to mid-July 1933*

The notion that there may be a link between the publication of Dr Arcangeli's medical report and the fact that Sraffa did not send the PCI any copy of Gramsci's letters to Tatiana Schucht written from end of April to mid-July 1933, hinges on two fundamental elements: (1) Sraffa learned of the effect the publication of the report had on the examination of Castellett's petition (and probably of the very publication of the report) in a letter from his father Angelo sent from Switzerland on 23 May, 1933;[33] (2) the first letter of the block that were not forwarded was dated 17 April 1933, thus probably received by Sraffa towards the end of that same month, when he was already in Cambridge. While we may assume that Sraffa intended to accumulate a number of copies of letters before forwarding them to the *centro estero,* or to hand them all over at once in the summer when he could deliver them personally in Paris, we may also speculate that once he had been apprised of the publication of the medical report and its ramifications, he began to wonder whether it was advisable to continue sending them at all. In fact, if we compare the list of letters compiled by Togliatti between 1937 and 1941, the list of documents of Gramsci-related material which Sraffa gave Giorgio Napolitano in 1972 and the lists of letters he gave Elsa Fubini in 1974 (see Table 18.1),[34] we

[32] See the letter from Angelo Sraffa to Piero Sraffa, 11 June 1933 (FIG, Fondo Sraffa; photocopy in SP C115/1/153). The reduction of Gramsci's sentence by 3 years and 4 months' detention, based on article 134 of the 1889 Penal Code, meant that his release date would have been January 19, 1936 instead of 19 May, 1939. The documentation on Castellett's petition to the Court is held in ACS, Tribunale Speciale, Esecuzioni, 1927–1943, 14, 394).

[33] The letter has been reproduced (albeit wrongly dated 29 May) in Spriano (1979, pp. 175–6). From a letter from Togliatti, dated 24 May (Spriano 1979, pp. 173–4) we also know that Sraffa had written to him on 19 May, not yet informed of the publication of Arcangeli's report (this letter from Sraffa to Togliatti does not seem to have been preserved).

[34] The documents Sraffa gave to Napolitano in 1972 included only one letter from Gramsci to Tatiana Schucht: the letter dated 5 June 1933.

**Table 18.1** Letters from Antonio Gramsci to Tatiana Schucht (December 1932–December 1933)

| *Letters from Antonio Gramsci to Tatiana Schucht (December 1932–December 1933)* | *Date (known or guessed) when Tatiana Schucht forwarded it to Piero Sraffa* | *Palmiro Togliatti's list (1937–1941)* | *Elsa Fubini's lists (1974)* | *First published in* |
|---|---|---|---|---|
| 5 December 1932 | 8–10 January 1933 | | • | 1965 |
| 12 December 1932 | 18 December 1932 | • | | 1947 |
| 19 December 1932 | December 1932–1933 | • | | 1964 |
| 26 December 1932 | January 1933 | • | | 1964 |
| 2 January 1933 | 8–10 January 1933 | • | | 1965 |
| 9 January 1933 | 30 January 1933 | • | | 1964 |
| 22 January 1933 | 30 January 1933 | • | | 1988 |
| 30 January 1933 | 13 February 1933 | • | | 1947 |
| 6 February 1933 | 13 February 1933 | • | | 1964 |
| 13 February 1933 | 22 February 1933 | • | | 1964 |
| 20 February 1933 | | • | | 1965 |
| 27 February 1933 | 3 March 1933 | | • | 1964 |
| 6 March 1933 | 10 March 1933 | • | | 1964 |
| 14 March 1933 | 22–23 March 1933 | • | | 1947 |
| 21 March 1933[a] | | | | 1964 |
| 27 March 1933 | 18 April 1933 | • | | 1964 |
| 3 April 1933 | 18 April 1933 | • | | 1947 |
| 10 April 1933 | 18 April 1933 | • | | 1947 |
| 17 April 1933 | | | • | 1965 |
| 23 April 1933[a] | | | | 1988 |
| 30 April 1933[a] | 4 May 1933 | • | • | 1965 |
| 8 May 1933 | 10 May 1933 | | • | 1964 |
| 16 May 1933 | 18 May 1933 | | • | 1965 |
| 22 May 1933 | | | • | 1964 |
| 29 May 1933 | 31 May 1933 | | • | 1964 |
| 5 June 1933 | 7 June 1933 | | [b] | 1965 |
| 11 June 1933 | 16 June 1933 | | • | 1965 |
| 18 June 1933 | 21 June 1933 | | • | 1964 |
| 2 July 1933 | 5 July 1933 | | • | 1964 |
| 6 July 1933 | 9 July 1933 | | • | 1964 |
| 10 July 1933 | 14 July 1933 | | • | 1964 |
| 17 July 1933 | 20–21 September 1933 | • | | 1947 |
| 24 July 1933 | 20–21 September 1933 | • | | 1947 |

(*continued*)

**Table 18.1** (continued)

| *Letters from Antonio Gramsci to Tatiana Schucht (December 1932–December 1933)* | *Date (known or guessed) when Tatiana Schucht forwarded it to Piero Sraffa* | *Palmiro Togliatti's list (1937–1941)* | *Elsa Fubini's lists (1974)* | *First published in* |
|---|---|---|---|---|
| 1 August 1933 | 20–21 September 1933 | • | | 1965 |
| 8 August 1933 | 20–21 September 1933 | • | | 1964 |
| 23 August 1933 | 27 August 1933 | • | | 1947 |
| 28 August 1933 | 11 September 1933 | • | | 1965 |
| 3 September 1933 | 11 September 1933 | • | | 1964 |
| 25 September 1933 | 22 October 1933 | | • | 1965 |
| 1 October 1933 | 22 October 1933 | | • | 1965 |
| 3 October 1933 | 16 October 1933 | | • | 1965 |
| 24 October 1933 | 28 October 1933 | | • | 1965 |
| 29 October 1933 | 5 November 1933 | | • | 1965 |
| 5 November 1933 | 17 November 1933 | | [c] | 1965 |
| 12 November 1933 | 17 November 1933 | | • | 1965 |
| 17 November 1933[d] | 17 November 1933 | | • | 1991 |
| 20 November 1933 | 1 December 1933 | | [c] | 1965 |
| 27 November 1933 | 1 December 1933 | | [c] | 1965 |
| 4 December 1933[e] | | | | 1965 |

[a]See note 35

[b]Sraffa handed in his copy of this letter to Giorgio Napolitano in March 1972

[c]These copies, even though not explicitly mentioned in Elsa Fubini's 1974 lists of copies of Gramsci's letters she had received from Sraffa, may be assumed to have also been given to her by Sraffa (for a detailed analysis of this issue see Naldi 2020, Tabella 1)

[d]Telegram (not preserved) sent by Gramsci to Tatiana Schucht on 17 November 1933 known to us from the transcription Schucht sent to Sraffa in her 17 November letter (given by Sraffa to Elsa Fubini in 1974)

[e]It is likely that Tatiana Schucht did not prepare a copy of this letter

can see that in the early 1970s all of Gramsci's letters from 17 April to 10 July, 1933 were in Sraffa's hands and that Togliatti could not have seen any of them neither through the *centro estero* nor at the Schucht residence.[35]

In short, we could maintain that Sraffa, passing through France on his way back to England following spring break in 1933, had consigned all the copies of Gramsci's letters he had received while in Italy, along with a copy of Dr Arcangeli's report, to the *centro estero,* but that once he returned to England, he held on to all the copies he received from Tatiana Schucht, not even handing them over when he next travelled through France the following summer. Clearly, the publication of information meant to be kept private kept Sraffa from forwarding letters containing equally sensitive information about Gramsci's dire health conditions and his request to be transferred to a different prison infirmary or to a clinic.[36] The same letters also contained references to two letters that Sraffa was

[35] One letter from Gramsci to Tatiana Schucht, the one of 30 April, 1933 (Gramsci 1994, p. 292), is on both the list compiled by Elsa Fubini in 1974 and that of Togliatti drafted between 1937 and 1941. The reason for this may be that, although Sraffa did not forward it to the *centro estero*, Togliatti may have been able to read and transcribe it at the Schucht residence. There is another notable element to be seen in comparing the two lists: neither of them contain Gramsci's 21 March, 1933 letter to Tatiana Schucht (Gramsci 1994, pp. 282–3). In it Gramsci spoke about the state of his health, following his health crisis on 7 March and Dr. Arcangeli's medical visit to him in prison. Schucht most likely received this letter while she was still in Turi, in the same period that Sraffa was there, or managed to show it to him when they met in Rome sometime between March and April. We may presume that this letter was absent from both lists because Schucht did not transcribe it, due to all the worrisome and arduous occurences and undertakings going on at the time. In fact, no transcription of the letter exists among the archives, only the original itself. Another letter missing from the lists, for obvious reasons, is that of 23 April, 1933 which Schucht never received, as it had been confiscated by the Turi prison authorities (Gramsci 1994, pp. 289–92—this letter is now housed in the Archivio Centrale dello Stato).

[36] We may provide two instances from letters written at the beginning and at the end of the relevant period: "My condition has somewhat improved, but the least little thing is enough to make me ill again. From one day to the next, due to the merest trifle, I fall back into serious prostraction" (Gramsci to Tatiana Schucht, 17 April 1933—Gramsci 1994, pp. 288–9); "You must inititate an urgent petition for my transfer as soon as possible from the Turi prison to the infirmary of another prison where there are specialists who can examine me thoroughly enough to establish what combination of ailments I am afflicted with and can take an x-ray of my lungs that will solve the doubts of both Professor Arcangeli and the prison inspector Dr. Filippo Saporito. I beg you to believe that I cannot bear it anymore. The pain in cerebellum and my cranium drives me wild. The difficulty in using my hands has increased and increases progressively, and this cannot simply be due to arteriosclerosis" (Gramsci to Tatiana Schucht, 6 July 1933—Gramsci 1994, p. 307).

expected to have written to Soviet diplomats,[37] about which the PCI was not be apprised, as per Gramsci's request, as well as to Greico's letter, which he had so harshly criticized.[38]

### 4.2 The Second Disaster and Gramsci's End of September to End of November 1933 Letters

In the period following this moratorium on forwarding Gramsci's letters, by comparing the list of Gramsci's letters prepared by Togliatti in 1937–1941 to the records of documents Sraffa gave to Istituto Gramsci in 1972 and 1974, we may affirm that between July 17 and September 3, 1933 the copies Sraffa received from Tatiana Schucht during his summer stay in Italy were once again delivered to the *centro estero*, as those received during the Easter holidays had been.

This delivery most likely took place in early October, when Sraffa was passing through France on his way back to England. Even though the letters mentioned Gramsci's request to be transferred to another prison's infirmary, the tone was sufficiently less dramatic than the previous ones, which may have contributed to allow Sraffa to reconsider his previous doubts about confiding this information to the *centro estero*. However, the following block of letters (from the end of September to the end of November 1933) were again withheld. The reason for this can be understood in the light of another letter Sraffa received from his father. The letter was sent on December the 1st, once again from Switzerland.[39] Sraffa was in Cambridge in early December, planning to return to Italy for winter break, and presumably planning to hand over the October-November letters as usual on his way. The letter from his father, however, informed him of the behind-the-scenes motivations for the Special Tribunal's refusal to grant Saverio Castellett's appeal. Although Sraffa already knew that *Procuratore Generale* Carlo Fallace's proposal to deny the appeal had been

[37] "I want to remind you again and urge you about the two letters to be written that I mentioned to you already in January. I don't understand why the attorney, who had not rejected the idea, was then so dilatory" (Gramsci to Tatiana Schucht, 10 July 1933—Gramsci 1994, p. 311).

[38] "In 1927–1928 [...] the pretrial judge was right to say that it really seemed as though my friends were collaborating to keep me in prison as long as possible" (Gramsci to Tatiana Schucht, 16 May 1933—Gramsci 1994, p. 295).

[39] The 1 December, 1933 letter from Angelo Sraffa to Piero Sraffa is kept in SP C300/4 and partially reproduced in Spriano (1979, p. 89, n.17), cfr. de Vivo (2017, p. 49).

accepted by the Special Tribunal on 13 October,[40] he was probably unaware that during the summer Fascist Police had confiscated a document instructing Communist political prisoners on how to behave should they have an opportunity to appeal for parole,[41] and that this document had been instrumental in definitively blocking Castellett's efforts. Before that incident, thanks to the intervention of Sraffa's uncle, Mariano D'Amelio, First President of the highest court of justice in Italy (*Corte di Cassazione*) at the time, there was every indication that the outcome would have been positive, despite the setback caused by the publication of Arcangeli's medical report. We have no doubt that Sraffa considered it obvious that the PCI needed to handle the flow of information regarding such a delicate matter with the utmost protection, in particular in view of how extensively the Fascist Police had been able to infiltrate anti-fascist organizations. And we may surmise that the news he received from his father deeply disappointed him. The fact remains that evidence suggests that on his trip through France in mid-December, Sraffa did not deliver the copies of the letters he had been receiving in Cambridge since October. These copies were also among his papers in 1974 and once again we can see that they touched on subjects which could justify his decision to withhold them, out of fear that the *centro estero* might have shared information better kept under wraps. These letters informed about the Special Tribunal's rejection of Castellett's petition and, more importantly, that the request to be transferred to a nursing facility had been granted to Gramsci, and that many obstacles to this outcome had been surmounted.

## 5 Looking Out for Gramsci's Best Interests

In recapitulation, our study of the circumstances surrounding Sraffa's neglect to forward to the PCI two substantial groups of Gramsci's letters to Tatiana Schucht, written between 17 April and 4 December, 1933, allows us to see how the dates of these letters correspond to Sraffa's realization that the PCI's *centro estero* could not guarantee that delicate information would be treated with all the necessary caution. Publication of information contained in these letters could have been particularly

[40] See the letters from Tatiana Schucht to Piero Sraffa of 5 November, 1933 (Gramsci and Schucht 1997, pp. 1370–1) and 9 November, 1933 (forthcoming) and those from Gramsci to Tatiana Schucht of 29 October and 5 November, 1933 (Gramsci 1994, pp. 330–2, 334–5).

[41] See Naldi (2013, pp. 383–4).

damaging to Gramsci's situation. It could have interfered with his chances of improving his conditions as an inmate or of getting even adequate medical care. We may therefore conclude that Sraffa, making a distinction between fulfilling agreements with the PCI and actually looking out for Gramsci's best interests, decided that withholding these letters was the best option for protecting his friend.

These conclusions have been supplemented by other observations related to the letters of 5 December, 1932 and 27 February, 1933. Gramsci specified that these letters were to be read only by Tatiana Schucht and Piero Sraffa, thus were not to be communicated to the *centro estero*—and Sraffa adhered to this request. However, the information available suggests that Sraffa also transmitted the letters he himself was receiving from Tatiana Schucht to the *centro estero*, in addition to the copies of Gramsci's letters to her. This leads us to deduce that he did not respect Gramsci's request to prevent the PCI from learning about his petition to the Soviet government to try to secure his release by negotiating with the Italian government. In fact, while he did not forward to *centro estero* Gramsci's 27 February, 1933 letter, he seems to have submitted Schucht's 5 March, 1933 letter,[42] which allowed Party leaders to get wind of the potential negotiations pressed for by Gramsci. This decision may be seen under different perspectives. On the one hand, Sraffa probably made it with the understanding that it would have been effectively impossible for PCI leaders to be kept in the dark on such matters. On the other hand, we may see it as a measure of Sraffa's trust in Party leaders, with whom he maintained direct contact on Gramsci's behalf—trust which between March and April 1933 had not yet been shaken by the aforementioned "two first class disasters".

We may also add that nothing of what we have said is meant to imply that either Gramsci's or Sraffa's or Tatiana Schucht's faith in socialist ideals and in the parties and international movements they inspired was shaken by these experiences. There is no doubt that the lack of trust each of them (in different occasions and degrees) may have felt was only directed towards some of the people in the structures which governed those parties and movements, not towards those very parties and movements. This can be appreciated by studying their extant correspondence after 1933 up to Gramsci's death and other documents and testimonies dating to subsequent years.

[42] See Naldi (2020, pp. 258–66).

## References

Canali, M. (2013) *Il tradimento. Gramsci, Togliatti e la verità negata*, Padova: Marsilio.

Daniele, C. and Vacca, G. (2005) *Togliatti editore di Gramsci*, C. Daniele and G. Vacca eds, Roma: Carocci.

de Vivo, G. (2009) Gramsci, Sraffa e la «famigerata lettera» di Grieco, *Passato e presente*, n.77, pp. 79–91.

de Vivo, G. (2017) *Nella bufera del Novecento. Antonio Gramsci e Piero Sraffa tra lotta politica e teoria critica*, Roma: Castelvecchi.

Fabre, G. (2015) *Lo scambio. Come Gramsci non fu liberato.* Palermo: Sellerio.

Gramsci, A. (1964) *2000 pagine di Gramsci*, G.Ferrata and N.Gallo eds., Milano: Mondadori.

Gramsci, A. (1992) *Lettere 1908–1926*, A.A.Santucci ed., Torino: Einaudi.

Gramsci, A. (1994) *Letters from Prison*, volumes 1 (1926–30) and 2 (1931–37), edited by F.Rosengarten, translated by R. Rosenthal, New York: Columbia University Press.

Gramsci A., Schucht, T. (1997) *Lettere 1926–1935* (C. Daniele and A. Natoli eds.), Torino: Einaudi.

Lattanzi, E., Naldi, N. (2018) Documents on Piero Sraffa at the Archivio Centrale dello Stato and at the Archivio Storico Diplomatico, *Research in the History of Economic Thought and Methodology*, Vol. 35B, pp. 61–100.

Naldi, N. (1998a) Some notes on Piero Sraffa's biography: 1917–1927. *Review of Political Economy*, 10(4): 493–515.

Naldi, N. (1998b). Dicembre 1922: Piero Sraffa e Benito Mussolini. *Rivista Italiana degli Economisti*, 3(2): 271–99.

Naldi, N. (2000) The friendship between Piero Sraffa and Antonio Gramsci in the years 1919–1927, *European Journal of the History of Economic Thought*, vol.7, n.1, pp. 79–114.

Naldi, N. (2012) Two notes on Piero Sraffa and Antonio Gramsci, *Cambridge Journal of Economics*, Vol.36, n.6, pp. 1401–15.

Naldi, N. (2013) La liberazione condizionale di Antonio Gramsci, *Studi Storici*, April-June, vol.54, n.2, pp. 379–92.

Naldi, N. (2020) Piero Sraffa intermediario delle lettere dal carcere, in *Un nuovo Gramsci. Biografia, temi, interpretazioni* (G.Francioni and F.Giasi eds.), Roma: Viella, 2020, pp. 241–78.

Pons, S. (2004) L'«affare Gramsci-Togliatti» a Mosca, *Studi storici*, XLV, n. 1, pp. 84–116.

Ravera, C. (1973) *Diario di trent'anni. 1913–1943*, Roma: Editori Riuniti.

Rossi A., Vacca, G. (2007) *Gramsci tra Mussolini e Stalin*, Roma: Fazi Editore.

Spriano, P. (1979) *Antonio Gramsci and the Party: The Prison Years*, London: Lawrence and Wishart.

Sraffa, P. (1924) Problemi di oggi e di domani, *L'Ordine Nuovo*, 1–15 aprile, p. 4.
Sraffa, P. (1991) *Lettere a Tania per Gramsci*, edited by V.Gerratana, Roma: Editori Riuniti.
Vacca, G. (2012) *Vita e pensieri di Antonio Gramsci 1926–1937*, Torino: Einaudi.

# Index[1]



[1] Note: Page numbers followed by 'n' refer to notes.



A. Sinha (ed.), *A Reflection on Sraffa's Revolution in Economic Theory*, Palgrave Studies in the History of Economic Thought, https://doi.org/10.1007/978-3-030-47206-1

## Q

The manufacturer's authorised representative in the EU is Springer Nature Customer Service Centre GmbH, Europaplatz 3, 69115 Heidelberg, Germany. If you have any concerns regarding our products, please contact ProductSafety@springernature.com

Printed and bound by CPI Group (UK) Ltd, Croydon, CR0 4YY
30/09/2026
02225246-0006